Mathematics in the Making

Editor: M. H. Chandler
Art Editor: M. Kitson
Assistant: A. T. Lockwood
Consultant Artists: Andre Czartoryski, Richard Jones
Research: A. F. Walker

First published in 1960 by Macdonald & Co. (Publishers) Ltd.,
16 Maddox Street, London W.1.
in association with Rathbone Books Limited.

Text set by Purnell & Sons Ltd., Paulton, Bristol.
Printed in Great Britain by L. T. A. Robinson Ltd., London.

Mathematics in the Making

Lancelot Hogben

Macdonald : London

Foreword

When I wrote *Mathematics for the Million* during a prolonged sickness in hospital, and without anticipating its publication two or three years later under pressure from a euphoric American publisher, I certainly did not flatter myself with the hope that it would be a best-seller or that it would justify itself, if only by convincing other publishers that there is a sale for mathematical books written for a large reading public by authors far better endowed to undertake the same task.

At the time of writing it, I had read most of the composite histories of mathematics, but had studied few available original sources. Accordingly I did not then sufficiently realise how often the task of charitably interpreting the achievements of the individual in the idiom and notation of a later period magnifies them beyond recognition. During the past twenty years, the work of later scholars, in particular Otto Neugebauer and Joseph Needham, has given us good reason to revise much of what we learned from the writings of D. E. Smith, Cajori, Rouse Ball and others of the same vintage. Till we know far more than we know as yet, we shall not be ready to survey the history of mathematics in its entirety as a facet of the history of the technique of human communications.

In short, the historical matrix of *Mathematics in the Making* is not in any sense an authoritative guide to the history of mathematics. It merely signifies an overdue aspiration to bring to being a new humanistic approach to learning at every level. I therefore hope that my critics and readers will judge this book less by its verbal content than by what is the outcome of a co-operative undertaking, sustained by the belief that we can immeasurably speed up the process of assimilating knowledge by exploiting visual aids to an extent as yet seriously undertaken by no text-books.

In this persuasion I have been highly fortunate because I have now had the opportunity to work with a team of artists determined both to understand what every charity is in aid of and to confront the reader with an aesthetic challenge of sustained interest. Maybe the time will come when television can take over the job and do it better. Even so, a broad curricular framework in a static medium will not be without relevance to the efforts of those who exploit the possibilities of the moving image.

About the choice of contents in the text, as also of the visual material, let me say this. The professional mathematician who writes for a large reading public is too apt to forget how few intelligent people actually continued their mathematical studies at high school or college far enough to follow what he (or she) has to say. It is also easy for the expert to assume wrongly that those who did so still recall their early teaching. If this book has no other useful purpose, I therefore hope that it will make it easier for a wider reading public to get the best out of a large number of available contemporaneous sources of which I list a few at the end of this book. If any author finds himself (or herself) unjustly excluded from so brief a catalogue, let me say that a book which helps the reader to *brush up* (and perhaps augment) his (or her) stock in trade of mathematical lore will add to the potential reading public of any author who writes at a more sophisticated level with less facilities for stimulating interest than those which my colleagues in a team enterprise have put at my disposal.

Lancelot Hogben

Contents

Chapter 1 Counting and Measurement

Fifty years ago, few if any would have anticipated the possibility of witnessing within a lifetime the political emergence of the peoples of Asia and Africa to the status of first-class powers, the cultural renaissance of the Far East, the possibility that one person could be visible and audible to persons of all nations at one and the same time, that unlimited sources of power would be henceforth available to destroy all life on this planet or to release man from the need for irksome toil, that it would now be possible to design machines which dispense with the need for nearly all routine tasks hitherto carried out by human beings or beasts of burden, that the hazards of death during infancy and childbirth would fall to about a tenth of what was customary in civilised communities at the turn of the century, still less that we should now accept as commonplace the imminent possibility of a human return voyage to and from the moon.

Though this catalogue is now commonplace to most of us, few highly educated persons, and among them very few whose vocation is politics, have started to take stock of its implications, only three of which concern the theme of this book. One is that Western civilisation will increasingly have to undertake a revaluation of a cultural debt it has hitherto amply acknowledged only to the Greco-Roman world. A second is that mathematics has invaded every facet of our everyday lives to an extent which excludes anyone without some acquaintance with mathematical techniques from any avenue to a deep understanding of the circumstances that shape them. The third is that the most challenging intellectual task of our world which must either unify or perish is the perfection of our means of communication. Among such, mathematics is pre-eminent as an instrument of intelligent planning at a level which transcends the Babel of native speech.

In this setting, traditionalists, unwilling to re-adjust their sights, envisage the bleak prospect of a community of robots who have relinquished the luxury of thinking to the electronic brain; but the prospect is forbidding only if we prefer to identify the luxury of thinking with art criticism or other inconclusive disputation. Otherwise, the breathless tempo of current events is an invitation to anyone over forty to embark on a rejuvenating course of self-education, and its challenge is exhilarating to those of us who retain the capacity to learn. Admittedly, it is one which imposes on us a more exacting effort of self-discipline than any previous generation has faced. Three hundred years ago, when the universities of the western hemisphere inculcated a familiarity with the learning of the Greek-speaking world, such knowledge was a passport to all mathematics held in high esteem; but mathematics made vast strides between 1650 and 1850 A.D. Current school teaching takes cognisance of comparatively little of the outcome, still less of the very considerable innovations of the last hundred years.

In short, those who now aspire to understand

These scenes from a 3400-year-old mural painting at Abd-el-Qurna show two quite different processes which call for the use of numbers. To be clear about the difference between using numbers as labels for counting discrete objects and as labels to record the results of measuring is to take an important step towards mathematical literacy.

the mathematical techniques which most intrusively influence our daily lives, and more especially those of us who dropped the study of mathematics with relief at the end of our schooldays, have to undertake an extensive programme of self-education beyond the scope of college courses in the mid-nineteenth century. One difficulty which then confronts us arises from the circumstance that professional mathematicians are rarely, if ever, good salesmen. Indeed few of them would wish us to regard themselves as such. By the same token, very few of them from the days of Euclid onwards are good educationists, if the criterion of a good educationist in this context is ability to elicit intelligent understanding from pupils with no native inclination for the speciality of the teacher.

The viewpoint of this book is that the best therapy for the emotional blocks responsible for the defeatist frame of mind in which many highly intelligent people face a mathematical formula is the realisation that the human race has taken centuries or millennia to see through a mist of difficulties and paradoxes which instruction now invites us to solve in a few hours or minutes. Thus the record of mathematical discovery is more human than some of its expositors recognise; and many of us who start with no inclination for mathematical study as a pastime can come to some appreciation of its intrinsic entertainment value by first acquainting ourselves with what the teaching of our schooldays failed to disclose about how material needs and the intellectual climate of earlier times shaped the making of mathematics. This being so, it is not the writer's intention to bring the story of mathematics in the making beyond the fringe of controversies which have raged somewhat inconclusively during the last hundred years, nor to touch on many recently developed branches which have had, to date, little pay-off in the world's work.

Since many eminent applied mathematicians of the past generation, including Albert Einstein, have expressed doubts about how far some of the themes of the contemporary debate concerning the basic concepts of mathematics are meaningful, it is too early to forecast with confidence how many of the current preoccupations of pure mathematics will survive, unless we assume that there is a conceivable finality about the verbal definitions which satisfy the most acute intellects of a particular historical *milieu.* Because we can teach an electronic brain to answer only questions stated in the language to which it responds, mechanical progress of the last two decades has raised questions which those previously in search of a new rationale for our confidence in mathematical techniques assuredly did not anticipate. What the consequences will be when our definitions have to be meaningful in the dictionary of the language of the machine remains to be seen.

Much of the debate of the past century has proceeded at a purely verbal level with the implication that the field of mathematics embraces any form of reasoning worthy of the name of philosophy or logic; but this is a burden which few applied mathematicians would wish to shoulder. Few, if any, biologists, geologists or archaeologists would concede the claim. Most chemists would be hesitant to endorse it, and by no means all physicists would do so. Indeed, those who make it sometimes write as if they want to have it both ways. On the one hand, they invite us to venerate, as we may well venerate, the achievements of an Archimedes or a Newton. On the other, they ask us to believe that there was no rational basis for what Archimedes and Newton asserted before the time of Gauss, Cantor, Dedekind, Weierstrass and other writers of the last hundred years.

Few of us can hope to gain any insight into the more recent branches of mathematics without a firm foothold in the foundations laid during Newton's lifetime. Since no one disagrees about the fact that Newton's mathematics is the basis of how we are now able to calculate the orbit and requisite initial speed of a Sputnik, we are in less uncharted territory if we survey mathematics in the making against the background of what indisputably useful work mathematicians have accomplished. On that understanding, we may *provisionally* adhere to the following formula:

> mathematics is the technique of *discovering* and *conveying* in the most *economical* possible way *useful* rules of *reliable* reasoning about *calculation, measurement* and *shape*.

Before we start the story of the making of mathematics, it will accordingly dispose of some misunderstandings if we examine carefully three sets of key words in the above: (a) the processes of *discovering* and *conveying*; (b) the qualifications *useful*, *economical* and *reliable*; (c) the three topics for discussion, *calculation*, *measurement* and *shape*. What is common to the last three in the compass of our programme is in fact the different ways in which we use numbers. In this chapter we shall speak of numbers only in a very restricted and in the most primitive sense, *i.e.* as labels for counting discrete objects or for the result of matching a tangible object against the scale divisions of a measuring rod. In short, our concern at this stage is primarily with *integers*.

It is difficult to do justice to the distinction between *discovering* and *conveying* a reliable rule unless we are first clear about what we here mean by *reliable*. This will emerge if we are also clear about the special sense in which we can speak of a rule as *useful* without intentionally giving offence to the sensibility of a very pure mathematician. Ordinarily, we think first and foremost of a rule as useful if it helps us to solve a practical problem; but here we shall sidestep whether the immediate problem is of practical interest by merely assuming the possibility that a rule may eventually be useful in that sense. If indeed that is so, the only sense in which one rule may be more useful than another for solving a particular problem depends on the answer to the question: does it save us more or less time and/or effort, if we do have to tackle it? It will therefore serve our purpose if we examine a problem which has no self-evident pay-off in hard cash. We may state it simply as follows: *if you add up any number of odd numbers starting from the first, what is the total?*

If you toy with this conundrum, you may proceed on some such plan as this: $1=1$; $1+3=4$; $1+3+5=9$; $1+3+5+7=16$; $1+3+5+7+9=25$ *etc.* Likely enough, you will then notice that the sum of the first two ($1+3=2^2$), first three ($1+3+5=3^2$), first four ($16=4^2$), first five ($25=5^2$) odd numbers conforms to a pattern which is briefly expressible if we use n for any number. We may then state the rule in the form: the sum of the first n odd numbers is n^2. If this discovery is always true, the rule is a useful rule because it saves you effort. Clearly it does so. It would take you many minutes to add up the first hundred odd numbers; but a 10-year-old child of normal intelligence can compute mentally in less than two seconds what the rule tells you, *i.e.* the first 100 odd numbers add up to $100^2=10{,}000$.

When not employed as meteorologists, mathematicians, and rightly so, are not content with the discovery that this rule seems to be O.K. as far as you care to take it. They seek assurance that it will still be true if you happen to use it outside the range in which you have already checked it; and it is easy to cite many examples of seemingly satisfactory ones which can let us down badly. For instance, the formula $n^4-10n^3+36n^2-49n+24$ for the sum of the first n even integers is true when $n=1$, 2, 3, 4, but otherwise false. Likewise, the formula $n^2+9n-22$ for the sum of the squares of the first n integers is true when $n=3$ or 4. Otherwise it is not.

Here you have a clue to one reason why mathematicians who write about mathematics in the making are prone to speak with two voices. The truth is that periods in which great discoveries happen are not always, if indeed ever, those in which mathematicians are most punctilious about what later generations call *proof*. What they may mean by the word may not be easy to define in a foolproof way; but it will here suffice to say that we have not proved that the n^2 rule for the sum of the first n odd numbers is reliable till we have shown either of two things:
(a) it is true however large we make n;

(b) it is true so long as n does not exceed or fall short of some number, say $10,000=N$.

Before we look more deeply into what we signify by *reliable* in our definition of what we here mean by mathematics, it will be best to examine another epithet which occurs in it. We have spoken of *conveying* a rule in the most economical way possible. To say this signifies that mathematics is a means of communication. In short, it is a universal written language of which the signs are meaningful in the same sense that the characters of the Chinese script are meaningful. That is to say, they convey the same meaning to all who have learned to read them, but they have no relation to the particular sounds people of different speech communities utter when they convey the meaning of a sign by word of mouth. At the most primitive level this is clearly true of number signs. Thus 5 or v convey the same meaning to a Swede who utters *fem*, to a Frenchman who utters *cinq* or to a Welshman who utters *pump*. Similarly, ∵ (*because*) and ∴ (*therefore*) convey the same to a Norwegian who utters *fordi* for the first and *derfor* for the second, to a Frenchman who utters *parce que* and *donc* and to a Welshman who utters *am* and *felly*. Thus it is that people from North and South China, though unable to converse with one another, may nevertheless be able to read the same classics written in the same characters.

The analogy is pertinent from another viewpoint. The characters of the Chinese sign script started as pictures though they have long since lost any recognisable trace of their origin as such. Likewise, in ancient times, a mathematical proof was primarily a pictorial representation, either: (a) one that employed line figures of circles, triangles, squares and the like; or (b) figurate patterns of dots as seen in the accompanying illustrations. To grasp how such a figurate lay-out may disclose the reliability of a rule, it will be needful to be clear about what a mathematician means by a *nonrecurrent series*. A collection of numbers constitutes such a series when we can place them in such a sequence that the relation of any term (*i.e.* member of the sequence) to its immediate predecessor is expressible in the same way. To make this relation explicit, we label the place (so-called *rank*) of a term in the sequence by one of the sequence of integers (so-called *natural numbers*). Below we see the first even and odd numbers set out in this way:

FIGURATE REPRESENTATION OF ODD AND EVEN NUMBERS

rank	*1*	*2*	*3*	*4*	r
Even	$E_1=2$ $=2(1)$	$E_2=4$ $=2(2)$	$E_3=6$ $=2(3)$	$E_4=8$ $=2(4)$	$E_r=2r$

rank	*1*	*2*	*3*	*4*	r
Odd	$U_1=1$ $=(E_1-1)$	$U_2=3$ $=(E_2-1)$	$U_3=5$ $=(E_3-1)$	$U_4=7$ $=(E_4-1)$	$U_r=2r-1$

You here notice that successive terms of the two series (*even* and *odd*) are both formed from their predecessors by adding 2; we may write this in the form:

$$E_{r+1}=E_r+2=2r+2 \quad \text{and} \quad U_{r+1}=(U_r+2)=2r+1$$

We may likewise represent the *sum* of the first n odd numbers thus: $S_1=1$; $S_2=(1+3)$; $S_3=(1+3+5)$ *etc.* These totals also form a series which we may lay out as follows:

rank	*1*	*2*	*3*	*4*	
Terms	$S_1=1$	$S_2=4$	$S_3=9$	$S_4=16$	. . .
	. . .	$=(S_1+3)$	$=(S_2+5)$	$=(S_3+7)$	. . .
	. . .	$=(S_1+U_2)$	$=(S_2+U_3)$	$=(S_3+U_4)$	. . .

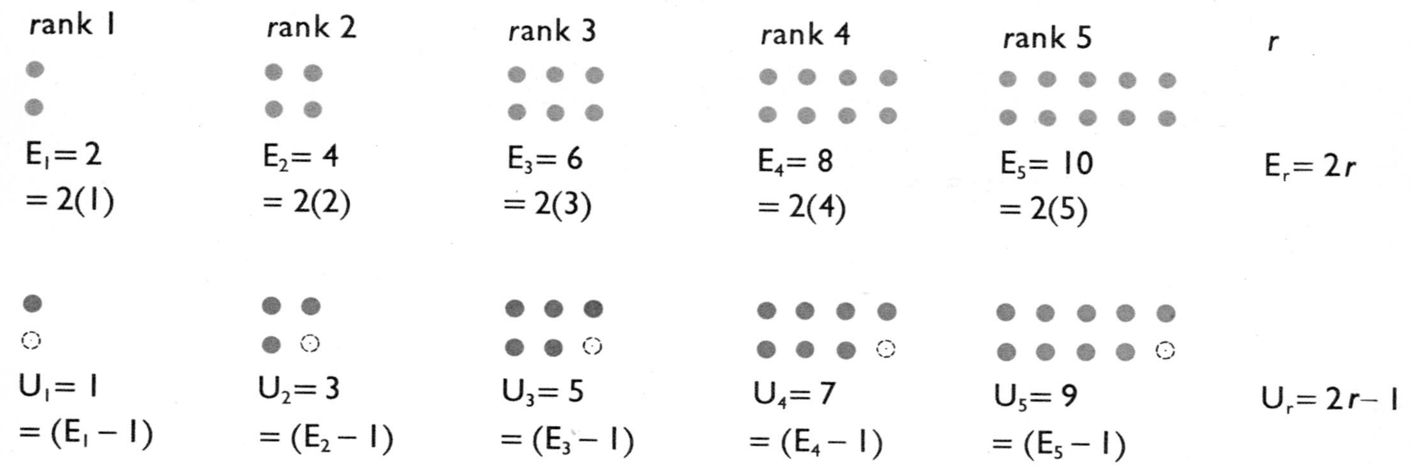

In this lay-out, each term has the same relation to its predecessor, *i.e.*:

$$S_{r+1} = S_r + U_{r+1} = S_r + 2r + 1$$

The blank spaces under Rank 1 in the schema above pinpoint the second part of our provisional definition of proof. Even and odd numbers have this in common: (a) each term of such series is a *whole* number; (b) each is 2 more than its predecessor. If we continue our series of odd numbers backwards from Rank 1 ($U_1 = 1$), we therefore get $U_0 = -1$, $U_{-1} = -3$, $U_{-2} = -5$ *etc.* Now $(S_0 + U_1) = (-1+1) = 0$; but $S_1 = 1$. So the n^2 rule, if true, can be true only for whole numbers which are either *all positive* or *all negative*.

We are now ready to examine how the figurate pattern exposes the general truth of the rule $S_n = n^2$. If we look at any one of the squares, say the third ($S_3 = 9 = 3^2$), we see that we can make the next square ($4^2 = 16$) by putting three dots on one side, three on the adjacent side and one other in the corner, thus adding in all $2(3)+1 = 7$, which is the fourth odd number. Evidently, we can make

In ancient times a mathematical proof was primarily a pictorial representation. Here figurate patterns of dots reveal the reliability of the rules by which we can calculate the total of consecutive even or odd numbers provided they are all positive or all negative.

To build up the sum of consecutive odd numbers beginning with 1, we first add 3, then 5, then 7, and so on. We build up the first consecutive square numbers in precisely the same way. The figurate pattern exposes the general truth of the rule $S_n = n^2$.

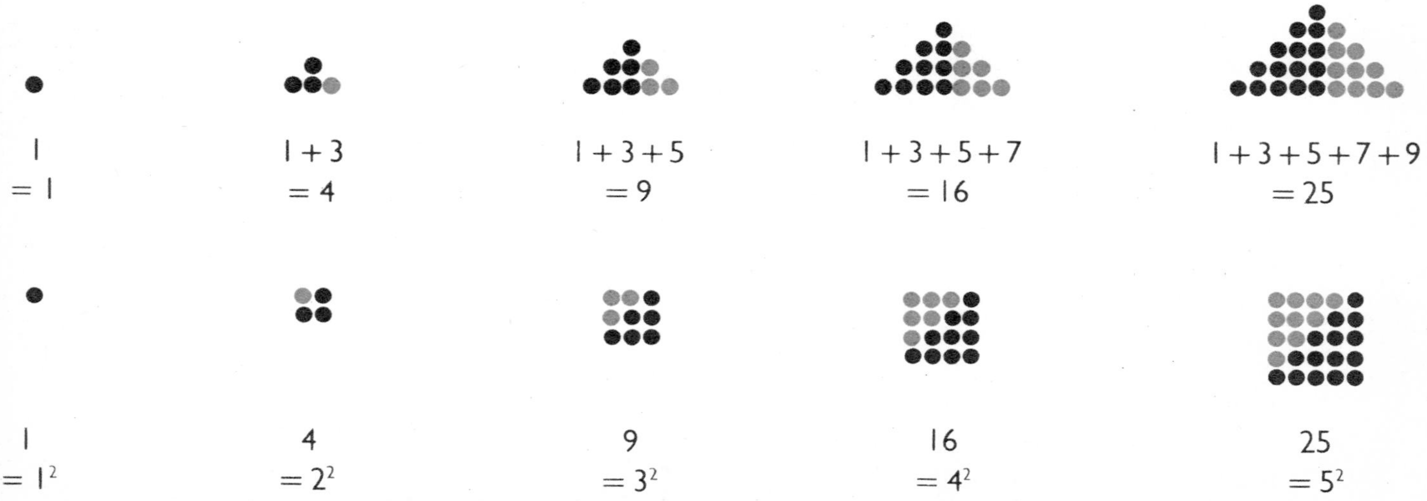

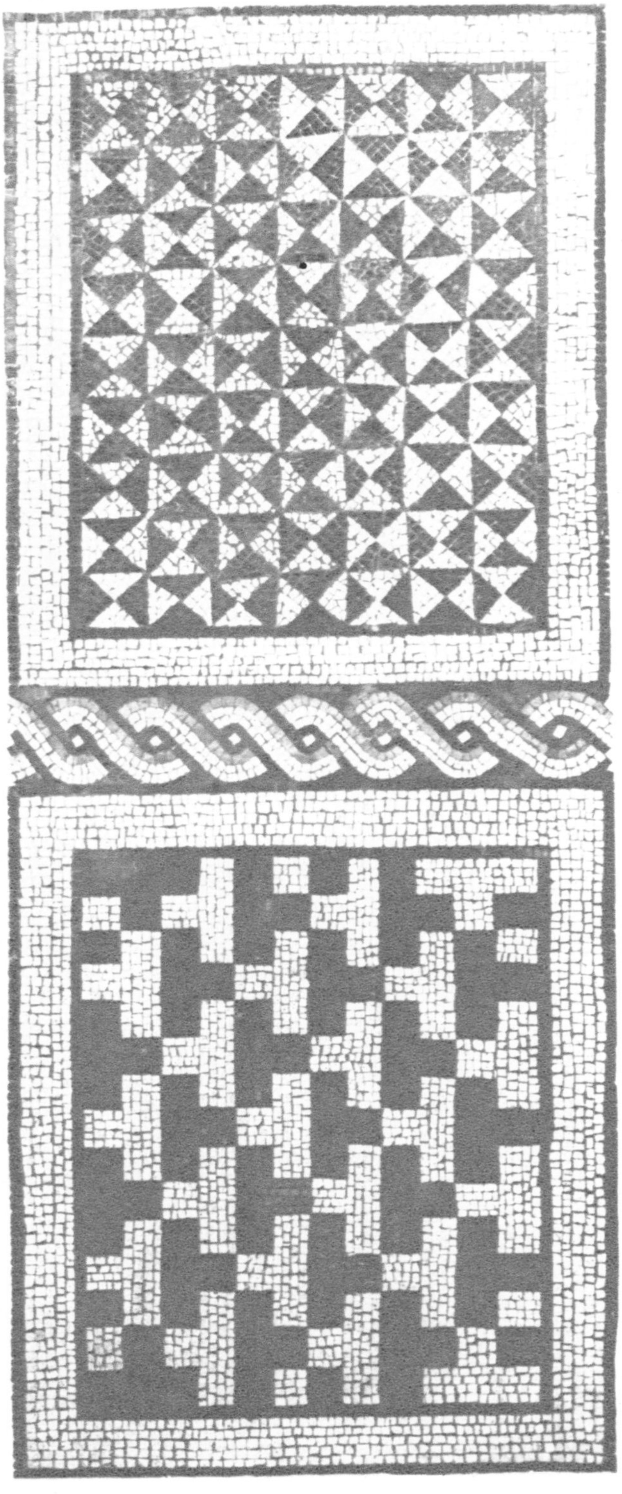

Part of a Roman mosaic floor found in a villa in Dorset, England. The making of mosaics and tiled floors, both of great antiquity, may well have given the clue to figurate representation of whole numbers and to the scale diagram – parent of the geometrical figure.

the next square (*i.e.* 5^2) by adding $2(4)+1=9$ which is the fifth odd number. Thus every term of the series of squares is obtainable by adding the odd number of the succeeding rank. If we can make *both* the sum of the first three odd numbers and the third square number by adding the third odd number to the second square (which is itself the sum of the first two odd numbers), we can thus repeat the process of generating the series of squares and corresponding sums of positive odd numbers indefinitely. For any member (S_n) of rank n in the series corresponding to the sum of the odd numbers, we may therefore write $S_n=n^2$, if we do so with the proviso that n is always positive or always negative.

Figurate representation, which is of great antiquity, is adaptable only to reasoning about series whose members are integers; but the pictorial demonstration of this simple and highly economical rule is a pattern for a less restricted principle of mathematical proof called *induction.* When we can assemble numbers in series form, whether picturable or not, trial and error may suggest a formula expressing the value of a term of rank r in terms of r itself and some fixed numbers, *e.g.* 2 and 1 in $U_r=(2r-1)$; but such fixed numbers, and hence the value of the terms, need not themselves be integers. For instance, $t_{r+1}=(t_r+0.75)$ if $t_r=0.75r+0.25$; but t_r will not then be an integer when $r=2, 3, 4, 6, 7, 8$ *etc,* though it will be so if $r=1, 5, 9$ *etc.*

If we do know the relation of each term to its predecessor or successor (*e.g.* $U_{n+1}=U_n+2$), there is a very simple criterion of whether a formula for finding any term in the series is reliable. If our formula tells us what the rth term is, we can adapt it to show what the $r+1$th term would be in virtue of what we know about the relation of any one term to its predecessor. The result should be that the formula remains unchanged except that $r+1$ takes the place of r in it. The application of this method of testing the n^2 formula for the sum (S_n) of the odd numbers from rank 1 to rank n inclusive is very simple if we know what is easy to represent in a figurate way, *viz.* that $(n+1)^2$

$=n^2+2n+1$. We can then proceed as follows. Since $S_{n+1}=S_n+U_{n+1}$ and $U_{n+1}=U_n+2=2n+1$:

$$S_{n+1}=S_n+2n+1$$

If our formula for S_n is correct, this means that

$$S_{n+1}=n^2+2n+1=(n+1)^2$$

Thus the formula is true for the sum of any number of positive odd numbers, if it is true for its predecessor; but we know that it is true for $n=1(S_1=1)$. So it must also be true for $n=2$, and if so for $n=3$ and so on, indefinitely.

Though this way of conveying the reliability of the n^2 rule merely expresses what is implied in figurate representation by recourse to a more compact battery of signs, it has the great advantage that the method enables us to discuss series which we cannot express in a figurate way. For instance, the reader may apply it to test the rule for the sum (S_n) of the terms of rank 1 to rank n inclusive for the series $t_r=\frac{1}{4}(3r+1)$. As stated, $t_{r+1}=t_r+\frac{3}{4}$; and the correct formula for S_n is:

$$S_n=\frac{n(3n+5)}{8}$$

To carry out the check last mentioned, we have to recall a few of the simplest tricks we learned about manipulating the sign language of mathematics in our schooldays. The exploitation of the so-called method of induction is indeed possible only when the sign language of mathematics has progressed beyond the crudely pictorial level. This has been true only during the past 350 years, and even such familiar items of the sign language of mathematics as + and − have been in use less than 700 years.

When pictorial representation of one sort or another is excessively cumbersome or otherwise impracticable, a battery of signs, which have no pictorial significance as such, now makes it possible for us both to convey rules and to reason about their reliability without losing our way in a maze of words. This sufficiently explains only one reason why the word *economical* comes into our definition of what we mean by mathematics.

$(3+1)^2=3^2+2(3)+1$

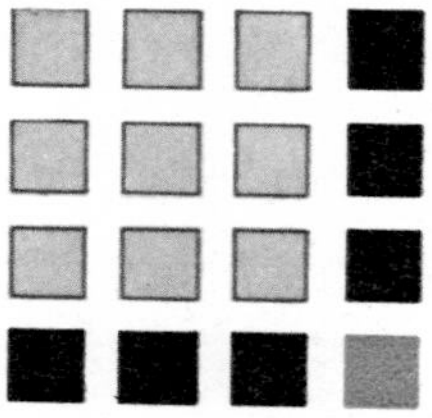

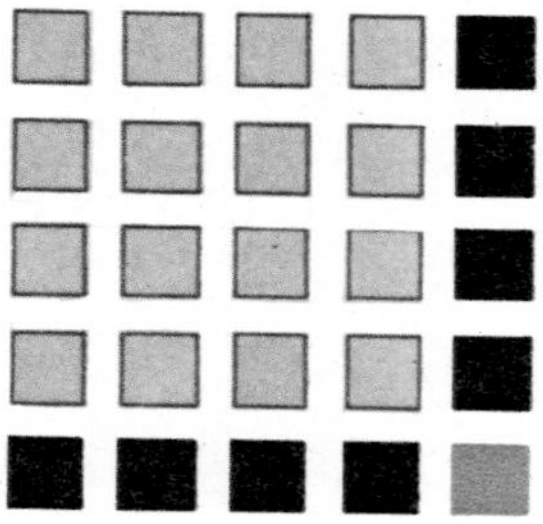

$(4+1)^2=4^2+2(4)+1$

$(n+1)^2=n^2+2n+1$

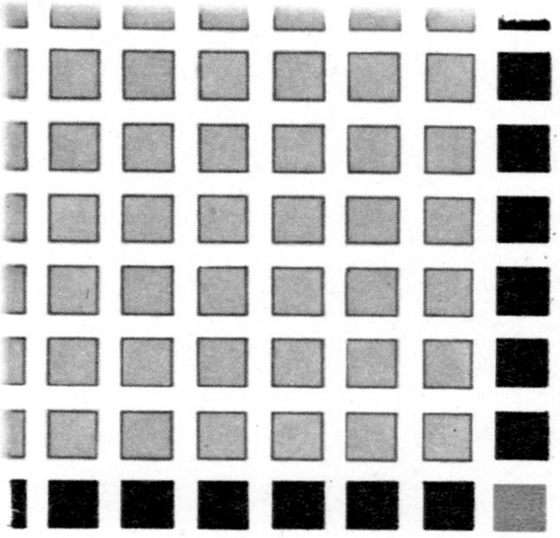

Tile patterns make it clear that to increase a square 3 tiles wide to one 4 tiles wide we must add $(2\times3)+1$ *tiles, making the total* $3^2+2(3)+1$. *If the original square has a width of an unspecified number of tiles, n, the square a tile wider will contain* n^2+2n+1 *tiles.*

A car travels at 30 m.p.h.
On braking it loses speed
at the rate of 22 feet
per second each second.
What is its stopping
distance?

In the sign language of today, the translation is as follows. In terms of initial speed (s_0) *and final speed* (s_t) *at fixed acceleration* (a) *in a time interval* (t), *the mean speed is* $s_m = \frac{1}{2}(s_0 + s_t) = \frac{1}{2}s_0$ *if* $s_t = 0$ *as above. In terms of time and distance* (d) *traversed,* $s_m = d \div t$ *by definition, whence* $d = s_m . t = \frac{1}{2}s_0 . t$. *Also by definition,* $a = (s_0 - s_t) \div t = s_0 \div t$ *so that* $t = s_0 \div a$ *and* $d = \frac{1}{2}s_0 . t = s_0^2 \div 2a$. *Here* $s_0 = (5280)30 \div 60^2 = 44$ *in feet per sec. and* $2a = 44$, *so that* $d = 44^2 \div 44 = 44$ *feet.*

The sign language of modern mathematics is indeed economical in other ways, one of which we may here illustrate by restricting ourselves to series of numbers as hitherto. To discuss series of terms of any sort, we may label them as t_0, t_1, t_2, t_3 . . . t_r *etc* in which the subscript r of t_r signifies its rank. When we are talking about a sum, we might write as below with little economy of space the sentence: *add up all the terms of the series from the term of rank a to the term of rank b*:

$$(t_a + t_{a+1} + t_{a+2} \ . \ . \ . \ t_{b-1} + t_b)$$

Nowadays we usually say the same thing in far less space by giving Σ, the Greek capital letter for s (in *sum*), a special meaning, *viz.*

$$\sum_{r=a}^{r=b} t_r = (t_a + t_{a+1} + t_{a+2} \ . \ . \ . \ t_{b-1} + t_b)$$

Thus we may write the sum of the first n odd numbers starting with the term of rank 1 alternatively as

$$\sum_{r=1}^{r=n} U_r = \sum_{r=1}^{r=n} (2r - 1)$$

A simple trick adapts this sign to the use of series whose terms are alternatively positive and negative, for instance the two below:

r	*1*	*2*	*3*	*4*	*5*	*6*
a_r	$+1$	-3	$+5$	-7	$+9$	-11
b_r	-2	$+4$	-6	$+8$	-10	$+12$

Regardless of sign, the terms of the series above respectively correspond to the odd numbers $(2r-1)$ and the even numbers $(2r)$. To get the signs right, all we have to do is to multiply each term in the series a_r by $(-1)^{r-1}$ and each term in b_r by $(-1)^r$. For the sum of the terms from rank 1 to rank n inclusive, we can therefore write:

$$\sum_{r=1}^{r=n} a_r = \sum_{r=1}^{r=n} (-1)^{r-1}(2r-1)$$

and

$$\sum_{r=1}^{r=n} b_r = (-1)^r . 2r$$

From these examples, it should be clear that the sign language of mathematics has now become highly economical in the sense that it expresses in a very compact way what would take very many words to convey; but the language of modern mathematics is economical in another sense which we may convey by saying that the dictionary of signs contains very many synonyms.

So far, we have used the sign $=$ as we have learned to use it in our schooldays. In higher mathematics, it is now common to use a special sign to indicate that two expressions which contain ordinary number signs are synonymous. Thus $\equiv$ will henceforth stand for *means the same as* in the sentence $a^2+2ab+b^2 \equiv (a+b)^2$. On the other hand, we shall translate $=$ as *is equivalent to the particular numerical value* in $y=3$.

Those of us who are still under thirty have now learned, as few of us over fifty did learn at school, to think of mathematical reasoning as a process of translation. What we customarily call the mathematical solution of a practical problem then involves two steps. The first is to translate the language of everyday life into the sign language of mathematics. The second is to replace different parts of the statement by simpler synonyms until we reach a numerical equality such as $y=3$. This is equivalent to saying: the number sign on the right is the numerical value of the sign on the left, if what we have previously assumed to be a true statement of the problem is correct. To carry out this process of translation by substitution of synonyms, we treat signs within brackets as blocks, the meaning of which we interpret by recourse to certain familiar tricks. The most elementary ones are:

(1) $(+a)\ (+b) \equiv +ab$
$(-a)\ (+b) \equiv -ab \equiv (+a)\ (-b)$
$(-a)\ (-b) \equiv +ab$

(2) $\left(\frac{a}{b} \equiv \frac{c}{d}\right) \equiv (ad \equiv bc) \equiv \left(\frac{a}{c} \equiv \frac{b}{d}\right) \equiv \left(\frac{b}{a} \equiv \frac{d}{c}\right)$

(3) $\frac{a}{b} \pm \frac{c}{d} \equiv \frac{ad \pm bc}{bd}$

(4) $\frac{a}{b} \times \frac{c}{d} \equiv \frac{ac}{bd}$

(5) $\frac{a}{b} \div \frac{c}{d} \equiv \frac{ad}{bc}$ whence $1 \div \frac{c}{d} \equiv \frac{d}{c}$

(6) $a^n \equiv a \times a \times a \ldots$ (n factors each a)

(7) $\sqrt{a} = b \equiv a = b^2$ and $\sqrt[n]{a} = b \equiv a = b^n$

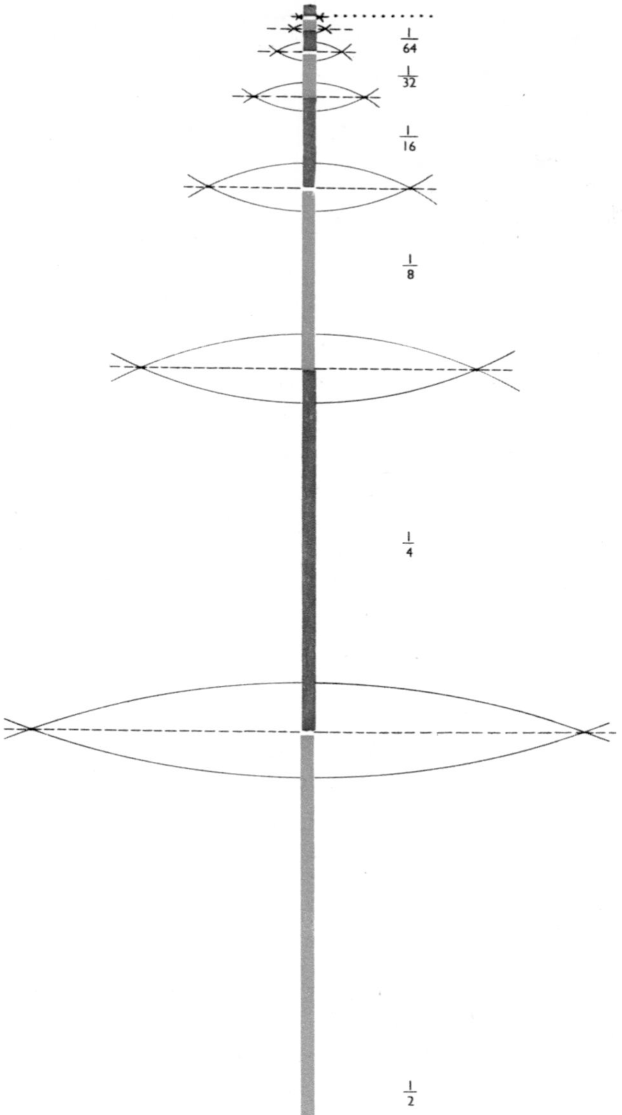

By use of the compass, mathematicians of antiquity could repeatedly bisect a straight line of unit length into 2, 4, 8, 16, 32, etc., equal segments. Nevertheless it took them long to grasp that $\frac{1}{2}+\frac{1}{4}+\frac{1}{8}+\frac{1}{16}+\frac{1}{32}$, *etc., is a convergent series. It can never exceed unity.*

Some of the synonyms which we recognise as such by applying now commonplace rules of the sign language of mathematics are likewise recognisable at a pictorial level by looking at figures one can draw on a floor of mosaic tiles and by tracing lines on sand with a straight-edge or with a cord and a couple of pegs to do the job of a compass when there is no compass to hand. Here are three which have been on record for 4000 years at least. Our pictures show why they are reliable rules of reasoning:

$$a^2+2ab+b^2 \equiv (a+b)^2$$
$$a^2-2ab+b^2 \equiv (a-b)^2$$
$$a^2-b^2 \equiv (a+b)\ (a-b)$$

By solid figures we can likewise exhibit such a relation as $a^3-b^3 \equiv (a-b)\ (a^2+ab+b^2)$; but we cannot vindicate in any corresponding pictorial form $(a^4-b^4) \equiv (a-b)\ (a^3+a^2b+ab^2+b^3)$ or indeed the breakdown of a^n-b^n except when n is a positive integer not exceeding 3. What rules we can indeed represent by the scale diagram form may be also expressible in figurate form; but when we can make a scale diagram of them instead of a mosaic – which is merely a figurate pattern – we take a decisive step forward in the art of calculation. We are no longer hamstrung by having to work with whole numbers only.

By taking this step, we approach an issue which has kept mathematicians since about 500 B.C. fully employed in the attempt to clarify what is or is not a reliable rule of mensuration; and we have committed ourselves to the use of numbers as labels in a new way. So far, we have been talking only about whole numbers as labels for *counting* discrete objects, and only about how we can manipulate them reliably when performing simple calculations. In this domain we may schematise what counting implies by imagining that we have at our disposal as many boxes as we require, each containing a particular number of balls and each labelled accordingly by one of the integers 1, 2, 3 ... On the assumption that no two boxes contain the same number, the process of counting, say, A apples is that of finding the box containing balls which we can singly pair off with each apple, leaving no remainder. If the label on the box is 67, we say that $A=67$.

In one way, this is like what we do when we use numbers as labels of *measurement*, since the scale divisions of our ruler or dial are discrete like the balls of our parable. There is, however, an essential difference between the *matching process* of putting balls with apples in one-to-one correspondence and the matching process of assigning a figure for the length of a wall or for the angle of elevation of a flagstaff at ground level. Before we begin to discuss proof in connexion with measurement, we should therefore be clear about what we really mean by measurement. To convey it economically, we shall need to enlarge our dictionary of signs as below:

Sign Language	*Meaning*
$a>b$	a is greater than b
$b<a$	b is less than a
$a \geqslant b$	a is greater than or equal to b ($\equiv a$ is not less than b)
$b \leqslant a$	b is less than or equal to a ($\equiv b$ is not greater than a)
$b \simeq a$	b is approximately equal to a, *i.e.* so near to a that the difference between them is of no practical importance in the context.

Today we can recognise mathematical synonyms as such by applying what are now commonplace rules of the sign language of mathematics. Before the formulation of such rules it was difficult to recognise them without the aid of mosaic or scale diagrams. The diagrams here and opposite are recognition-aids to three synonyms:
$(a+b)^2 \equiv a^2+2ab+b^2$; $(a-b)^2 \equiv a^2-2ab+b^2$; $a^2-b^2 \equiv (a+b)\ (a-b)$.

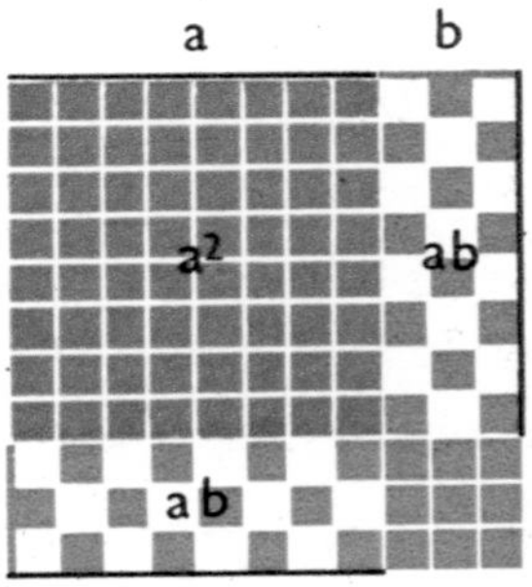

$(a+b)^2 \equiv a^2+2ab+b^2$

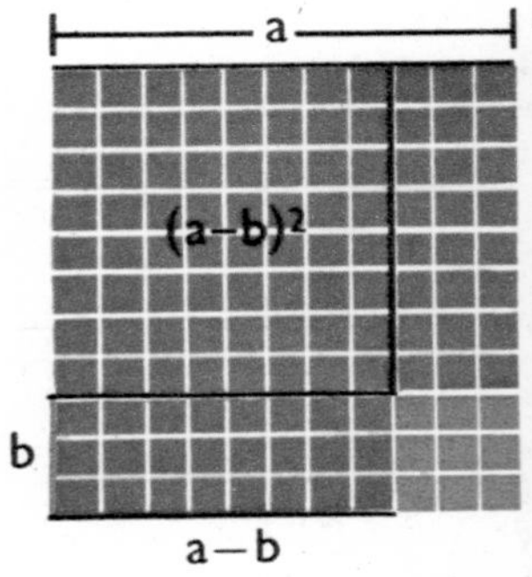

$(a-b)^2 \equiv a^2-2(a-b)b-$
$\equiv a^2-2ab+2b^2-$
$\equiv a^2-2ab+b^2$

These signs make it possible to discuss without tiresome circumlocution the use of numbers in a practical situation when an exact specification is irrelevant or impossible. For instance, it may serve our purpose to define the ratio of the circumference to the diameter of a circle ($\pi = 3.14159\ldots$) correct to 5 significant figures, in which event we shall write alternatively $\pi \simeq 3.1416$, or more explicitly $3.14159 < \pi < 3.14160$. However, the reason for introducing such signs here is that measurement in the practical sense of the term is a process of *matching as best we can*. When we use a number (n_m), whether a whole one or not, to label the result of measuring something, the most we can therefore say about it honestly is a statement of the type $a < n_m < b$, i.e. a is less than n_m which is itself less than b.

This will be clear if we examine what we mean by measuring the length (L) of a wall. At the most primitive level, we may adopt a stride as our unit and count the paces required to reach opposite ends. In the microcosm of drawing-board procedure this multiplicative method corresponds to repeated rotation of a compass set at a fixed angle through a half circle. However carefully we do this, we have no guarantee that the tip of the toe or the point of the compass will correspond as exactly as the eye can judge with the end of the wall or with the line which represents it in our scale diagram. If we say that the wall is 20 paces long, we may mean either of two things:

(a) it is at least twenty but not twenty-one paces long, *i.e.* $20 \leqslant L < 21$;

(b) it is nearer to twenty than to any other number of paces, *i.e.* $19.5 < L < 20.5$.

Nowadays, we usually perform the matching process by some sort of *calibrated* device or scale, *i.e.* a tape, rod or chain lined up with marks equally far apart – as far as the eye can judge. We then place the zero mark beside one end of the wall (or line) and read off the number of the mark nearest the other end. To be sure, the end may – as far as the eye can judge – exactly correspond to a particular mark ($k\triangle$). Even so, the mark itself is recognisably thick, and our recorded measurement may then be out by as much as half its thickness. If $\triangle$ is the smallest division on the scale, the most exact statement we can ever guarantee the possibility of making is therefore:

$$k\triangle - \tfrac{1}{2}\triangle < L < k\triangle + \tfrac{1}{2}\triangle$$

For instance, our smallest scale division may be 1 mm., and the nearest mark we can match against one end of a line when we have set our scale with its zero mark against the other may be 45 mm. The most precise way of stating the length of this line is then $44.5 < L < 45.5$.

To those of us who take measurement seriously, the possibility of making a scale to sidestep the more laborious (and primitive) multiplicative process of matching needs justification. If we have no concern with the unit, we can mark off with as much accuracy as we require by the compass as many equal divisions as we like (*e.g.* 100) in the multiplicative way. On the other hand, we may decide to fix a unit (*e.g.* the standard metre in Paris). We can then bisect it repeatedly by use of the compass into 2, 4, 8, 16, 32, 64, 128, 256, 512, 1024 equal segments if our compass is sufficiently sensitive; but we shall need a different recipe to divide a pre-assigned distance into 10, 100, or 1000 equal intervals. What is perhaps the most useful

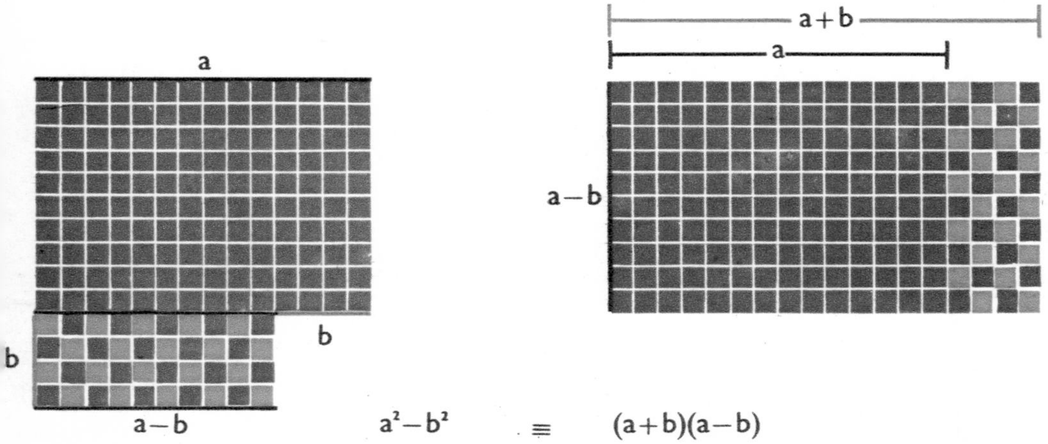

$a^2 - b^2 \equiv (a+b)(a-b)$

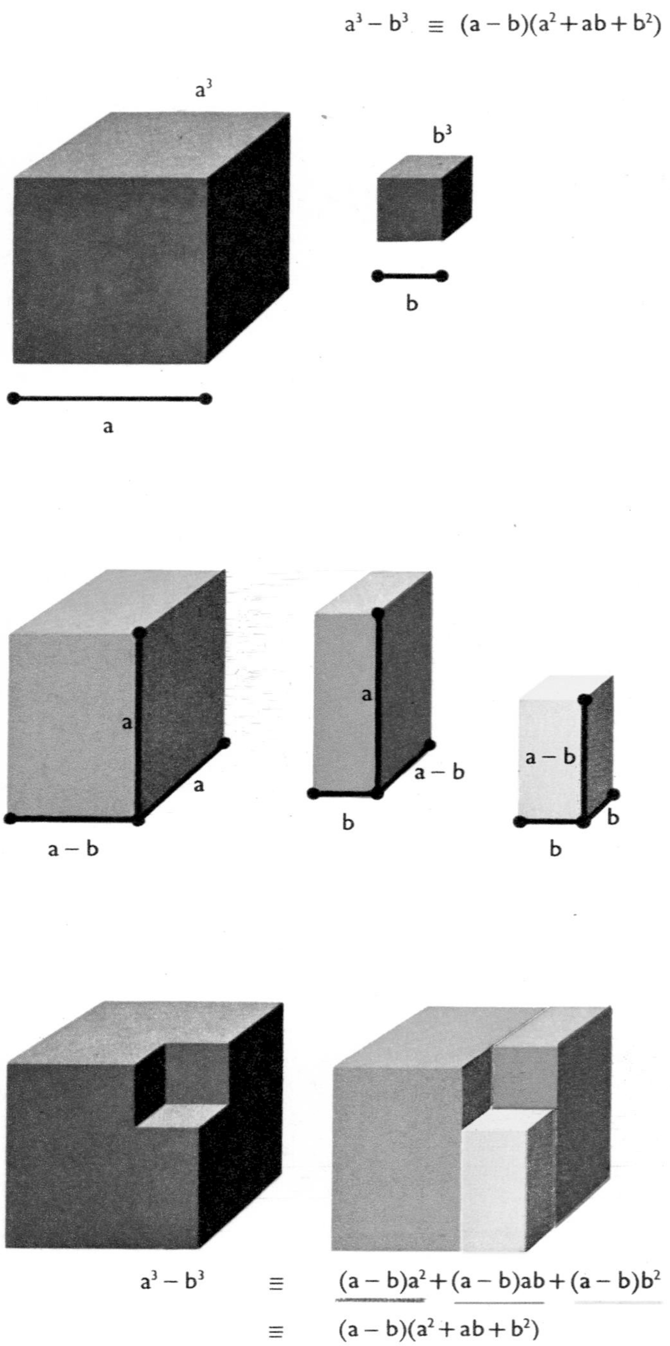
$a^3 - b^3 \equiv (a-b)(a^2+ab+b^2)$
a^3
b^3
b
a
a
a
a − b
a
a − b
b
a − b
b
b
$a^3 - b^3$
$\equiv (a-b)a^2 + (a-b)ab + (a-b)b^2$
$\equiv (a-b)(a^2+ab+b^2)$

A diagram can likewise exhibit such a relationship as $a^3 - b^3 \equiv (a-b)\ (a^2+ab+b^2)$; *but since we cannot intelligibly picture more than three dimensions, no scale diagram can vindicate the breakdown of* $a^n \pm b^n$ *when n is a positive integer greater than 3.*

and far-reaching rule of Euclid's geometry, though Euclid himself clearly did not regard it as such, supplies us with such a recipe. The rule itself is that:

> *the ratios of the lengths of corresponding sides (sides opposite equivalent angles) of right-angled triangles whose other two angles are equal are also equal.*

This is another way of saying that if we complete a right-angled triangle about any line inclined at a particular angle A to a base by erecting a vertical line on the latter, so that the side inclined to the base is of length h, that of the base of length b and that of the perpendicular of length p, the ratios $p:h$, $b:h$ and $p:b$ (or their reciprocals) are the same however large we make the triangle. We now have names for these ratios. Unless we have already done so, it will be useful to add them to our dictionary at an early stage, and to notice that we can represent each as a single line in a circle of unit radius ($h=1$):

$\sin A$ (sine of A) $p \div h$ so that $p = h \sin A$
$\cos A$ (cosine of A) $b \div h$ so that $b = h \cos A$
$\tan A$ (tangent of A) $p \div b$ so that $\tan A = \sin A \div \cos A$

Since the possibility of subdividing a scale in whatever way we choose depends on the use of the foregoing rule, a guarantee of its reliability is of the utmost practical importance. What we shall then regard as an adequate justification will depend partly on: (a) whether we want to claim no more precision of statement than is realisable in terms of the matching process of mensuration in the real world; (b) how much we are willing to assume about figures by common agreement. This brings us face to face with two differences between the viewpoint of Euclid and that of the surveyor or engineer. In terms of what we regard as proof in the context of measurement, the demands of the surveyor or of the engineer part company with the Greek tradition on two issues.

We may speak of the first as *commensurability*. Euclid devoted much of his system to the discussion of what we can say convincingly about two lines, if not more than one of them exactly matches a whole number of scale divisions on any one scale, however sensitive we conceive it to be. Of this, much more anon. In view of what we have seen to be true about the matching process in practice, it here suffices to say that we can never compare the measurement of two lines with the assurance that they are commensurable in Euclid's sense. Consequently, the search for a proof which satisfies the demand for reliability in the real world commits us in advance to no such exacting demands on the use of numbers.

In what follows next, we shall also part company with Euclid *vis-à-vis* what we can agree to assume as a basis for further discussion. Euclid believed, as did almost all Western mathematicians till less than 150 years ago, that: (a) his definitions of figures supplemented by a statement of self-evident principles were consistent with what one can construct with a straight-edge and a compass; (b) the instruments last named had some peculiar merit which other drawing devices lack. Among the seemingly self-evident principles he propounded was a criterion of when lines are *not* parallel. As we shall see later, we cannot infer the fundamental properties of parallel lines from any rule and compass recipe for making one line parallel to another without its aid, unless we invoke some other principle by no means self-evident. Nevertheless, geometers have been uneasy about Euclid's axiom for many centuries; and its discussion has led during the past 150 years to the construction of geometrical systems which treat, in effect, all possible lines as curved, albeit of curvature indistinguishable from any intelligible definition of a straight line throughout a very large span of the visible universe.

Though his definitions of a circle and of a circular arc are fully consistent with the properties of a compass, the truth is that Euclid's definition of a straight line has no bearing on the construction of a straight-edge, and his system discloses no reason to believe it is possible to make one. In short, there is nothing sacrosanct about a ruler as such. Consequently, we shall here adopt as our

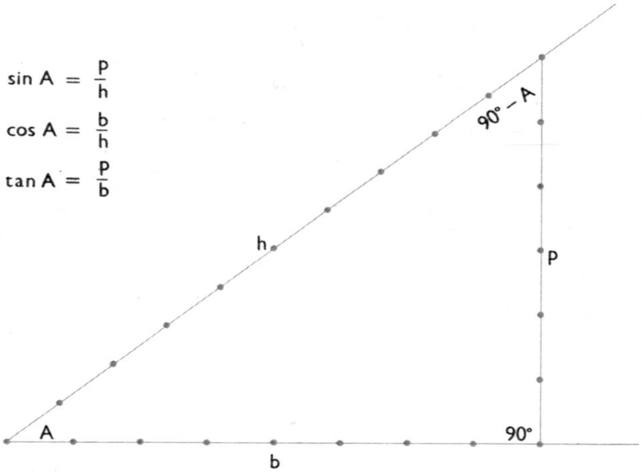

If (as we shall later see) the angles of a triangle add up to two right angles, the third angle of a triangle whose other two are A and 90° is $180° - (A + 90°) = 90° - A$. *From this it follows that:*

$$\sin(90 - A) = \frac{b}{h} = \cos A;\ \cos(90 - A) = \frac{p}{h} = \sin A.$$

In a circle of unit radius $(h = 1 = r)$:

$$\sin A = p \div h = p \quad \text{and} \quad \cos A = b \div h = b$$

The figure shows that tan A is actually the length of the tangent which subtends A at the base. It also shows why some writers of the past called the sine a semichord.

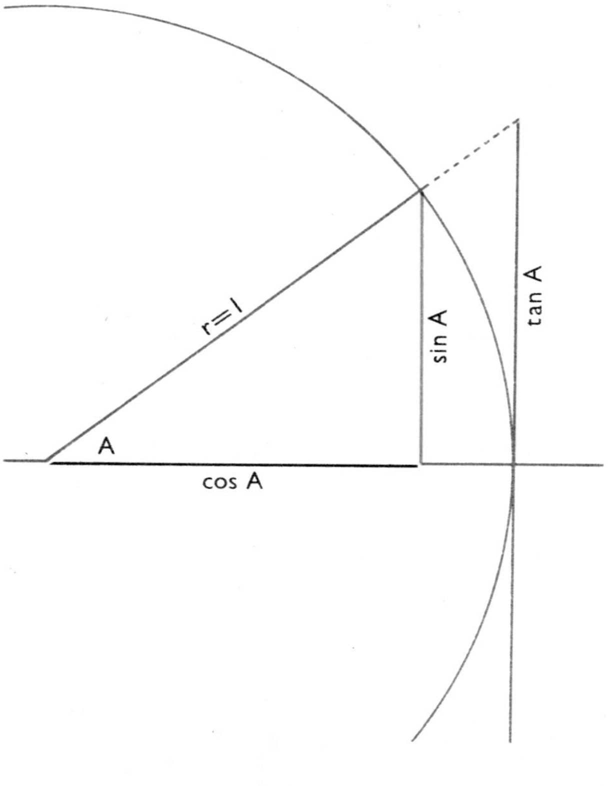

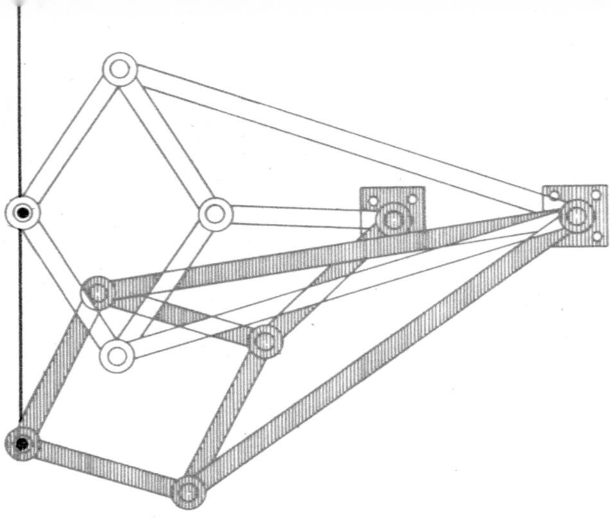

Euclid's definition of a straight line has no bearing on the construction of a straight-edge. The Peaucellier Linkage, invented in 1864, is the first constructed drawing instrument which truly traces a straight line.

definition of parallelism one which is consistent with the assumed properties of any drawing instrument designed to trace parallel lines, *viz.*:

> two straight lines are parallel if equally inclined to any one straight line which crosses them, and if so inclined are equally inclined to every other straight line which does so.

To proceed further, we shall not regard it as necessary to prove two statements which need no dissection to disclose their truth:

(1) vertically opposite angles of two intersecting straight lines are equal, whence alternate angles which a straight line traces across two parallel straight lines are also equal;

(2) two triangles are equivalent in every way other than orientation and position in space, if *either* the lengths of two sides and the angle between them are equivalent *or* the two angles at the extremities of a side of fixed length are equal.

The last statement follows from the fact that either of the two properties mentioned suffices to specify a way of making a three-sided figure whose metrical properties are unique. If we assume the first rule and are willing to adopt the foregoing definition of parallelism without more ado, a simple construction then shows what we may call CLAIM 1 of the case law of surveying, *viz.*: *the three angles of a triangle add up to* 180°.

From this, it follows that right-angled triangles have a property of peculiar interest. If A, B, C are the three angles of a triangle and $C=90°$, the foregoing assertion implies that $A+B=90°$, so that $B=90°-A$ and $A=90°-B$. On this understanding, we are now ready to dispose of the already stated rule which we shall call CLAIM 2, *viz.*: *the ratios of the lengths of corresponding sides of right-angled triangles whose other two angles are equal are also equal.*

The demonstration we shall now follow is one which Euclid would have rejected for reasons we look into at a later stage, when we ask ourselves the question: was Euclid's geometry a science of measurement? In our picture (page 22) we have fitted three right-angled triangles into a fourth. They fit because they are equi-angular; and we have chosen them so that the upright side of the first is a quarter of that of the fourth, the upright side of the second one half, and the upright side of the third three quarters. Thus the rule we wish to vindicate is so far true. We shall now make two assertions. To the first of these, Euclid would have taken no exception, if Cantor rightly interprets the implications of the first proposition of his tenth book. The second he would have been reluctant to admit.

(a) With a compass we can divide the upright side of any right-angled triangle into an even number of as many segments as we choose till they are indistinguishable for the purposes of assigning any number to the matching operation of measurement;

(b) we can do this with as much precision as we can ever hope to achieve, and nothing useful we can say about what we are doing in measurement can go beyond statements of this sort.

So far as anything we can prove by drawing figures has any relevance to what we are doing in measurement, the case for the defence is then complete. Everything we can say about super-imposing a right-angled triangle whose side is one quarter of a unit, one half of a unit or three quarters of a unit on a right-angled triangle of one unit, is equally applicable to super-imposing a triangle of $\frac{1}{1024} = (\frac{1}{2})^{10}$ of a unit – or any fraction of a unit which is some power of $\frac{1}{2}$ – till we have reached the limit at which we can distinguish a half scale division from a whole one. This is sufficiently exacting for the purpose in hand, *i.e.* if the purpose in hand is measurement undertaken with the explicit safeguard that the number of scale divisions involved is large enough to ensure a *zone of uncertainty* $(\pm \frac{1}{2} \triangle)$ proportionately small enough in comparison with the lengths (L_1, L_2, L_3) of the three sides. In numerical terms, this is achievable if we can set an upper limit to the error which may arise from cutting off all significant figures in both numerator and denominator of a ratio after a specified number (n) of them. As we shall later see, we can do so. Indeed, we should not be able to give instructions to a mechanical computer in the idiom it understands, if we could not do so.

Opposite angles of two intersecting straight lines are equal. If our definition of parallelism is consistent with the assumed properties of the parallel ruler, it follows that alternate angles which a straight line traces across parallel straight lines are also equal.

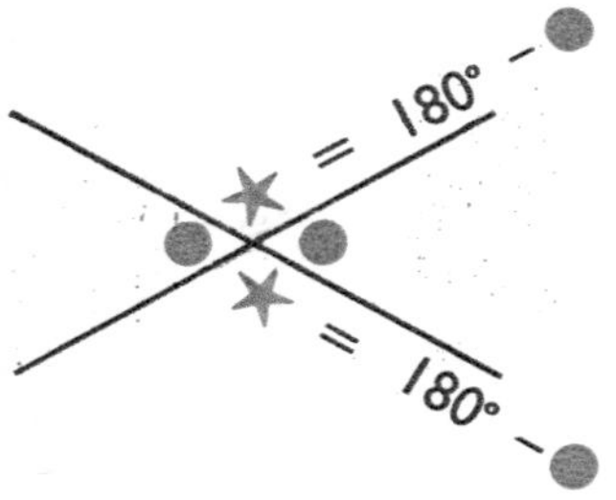

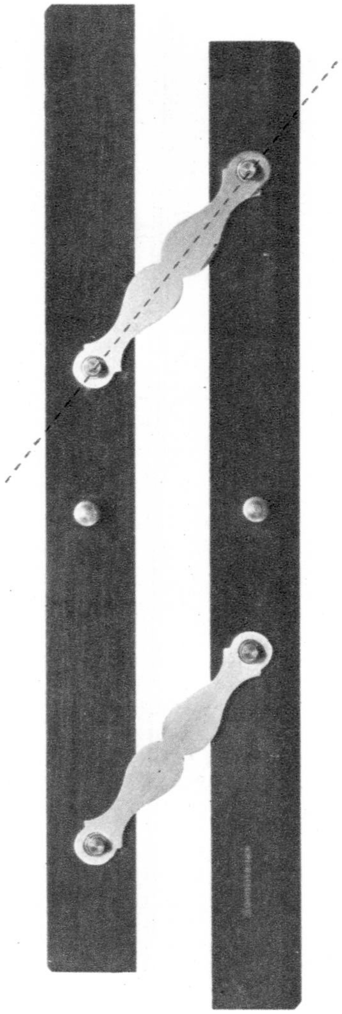

The making of a parallel ruler demands the assumption that two straight lines are parallel if equally inclined to any straight line which crosses them.

The three angles of a triangle add up to 180°. (From diagram opposite and definition of parallelism.)

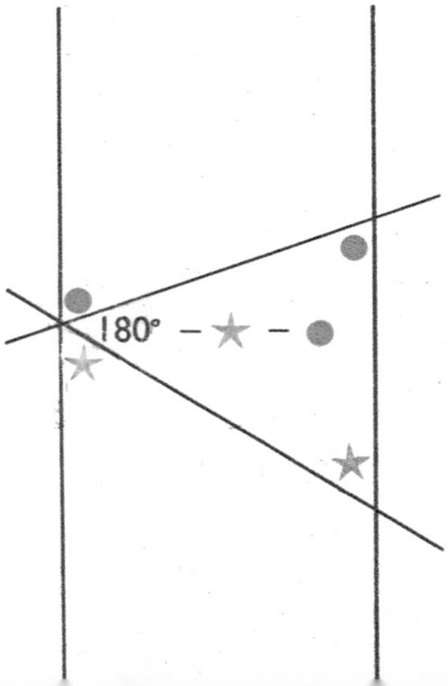

Cl.1

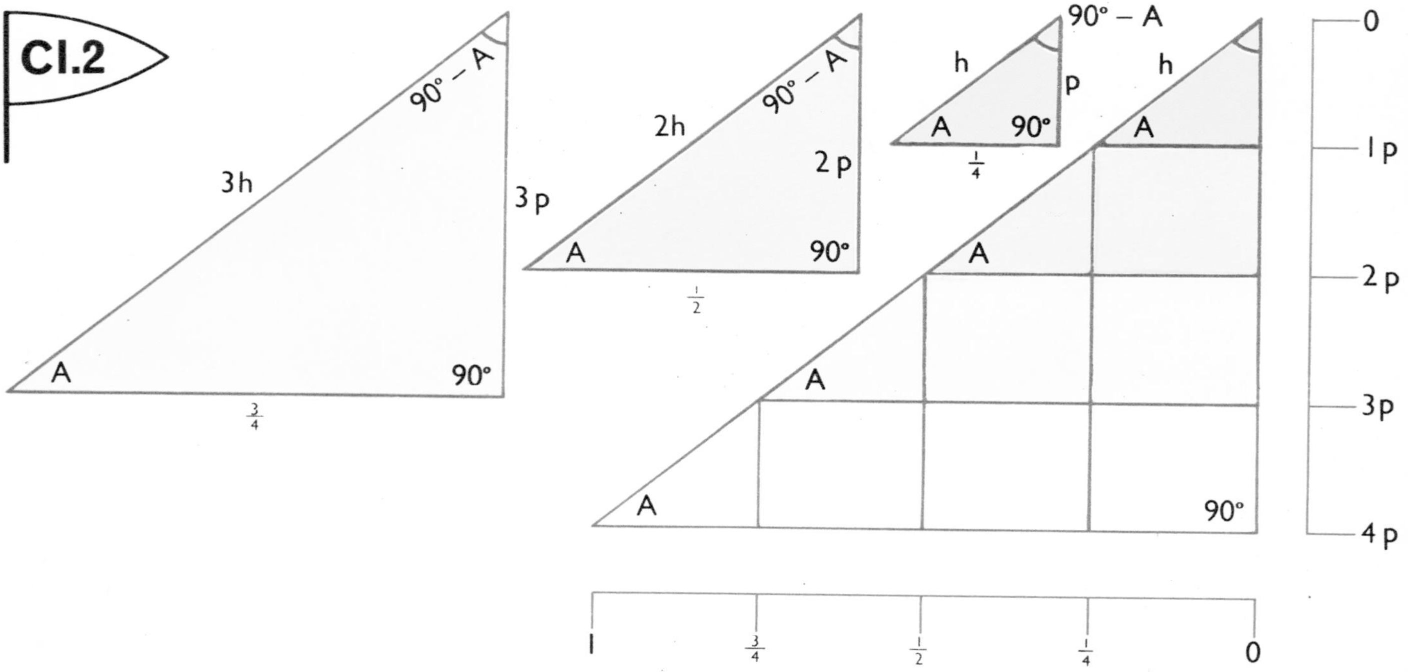

The ratios of the lengths of corresponding sides of right-angled triangles whose other two angles are equal are also equal: i.e. in a right-angled triangle, sin A, cos A and tan A do not depend on the length of the sides for fixed A.

From this it follows that if ae *is parallel with* AE, *the straight lines* OB, OC *and* OD *which cut* AE *into equal segments also cut* ae *into equal segments. Since we can mark off with a compass as many equal divisions as we like along any straight line* (AE), *this gives a recipe for dividing* a line of pre-assigned length (ae) *into as many equal segments as we like.*

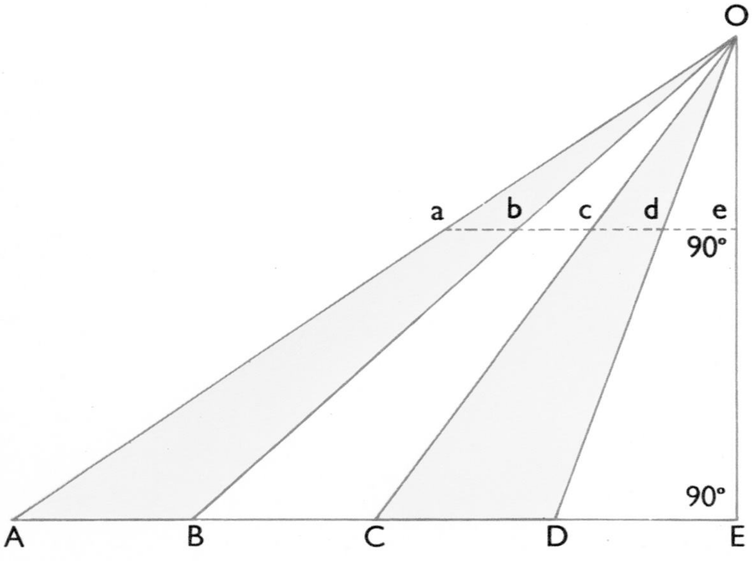

One pay-off appears in the picture below. If I want to divide a line matched against the standard metre at Paris into a hundred equal divisions, I can do so by: (i) first laying out on any convenient line one hundred equal segments by use of my compasses; (ii) then applying a rule which depends upon the one which we have last examined. The new one embraces two assertions:

(a) in a right-angled triangle, lines drawn from an apex to equidistant points along the opposite side (primary scale) divide any line parallel to that side into a corresponding number of equal segments (secondary scale);
(b) the ratio of the interval between two scale divisions of the secondary to that of the interval between two scale divisions of the primary is equal to the ratio of their vertical distances from the apex.

Apart from the fact that the last rule, of which the picture (left) exhibits the reliability, tells us how to divide a line into any number of equal segments with as much precision as we can hope to achieve within the domain of measurement, it provides us with the means of making measurement more precise in the sense that the zone of uncertainty is less. In the world of engineers and surveyors, what we mean by great accuracy in a statement such as $a < M < b$ is that $M - a$ and $b - M$ are very small fractions of M. For instance if we record 5 kilometres as the outcome of matching up to a scale whose smallest interval is 1 millimetre (=0.000001 km.), what we call briefly 5 km. is really something ($4.9999995 < M < 5.0000005$) which differs more or less from 5 km. by no more than 0.0000005 km., *i.e.* with a proportionate error not exceeding one in ten million. Till about 300 years ago there was only one method of getting great accuracy in this sense. We may call it the *pyramid* method.

The astronomer priests of antiquity accomplished what would otherwise seem to be incredible exploits of precision precisely because they made their instruments large enough. When we read that a Greek sea captain in about 400 B.C. was able to give the latitude of Marseilles correct to a tenth of a degree, it is therefore less remarkable if we also reflect upon the likelihood – indeed well-nigh certainty – that he did so by measuring the ratio of the length of the sun's noon shadow to the height of a very high tower which cast it. A calculation nearer to our time brings this home. To make the observations which provided so much new material for Kepler and Newton's theories of planetary motion, Tycho Brahe (1546–1601) used a quadrant (half-protractor with sights) with a radius of 19 feet. On this scale, one degree at the circumference is 4 inches, and 1 millimetre intervals, easily discernible to the eye, correspond to 100th of a degree. Till the time of Newton himself and after Kepler had completed his life-work, the pyramid principle was indeed the only way of achieving such accuracy.

We now have another way which makes it possible to match up intervals smaller than we can directly recognise with the eye. To do this, we need a second scale with divisions only slightly nearer than the finest ones of another. The invention of the secondary scale (*vernier*) depends on a neat application of the rule last stated. If $\triangle_1$ is the width of the smallest division of our primary scale, our rule tells us how to make a second scale with slightly smaller intervals $\triangle_2$ in a fixed ratio, *e.g.* $\triangle_1 \div \triangle_2 = 10 \div 9$ or $\triangle_1 \div \triangle_2 = 60 \div 59$. If the secondary scale slides along the graduated edge of the primary, we can place the zero mark of the secondary scale at a distance x

For part of the year the noon sun's elevation is less than the slope of the Great Pyramid. At such times the properties of similar right-angled triangles provide a means of measuring the pyramid's height. Ratio of stick's height to stick's shadow is the same as ratio of pyramid's height to pyramid's shadow measured from middle of base.

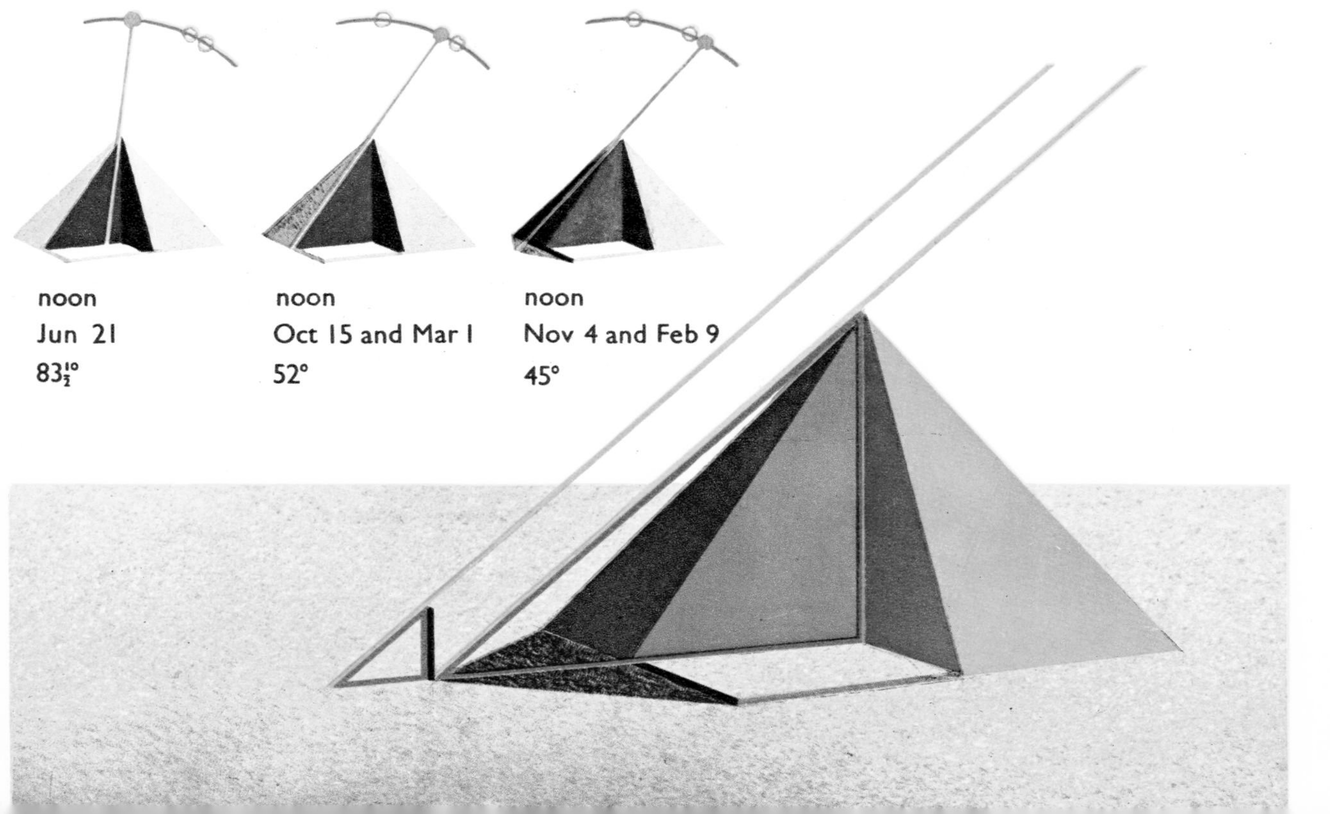

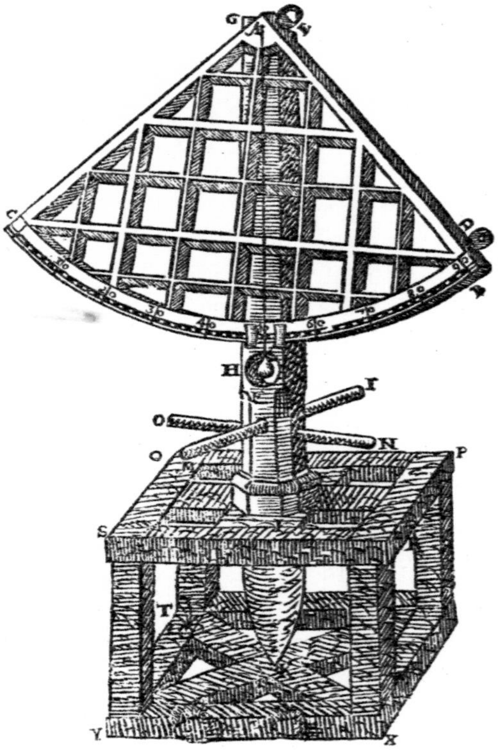

One of Tycho Brahe's quadrants had a radius of 19 feet. At the circumference, 1° ≃ 4 inches, and 0.01° ≃ 1 mm. (Sixteenth-century woodcut.)

To produce a secondary scale on which the divisions are nine tenths as great as those on the primary scale, we invoke Claim 2 (page 22). If the base of the smaller triangle is nine tenths as far from the apex as is the base of the larger triangle, it is also nine tenths as long as the base of the larger triangle. Straight lines divide both into ten equal segments.

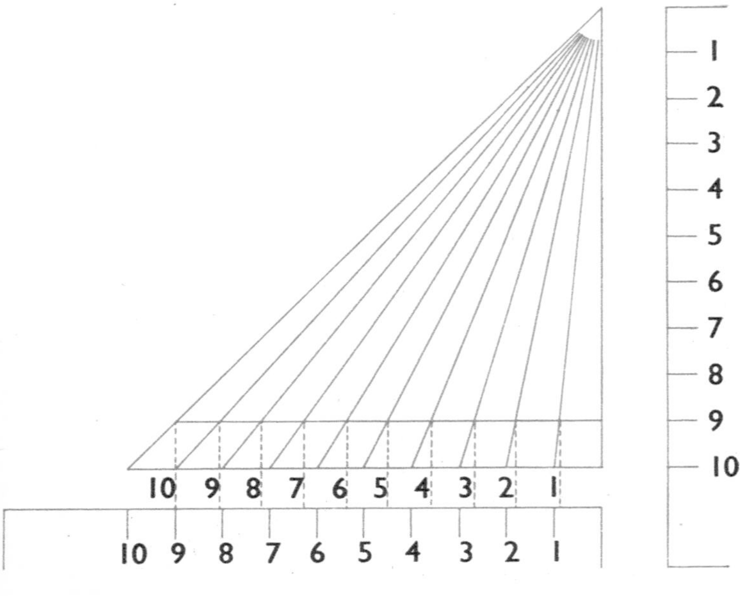

from some mark on the primary and note the first of its scale divisions in line (so far as the eye can judge) with a scale division on the primary. If we represent such a scale division on the secondary by the number a, our picture (right) shows that:

$$x + a\,.\,\triangle_2 = a\,.\,\triangle_1 \text{ so that } x = a(\triangle_1 - \triangle_2)$$
$$\text{If} \quad \triangle_2 = 0.9\,\triangle_1,\ a(\triangle_1 - \triangle_2) = (0.1)a.$$

To make a vernier for an angular scale, we have to exploit another Euclidean rule the reliability of which we shall vindicate at a later stage, *viz.* the boundaries of circles are proportional to their radii, whence the lengths of equi-angular concentric arcs are likewise proportional to their radii. We can thus mark off on a steel tape applied to the edge of a circle whose radius is fifty-nine sixtieths of that of the primary scale a secondary scale of equal segments ($\triangle_2$) of an arc each fifty-nine sixtieths of a segment ($\triangle_1$) on the primary scale. If we do so, $a(\triangle_1 - \triangle_2) = (\frac{1}{60})a$. The thing to note about this is that the ruling of a secondary scale to read off minutes $(\frac{1}{60}^\circ)$ from a primary with scale divisions for degrees a little wider than they have to be, if visually distinguishable, needs to be only slightly more sensitive than the primary.

If what has gone before has helped you to be clear about the difference between discovery and proof, time spent on it will be worth while because the emotional block of many intelligent people when confronted with a formula or a page of mathematical reasoning is more often than not due to a misunderstanding. The tradition of teaching mathematics places great emphasis on proof and rarely gives any clue to the way in which we decide what is worth while proving. Thus many of us realise too late, if at all, how much explorative work with numbers and scale diagrams goes into the making of mathematics. In short, few realise how much seemingly obscure but reliable reasoning about counting and measurement draws on a fund of natural history accessible to anyone who rambles aimlessly around the number landscape. Still fewer of us experience the uplift one's morale gets from realising that things we ourselves need a few hours to take in took the

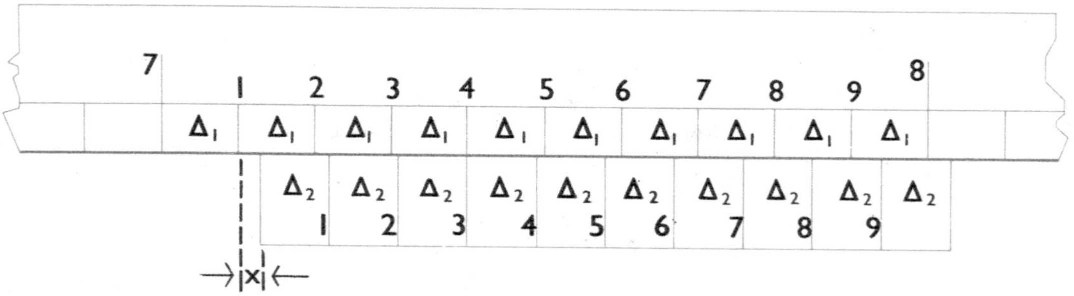

wisest men of ancient times the same number of centuries or millennia to get clear about.

What we have so far dealt with may also throw some light on a difference between the viewpoint of the pure mathematician and the consumer. It should give the reader who needs it a clue to one sense in which the former uses the term *rigorous*, in contradistinction to what we here call *satisfactory*, when speaking of proof. When we speak of the reliability of a rule in a practical sense, all that concerns us is whether it is valid within the whole range of situations in which we propose to use it. So long as we refrain from using it in other situations, the adequacy of its demonstration is then final. The mathematical purist who is eager to explore and to prescribe every possible situation in which it is applicable will not be content with finality in this sense. If it happens that events invite us to apply the rule to situations which do not fall within the scope of our demonstration, we have then reason to be grateful to those who have done a more thorough job.

The invention of the secondary scale (vernier) enables us to achieve greater precision with far smaller instruments. Here the divisions on the primary scale ($\triangle_1$) are 1 cm, those on the secondary scale ($\triangle_2$) are 0.9 cm. We can see that $x+3\triangle_2=3\triangle_1$, so that $x=3(\triangle_1-\triangle_2)=3(0.1\ cm)=3\ mm$. Thus, though our smallest scale divisions are 0.9 centimetres, we can measure in millimetres.

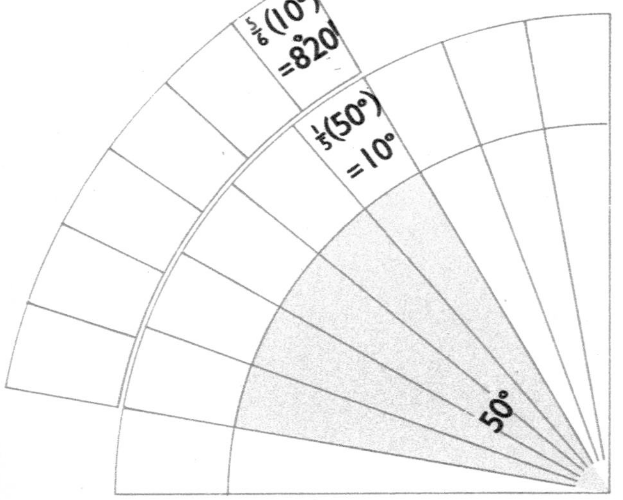

Here the divisions on the secondary scale are five sixths as great as those on the primary. If the primary scale is marked off in degrees we can thus measure in sixth parts of a degree. The secondary scale divisions are commonly fifty-nine sixtieths as great as those of the primary, and we can then measure in minutes.

Chapter 2 Our Hindu Heritage

So far we have spoken only of the most primitive use of numbers, *i.e.* as labels for enumeration of discrete objects including scale divisions. If we take the view that mathematics is pre-eminently a constructed written language in which we can reason about the various uses of number without getting bogged down in the morass of paradoxes and circumlocutions with which the use of vulgar speech besets our efforts to do so, we must now look at a use of number other than as mere labels of counting and measurement. Provisionally, we may define our theme in this chapter as some of the ways in which we use numbers as labels for manipulating numbers.

This does not mean that we are yet ready to talk about all the ways in which mathematicians classify numbers as cardinal and ordinal, negative and positive, fractional and integral, rational and irrational, algebraic and transcendental, real and (until it lately became more fashionable to replace them by number couples) complex. If we ask what they all have in common, it here suffices to say that the major concern of a considerable output of mathematical enquiry during the last century, and indeed during the last half century, has been to justify the use of a single word applicable to all of them by formulating a general pattern consistent with, but of wider terms of reference than, the rules of addition, subtraction and multiplication as we learn them at school in conformity with the requirements of dealing with numbers as labels of enumeration. It may help us to understand the need for classifying numbers and the need for prescribing a common pattern of rules we use when manipulating all we name as such, if we first dig into the history of the attempts mankind has made to devise signs for numbers.

Our story then starts with the dawn of human life on this planet. The several inter-fertile geographical varieties of the single species *Homo sapiens* have existed on this planet for 25,000 years. Nothing to which we can generously apply the term mathematics has existed much more than 5000 years; but the preceding 20,000 years, during the greater part of which human beings were hunters and food gatherers without a fixed abode, contributed a vast body of data to the manipulation of which a system of numeral signs became a social necessity when settled life began about 10,000 years before our own time. Concerning all we can conjecture plausibly before then, we have to rely on cave paintings of remote antiquity and facts about preliterate communities assembled comprehensively in Nilsson's monograph on primitive time reckoning.

From these sources we can get a picture of one of the earliest human preoccupations. The nomads of the Palaeolithic had neither clocks nor maps to guide them to the territories where berries would be ready for picking, herds gathering, birds nesting and edible wild grasses in seed each in its own season. Their only fixed landmarks were the rising and setting positions of

This early sixteenth-century picture shows one man computing with a form of abacus, the other with Hindu-Arab numerals. These numerals, consisting of nine signs to label the number of beads in any column of the abacus and a zero to label any column as empty, made it possible to compute on paper.

A Mayan stele commonly includes pictograms of deities, numerals like those shown below, and "face numerals", as above. The simple numerals consist of nineteen signs and a zero. If the Mayas had consistently used a fixed base of 20, would represent *20,* *400,* *8000, etc. Their preoccupation with the length of the solar year (witnessed by the face numerals), however, led them to use* *to represent 360,* *to represent 7200, and so on.*

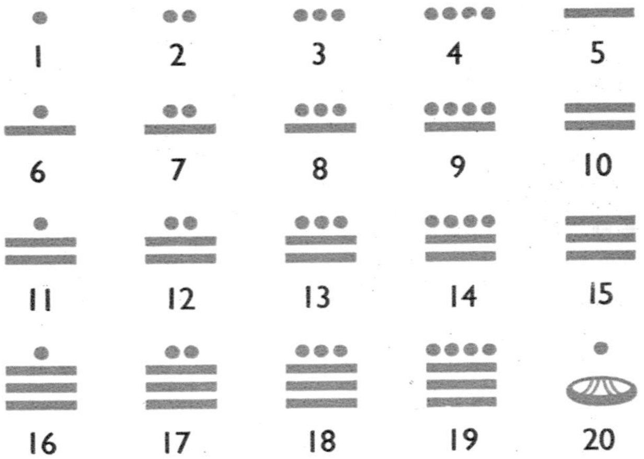

stars or star clusters on the boundary of the horizon, and their most sensitive check on the time of the year was the occasion when a particular bright star or star cluster rose or set just before dawn or just after sundown. The wisest old woman of the tribe would be able to recall how many full moons had elapsed since the birth of a child as the date for its initiation rite approached; and at this stage we may plausibly assume that folk recognised a month of 30 days as a gross unit of time.

The diurnal traffic of the celestial bodies over the seemingly hemispherical vault of heaven, the nightly half circle of the circumpolar stars, the rising and setting positions of the brightest stars, their relations to dawn or to sundown and the approximate length of the lunar month must have circumscribed man's astronomy and chronology till settled communities of herdsmen and grain growers make their appearance. Motive, means and opportunity then alike make manifest the fact and the consequences of a momentous human need.

To regulate a seasonal economy a solar calendar is an imperative requirement; and only where life is settled can there exist occasion to observe the variations of the sun's shadow cast by a pole or the variations of the sun's rising and setting positions throughout the seasons. Thenceforth, the challenge of a new environment is how to accommodate time reckoning in lunar with time reckoning in solar units, a challenge none the less material because misconceptions about the need to propitiate the celestial occupants of the heavens and to preserve intact the ceremonies of propitiation or of initiation into adult life reinforced a more intelligible need for a seasonal

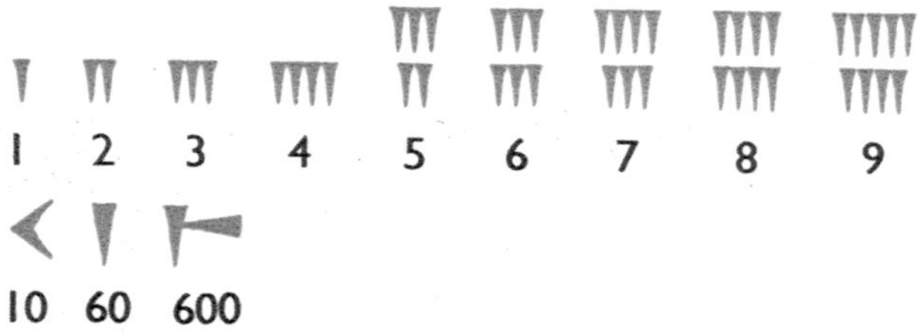

Numerals appear at the left side of each column of this Babylonian calendar listing lucky and unlucky days. Astronomers consistently employed a base of 60, but used number signs repetitively up to nine times. Each sign represented different values in different contexts.

Three of these Chinese numerals of the kind commonly used in accountancy still exemplify the repetitive principle. Some appear in the document below (left).

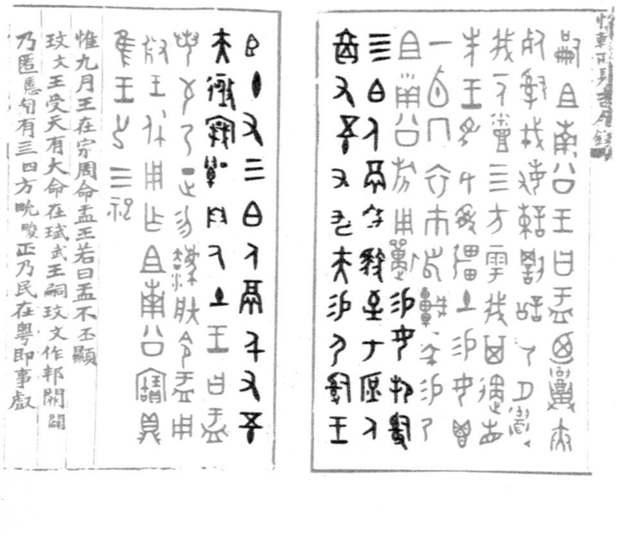

Black part of this early Chinese document reads: I bestow on you also 4 po (feudal chiefs) and of serfs, from charioteers to commoners, 659 men. I bestow on you 13 po from among the royal officials with 1050 serfs . . . from their land.

The Egyptian scribe of about 2500 B.C., immortalised in the statue on the right, would have recorded receipt of taxes in number signs like those shown above. In the sandstone stele of 1450 B.C. (top) *the higher-value signs are inscribed to the right of the lower-value signs. The number seven hundred and forty-three appears in the same sequence as if we were to write it 347.*

calendar. When there was yet no private property, the need for counting members of a flock had not been exacting. Besides, one can count them simultaneously. In this respect days are not like sheep or goats or cattle. Time flies, and there is a limit to what the recollections of even the wisest among the aged members of the tribe can subsume without an *aide memoire*.

From three circumstances, we can discern with something more than plausibility how this started. One is that the only ingredients of what we call the writing of the Maya culture are number signs and pictograms of calendrical deities. A second is the prominent role of primitive astronomy in the earliest writings of the older Mediterranean civilisations. A third, perhaps most telling, is the repetitive principle, illustrated by the familiar Roman symbols mmmm (4000), ccc (300), xx (20) or iii (3), manifest in the Egyptian, Mesopotamian, Phoenician and Mayan numeral signs. It persists in the Chinese *rod* numerals and in early Hindu forms of the ciphers.

If our reconstruction of the dim past is so far correct, it is not trivial to remark at this stage that the ordinal use of whole numbers as labels to record the traffic of time is at least more imperative primitively than their cardinal use as labels to signalise the tribally-shared capital gain of the hunt. Be that as it may, all we can say with assurance is consistent with what we may surmise as the only way in which it was initially possible to accomplish the end in view with the means at man's disposal. The first step is to chip marks daily on a pillar or tree trunk to check up on the passage of time. Here again, study of Mayan and Toltec monuments, though far later than the Egyptian or Babylonian of 3000 b.c., reinforces what we might surmise on other grounds. Every 52 years the temple builders added another peel to the onion step-pyramid of the Serpent, near Mexico City.

We may assume that the next stage in the emergence of a written record, into which seals primitively identified with the totem animal sacrificed on some key date of the primitive

calendar may have already intruded, was how to group the notches of the permanent tally to economise space. One may guess that the custodians of the calendar would find it convenient to group marks chipped on a stone or wooden pole in some such way as the following record of the lapse of 27 days:

$$\text{卌卌 卌卌 卌卌 卌卌 卌卌 ||}$$

In course of time, a secondary record of such an event might assume a more stylised pattern such as the following which is very much like the way in which Caesar, or rather Caesar's slave, would record the same number several millennia later:

$$\boxtimes \quad \boxtimes \quad \triangledown \quad ||$$

Here two claimants are manifest in the solution. Most commonly the group, or as we now say *base*, is a multiple of five — the number of fingers of one hand. Most peoples of the world, like the Egyptians, use 10; but the Romans, as we all know, used both 5 (v) and 10 (x). The Mayas used 20, *i.e.* finger and toe counting, with a glaring anomaly

I	Γ	Δ	H	⊓	X
1	5	10	100	500	1000

This fragment of an Athenian calendrical inscription dating from the fifth century B.C. shows early Greek number signs, used repetitively, like Roman numerals.

Concern with the calendar at the dawn of civilisation is nowhere more apparent than in Central America. Above we see the Temple of Tenayuca as it is today, below as it was at the time of the Conquest. The bottom picture shows how new layers were added at intervals of 52 years (probably A.D. *1351, 1403, etc.).*

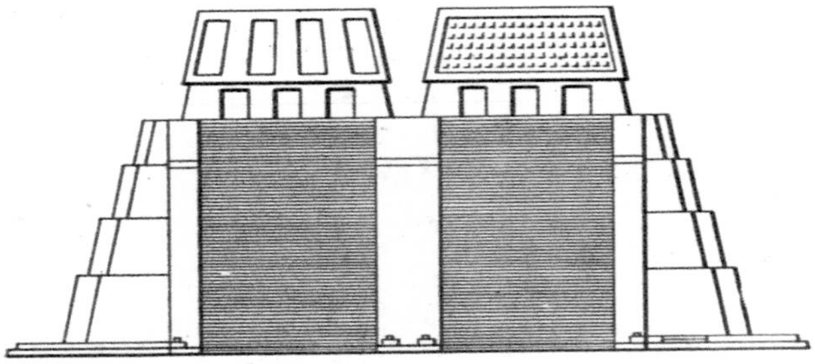

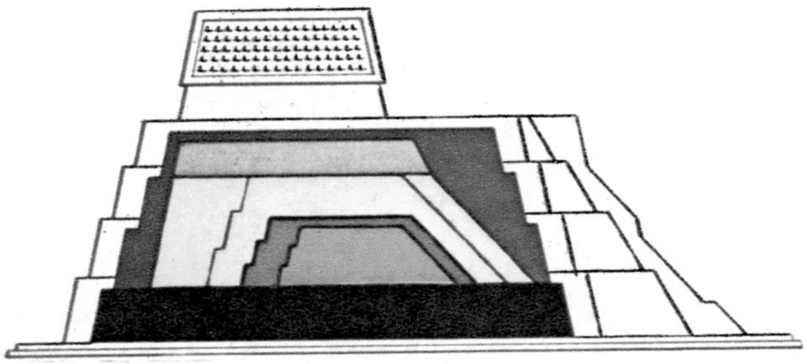

which bears further testimony to the most pressing social need which prompted people at a primitive level of social life to record the passage of time. In terms of the values of the beads of an abacus, which we shall next see the need for, the Mayan values run 1, 20, 360, 7200, 144000 and so on. Here the one out of step is 360, *i.e.* 12 lunar months of 30 days, a widespread early estimate of the solar year, being the best of a bad bargain, if one assumes in advance that it is possible to square the more primitive calendar of full moons with the cycle of the sun throughout the seasons.

In their astronomical calculations the Babylonians consistently used a base of 60; and we shall later see that this had not inconsiderable advantages for dealing with fractions. However, the signs they took over from their Sumerian teachers group strokes in tens and units to make up a multiple of a particular power of the base. Though lucid in comparison with the Greek and Hebrew systems, this repetitive way of writing was extremely bulky; and it may well be that the V, L, D in the Etruscan-Roman system were not relics of an earlier base 5 but devices like the comparatively late trick of inversion (e.g. IV for IIII or XLIV for XXXXIIII) to economise space.

For calculations, all the literate peoples of antiquity, and indeed till well after the beginning of our era, relied largely on the nursery device known as the abacus. In its most sophisticated form, it is a frame on which beads slide as required. In its earliest, it was probably a series of grooves on the sand with pebbles which the computer could remove or replace up to a prescribed limit. Actually the medical term *calculus* for a stone in the bladder and the same root in *calculation* have their origin in the Latin word for a pebble. We may operate an abacus in one of two ways. The groove may hold as many pebbles as the base. We then empty the groove when full, put one on the next one and reject the remainder. Alternatively, the groove may accommodate one less than the base. We then put the next pebble added to a full groove on the next one before we empty it. The Babylonians, and later the Greeks who learned from them what little they knew of the art of calculation, had various *ad hoc* devices to shorten the labour by recourse to tables too long to commit to memory. What the Greeks called *arithmetic*, the study of prime numbers and of series, such as the sum of the squares or of the natural numbers, had no affinity with our simple rules of calculation on paper called more appropriately by their Moslem name *algorithms*.

That the initial impetus to a written record of any sort arose from the social necessity of primitive time reckoning when the Neolithic revolution signalised the emergence of settled communities in the warmer parts of the northern hemisphere is a conclusion which would be sufficiently plausible even if not embalmed in the impressive relics of statuary calendrical tallies of the jungle temples in Central America. In any event, the repetitive character of the most ancient number scripts, whether of Egypt, Mesopotamia, the Phoenician colonies or China, preserved to this day in the Etruscan-Roman numerals of chapter headings or watch dials, bears eloquent testimony to the origin of number signs as strokes to record successive events.

The medical use of the term *calculus* for a stone

The need for numbers as labels of measurement arises when private property emerges. In Mesopotamia traders needed an authorised system of just weights as long ago as early Sumerian times, though the Babylonian duck-weights pictured above are of far later date. As soon as people begin to use numbers for measurement, fractions take on new importance. The Egyptian measuring vessel below is marked $8\frac{1}{6}$ hins.

in the bladder recalls the antiquity of the already-mentioned most primitive form of a more mobile tally than a tree trunk or stone pillar; but the need for such could not arise till there was also a need for the use of numbers as labels of measurement. We may date the need for such with the beginning of the historic record, when a quasi-narrative form of writing became possible through incorporation of pictograms, possibly when seals with totemistic signatures of sacred animals as symbols of authority, ownership or craftsmanship, drew on a fund of ingredients whose origins are traceable to the cave paintings of the Aurignacian hunters.

The need for numbers as labels of measurement arises only when private property emerges; that is to say, when the astronomer priesthood as custodians of the ritual calendar of the seasons with an attendant privileged caste of surveyors and architects can levy tribute on the produce of the tillers of the soil. According to tradition, the Egyptian priesthoods, at least at one stage, assessed tribute on the size of the small-holding; and the frequent removal of landmarks by inundations of the Nile called for frequent re-assessment of tribute due, whence also both for some standards of area and for some standards of volume for the amount of produce delivered. In Sumeria the early emergence of a merchant class trading in metals and other hardware necessitated an authorised system of just weights.

It may seem a far cry from pure mathematics to the creation of some system of weights and measures; but the event was a milestone in the story of numbers. So long as we use numbers only as labels for counting days and full moons, or the solstices, or sheep in a flock, a fraction is a concept with very limited relevance or reference. Thus we can divide a flock of 31 sheep in only 30 ways, *viz.* $\frac{1}{31}$, $\frac{2}{31}$, $\frac{3}{31}$, etc.; and it is meaningless to speak of two fifths or 0.1 of the flock. Once we start to measure land or tribute and to barter wares, we are free to divide in any convenient way we choose our largest unit (*e.g.* a ton) into smaller ones (*e.g.* cwt., lb., oz.). Vestiges of this concrete

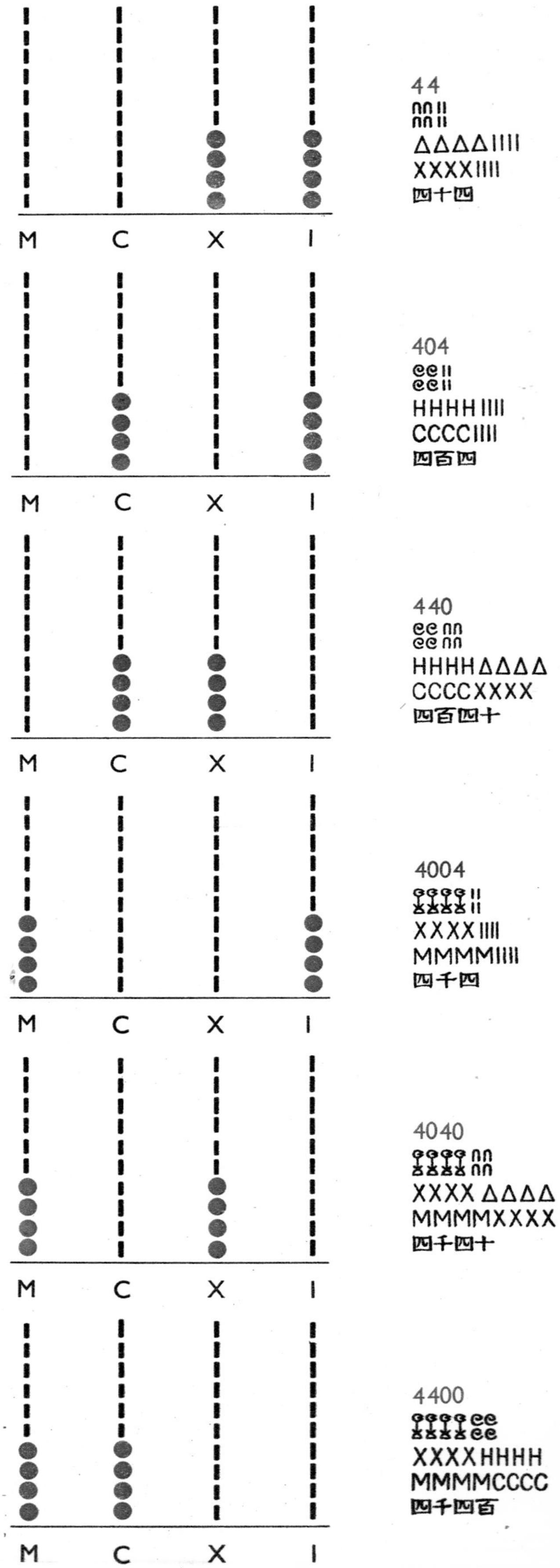

Left: Hindu-Arab numerals reflect economically what the abacus registers. Egyptian, early Greek and Roman numerals do so in a more cumbersome way. All use different signs for beads in different columns, repeating the sign as many times as there are beads. All lack a sign to indicate that a column is empty. Standard Chinese numerals simply state what the abacus shows, e.g. "four tens four."

Α	Β	Γ	Δ	Ε	F	Ζ	Η	Θ
א	ב	ג	ד	ה	ו	ז	ח	ט
1	2	3	4	5	6	7	8	9
Ι	Κ	Λ	Μ	Ν	Ξ	Ο	Π	Ϙ
י	כ	ל	מ	נ	ס	ע	פ	צ
10	20	30	40	50	60	70	80	90
Ρ	Σ	Τ	Υ	Φ	Χ	Ψ	Ω	λ
ק	ר	ש	ת	ך	ם	ן	ף	ץ
100	200	300	400	500	600	700	800	900
/Α	/Β	/Γ	/Δ	/Ε	/F	/Ζ	/Η	/Θ
1,000	2,000	3,000	4,000	5,000	6,000	7,000	8,000	9,000

To economise space, the Hebrews and later Greeks used the whole alphabet as a battery of number signs. In so doing they robbed the signs of any operational meaning. The Hebrews used the same sign for 1000 as for 1.

The task of learning Alexandrian addition tables, even for work with comparatively small numbers, was almost superhuman. Right: An imaginary mouse has nibbled away the parts shown blank on such a table. It may amuse you to complete it. Answers near end of book.

Α	Β	Γ	Δ	Ε	F	Ζ	Η	Θ	Ι	
Β	Γ	Δ	Ε	F	Ζ	Η	Θ	Ι	ΙΑ	Α
	Δ	Ε	F	Ζ	Η	Θ	Ι	ΙΑ	ΙΒ	Β
		F	Ζ	Η	Θ	Ι	ΙΑ	ΙΒ	ΙΓ	Γ
			Η	Θ	Ι	ΙΑ			ΙΔ	Δ
				Ι	ΙΑ	ΙΒ		ΙΔ	ΙΕ	Ε
					ΙΒ		ΙΔ	ΙΕ	ΙF	F
						ΙΔ	ΙΕ	ΙF	ΙΖ	Ζ
							ΙF	ΙΖ	ΙΗ	Η
								ΙΗ	ΙΘ	Θ
									Κ	Ι

Ι	Κ	Λ	Μ	Ν	Ξ	Ο	Π	Ϙ	Ρ	
Κ	Λ	Μ	Ν	Ξ	Ο	Π	Ϙ	Ρ	ΡΙ	Ι
	Μ	Ν	Ξ	Ο			Ρ	ΡΙ	ΡΚ	Κ
		Ξ	Ο	Π			ΡΙ	ΡΚ	ΡΛ	Λ
			Π	Ϙ	Ρ		ΡΚ	ΡΛ	ΡΜ	Μ
				Ρ	ΡΙ	ΡΚ	ΡΛ	ΡΜ	ΡΝ	Ν
					ΡΚ	ΡΛ	ΡΜ	ΡΝ	ΡΞ	Ξ
						ΡΜ	ΡΝ	ΡΞ	ΡΟ	Ο
							ΡΞ	ΡΟ	ΡΠ	Π
								ΡΠ	ΡϘ	Ϙ
									Σ	Ρ

Ρ	Σ	Τ	Υ	Φ	Χ	Ψ	Ω	λ	/Α	
Σ	Τ	Υ	Φ	Χ	Ψ	Ω	λ	/Α	/ΑΡ	Ρ
	Υ	Φ	Χ	Ψ	Ω	λ		/ΑΡ	/ΑΣ	Σ
		Χ	Ψ	Ω				/ΑΣ	/ΑΤ	Τ
			Ω	λ	/Α	/ΑΡ	/ΑΣ	/ΑΤ	/ΑΥ	Υ
				/Α	/ΑΡ	/ΑΣ	/ΑΤ	/ΑΥ	/ΑΦ	Φ
					/ΑΣ	/ΑΤ	/ΑΥ	/ΑΦ	/ΑΧ	Χ
						/ΑΥ	/ΑΦ	/ΑΧ	/ΑΨ	Ψ
							/ΑΧ	/ΑΨ	/ΑΩ	Ω
								/ΑΩ	/Αλ	λ
									/Β	/Α

Hindu numerals, including the zero, came into use certainly not later than A.D. *850, and very probably some time before* A.D. *400. The number at the bottom centre of this fragment of the twelfth-century Bakshali MS is 109350. The zero sign is still a dot, and only the 3 bears much resemblance to its modern form. Below we see how other signs slowly changed shape.*

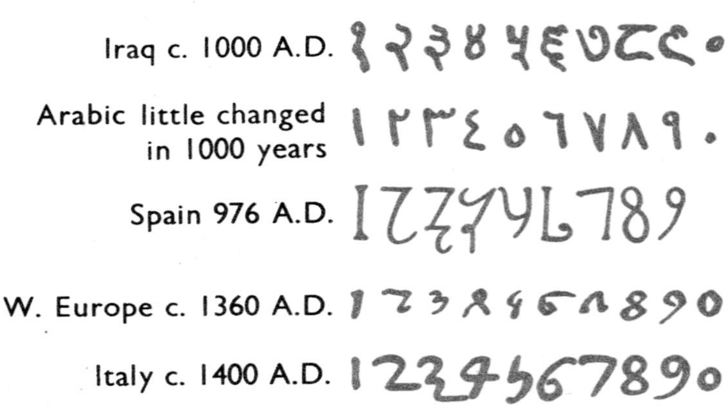

Sundial of A.D. *1453 bearing Hindu-Arab numerals.*

way of looking at fractions bogged down much thinking about number in the world of antiquity. As children, we still make our first approach to a fraction in terms of so many (*m*) out of a total (*n*) equal parts of something; and the first great milestone in pure mathematics was the dilemma faced by the early Greek geometers when they came across fractions which are not expressible in this way.

They might well have faced the dilemma in a different way if they had not taken a decisive step backward at a crucial stage in the history of human communications. The narrative sign-writing of the calendar priesthoods was an art to which few could hope to aspire. It was therefore a momentous step forward when Semitic slaves in Egypt stumbled on the first crude form of alphabetic script about 1800 B.C. The new way of writing had two merits. It was easier to learn; and it was vastly less cumbersome than that of the Temple archives. None the less, its emergence to full stature was a slow process. Till the beginning of Greek literature a thousand years later, what alphabetic writing the Semitic peoples of the Orient transmitted by western sea-routes and eastwards overland had a modest usefulness among traders and master mariners for short memoranda.

At a time when mankind had no prevision of its full usefulness and eventual supremacy as a means of storing knowledge, the users thus blundered into the habit of representing numbers by letters. The transparent meaning of M and C in the Etruscan-Roman signs which preserve the repetitive principle, as do the earliest Greek number signs, suggests that this started with no deliberate intention of economising space; but the Hebrews and the later Greeks enlisted the whole alphabet to make up a new battery of number signs with different values for the same number of units on different columns of the abacus. It is difficult to exaggerate the pernicious consequences of this step. It divorced the sign language of number from any operational meaning the repetitive system conveys. It added to the difficulty

$$60^2 = 3600 \qquad 60^1 = 60 \qquad 60^0 = 1$$

The merits of the numeral system we use has nothing to
do with the circumstance that we employ a base of 10.
They depend solely on the fact that the position of the
signs in a sequence indicates the number of beads on
successive bars of a counting-frame with a fixed base.
A larger fixed base, say 60 or 20, would merely involve
learning longer tables of addition and multiplication.
A smaller one, say 7 or 2 (that of the electronic brain),
would mean taking more space to record numbers.

$$7^4 = 2401 \qquad 7^3 = 343 \qquad 7^2 = 49 \qquad 7^1 = 7 \qquad 7^0 = 1$$

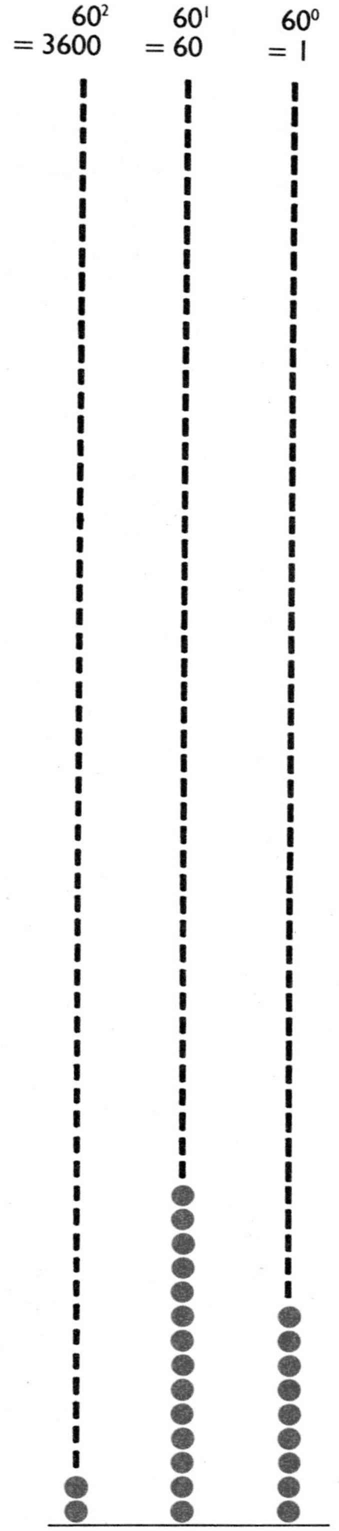

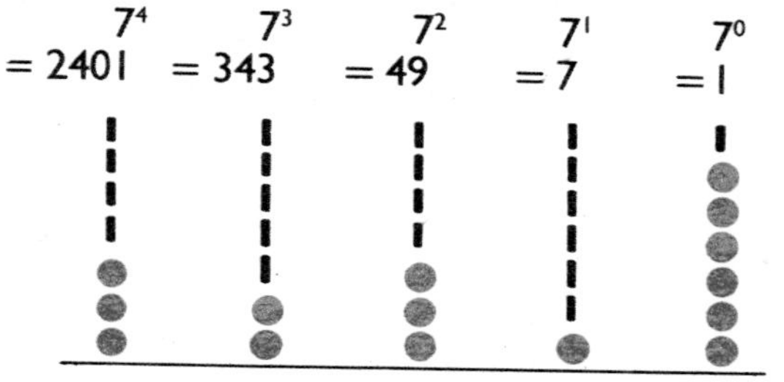

$$3(7^4) + 2(7^3) + 3(7^2) + 1(7^1) + 6(7^0) = 8049$$

$$10^3 = 1000 \qquad 10^2 = 100 \qquad 10^1 = 10 \qquad 10^0 = 1$$

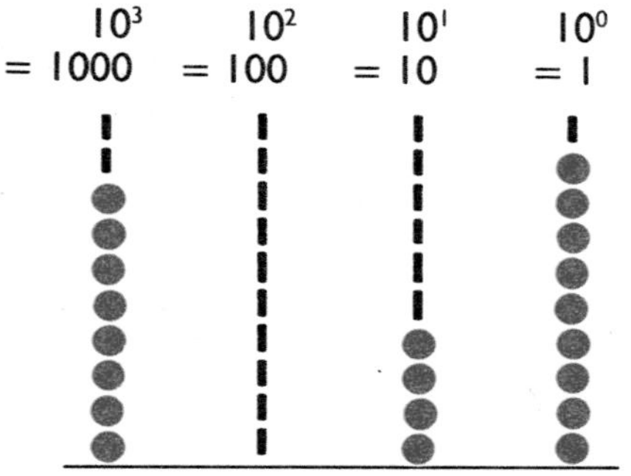

$$8(10^3) + 0(10^2) + 4(10^1) + 9(10^0) = 8049$$

$$20^3 = 8000 \qquad 20^2 = 400 \qquad 20^1 = 20 \qquad 20^0 = 1$$

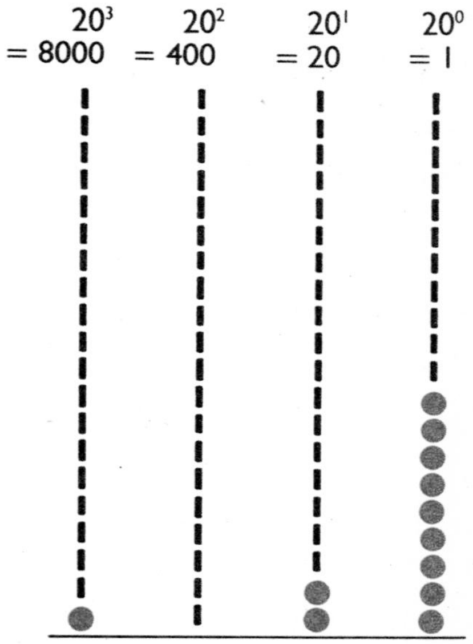

$$1(20^3) + 0(20^2) + 2(20^1) + 9(20^0) = 8049$$

$$2(60^2) + 14(60^1) + 9(60^0) = 8049$$

$$2^{12} = 4096 \quad 2^{11} = 2048 \quad 2^{10} = 1024 \quad 2^9 = 512 \quad 2^8 = 256 \quad 2^7 = 128 \quad 2^6 = 64 \quad 2^5 = 32 \quad 2^4 = 16 \quad 2^3 = 8 \quad 2^2 = 4 \quad 2^1 = 2 \quad 2^0 = 1$$

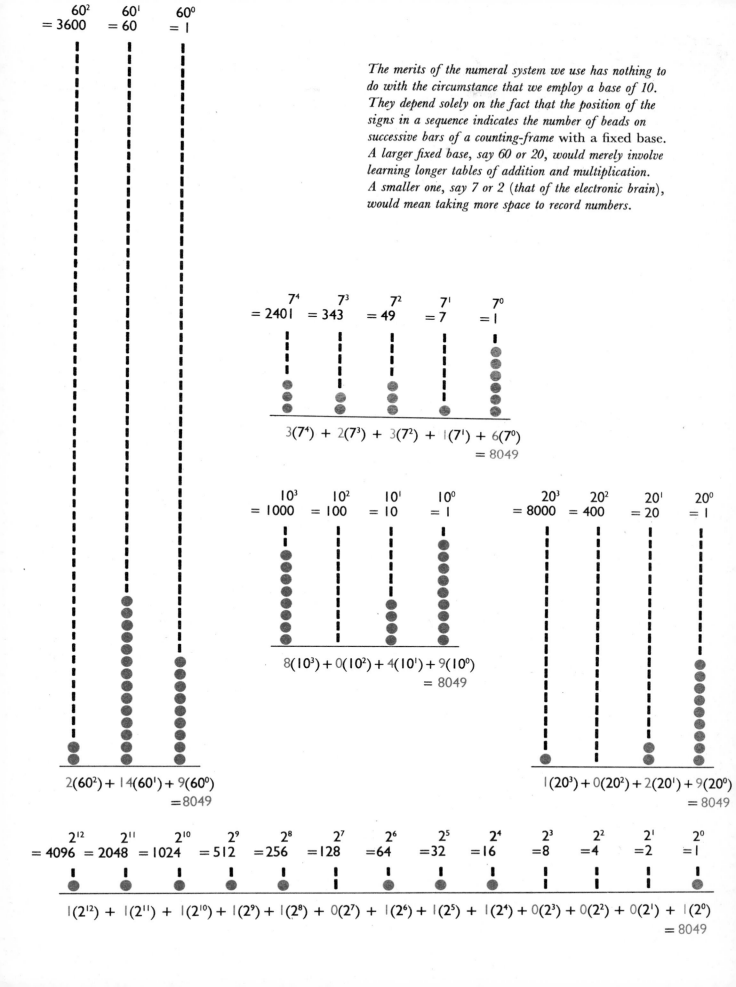

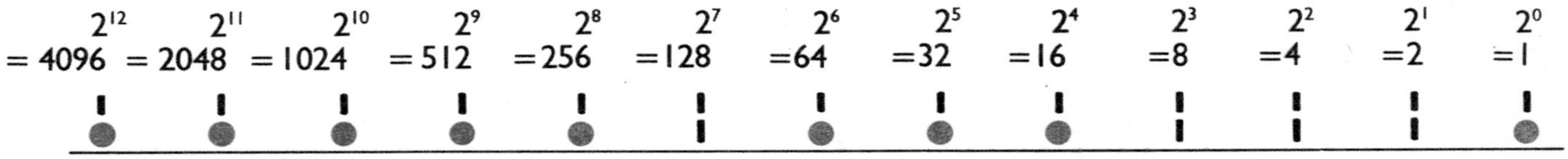

$$1(2^{12}) + 1(2^{11}) + 1(2^{10}) + 1(2^9) + 1(2^8) + 0(2^7) + 1(2^6) + 1(2^5) + 1(2^4) + 0(2^3) + 0(2^2) + 0(2^1) + 1(2^0) = 8049$$

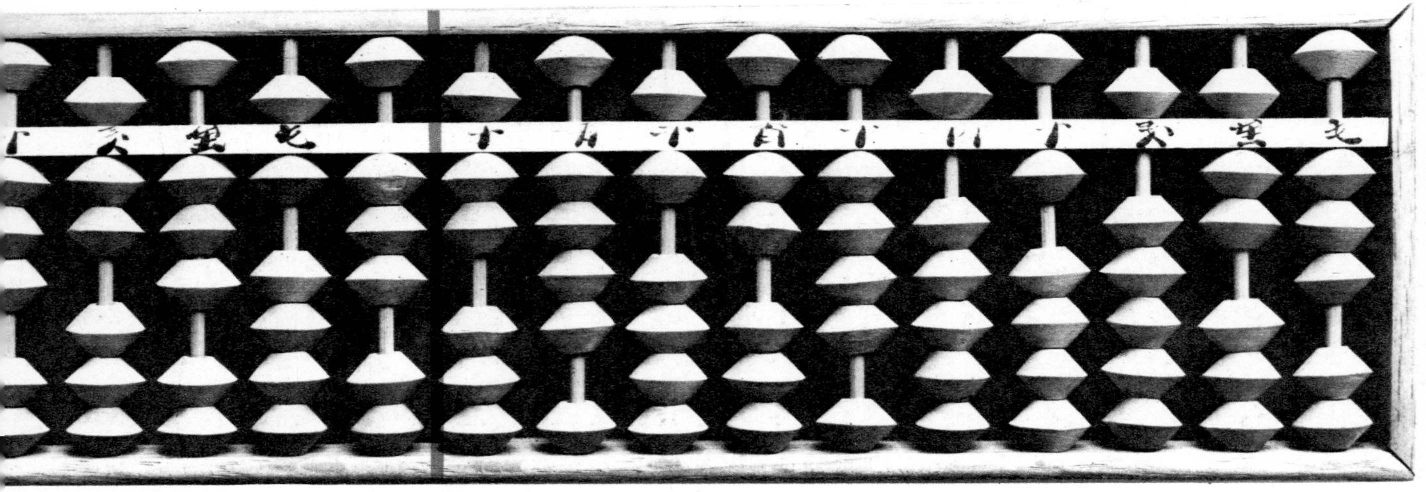

On this Japanese abacus (above) beads above the crossbar have a value of 5, beads below it a value of 1. Only beads pushed to the centre count. The number shown here is 37363.7462451573. (Bars left of the brown line show positive powers of 10, right of it negative powers.)

of thinking about numbers too large for the signs of the alphabet to accommodate. It made the task of memorising tables of addition, subtraction and multiplication to facilitate work with comparatively small numbers on the abacus almost superhuman. It thus made the astronomer more dependent on tables or on mechanical aids to perform laboriously calculations we now learn to carry out effortlessly before we reach our teens. It doubtless delayed the emergence of the sort of sign writing we customarily call algebra. *Inter alia*, it also encouraged the persistence of superstitious beliefs about lucky, unlucky or sacred numbers if the alphabetical sequence had a verbal meaning, as is true of the magic square which spells out the name of the great Unmentionable in Hebrew.

That we no longer need to limp along with the bars of the counting frame as our crutches depends on the fact that at some time not much earlier than A.D. 300 and not much later than A.D. 700 it became the custom in the counting houses of India to represent the empty column on the abacus by a dot corresponding to what later became our zero sign. Its original name *sunja*, signifying empty, proclaims its origin. One implication of having such a sign is that we can convey the position of a column of the abacus without recourse to new symbols such as M, C, X in the Roman system. In short, the number of necessary signs other than zero is one less than the base, *i.e.* 9 if the base is 10. The same easily memorised tables of addition, subtraction and

$18(20^3)$
$= 144000$

$18(20^2)$
$= 7200$

$18(20^1)$
$= 360$

20^1
$= 20$

20^0
$= 1$

$0 \times 18(20^3)$
$+$
$1 \times 18(20^2)$
$+$
$2 \times 18(20^1)$
$+$
6×20^1
$+$
9×20^0
$= 8049$

↑

The Mayan inconsistency

$\bullet = 1 \times 18(20^2)$

$\bullet\bullet = 2 \times 18(20^1)$

$\underline{\bullet} = 6 \times 20^1$

$\underline{\bullet\bullet\bullet\bullet} = 9 \times 20^0$

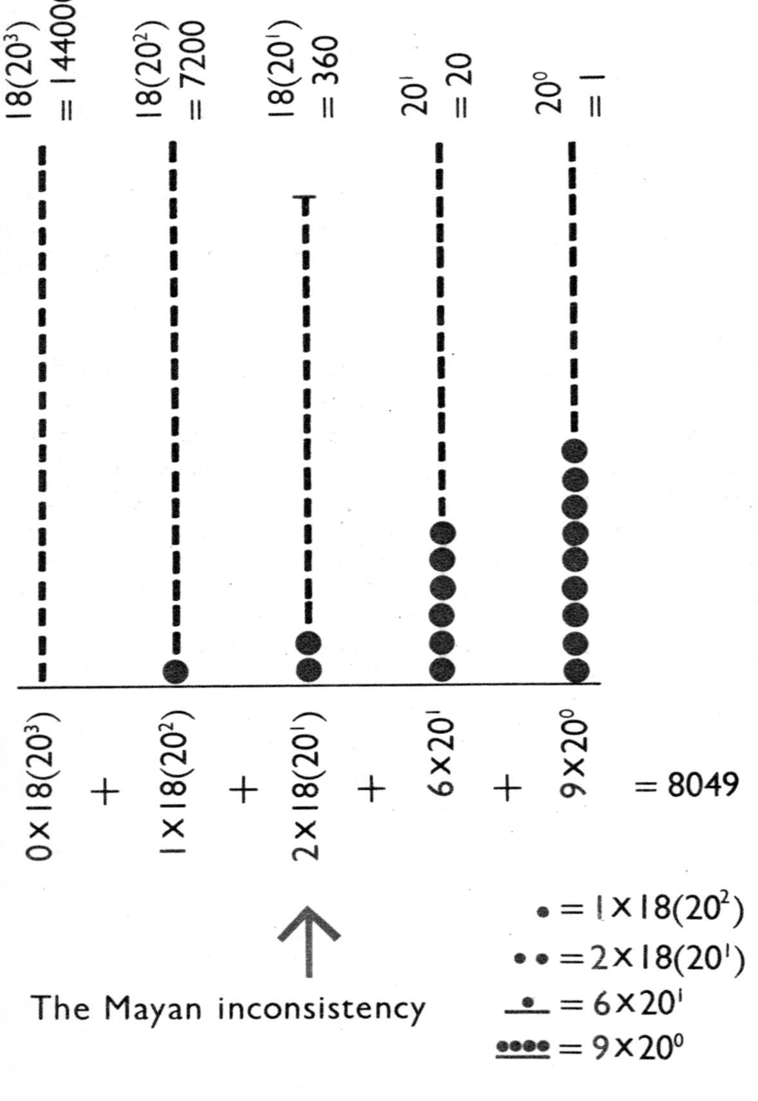

In finding a simple way of representing 360, the Mayas forfeited the benefits of a fixed base. In their system the third position is out of step with all the rest.

Each time we move one step from left to right along the bars of an abacus, we divide the value of each bead on a bar by 10. We also reduce the index-label by 1. Hence if we label 10 as 10^1, and 1 as 10^0, it is consistent to label $\frac{1}{10}$ as 10^{-1}, $\frac{1}{100}$ as 10^{-2}, and so on. The abacus can thus provide the ideal model of a basal fraction.

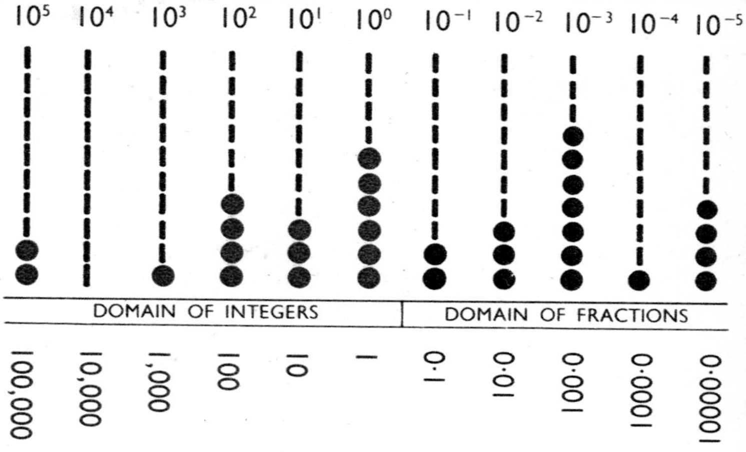

$= 201436{\cdot}23714 = 10^5(2{\cdot}0143623714)$

multiplication then suffice for manipulating any column; and when we have learned them we can perform the *carry over* operation on paper without mechanically rearranging the beads on a counting-frame.

The Mayan numeral script also had a zero sign; but the anomalous change of base, by giving the place value next to 20 as 360 instead of 400, deprived it of what would otherwise have been its supreme usefulness. The great advantages of the numeral system we owe to Hindu civilisation depend solely on the fact that the position of the signs in a sequence indicates the number of units on successive bars of a counting-frame with a *fixed base*, *e.g.* 423 signifies $4(100)+2(10)+3(1)$ if the base (b) is 10. The rule for operating an abacus with bars (or grooves) able to accommodate the same number of beads does not depend upon how many it can carry. Thus the very great merits of the system we now use have nothing to do with the circumstance that our base is 10. A larger base (*e.g.* 20 or 60) would entail the need to memorise longer tables of addition, subtraction and multiplication, and a smaller base would make more demands on space. With $b = 3$, for instance, what we write as 423 would be $1(243)+2(81)+0(27)+2(9)+0(3)+0(1)$, which we may write as $(120200)_3$ to distinguish it from $(423)_{10}$. The addition, subtraction and multiplication tables for $b = 3$ would be as below, where ciphers shown in italics carry the instruction which we usually express by "borrowing."

Addition

	0	**1**	**2**	**10**
0	0	1	2	10
1	1	2	10	11
2	2	10	11	12
10	10	11	12	20

Subtraction

	0	**1**	**2**	**10**
0	0	1	2	10
1	*2*	0	1	2
2	*1*	*2*	0	1
10	*0*	*1*	2	0

Multiplication

	0	**1**	**2**	**10**
0	0	0	0	0
1	0	1	2	10
2	0	2	11	20
10	0	10	20	100

Needless to say, calculations with the base 2 involve least load to the memory, and are most space-consuming. Thus $(8192)_{10} = (10{,}000{,}000{,}000{,}000)_2$. The only tables to memorise are

$1+0 = 1$	$1-0=1$	$1\times 0 = 0$
$1+1 = 10$	$1-1=0$	$1\times 1 = 1$
$1+10=11$	$10-1=1$	$1\times 10=10$
$1+11=100$	$100-1=11$	$10\times 10=100$

If a machine which uses no paper for the calculation and visibly records only the last step translated into our decimal ciphers, the bulkiness of the representation of the stages of a calculation to base 2 is not disadvantageous and the simplicity of its *modus operandi*, in particular the algorithm for extracting a square root, is highly meritorious. Thus the most modern computing machines carry out their calculations in this way.

If readers would like to indulge more profitably in as much fun as an intelligent person can hope to get from a newspaper crossword puzzle, they can construct tables of addition, subtraction and multiplication for the base 5, first check the following computations and then make up examples – especially in division – which they can check in the same way.

$(256)_{10} = (2011)_5 : (41)_{10} = (131)_5 :$
$(10496)_{10} = (313441)_5 : (297)_{10} = (2142)_5$

Addition

256 *on ten*	2011 *on five*
41	131
$(297)_{10}$	$(2142)_5$

Multiplication

256 *on ten*	2011 *on five*
41	131
256	2011
1024	11033
10496	2011
	313441

A thousand years separate the beginnings of Greek geometry from this momentous innovation. It speedily bore fruit in providing a prescription of the familiar algorithms which vastly simplified the art of computation; but fully another thousand years elapsed before it laid bare all the lessons the abacus can teach us. Foremost among these – though last in historical sequence – was the possibility of putting numbers to a new use. If we illustrate this by recourse to the range of ciphers we use in connexion with the base ten, it goes without saying that 10 could equally well stand for 2, 3 or *any other* number we might choose as a base. First notice the sequence of values of a bead on successive bars of the counting frame when written in the usual way, *viz.*

. . . 10,000 . 1000 . 100 . 10 . 1

In this sequence $10{,}000(=10^4)$ is 1 followed by 4 zeros, $1000(=10^3)$ is 1 followed by 3 zeros, $100(=10^2)$ is 1 followed by 2 zeros. It is therefore appropriate to write 10 as 10^1 and 1 as 10^0, *i.e.* 1 followed by no zero. So far this is merely a snappier way of writing a number as a label for the outcome of counting; but we can look on it in a different way, *viz.* 10^5 represents an *instruction* to multiply 1 by 10 five times, and 10^0 as an instruction *not* to multiply it by 10. That $10{,}000 \times 100 = 1{,}000{,}000$ in this notation means $10^4 . 10^2 = 10^6$, which we may generalise for integers b, m, n as:

$$b^m . b^n = b^{m+n}$$

So far we have been speaking of positive integers. So long as we use numbers as labels for enumeration or measurement, we can think of -6 or -16 only in terms of the abomination of the short weight (*i.e.* a *deficit* of what is due) or as an arbitrary consequence of naming our scale divisions, *e.g.* -273°C as absolute zero and 0°C as 273 on the absolute scale; but inspection of an abacus labelled in the foregoing way endorses a new use for negative integers. Each time we go from left to right, we divide the value of each bead of the column by 10 and reduce the index by 1. If we label 10 as $10^1(=10^2 \div 10^1 = 10^{2-1})$ and 1 as $10^0(=10^1 \div 10 = 10^{1-1})$, it is consistent with the overall plan to label $\frac{1}{10}$ as $10^{-1}(=10^0 \div 10)$, $\frac{1}{100}$ as $10^{-2}(=10^{-1} \div 10)$ and so forth.

Long before the convenience of this improvement in the means of human communication became apparent, it was evident to at least one Hindu-Moslem mathematician that we have now a new way of writing down a fraction consistent with the principle of position, *i.e.* as a decimal if the base is ten, or what we may more generally designate as the *basal* way. We shall discuss later the considerable theoretical implications of this step; here it suffices to point out that it immensely simplifies the problem of designing a computing machine of any sort.

We have now left far behind us the concrete way of thinking of fractions in terms of subunits such as minutes and seconds; and we have also extricated ourselves from a difficulty which was very real to the ancient world. Multiplication on the traditional abacus is an exact operation; but division usually leaves us with a remainder on our hands. The world of antiquity could conceptually extend the bars of the counting frame as far as it chose to do so with the reservation that its limited battery of number signs made this an effort. Though the Babylonians came near to doing so, what no one else could clearly conceive was the possibility of limitless division by equipping the counting frame with an unending sequence of columns to represent fractions with denominators referable to successive powers of the same base.

Just as we interpret 5 in 10^5 as an instruction to multiply 1 by 10 five times, we interpret 10^{-5} as an instruction to divide 1 by 10 five times, and we may generalise the instruction for positive integers b, m, n in the form

$$b^m \div b^n = b^{m-n}$$

When the performance of one operation cancels the performance of a second, we speak of the latter as *inverse* to the former. Thus subtraction is the inverse operation with respect to addition, since subtracting 3 from 8 or 14 from 23 *etc.* cancels the effect of adding 3 to 8 or 14 to 23 *etc.* By the same token, division is the inverse operation with respect to multiplication. Thus we interpret m in 10^m to mean: perform the operation of multiplying unity m times by 10. We therefore interpret $-m$ in 10^{-m} as an instruction to perform the corresponding inverse operation of *dividing* unity m times by 10.

This is a useful addition to our number language, as shown by the following exercise in translation:

$\boldsymbol{sin\ A = x} \equiv$ the semichord subtended by the angle A in a circle of unit radius is x.

$\boldsymbol{sin^{-1}x = A} \equiv$ the angle which subtends the semichord x in a circle of unit radius is A.

Alas, college and high school textbooks still cling to such inconsistencies as $\sin^2 A$ for what we should, and in this book shall, write as $(\sin A)^2$, meaning in Euclid's idiom that it is the square on the semichord.

A	B	Γ	Δ	E	F	Z	H	Θ	I	K	Λ	M	N	Ξ	O	Π	Ϙ	P	Σ	
A	B	Γ	Δ	E	F	Z	H	Θ	I	K	Λ	M	N	Ξ	O	Π	Ϙ	P	Σ	A
	Δ	F	H	I	IB	IΔ	IF	IH	K	M	Ξ	Π	P	PK	PM	PΞ	PΠ	Σ	Y	B
		Θ	IB	IE	IH	KA	KΔ	KZ	Λ	Ξ	Ϙ	PK	PN	PΠ	[redacted]	ΣM	ΣO	T	X	Γ
			IF	K	KΔ	KH	ΛB	ΛF	[redacted]	[redacted]	PK	PΞ	Σ	ΣM	[redacted]	[redacted]	TΞ	Y	Ω	Δ
				KE	Λ	ΛE	M	ME	[redacted]	P	PN	Σ	ΣN	T	[redacted]	[redacted]	YN	Φ	/A	E
					ΛF	MB	MH	NΔ	Ξ	[redacted]	PΠ	ΣM	T	[redacted]	[redacted]	[redacted]	ΦM	X	/AΣ	F
						MΘ	NF	ΞΓ	O	PM	ΣI	ΣΠ	TN	YK	[redacted]	[redacted]	[redacted]	ψ	/AY	Z
							ΞΔ	OB	Π	PΞ	ΣM	TK	Y	YΠ	ΦΞ	XM	[redacted]	[redacted]	/AX	H
								ΠA	Ϙ	PΠ	ΣO	TΞ	YN	ΦM	XΛ	ψK	ΩI	ϡ	/AΩ	Θ
									P	Σ	T	Y	Φ	X	ψ	Ω	ϡ	/A	/B	I
										Y	X	Ω	/A	/AΣ	/AY	/AX	/AΩ	/B	/Δ	K
											ϡ	/AΣ	/AΦ	/AΩ	/BP	/BY	/Bψ	/Γ	/F	Λ
												/AX	/B	/BY	/BΩ	/ΓΣ	/ΓX	/Δ	/H	M
													/BΦ	/Γ	/ΓΦ	/Δ	/ΔΦ	/E	/I	N

Part of our debt to India is that we no longer have to learn Alexandrian tables of multiplication. The mouse we met on page 35 has nibbled this one. You can check your effort to complete it near end of book.

In learning to think of a number as a label for an operation in contradistinction to a label for counting days, sheep, weights or scale divisions, and thereby vastly widening our fractional horizon, we have still the visual model of the counting-frame as a guide to exploration. It is now tempting to take a step into the unknown. Medieval mathematicians used the sign $\sqrt{x}$ to represent the operation of finding a number which yields x when multiplied by itself. Now we can tentatively write 10^1 in the form $10^{\frac{1}{2}+\frac{1}{2}}$. If we make the rule that $10^m . 10^n = 10^{m+n}$ is true when n and m are vulgar fractions, we must interpret $\sqrt{10}$ as $10^{\frac{1}{2}}$, $\sqrt[3]{10}$ as $10^{\frac{1}{3}}$ *etc.*; and this convention leads us into *no inconsistencies*. It is easy to see that $(b^m)^n = b^{mn}$ if m, n and b are integers. Likewise $(b^{-m})^n = b^{-mn} = (b^m)^{-n}$, whence we must interpret:

$$(100^3)^{\frac{1}{2}} = 1000 = 10^3 = (100^{\frac{1}{2}})^3$$

In the same way, we must interpret $b^{-\frac{1}{2}}$ as the reciprocal of $\sqrt{b}$, *e.g.* $100^{-\frac{1}{2}} = 0.1$. If we now factorise the base, *e.g.* $10 = 2(5)$, we see that $10^2 = 100 = 4 \times 25 = 2^2 . 5^2$, or more generally if $b = f_1 . f_2 . f_3$ .. we can write $b^m = f_1^m . f_2^m . f_3^m$ .. and $b^{-m} = f_1^{-m} . f_2^{-m} . f_3^{-m}$ .., whence we can extend our sign language to include a base which is a vulgar fraction, as when we write $b = (49 \div 400) = 7^2 . 20^{-2}$. Again, we encounter no inconsistency when we write

$$\left(\frac{49}{400}\right)^{\frac{1}{2}} = 7^1 . 20^{-1} = \frac{7}{20}.$$

We have now roamed far enough into a domain where numbers are no longer labels for operations which we can visualise obliquely against the background of the abacus model. Indeed, we are ready to scrutinise some undisclosed difficulties of number lore in antiquity. Foremost of these is that of entertaining the possibility that a limitless series of positive fractions may be *convergent* in the sense that the sum of all its terms, however many, may be finite and indeed a manageably small number. As far as we know, nobody did clearly grasp this before Archimedes used the following:

$$\left(1 + \frac{1}{4} + \frac{1}{16} + \frac{1}{64} + \frac{1}{256} \ldots \textit{ad infinitum}\right) = S$$

$$= \left(\frac{1}{4^0} + \frac{1}{4^1} + \frac{1}{4^2} + \frac{1}{4^3} + \frac{1}{4^4} \ldots \textit{ad infin.}\right)$$

Here $\frac{1}{4} S = \left(\frac{1}{4^1} + \frac{1}{4^2} + \frac{1}{4^3} \ldots \textit{etc.}\right)$,

whence by subtraction:

$$\left(1 - \frac{1}{4}\right) S = \frac{3}{4} S = \frac{1}{4^0} = 1 \text{ and } S = \frac{4}{3}$$

If one specialises in being wise after the event, one may think it odd that the passion of the early Greeks for the regimen of the compass did not lead them to infer that successively bisecting the remainder of a line of unit length after the first bisection is equivalent to writing:

$$1 = \frac{1}{2} + \frac{1}{4} + \frac{1}{8} + \frac{1}{16} + \frac{1}{32} + \frac{1}{64} \ldots \textit{etc.}$$

However, one may hold, as Francis Bacon held, that it is unwise to extol the powers of the human mind when we should rather seek out its proper helps. If so, one may be thankful that our Hindu heritage makes the following example of the existence of a convergent series self-evident:

$$\tfrac{1}{3} = 0.\dot{3} = 0.33333 \ldots \textit{ad infinitum}$$

$$= \frac{3}{10^1} + \frac{3}{10^2} + \frac{3}{10^3} + \frac{3}{10^4} \ldots \textit{ ad infin.}$$

In contrariety to what many eminent nineteenth-century mathematicians have said, E. J. Bell, a contemporary pure mathematician who writes with a matchless combination of formidably extensive erudition and brilliantly incisive wit, warns us that we shall not exaggerate our debt to the great Hindu number reform if we recall how large tracts of late nineteenth-century and twentieth-century mathematics have no explicit connexion with number as such. It suffices to say that we may well leave to posterity to decide whether so formidable an output of verbal dispute, upholstered or otherwise with non-verbal symbolism, has any lasting function other than as an as yet uncompleted programme for disclosing inconsistencies in the sign writing of a prolific period of discovery during three preceding centuries. If it has not, it will justify its claims to importance in the long run only if it produces, like the substitution of number couples for complex numbers as already mentioned, a more consistent and coherent, whence also more communicable, symbolism for techniques whose numerical usefulness stands the test of time.

The mental block which prevented the best brains of the early Greek period from coming to terms easily with the notion of a convergent series is one among many examples of mists dispelled by the privilege of reliance on a more convenient notation for numbers. One of the more sophisticated by-products of Greek arithmetic introduces us to an illuminating way of representing any integer and gives us a clue which will help us to dispel a Greek dilemma already mentioned. Euclid gives a proof of the assertion that there is no limit to the number of prime numbers. In effect, he proceeds in this way. Suppose that we knew no prime greater than 17. Let us then multiply *all* primes up to and including 17, thus obtaining the number:

$$1.2.3.5.7.11.13.17=510510$$

The only numbers by which this is divisible are its prime factors and their products, *e.g.* 10, 21, 39 *etc.* By adding 1 we obtain 510511 which will leave a remainder of 1 if we divide it by any of its prime factors, whence also by any product of its prime factors. Hence it is not factorisable and is a prime greater than 17. The argument is easy enough to generalise. However large may be the largest prime number we can name as yet, we can always find another larger than itself by adding 1 to the product of all the primes up to and including itself.

With this clue at our disposal, the use of the index notation makes manifest at least one property of numbers much more puzzling to the best brains of antiquity than it need be to an intelligent teen-ager of our own time if already equipped to perform the following elementary exercises in translation. If a, b, c *etc.* may each correspond to any positive integer or to zero, we can clearly express any positive whole number (n) as the product of the powers of all prime numbers less than (or equal to) itself in the form:

$$n=2^a.3^b.5^c.7^d.11^e.13^f \ldots . \text{ etc.}$$

$$\text{e.g. } 1260=2^2.3^2.5^1.7^1.11^0.13^0.17^0.19^0 \ldots \text{etc.}$$

This is clearly so: (a) because any factorisable number is expressible as the product of primes (*e.g.* $200=2^3.5^2$); (b) because $b^0=1$ for all b, any prime whose power is zero being replaceable by unity makes no difference to the result, *e.g.* $2^3.3^0.5^2=2^3.5^2$. A prime itself is, of course, expressible in the form illustrated by:

$$2^0.3^0.5^0.7^0.11^0.13^0.17^1=17$$

Now suppose that p is an integer and that another integer N is the pth power of n expressed as above, so that

$$N=2^{ap}.3^{bp}.5^{cp}.7^{dp} \ldots \text{etc.}$$

$$N^{\frac{1}{p}}=2^a.3^b.5^c.7^d \ldots \text{etc.}=n$$

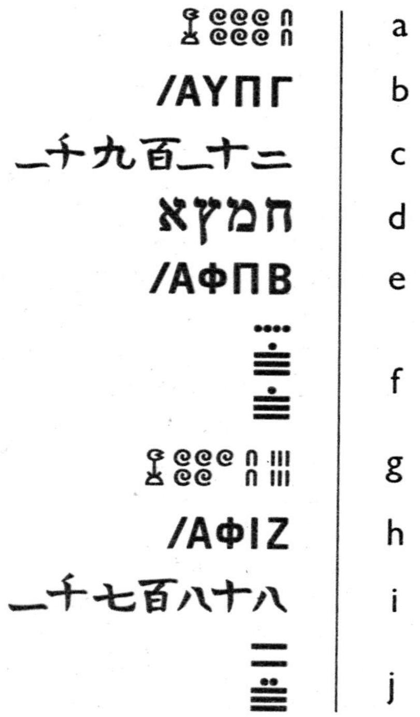

Here are ten dates notable in world history written in various number scripts of antiquity. Can you re-write them in modern form and tie each with a great event? All necessary clues for re-writing appear in this chapter. Answers are given near end of book.

To say that the pth root of an integer (N) is itself an integer (n) therefore signifies that the power (other than zero) of any prime factor of N is either p or an exact multiple of p.

Let us now look at how we can factorise such an integer N. For illustrative purposes we may assume that N $(=2^{ap}.3^{bp}.5^{cp}.7^{dp})$ is the pth power of an integer n $(=2^{a}.3^{b}.5^{c}.7^{d})$ of which only the first four prime factors have indices (a, b, c, d) greater than zero. All possible ways of resolving N into two factors of which one is the pth power of any integer are then

$$(2^a.3^b)^p.(5^c.7^d)^p = (2^a.5^c)^p.(3^b.7^d)^p$$
$$= (2^a.7^d)^p.(3^b.5^c)^p = (2^a.3^b.5^c)^p.7^{dp}$$
$$= (2^a.3^b.7^d)^p.5^{cp} = (2^a.5^c.7^d)^p.3^{bp}$$
$$= (3^b.5^c.7^d)^p.2^{ap}$$

It is now easy to see that the following assertion is always true:

> if we can resolve an integer (N_1) which is the pth power of another into two factors of which one (N_2) is also the pth power of an integer, the other factor must likewise be the pth power of an integer.

It is here important to be clear what this assertion does *not* mean, *viz.* that it is impossible to resolve the pth power of an integer into two factors unless *each* is expressible as the pth power of some integer. For instance, we can resolve into two factors $576 = (2^3.3)^2$ in either of two ways:

(a) *both are perfect squares*:

$$4(144) : 9(64) : 16(36)$$

(b) *neither is a perfect square*:

$$2(288) : 3(192) : 6(96)$$

Let us now examine what meaning we may give a quantity such as $\sqrt[3]{5}$ in the older, and $5^{\frac{1}{3}}$ in our more informative, notation against the background of this argument. Since $1^3=1$ and $2^3=8$, the cube root of any number between 1 and 8 presumably lies between 1 and 2. So it is tempting to assume that $\sqrt[3]{5}$ is expressible as an improper vulgar fraction in terms of integers a and b $(>a)$ as:

$$\sqrt[3]{5} = 1 + \frac{a}{b} = \frac{a+b}{b}$$

If we replace $(a+b)$ by the integer N_1 and b by N_2 to keep in step with the foregoing argument, we may therefore write:

$$\sqrt[3]{5} = \frac{N_1}{N_2} \text{ and } 5 = \frac{N_1^3}{N_2^3}$$

$$\therefore\ N_1^3 = 5N_2^3$$

Now we have seen that this can be true only if 5 is the cube of an integer; but we know that it is not. Therefore we have to conclude that we cannot express $\sqrt[3]{5}$ as the ratio of two integers N_1, N_2, *i.e.* as a vulgar fraction. The reader should be able to apply the foregoing argument to the simpler case $\sqrt{2}$. It follows that $\sqrt{2}$ is not expressible as a vulgar fraction, whence also a ratio such as $\sqrt{5}:\sqrt{2}$ or $\sqrt{2}:1$ is not expressible in terms of a vulgar fraction.

None the less, we know by trial and error that we can indeed find a number between 1 and 2 whose cube is as near to 5 or whose square is as near to 2 as we like to make it; and we can repeatedly narrow the boundary of error in evaluating such a number. Thus:

$$(1.4)^2 = 1.96 < 2 < 2.25 = (1.5)^2, \text{ whence}$$
$$1.4 < 2^{\frac{1}{2}} < 1.5$$

Proceeding in this way, we find:

$$1.41 < 2^{\frac{1}{2}} < 1.45$$
$$1.414 < 2^{\frac{1}{2}} < 1.415$$
$$1.4142 < 2^{\frac{1}{2}} < 1.4143$$
$$1.4142135 < 2^{\frac{1}{2}} < 1.4142136$$

Before discussing how we can tailor our representation of fractions to accommodate the seemingly paradoxical conclusion that such, so-called *irrational*, numbers are not expressible as the ratio of two integers, let us pause to notice that they are by no means exceptional. Since $2^2=4$, $2^3=8$, $2^4=16$, $2^{13}=8192$, there are two integers (2, 3) whose irrational square roots lie between 1 and 2 (or -1 and -2), six (2, 3, 4, 5, 6, 7) whose cube roots, fourteen whose fourth roots and 8190 whose 13th roots also lie between the same limits positive or negative. What we may speak of as the *packing* of such numbers between successive integers increases uniformly. There are 126 integers whose irrational seventh roots lie between 1 and 2, but the number of integers whose seventh roots lie between 9 and 10 is in fact over 5 million. By trial and error before attempting a general proof, the reader should be able to show that the number of integers whose irrational pth roots lie between N and $N+1$ or $-N$ and $-(N+1)$ is $(N+1)^p-(N^p+1)$.

That the crude notion of a real number as an integer or as the ratio of two integers cannot accommodate a vast class of fractions would be very disturbing if our only means of recording a fraction were the vulgar form such as $\frac{5}{9}$, $\frac{12}{27}$ or $\frac{203}{481}$. It ceases to worry us when we have fully explored the implications of basal notation. Though all vulgar fractions expressible as the ratio of two integers are also expressible as basal fractions, the converse is happily not true. Fractions expressible as the ratio of two integers are divisible into two classes:

(i) if the denominator contains no factor other than a power of one of the prime factors of the base (*i.e.* is of the form $2^a.5^b$ if the base is 10), the corresponding basal fraction terminates at the ath place if $a > b$ or at the bth place if $b > a$ as is true in the decimal system of

$$\frac{3}{16}=\frac{3}{2^4.5^0}=0.1875 \text{ and } \frac{2}{25}=\frac{2}{2^0.5^2}=0.08$$

(ii) if the denominator contains a factor which is a prime not itself a factor of the base, the same sequence of one or more ciphers repeats itself indefinitely, as is true of

$$\tfrac{2}{3}=0.\dot{6}=0.66666\ldots$$

$$\tfrac{3}{7}=0.\dot{4}2857\dot{1}=0.428571428571428571428571\ldots$$

Basal fractions of the second type, *i.e.* periodic (recurrent) and nonterminating, are of dual interest to our theme. As we have seen, they familiarise us with the notion of a *convergent* infinite series, *i.e.* the fact that we may be able to go on adding smaller and smaller fractions indefinitely without exceeding a definite sum. For instance, we see at once that $0.\dot{1}$ lies between 0.11 and 0.12, between 0.111 and 0.112, between 0.1111 and 0.1112 and so on. Any periodic nonterminating basal is indeed equivalent to an infinite series of which the denominators of successive terms are increasing powers of the base; but there is no conceivable reason why the denominators of successive terms of an infinite series of vulgar fractions should increase in this way only.

When we have dismissed vulgar fractions of the two types mentioned, we have not indeed disposed of all conceivable basal fractions. We can conceive and construct convergent infinite series in which no regular periodicity emerges in the sequence of ciphers of the basal fraction, for instance 0.23223222322223... An important example is Euler's number ($e=2.7182818285\ldots$) of which the part beyond the decimal point is the sequence

$$\frac{1}{1.2}+\frac{1}{1.2.3}+\frac{1}{1.2.3.4}+\frac{1}{1.2.3.4.5}\ \textit{etc.}$$

$\sqrt[2]{1}=1{\cdot}00\ldots$	$\sqrt[3]{1}=1{\cdot}00\ldots$	$\sqrt[4]{1}=1{\cdot}00\ldots$
		$\sqrt[4]{2}=1{\cdot}18\ldots$
	$\sqrt[3]{2}=1{\cdot}25\ldots$	$\sqrt[4]{3}=1{\cdot}31\ldots$
		$\sqrt[4]{4}=1{\cdot}41\ldots$
	$\sqrt[3]{3}=1{\cdot}44\ldots$	$\sqrt[4]{5}=1{\cdot}49\ldots$
$\sqrt[2]{2}=1{\cdot}41\ldots$		$\sqrt[4]{6}=1{\cdot}56\ldots$
	$\sqrt[3]{4}=1{\cdot}58\ldots$	$\sqrt[4]{7}=1{\cdot}62\ldots$
		$\sqrt[4]{8}=1{\cdot}68\ldots$
		$\sqrt[4]{9}=1{\cdot}73\ldots$
	$\sqrt[3]{5}=1{\cdot}70\ldots$	$\sqrt[4]{10}=1{\cdot}77\ldots$
$\sqrt[2]{3}=1{\cdot}73\ldots$		$\sqrt[4]{11}=1{\cdot}82\ldots$
	$\sqrt[3]{6}=1{\cdot}81\ldots$	$\sqrt[4]{12}=1{\cdot}86\ldots$
		$\sqrt[4]{13}=1{\cdot}89\ldots$
	$\sqrt[3]{7}=1{\cdot}91\ldots$	$\sqrt[4]{14}=1{\cdot}93\ldots$
		$\sqrt[4]{15}=1{\cdot}96\ldots$
$\sqrt[2]{4}=2{\cdot}00\ldots$	$\sqrt[3]{8}=2{\cdot}00\ldots$	$\sqrt[4]{16}=2{\cdot}00\ldots$

Between 1 and 2 lie the irrational square roots of $2=(2^2-2)$ integers, the irrational cube roots of $6=(2^3-2)$ integers, the irrational fourth roots of $14=(2^4-2)$ integers, and the irrational nth roots of (2^n-2) integers.

It is easy to see that this series is convergent. When we reach the ninth term, its value is

$$\frac{1}{1.2.3.\,.\,.10}=\frac{1}{362880}$$

The remainder of the series then simplifies to

$$\frac{1}{362880}\left(\frac{1}{11}+\frac{1}{11.12}+\frac{1}{11.12.13}\cdots\right)$$

Thus the remainder after 9 terms is less than

$$\frac{1}{362880}\left(\frac{1}{10^1}+\frac{1}{10^2}+\frac{1}{10^3}\cdots \textit{ad infin.}\right)$$

$$=\frac{1}{9(362880)}$$

Prima facie, we have no reason to believe that, and good reason to doubt whether, we could ever express as the exact ratio of two nameable integers a nonterminating nonperiodic basal fraction built up of terms which embrace as factors of the denominator a limitless sequence of different prime numbers; and we shall later see that any irrational number is expressible as a sum involving a limitless succession of terms the denominators of which involve multiples of more and more different primes. However, the impossibility of representing an irrational number as the ratio of two integers need puzzle us less at this stage, if we now acquaint ourselves with one way of writing a fraction which can in fact represent an irrational number in a rational way.

As an example $5^{\frac{1}{2}}$ will serve well enough. We know that $2^2=4$ and $3^2=9$. If we can give a meaning to $5^{\frac{1}{2}}$ it therefore lies between 2 and 3, and we may write it as $2+r$ in which r is the fractional remainder. Thus $(2+r)^2=5$, so that

$$4+4r+r^2=5 \quad \textit{and} \quad r(4+r)=1$$

$$\therefore r=\frac{1}{4+r}$$

$$\therefore r=\cfrac{1}{4+\cfrac{1}{4+r}}$$

$$\therefore \boldsymbol{r}=\cfrac{\mathbf{1}}{\mathbf{4}+\cfrac{\mathbf{1}}{\mathbf{4}+\cfrac{\mathbf{1}}{\mathbf{4}+\boldsymbol{r}}}}$$

and so on, indefinitely.

By inspection, we see that

$$\frac{1}{4}>r>\frac{1}{4+\frac{1}{4}}$$

In short, we can bound successively more refined limits between which r lies, by using as approximations:

$$\frac{1}{4};\ \frac{1}{4+\frac{1}{4}};\ \cfrac{1}{4+\cfrac{1}{4+\frac{1}{4}}};\ \cfrac{1}{4+\cfrac{1}{4+\cfrac{1}{4+\frac{1}{4}}}}\ \textit{etc.}$$

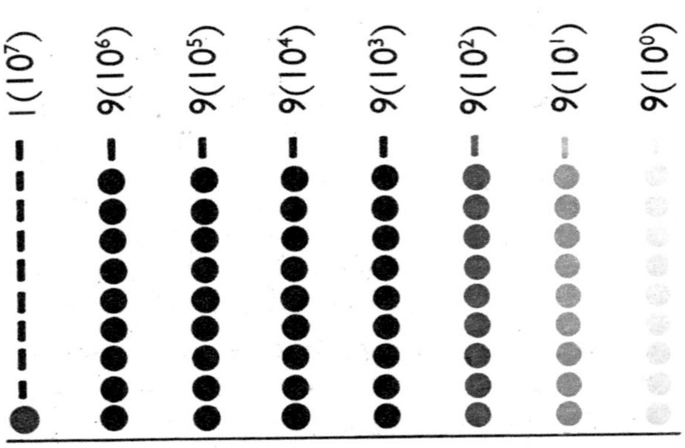

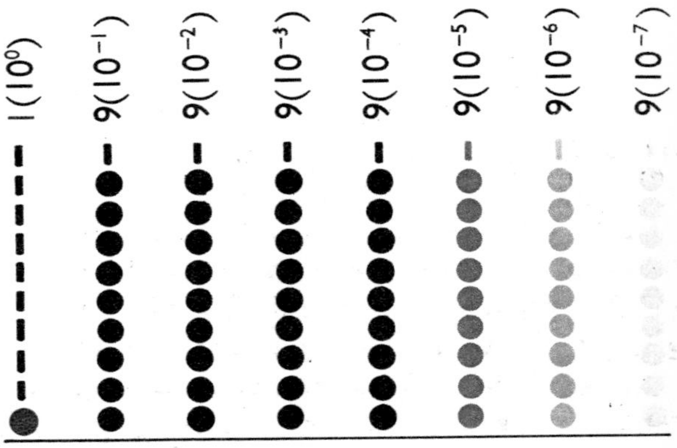

Early mathematicians found it hard to grasp that the sum of an infinite series of positive fractions may be a finite number. Modern notation makes it obvious that $1(10^7)>9(10^6)+9(10^5)\ldots+9(10^0)$. *Use of basal fractions makes it obvious that* $1(10^0)>9(10^{-1})+9(10^{-2})+9(10^{-3})$ *to infinity.*

In decimal notation, this gives

$$0.25 : 0.2353 : 0.2361 : 0.2360\ldots$$

By multiplying out $(2+r)$ based on the mid-value of the last pair, we get a number differing from 5 by less than 1 in 50,000, *viz.* $(2.23605)^2 = 4.99992$ (correct to 6 significant figures).

Some readers will here find it a useful exercise to apply the same method to evaluate successive approximations to $\sqrt{2}$ as:

$$1\tfrac{1}{2} : 1\tfrac{2}{5} : 1\tfrac{5}{12} : 1\tfrac{12}{29} \textit{ etc.}$$

One must not infer from the foregoing remarks that a continued fraction necessarily corresponds to a nonterminating nonperiodic basal fraction. A single example suffices to show that this is not so. It is indeed a more general pattern which embraces the vulgar fraction as a particular case. If we choose the approximation to $(2+r)$ for $4^{\frac{1}{2}}$, we derive $r=0$ directly; but if we choose $(1+r)$ we obtain

$$r = \frac{3}{2+r} \equiv r = \cfrac{3}{2+\cfrac{3}{2+r}} \equiv r = \cfrac{3}{2+\cfrac{3}{2+\cfrac{3}{2+r}}} \textit{ etc.}$$

Successive approximations by neglecting r in the bottom line are:

$$r_1 = \frac{3}{2} : r_2 = \frac{3}{2+\frac{3}{2}} = \frac{6}{7} : r_3 = \frac{3}{2+\frac{6}{7}} = \frac{21}{20} \textit{ etc.}$$

The eleventh and twelfth approximations are respectively:

$$r_{11} = \frac{132681}{132680} = 1 + \frac{1}{132680}$$

$$\text{and } r_{12} = \frac{398040}{398041} = 1 - \frac{1}{398041}$$

Thus r approaches more and more nearly to the limit $r \simeq 1$ in agreement with the identity $(1+r)^2 = 4$.

Though very space-consuming, the *continued fraction* has one merit the rationale of which the reader may explore as a puzzle. Alternate approximations exceed and fall short of the ideal value. Thus we have derived for $\sqrt{5}$, the values 2.25, 2.2353, 2.2361, 2.2360 . . . *etc.* At any stage in the sequence, the ideal value lies between two adjacent estimates, so that

$$2.25 > 5^{\frac{1}{2}} > 2.2353 < 5^{\frac{1}{2}} < 2.2361 > 5^{\frac{1}{2}} > 2.236\ldots \textit{ etc.}$$

When we have thus disposed of the paradox that some fractions are not expressible as the ratio of two integers, we have still to resolve a very legitimate doubt prompted by the reflection that we can neither multiply exactly, nor divide exactly, a number such as $\sqrt{5}$ by a number such as $\sqrt{3}$. We shall later see that the Greek mathematicians of antiquity reacted to this challenge by substituting the notion of *magnitude* for the notion of number. In effect, this meant that they abandoned the attempt to interpret a ratio such as $\sqrt{5}:\sqrt{7}$ in any terms other than an idealised scale diagram. Having no prevision of such a possibility, they felt no obligation to communicate in the only language which the electronic brain understands. We ourselves cannot afford to shirk the obligation.

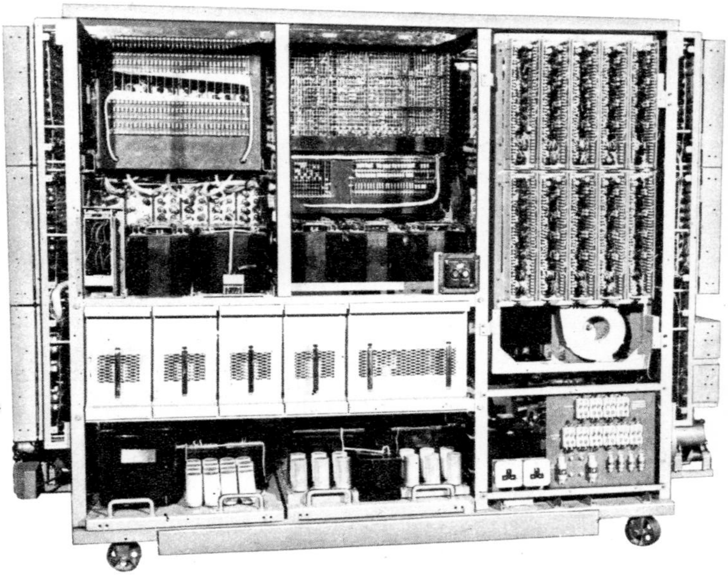

Unit of an electronic computer. All such devices work initially *to a base of 2, though many can now translate into any other base required.*

We have seen that we can express numbers such as $\sqrt{5}$ and $\sqrt{3}$ to any order of precision suitable to the end in view; but we have not shown that we can compute with them as confidently. We have given a meaning adequate to practical needs to a ratio such as $\sqrt{2:3}$; but we have not given any meaning to a ratio such as $\sqrt{2}:\sqrt{3}$. The issue is of great importance when we hand over our computations to an electronic brain, unless prepared to give it a blank cheque. Happily, our Hindu heritage which makes the existence of a convergent series so easy to grasp helps us here, and for the same reason. Let us then look at the two following numbers:

$$\text{(a) } 1000.\dot{9} \qquad \text{(b) } 9999.\dot{0}$$

If one retains only the first 4 ciphers of the unending sequence (a), the proportionate error will be less if we substitute any other combination of ciphers in the residue or any ciphers other than zero in the part retained; but if one substitutes any other combination for the first 4 ciphers of (b) and replaces at any finite stage one or more zeros by any other cipher in the unending sequence of zeros, the proportionate error will be greater. More generally, we may say that the result of rejecting all significant figures after the nth one results in the greatest possible *proportionate* error if a real number is expressible in the form

$$N_1 = 10^p(10^{n-1}+1)$$

$$\text{e.g. } 1.000\dot{9} = 10^{-3}(1001)$$

If we write $99 = 10^2 - 1$, $999 = 10^3 - 1$ *etc.*, we may likewise state the condition that the proportionate error must be least when we can express a number in the form

$$N_2 = 10^p(10^n - 1)$$

$$\text{e.g. } 99.999 = 10^{-3}(100000 - 1)$$

It should be clear that the maximum proportionate error in a product resulting from retaining only n significant figures in each of the factors is greatest when each is expressible like N_1 above. Let us therefore look at the situation when:

$$a = 10^p(10^{n-1}+1) : b = 10^q(10^{n-1}+1)$$

$$ab = 10^{p+q}(10^{n-1}+1)^2$$

If we reject all significant figures after the nth, our approximations will then be expressible as:

$$A = 10^p.10^{n-1} : B = 10^q.10^{n-1} : AB = 10^{p+q}.10^{2n-2}$$

Since $(10^{n-1}+1)^2 = 10^{2n-2} + 2.10^{n-1} + 1$, the gross error involved is:

$$ab - AB = \Delta = 10^{p+q}(2.10^{n-1}+1)$$

Whence the proportionate error is:

$$\frac{\Delta}{ab} = \frac{2.10^{n-1}+1}{(10^{n-1}+1)^2} < \frac{2.10^n+1}{10^{2n-2}}$$

This is the greatest proportionate error we can conceivably incur in computing a product when we reject all except the first n significant figures of both factors. Suppose we reject all after the sixth, the formula therefore means that the proportionate error (deficit) in the product can never exceed:

$$\frac{2.10^5 - 1}{2.10^{10}} = 0.0000100005$$

When we divide one number by another, four situations suggested by the foregoing may arise; but it will suffice to consider two; the proportionate error in the numerator is as great as can ever happen and the proportionate error in the denominator as small as can ever happen, or *vice versa*. For the first situation we may write:

$$\frac{a}{b} = \frac{10^p(10^{n-1}+1)}{10^q(10^n - 1)}$$

Our approximate value will then be

$$\frac{A}{B} = \frac{10^p.10^{n-1}}{10^q(10^n - 1)}$$

$$\frac{a}{b} - \frac{A}{B} = \Delta = \frac{10^{p-q}}{10^n - 1}$$

The proportionate error is $\triangle \div \frac{a}{b}$, so that

$$\frac{1}{10^{n-1}+1} < \frac{1}{10^{n-1}}$$

Alternatively, we may make the proportionate error greatest in denominator, so that

$$\frac{a}{b} = \frac{10^q(10^n-1)}{10^n(10^{n-1}+1)} \text{ and } \frac{A}{B} = \frac{10^q(10^n-1)}{10^n . 10^{n-1}}$$

$$\triangle = \frac{10(10-1)}{10^n}\left[\frac{1}{10^{n-1}(10^{n-1}+1)}\right]$$

and the proportionate error is

$$\frac{1}{10^{n-1}}$$

This is the greatest proportionate error we can make if we interpret the ratio a:b as a vulgar fraction whose numerator is equivalent to the first n significant figures of a and whose denominator is equivalent to the first n significant figures of b. If $n=6$, the proportionate error can never exceed 0.00001.

When a mechanical device for computation cuts off all significant figures after the first n in a particular operation of multiplication or a particular operation of division, we therefore know that the proportionate error can never exceed a specifiable number. We can evaluate the product or ratio of two nonterminating basal fractions, knowing confidently that our error cannot exceed a certain limit. In practice, this is all the assurance we need for trusting the outcome of machine computation. If the electronic brain could answer back, it might well do so as follows. *When you talk to me about multiplying or dividing one nonterminating basal fraction by another, I can carry out your instructions to your own satisfaction; but when you talk about Euclidean magnitudes I have no clue to what you mean.*

Chapter 3 Our Debt to Egypt and Mesopotamia

Few things seem more certain at the present time than the fact that the earliest recorded evidence of mathematical reasoning, as some pure mathematicians now use the term, is the contribution of Greek-speaking communities of the Mediterranean about 600 B.C. Few things are also more certain than that they themselves did not exaggerate their self-confessed and considerable indebtedness to the temple cultures of Egypt and Mesopotamia, that the priestly custodians of the calendar jealously guarded the secrets of how they arrived at the discoveries for which their records disclose no proof and that they had some good reasons for confidence in the reliability of at least some of them. What level of mathematical skill the earliest settled communities which have left remains elsewhere may have attained is a question to which we have as yet no answer; but it is unlikely that their achievements were comparable to those of Egypt and Mesopotamia. We can say with assurance that Egypt and Mesopotamia emerge first in the history of the written record; and that the record of their progress covers more than two millennia B.C.

In so far as the emergence of a written record first crystallises from the need for a calendar, they are the first centres of civilisation which have left their impress on our habits of using and of reasoning about numbers. To date the beginnings of either, we have still to rely mainly on astronomical considerations with a bearing on the calendar itself. On the most plausible view, we may infer that some sort of tally of the solar year began in Egypt about 4241 B.C. and perhaps 1500 years earlier in what is now Iraq. By 3000 B.C. in both regions there were cities of which the temple, with its divine overlord and an official caste of calendar makers, surveyors, architects and irrigation engineers, was the focus. In Egypt, the lower castes were the infantry and a vast labour force of slaves; but Mesopotamia was the centre of a flourishing overland trade, conducted by merchants who were also freemen well before 2000 B.C. In contemporary Crete, the common pattern seems to have been much the same; but its lasting contribution to the intellectual outlook of mankind will remain problematical till the decipherment of the two earlier of the three Minoan scripts.

On the other hand, we have abundant documentary evidence concerning the level attained by about 2000 B.C. both in Egypt and in Mesopotamia. At the date last mentioned, Chinese culture, the antiquity of which was till lately grossly exaggerated, was in its infancy; and we may now date its first intellectual efflorescence at about the time when a Greek-speaking civilisation comes into our picture. Though we may well have to pay more attention to the influence of the civilisation which bestowed on the West silk, paper, gunpowder, block printing and lock gates for canals during the course of the Christian era, it is unlikely that it contributed much to the foundations of mathematical lore or indeed to

This colonnade of the temple of Karnak was oriented so that only the rising sun on the summer solstice shone straight along it, thus enabling the priests to record the length of a year.

mathematics of any sort before the latter phase of the Alexandrian culture.

If we subsequently speak of Babylonian mathematics, it is necessary to explain that the term is strictly appropriate only to a short period. Great cities of which the temple site was the core existed in Mesopotamia by 3500 B.C. Of the various inhabitants of the region, the northern Hittites spoke an Indo-European language, and the people of the coastal region of Sumer bordering the Persian Gulf spoke a language possibly allied to Turkish and certainly not Semitic. Others variously called Akkadians, Amorites, Chaldeans (in the south-west on the west bank of the Tigris) and Assyrians (in the north) spoke Semitic dialects; but the parent culture of the whole region was Sumerian. Its script, as one might expect, was primitively pictographic; but it became highly stylised at an early date, in the manner called *cuneiform*, by the custom of imprinting the signs with a wedge-like writing tool on soft clay tablets afterwards baked in the sun.

When the Akkadian dynasty of Sargon subdued Sumer about 2500 B.C., the usurpers took over the 4000-strong pantheon of the Sumerian priesthood, their language for ritual incantations and their script for official edicts. The dynasty lasted only about 100 years. Thereafter the various cities north of Sumer, notably Ur of the Chaldees, Nippur, Asshur and Babylon, came under the rule of Amorites, Chaldeans and other contenders for supremacy till Babylon gained (2169–1870 B.C.) leadership under a dynasty made memorable by the legal code of Hammurabi. To this period belong all the documentary remains which concern us in this chapter. The chief Egyptian papyri – the Moscow and the Rhind – with any bearing on our theme belong to the end of the same period or not much later.

It would be erroneous to hold that either Egypt or Babylon made no contribution to the Greek synthesis thereafter. Though we have little documentary evidence of sustained progress in Egypt after 1500 B.C., Babylonian astronomy continued to make progress throughout the Persian occupation of the surrounding territory after the invasions of Cyrus (539 B.C.) and his successors. About 330 B.C., Alexander the Great decreed the clear-

A fragment of the still-undeciphered Linear A script. Not until the two earlier of the three Minoan scripts are decoded shall we know much about Crete's contribution to man's intellectual outlook before 2000 B.C.

This map makes no attempt to pinpoint a single moment in history. Its purpose is to show the sites of some leading centres of culture during the immense period of Egyptian and Mesopotamian greatness.

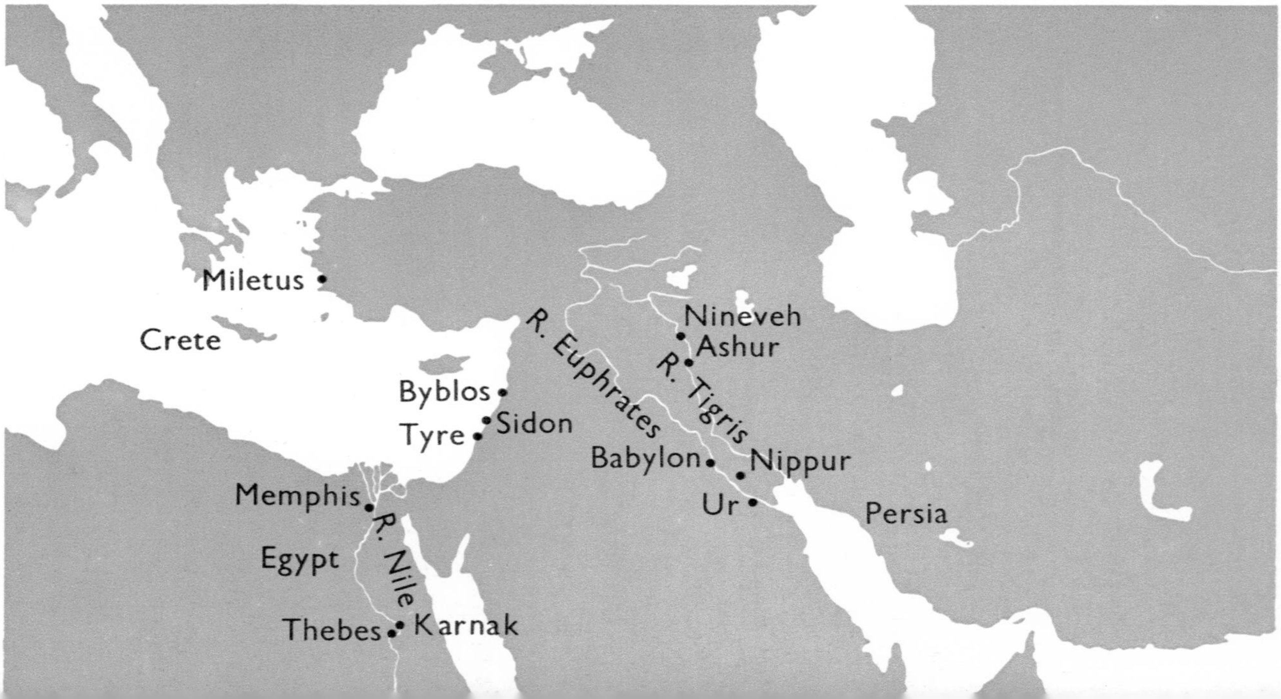

ance of the temple site of Babylon itself, whereafter it has been customary to speak of the successors of the Babylonians as Chaldeans. Doubtless, there were close contacts between the Greek-speaking peoples and those of the Near East when Alexandria became the centre of Greek culture under the dynasty of the Ptolemies, founded by one of Alexander's generals.

At least, we know that the Seleucid dynasty founded by another of Alexander's successors was sympathetic to the old religion and founded cities for Greek and Macedonian colonists throughout Babylonia. Shortly before the Alexandrian invasion, a Chaldean astronomer reputedly recorded the precession of the equinoxes, a 25,000-year cycle of the position of the fixed stars relative to the horizon plane now interpreted by saying that the North Pole, like a spinning top, wobbles about an axis at right angles to the plane of its solar orbit. This was a century before Hipparchus announced its discovery along with many other results referable to the moon's motion and to the sidereal periods of the planets, already incorporated in Chaldean temple lore, as the harvest of several millennia of patient observation.

When one writes about the science or mathematics of another age, it is almost impossible to steer a middle course between two erroneous extremes. On the one hand, one may underrate its achievements through misunderstanding of what were its needs and what its difficulties. On the other, it is a common fault to discern anticipation of later discoveries of which people of earlier times could have had no prevision unless gifted with superhuman subtlety or second sight. If this chapter puts forward suggestions of how the priestly mathematicians of the temple made what discoveries they did, the best that one can say about many of them is that they *may* be right. Their justification is that the discoveries themselves provide a useful peg on which to hang a little information which is intrinsically useful, regardless of its historical relevance.

Two thousand years before its rise to power, the temple astronomers of the Nile had far surpassed the most brutal and totally uncreative of all the early civilisations in the art of representing and using numbers vastly larger than the M of the Roman blackshirts. Unless the papyrus of the scribe Ahmes (*circa* 1600 B.C.) is merely the relic of a very elementary contemporary textbook, documentary evidence at our disposal indicates that the art of computation in Egypt was far behind the level attained well before 2000 B.C. in Mesopotamia. This is especially evident if we scrutinise the ham-handed way in which the Egyptians handled fractions for reasons of which the advantages are still by no means wholly transparent to ourselves. Presumably to shorten work on the abacus, they broke down a fraction

Babylon, shortly after excavations made at the turn of our own century. Here thousands of clay tablets have been unearthed. Written in known scripts, they throw a flood of light on Babylon's contribution to knowledge.

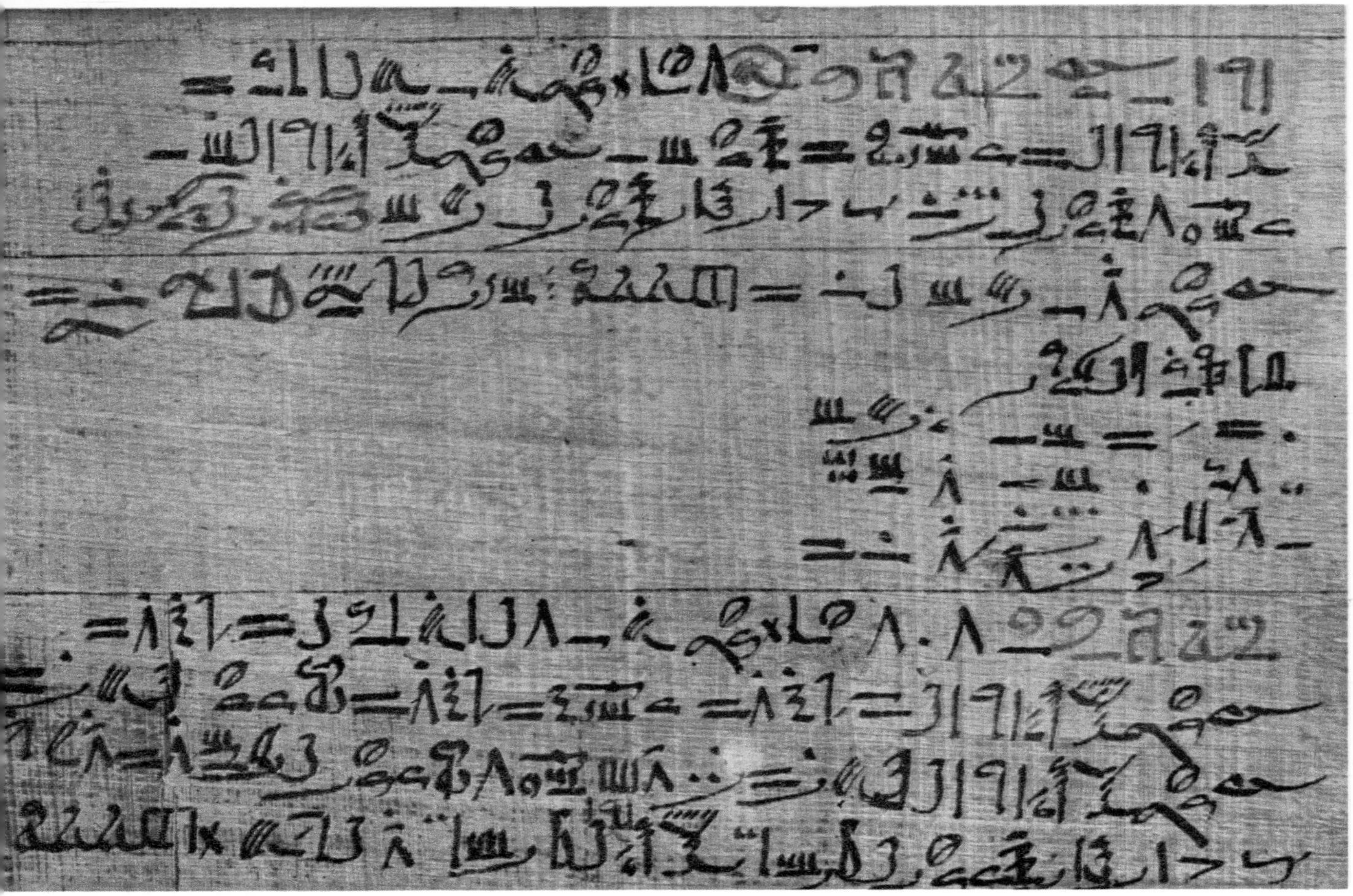

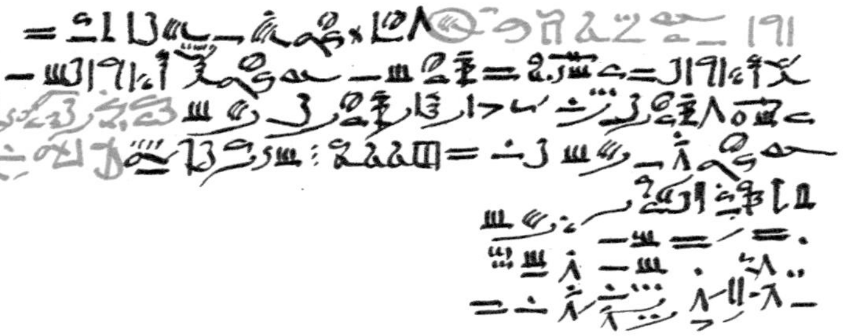

The first part of this extract from the Rhind Papyrus sets the problem of finding the volume of a cylindrical granary 9 units in diameter and 10 units high. It gives the solution in a form somewhat as follows: Take away $\frac{1}{9}$ of 9 which is 1. This leaves you with 8. Multiply this number by itself and you have 64. Multiply 64 by 10 and you have 640. To find the area of a circle of known diameter the scribe Ahmes thus used the formula $\left(\frac{8d}{9}\right)^2$*, equivalent to* $\frac{256}{81}r^2$ *or* $3.1605r^2$ *which gives a close approximation to* π.

whose numerator is greater than 2 into fractions whose numerator is unity (or 2) as the following examples illustrate:

$$\frac{149}{308} = \frac{1}{4} + \frac{1}{7} + \frac{1}{11}$$

$$\frac{23}{45} = \frac{2}{5} + \frac{1}{9}$$

$$\frac{13}{45} = \frac{2}{9} + \frac{1}{18} + \frac{1}{90} \text{ or } \frac{1}{5} + \frac{2}{45} + \frac{2}{45}$$

It is an amusing exercise to play with the trick of breaking down fractions in this way; and anyone who attempts it will doubt whether the feat was feasible unless the temple computers worked with tables embodying the experience of many generations of predecessors. On the other hand, the Babylonians transmitted to their Greek-speaking successors a system which had considerable merits and one which survives in our division of the hour or degree into minutes (*pars minuta*) and seconds (*pars secunda*) as the names have come down to us through Latin. The system was already used during the period of Akkadian supremacy. We may therefore conclude that it was a bequest of the Sumerian culture.

At least in their astronomical work, the Babylonians used at an early date the base $b=60$ consistently; and the diacritic marks which we now use when we write 54° 30′ 25″ would have provided, albeit in a cumbersome way, all that the introduction of the zero sign accomplishes for us, if the size of the base had not imposed an exhausting task on a beginner confronted with the invitation to memorise tables of addition, subtraction or multiplication. Indeed, it seems that the Chaldean astronomers of the Seleucid period did use a convention for the empty column of the abacus. Be that as it may, the Alexandrian astronomers who used the same sexagesimal fractions, *i.e.* fractions expressed as 60^{-n} in the same way as we can express fractions as 10^{-n}, had the worst of both worlds. They already expressed integers in terms of $b=10$.

Though the choice of so large a base as 60 puts a large burden on the memory unless one relies on tables as we now use logarithmic tables or tables of trigonometrical ratios, it has the merit that 60 is expressible in terms of powers of 3 prime factors, 2, 3, 5, being divisible by 2, 3, 4, 5, 6, 10, 12, 15, 20 and 30. If N contains no prime factors other than the above, N^{-1} is therefore expressible in minutes (60^{-1}), seconds (60^{-2}) and sub-seconds (60^{-3}) for any multiple of the fractions shown below:

$\frac{1}{1}$	60′	$\frac{1}{12}$	5′	$\frac{1}{32}$	1′ 52″ 30‴
$\frac{1}{2}$	30′	$\frac{1}{15}$	4′	$\frac{1}{36}$	1′ 40″
$\frac{1}{3}$	20′	$\frac{1}{16}$	3′ 45″	$\frac{1}{40}$	1′ 30″
$\frac{1}{4}$	15′	$\frac{1}{18}$	3′ 20″	$\frac{1}{45}$	1′ 20″
$\frac{1}{5}$	12′	$\frac{1}{20}$	3′	$\frac{1}{48}$	1′ 15″
$\frac{1}{6}$	10′	$\frac{1}{24}$	2′ 30″	$\frac{1}{50}$	1′ 12″
$\frac{1}{8}$	7′ 30″	$\frac{1}{25}$	2′ 24″	$\frac{1}{54}$	1′ 6″ 40‴
$\frac{1}{9}$	6′ 40″	$\frac{1}{27}$	2′ 13″ 30‴	$\frac{1}{60}$	1′
$\frac{1}{10}$	6′	$\frac{1}{30}$	2′		

Needless to say, this also permits one to express as *terminating* sexagesimals a large range of fractions whose denominator is less than 60. In addition to integers which we can represent as products of the powers of its prime factors within the interval $N=1$ to $N=60$ we have also:

(a) 14 primes, *viz.*
7, 11, 13, 17, 19, 23, 29, 31, 37, 41, 43, 47, 53, 59

(b) 20 multiples of the foregoing 14, *viz.*
14, 21, 22, 26, 28, 33, 34, 35, 38, 39, 42, 44, 46, 49, 51, 52, 55, 56, 57, 58

The integral values of N not included in the foregoing table do not lead to terminating sexagesimal fractions. For instance:

$$\frac{1}{7}=\frac{1(60)'}{7}=\left(8\frac{4}{7}\right)' : \frac{4'}{7}=\frac{4(60)''}{7}=\left(34\frac{2}{7}\right)'' : \frac{2''}{7}=\frac{2(60)'''}{7}=\left(17\frac{1}{7}\right)''' \ldots$$

From this point onwards, the procedure repeats itself, whence

$$\tfrac{1}{7}=8'\ 34''\ 17'''\ 8''''\ 34'''''\ 17''''''\ \textit{etc.}$$

By analogy with our way of writing one seventh as $0.\dot{1}4285\dot{7}$, we might nowadays write this briefly:

$$\tfrac{1}{7}=\dot{}(8'\ 34''\ 17''')\dot{}$$

In the same way, we find that:

$$\tfrac{3}{11}=\dot{}(16'\ 21''\ 49'''\ 5''''\ 27''''')\dot{}$$

$$\tfrac{5}{13}=\dot{}(23'\ 4''\ 36'''\ 55'''')\dot{}$$

$$\tfrac{1}{17}=\dot{}(3'\ 31''\ 45'''\ 52''''\ 56'''''\ 28''''''\ 14'''''''\ 7'''''''')\dot{}$$

$$=0.\dot{0}58823529411764\dot{7}$$

In such periodic sexagesimals we have the germs of several notions which were difficult for the cleverest people of antiquity to grasp, in particular the finite limit of the sum of an infinite number of terms; but we have no evidence that either the Alexandrians or their teachers fully appreciated this implication, possibly because: (a) it would rarely be necessary to continue to carry on the calculation till the recurrence was manifest; (b) people who relied on the abacus or on tables of reference to perform the computation would not lightly incur the labour of doing so for the fun of it.

Our sources of information concerning the level of Mesopotamian mathematics are profuse. Many hundreds of tablets in the cuneiform script deal with problems which we should now call algebraic, or with geometrical relations; but many thousands are tables for computation, including tables of reciprocals of sexagesimals for work with fractions, tables of squares and tables of cubes for interpolating cube and square roots, together with tables of the sums of squares and cubes (*e.g.* $2^3+3^3=35$ *or* $3^3+4^3=91$) for manipulating problems which involve the solution of a cubic equation in our own idiom. One of the richest finds – some 50,000 tablets – comes from a temple site at Nippur destroyed by the Elamites about 2000 B.C. If we use the term *algebra* in a metaphorical way, we cannot doubt that Babylonian algebra was well ahead of Egyptian at that date.

Two rules of Babylonian algebra – in a metaphorical sense of the term – are instructive in connexion with the greatest achievement of Egyptian geometry. They will help us to clarify the different ways in which people use the two terms. In common parlance, we use the terms algebra and geometry respectively to convey manipulation with signs and manipulation with plane or solid figures. When an historian speaks of Babylonian and Hindu algebra or when a mathematician speaks of the geometry of four dimensions, the terms have a less restricted connotation. No one consistently used symbols of the sort we now use in school algebra before the seventeenth century of our own era, and the mathematicians of antiquity leaned heavily on geometrical reasoning to justify rules of solution for the class of problems which constitute the subject matter of elementary algebra today. Conversely, the sign language of elementary algebra may intrude into the visual domain, where a common pattern for expressions which

Site of Nippur, near the step-pyramid temple (ziggurat). The site yielded some 50,000 clay tablets, many of which testify to a considerable knowledge of mathematics.

condense the properties of figures in one, two and three dimensions of space is one which we can symbolically and usefully extend into a domain beyond our power to visualise. Two families of series with which the Babylonians were familiar by 2000 B.C. illustrate this forcibly. One includes the triangular numbers (1, 3, 6, 10, 15 *etc.*) and the tetrahedral numbers (1, 4, 10, 20, 35 *etc.*). The other includes the square numbers (1, 4, 9, 16, 25 *etc.*) and the pyramidal numbers (1, 5, 14, 30, 55 *etc.*).

Such figurate numbers provide us with a simple example of the overflow of geometrical terms into an algebraic context. We may visualise the following series respectively as points (*zero* dimension), lines (*one* dimension), triangles (*two* dimensions) and tetrahedral figures (*three* dimensions):

1	1	1	1	1	1	.....
1	2	3	4	5	6	.....
1	3	6	10	15	21	.....
1	4	10	20	35	56	.....

The build-up of these series is as follows:

1	1	1	1
1	$(1+1)=2$	$(2+1)=3$	$(3+1)=4$
1	$(1+2)=3$	$(3+3)=6$	$(6+4)=10$
1	$(1+3)=4$	$(4+6)=10$	$(10+10)=20$

1	1
$(4+1)=5$	$(5+1)=6$
$(10+5)=15$	$(15+6)=21$
$(20+15)=35$	$(35+21)=56$

The build-up of the *r*th term in each of the 3 last sets is by adding its predecessor (*i.e.* $r-1$th) to the *r*th term of the preceding series. *e.g.* we get 56 by adding 35 to 21. By continuing this process we derive the "super-solid" figurates:

1	5	15	35	70	126	
1	6	21	56	126	252	
1	7	28	84	210	462	*etc.*

By trial and error, the reader may easily discover a plausible formula for the triangular and tetrahedral numbers, *e.g.*

This city plan of Nippur was made about 3500 years ago. Careful measurement of actual walls and buildings proves that it is drawn accurately to scale.

Some clay-tablet libraries may well have been carefully catalogued. The two sides of this early Sumerian tablet, only $2\frac{1}{2}$ in. long, list 62 literary works.

$$1=\frac{2.1}{2.1}:3=\frac{3.2}{2.1}:6=\frac{4.3}{2.1}:10=\frac{5.4}{2.1}\ etc.$$

$$1=\frac{3.2.1}{3.2.1}:4=\frac{4.3.2}{3.2.1}:10=\frac{5.4.3}{3.2.1}:20=\frac{6.5.4}{3.2.1}$$

By recourse to the rule of formation of the *r*th term, we may easily prove by the method of induction that the formulae suggested by these for the *n*th members of the triangular and tetrahedral numbers fall into line with the same pattern as the linear (*i.e.* natural) numbers:

Lines	**Triangles**	**Tetrahedra**
$\sum_{r=1}^{r=n} 1=\frac{n}{1}$	$\sum_{r=1}^{r=n} r=\frac{n(n+1)}{1.2}$	$\sum_{r=1}^{r=n}\frac{r(r+1)}{1.2}=\frac{n(n+1)(n+2)}{1.2.3}$

TRANSLATION

Add up the terms of the series whose members are units from that of rank 1 to that of rank *n* inclusive	Add up the terms of the series whose members are the natural numbers from that of rank 1 to that of rank *n* inclusive	Add up the terms of the series whose members are the triangular numbers from that of rank 1 to that of rank *n* inclusive

Likewise, the formula for what we may call the 4th dimensional series (1, 5, 15, 35, *etc.*) conforms to the same plan, being:

$$\frac{n(n+1)(n+2)(n+3)}{1.2.3.4}$$

Similar remarks apply to a second family of the same sort. Below, the 1-dimensional (first) sequence is the odd numbers, the 2-dimensional is that of the squares of the natural numbers, and the 3-dimensional sequence being that of the sums of squares, is visualisable as pyramids on a square base.

1	1	3	5	7	9	11
2	1	4	9	16	25	36
3	1	5	14	30	55	91

By trial and error, it is not hard to disclose the following pattern for the 3-dimensional sequence:

$$1=\frac{1.2.3}{1.2.3};\ 5=\frac{2.3.5}{1.2.3};\ 14=\frac{3.4.7}{1.2.3};\ 30=\frac{4.5.9}{1.2.3};$$

$$55=\frac{5.6.11}{1.2.3}$$

The law of formation of the *r*th term of this sequence is on all fours with that of the formation of the *r*th term of the tetrahedral numbers; and it

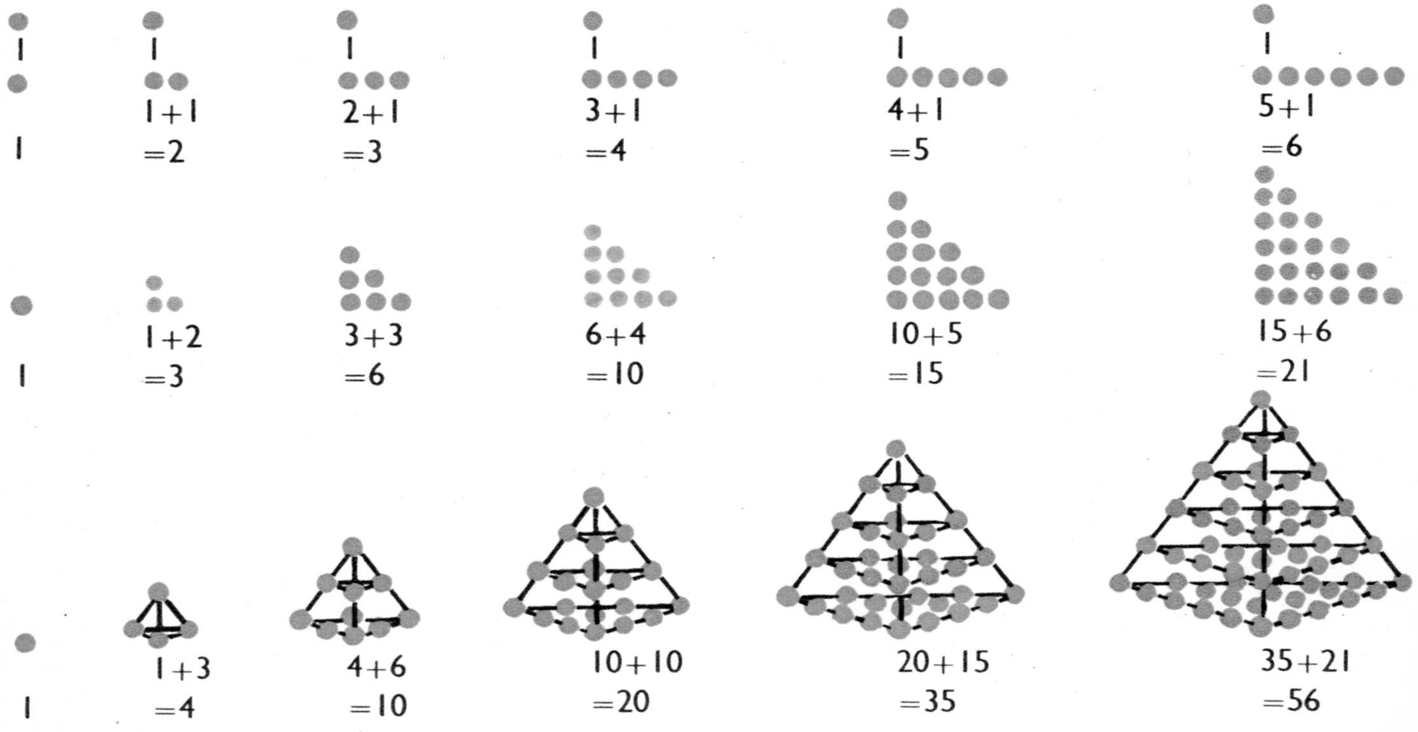

Left: Build-up of the family of series which includes the triangular numbers (1, 3, 6, 10, 15, etc.) and the tetrahedral numbers (1, 4, 10, 20, 35, etc.). Below right: The family which includes square numbers (1, 4, 9, 16, 25, etc.) and pyramidal numbers (1, 5, 14, 30, 55, etc.).

Mid-nineteenth-century ammunition store in Calcutta. Stacking ammunition in pyramids (second right) began before A.D. *1600. Knowledge of pyramidal numbers made it easy to work out how many balls in a pyramid of known height.*

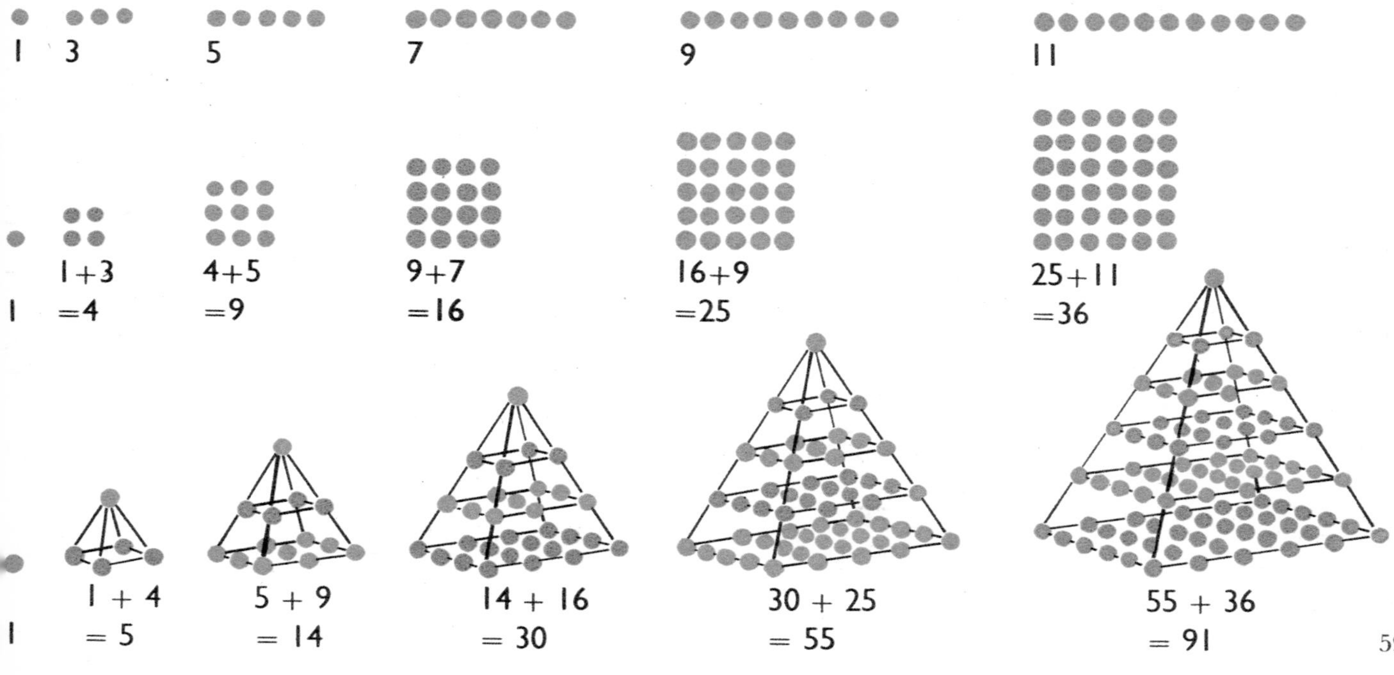

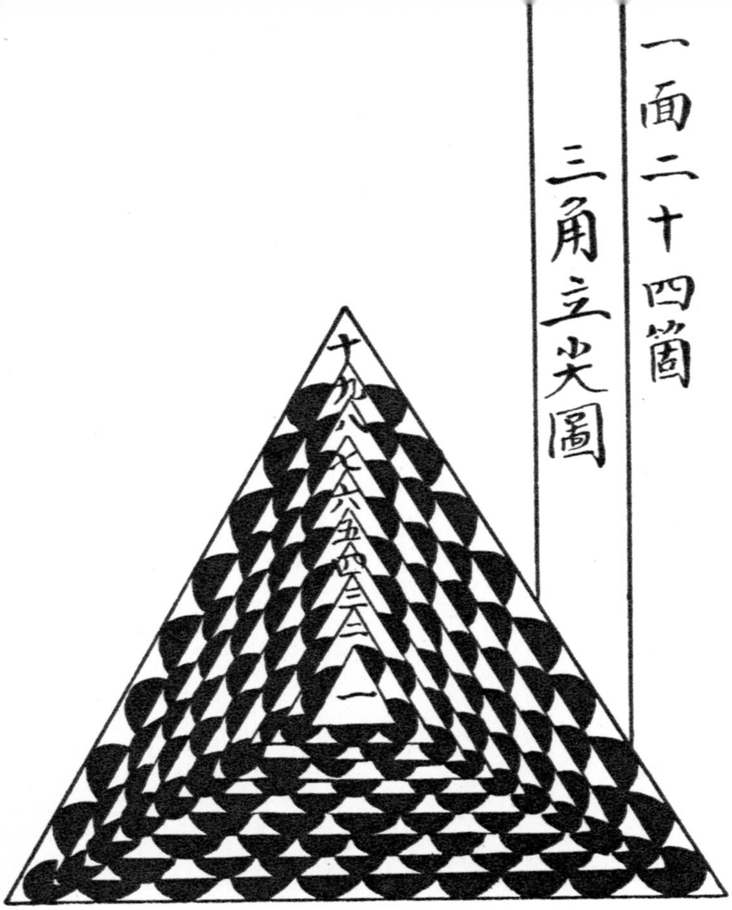

One can arrive at closer and closer approximations to the volume of a three-dimensional figure by packing it with standard solid units of smaller and smaller size. Here the sixteenth-century Chinese mathematician, Chou Shu-Hsüeh, illustrates the packing of a pyramid with ten layers of small spheres.

is a useful exercise to prove by induction the formula for the sum of the first n squares, *viz.*

$$\sum_{r=0}^{r=n} \mathbf{r}^2 = \frac{\mathbf{n(n+1)\ (2n+1)}}{\mathbf{6}}$$

This formula had a pay-off in the age of cannon-ball warfare and could well have had a pay-off at a far more remote time when numbers with peculiar properties had the portentous significance that 7 and the 7th odd number have for some of our contemporaries. Needless to say, our cuneiform tablets would not cite it in any such form as the above. The decipherment would read somewhat as follows:

> to get the total area of all the squares of (*i.e. whose sides are represented by*) the numbers from 1 to a certain other, first multiply this same last number by its successor and their product by the successor of its double and divide the result by 6.

Similarly, the rule for getting the nth triangular number would be verbal and somewhat as follows:

> to get a triangular number whose place in its sequence corresponds to a particular natural number, multiply the latter by its successor and divide by 2.

One use for a table of squares pinpoints the reliance of the mathematics of the ancient world on a scale diagram to justify operations we ourselves carry out in conformity with a symbolism which materialises the physical performance of calculating with an abacus. Thus the possibility of expressing the difference between the area of two squares on sides a and b units of length as a rectangle the lengths of whose adjacent sides are the sum and difference of a and b units now seems to us to be a tautology of computation in need of the services of no geometrical model to justify it as an application of the symbolic identity:

$$a^2 - b^2 = (a+b)(a-b)$$

To those of a vastly earlier vintage than ours the model embodied both a principle of practical land surveying and the rationale of a device for lightening labour on the abacus by replacing the repetitive addition of multiplying two numbers with a much shorter process if there was a table of squares to hand. If both integers are odd or both even, we may set out the operation as follows for (123) (27):

Add: $27 + 123 = 150$
Halve the sum: $\frac{1}{2}(150) = 75$
Subtract: $123 - 75 = 48$ *and* $75 - 27 = 48$
$\therefore (123)(27) = (75+48)(75-48) = 75^2 - 48^2$

If one integer is even and one odd, the procedure involves an additional step, thus:

$$(28)(123) = 27(123) + 123 = 75^2 - 48^2 + 123$$

In the same way, a verbal instruction for solving what we should call a quadratic equation in connexion with a problem of trade would rely on the idiom of line and area to convey the rules for partitioning a square field in accordance with the recipes we now convey in the form:

$$(a \pm b)^2 = a^2 \pm 2ab + b^2$$

Problems in the Egyptian documentary (papyrus) remains of the period 2000–1500 B.C. include examples of the summation of simple arithmetical series, to which class, needless to say, the triangular numbers and squares belong, though not mentioned as such. There is a rule for the sum of simple geometric series of integers (*e.g.* 1, 2, 4, 8, 16 *etc.*) but *not* of fractions (*e.g.* $1, \frac{1}{2}, \frac{1}{4}, \frac{1}{8}$ *etc.*). A few problems involve a quadratic solution. Otherwise, there is nothing on show but the simple class of problems we now solve by a linear equation such as $3x + 4 = 13$.

Babylonian algebra is immensely more sophisticated than this. There are verbal recipes for solving (by tables) what we should call simple simultaneous equations involving up to ten variables, and many examples of how to solve problems involving simple quadratic equations. Indeed, the recovered tablets record examples of verbal instructions for working out problems which entail simultaneous equations now expressible in a form such as

$$x - y + xy = 33 \text{ and } x + y = 8$$

However, it would be wrong to infer that the temple mathematicians envisaged what we should call a general solution of the quadratic. To recapture the temper of their times, we may distinguish five types illustrated by the following examples:

(1) $x^2 + 6 = 5x$ with two solutions ($x = +3$ or $+2$) both positive.

(2) $x^2 + x = 6$ with two solutions ($x = -3$ or $+2$) one positive, one negative.

(3) $x^2 = x + 6$ *ditto* ($x = +3$ or -2).

(4) $x^2 + 4x + 3 = 0$ with both solutions ($x = -3$ or -1) negative.

(5) $x^2 + 3x + 4 = 0$ with neither solution real.

If examined on their own merits, it is probable that (4) and (5) would be equally mysterious to a Babylonian or for that matter to a Greek of the fifth century B.C.; and the fact that at least one record cites both solutions for an equation of type (1) does not necessarily imply that the teacher had any conception of the square as the product alike of two equal negative or of two equal positive factors. It would be possible, as is consistent with the Babylonian treatment of a cubic equation, to solve type (1) by tabulating $x^2 + N$ for successive positive integers x and N as below:

x \ N	1	2	3	4	5	6
1	2	3	4	5	6	7
2	5	6	7	8	9	*10*
3	10	11	12	13	14	*15*
4	17	18	19	20	21	22
5	26	27	28	29	30	31

Inspection of the table shows that $x^2 + 6 = 5x$ if either $x = 2$ or $x = 3$.

A table of squares and square roots found at Nippur. One use of such a table may have been to eliminate the task of multiplication on the abacus.

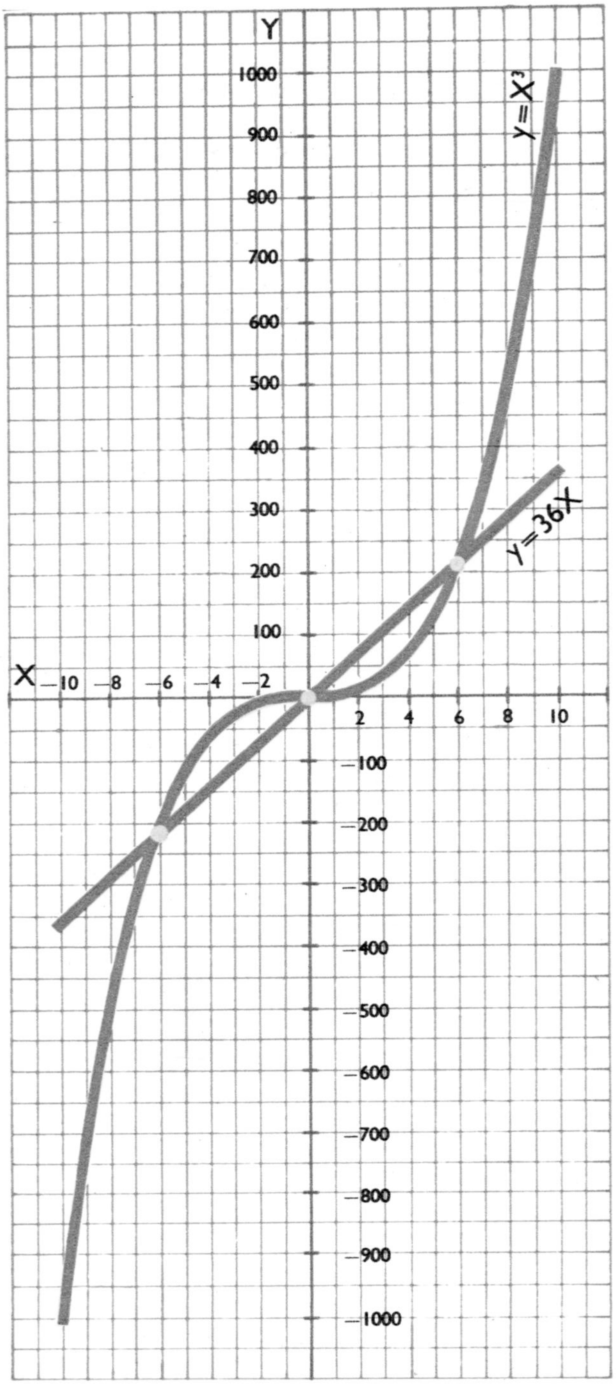

$y=x^3-6x^2-24x+64$
reduces to $y=X^3-36X$
where $X=x-2$
Solution : $X= -6$, 0 and +6
$\therefore x=+8$, +2 and −4

To arrive at this graphical solution of the above cubic equation, the first step is to reduce it to $y=X^3-36X$, where $X=x-2$. If we then plot $y_1=X^3$ and $y_2=36X$ the intersections give a good approximation to the real roots. Babylonian mathematicians may well have used the first step in their attempts to solve cubic equations.

Tables of sums of cubes and squares of integers (*e.g.* $3^3+2^3=35$) were an aid to finding a positive solution of a problem which we should express as a cubic equation, such as $x^3+5x^2=72$; but there is reason to believe that the temple mathematicians could tackle the more general case illustrated by $x^3+5x^2+10x=102$. We may plausibly reconstruct in our own sign writing the procedure for dealing with the general cubic $(x^3+ax^2+bx+c=0)$ as follows. First substitute $X+d=x$, and apply the volume rule:

$$(a+b)^3=a^3+3a^2b+3ab^2+b^3$$

$$\therefore\ (X+d)^3+a(X+d)^2+b(X+d)+c=0$$
$$=X^3+3dX^2+3d^2X+d^3+aX^2+2adX$$
$$+ad^2+bX+bd+c$$

$$\therefore\ X^3+(3d+a)X^2+(3d^2+2ad+b)X$$
$$+(d^3+ad^2+bd+c)=0$$

We can make the term involving X vanish, if we put

$$3d^2+2ad+b=0 \text{ so that } d=\frac{-a+\sqrt{(a^2-3b)}}{3}$$

Having found the required value of d, we may evaluate

$$(3d+a)=A \text{ and } (d^3+ad^2+bd+c)=-B$$

Whence we have:

$$X^3+AX^2=B$$

A table of X^3+AX^2 for different values of A and X gives the required solution if there is one.

Between 2000 B.C. and A.D. 1500 there was indeed no appreciable advance towards manipulating a cubic equation, and when a formal solution emerged in the sixteenth century of our era, it recalled the fundamental step in the Babylonian recipe. The modern solution is very laborious and for practical purposes of little use, since a graph gives us a good first approximation to the real roots, if any, whereafter an iterative procedure permits us to approximate as closely as required in accordance with the way in which the electronic brain solves such problems for us.

However, the first step in the modern general solution is worthy of comment because it discloses a useful recipe for the graphical procedure. In the foregoing argument we chose the value of d which transforms the equation into one which contains no term in X; instead, we might have eliminated x^2 by putting

$$3d+a=0 \text{ so that } d=\frac{-a}{3}$$

The equation then reduces to the form $X^3+pX=B$ in which

$$p=3d^2+2ad+b \text{ and } -B=(d^3+ad^2+bd+c)$$

The best graphical solution is to reduce it to this form so that $X^3=B-pX$. If we then plot $y_1=X^3$ and $y_2=B-pX$, the intersections $(y_1=y_2)$ disclose a good approximation to the real roots.

One other achievement of Babylonian verbal algebra is worthy of mention. In the social context of the earliest practice of usury, the problem of computing compound interest called for a solution; and the priestly mathematicians had a prescription for their pupils. One example deals with the time taken for a fixed sum to double itself at 20%. If $f(=0.2)$ is the fraction corresponding to the fixed interest, this involves finding for the lapse of time (t) the value of $(1+0.2)^t$, *i.e.* what we should now call evaluation of an exponential function. Doubtless there were tables constructed for the solution.

If only in the course of laboriously constructing tables of squares, the makers could not fail to notice that only certain integers (*e.g.* 4, 9, 16, 25) have square roots which are also integers. Indeed, the Babylonian mathematicians did make approximations of one sort and another to the irrationals. One example of great interest is the approximation $\sqrt{2}=1\frac{5}{12}$. The mere fact that $(1\frac{5}{12})^2$ differs from 2 by less than 0.01 is of little interest. What is intriguing about the result is that it is the third approximation for r in $(1+r)^2=2$ of the continued fraction sequence:

$$\frac{1}{2}:\frac{1}{2+\frac{1}{2}}=\frac{2}{5}:\frac{1}{2+\cfrac{1}{2+\frac{1}{2}}}=\frac{5}{12}$$

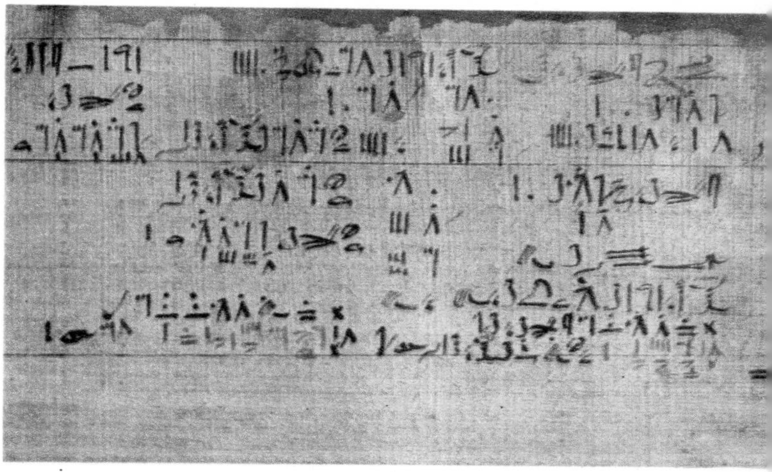

Top: A fragment of the Rhind Papyrus of about 1600 B.C. The part repeated below in hieroglyphic script may be freely translated thus:
$\frac{1}{4}, \frac{1}{8}, \frac{1}{10}, \frac{1}{30}, \frac{1}{45}$. *Bring the total to* $\frac{2}{3}$. *Our fractions represent* $11\frac{1}{4}$, $5\frac{1}{2}+\frac{1}{8}$, $4\frac{1}{2}$, $1\frac{1}{2}$ *and 1 [forty-fifths]. We must add* $\frac{1}{9}$ *and* $\frac{1}{40}$, *because* $\frac{1}{4}, \frac{1}{8}, \frac{1}{9}, \frac{1}{10}, \frac{1}{30}, \frac{1}{40}, \frac{1}{45}$ *and* $\frac{1}{3}$ *total 1. They represent* $11\frac{1}{4}$, $5\frac{1}{2}+\frac{1}{8}$, *5,* $4\frac{1}{2}$, $1\frac{1}{2}$, $1\frac{1}{8}$, *1 and 15 [forty-fifths] = 1.*

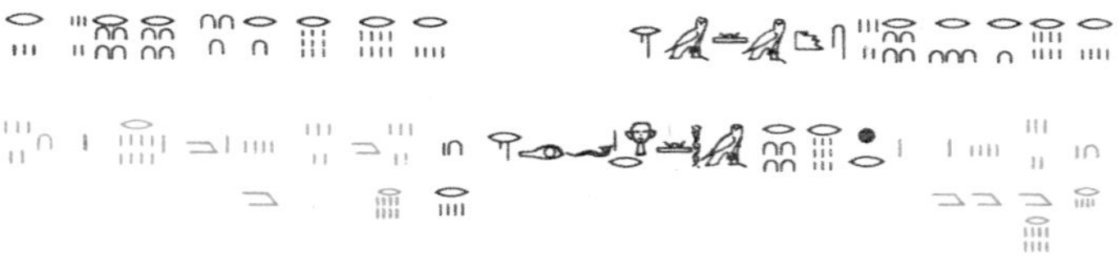

If the documentary evidence already available confirms the view that the art of computation in the temple sites of the first mercantile civilisation considerably surpassed that of the sacred observatories of the Nile, the relative importance of the enduring contributions of Egypt and Mesopotamia to the art of measurement is more difficult to assess. At the most primitive level, the custodians of the Egyptian calendar settled the length of the year as $365\frac{1}{4}$ days at a date plausibly put at 4241 B.C. or thereabouts. Like their Chinese contemporaries, and the Mayan priesthood before the latter added a five-day month to their year of eighteen 20-day months, the Babylonians obstinately clung to the solar-lunar compromise of 360 days, and interpolated leap years, as required, to keep the festivals in track with the seasons.

Beyond dispute, we know that Babylonian geometry embraced many conclusions of which we have no documentary relics from Egyptian sources; but the architectural and irrigation feats of the civilisation of the Nile encourage us to suppose that it had an elaborate code of reliable rules of mensuration. Moreover, two facts of which we have documentary evidence show that Egypt had outstripped Mesopotamia. The Egyptian value for the ratio (π) of the perimeter of a circle to its diameter as given in the Ahmes Papyrus is 3.16, which is much nearer the mark (3.14159...) than the Babylonian (and Biblical) value 3.0. The Egyptian formula for the volume of a pyramid as given in the Moscow Papyrus is correct, and the Babylonian is not.

The Egyptian rules for area (A) and perimeter (P) of the circle in terms of the diameter ($d = 2r$) are:

$$A = \left(\frac{8d}{9}\right)^2 = \frac{256}{81} r^2 \simeq \pi r^2$$

$$P = \left(\frac{8}{9}\right)^2 . 4d = 2\left(\frac{256}{81}\right) r \simeq 2\pi r$$

This makes $\pi \simeq 3.1605$. We have no certain information about how the temple geometers arrived at this figure. A possible clue is that it lies midway between the figures for the half perimeters of the equal-sided dodecagons enclosing and enclosed by a circle of unit radius or, what comes to the same thing, the mean of the areas of the inscribed 12-sided and escribed 6-sided regular polygons. The corresponding figures are 3.1058 and 3.2154, the mean being 3.1606.

If we look at the situation in terms of the imperative need for a regimen of mensuration, the salient issues other than the cultivation of astronomy in the service of the calendar, or as a means of maintaining authority over a slave population with superstitious awe when confronted with unusual celestial occurrences such as an eclipse, are classifiable under three headings:

(a) land measure – from which geometry takes its name – to meet the requirements of tax assessment and of irrigation;

(b) architecture, in particular the construction of monuments with a calendrical orientation;

(c) measures of quantity for tribute or merchandise.

In the context of surveying, we probably owe an equal debt to Egypt and to Mesopotamia in so far as the temple surveyors appreciated: (a) relations

This Babylonian tablet represents a crude attempt to evaluate the area of a circle in terms of the area of the escribed square. The Babylonians, however, never evaluated π as accurately as did the Egyptians.

between squares and rectangles used by the Babylonians and their Greek pupils as a rationale for solving problems which we should translate as quadratic equations; (b) the fact that the area of a triangle is equivalent to that of half the rectangle of the same height on the same base; (c) that the area of a circular enclosure is more or less closely obtainable by multiplying the square of the diameter by a fixed number.

The non-documentary evidence that the architecture forces on our notice is closely related to the requirements of calendrical astronomy. To fix the meridian, whence also to build a pyramid with faces so precisely due north, south, east and west as those of the Great Pyramid at Gizeh, it is necessary to bisect an angle between the sun's rising and setting positions at the equinoxes or to get the exact position when the sun's noonday shadow is least by bisecting the angle between its position of equal length before and after midday. We may therefore assume that the Egyptian rope-stretchers, as their Greek pupils spoke of the temple surveyors or architects, knew how to bisect an angle, probably by using peg and cord to trace circles and straight lines on the sand. Likely enough also, their experiences of the swing of the plumb-line grazing the plane of the horizon would acquaint them with what they presumably assumed to be true in their astronomical calculations, *i.e.* the line drawn from the centre of a circle to the tangent is at right angles to it.

The maximum error of the sides and of the corner angles of the Great Pyramid is a small fraction of 1%. Consequently, it is self-evident that the architects and ground surveyors could make a right angle in one way or another with remarkable precision. With the rope and peg procedure, their reliance on any one of three recipes is equally plausible:

(1) to employ the Euclidean construction which is also a recipe for bisecting a line;

(2) to take advantage of the fact that a rope knotted at lengths 5:4:3 when pegged out at the knots is the outline of a right-angled triangle;

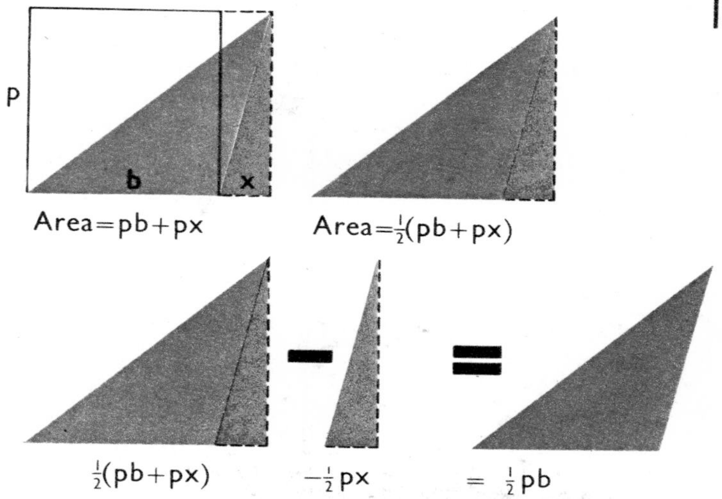

The area (S) of a triangle is equal to half the product of the length of any one side and that of its perpendicular distance from the opposite apex (i.e. $S = \frac{1}{2}pb$). If all angles of the triangle $< 90°$ the construction merely involves dropping a perpendicular from apex to base. Then each half of the triangle is half onc of the two rectangular segments enclosing the whole.

(3) to know that the angle joining the extremities of the diameter of a circle to any point on the circumference is a right angle.

With reference to these, it seems highly probable that the Egyptians and Babylonians knew the first trick, and that the Egyptians knew the second. There is documentary evidence that the Babylonians knew the third; and that they could also use their tables of squares to advantage by recognising 5:4:3 as a particular case of a rule illustrated by other numerical examples, *e.g.* 20:16:12 and 17:15:8. Needless to say, the Pythagorean rule ($h^2 = p^2 + b^2$) is that the square on the longest side of a right-angled triangle is equal to the sum of the squares on the other two sides. That the Babylonian geometers were familiar with many examples of the Pythagorean relationship other than the simplest (5:4:3)

The Gizeh pyramids face precisely north, south, east and west. Below are two geometrical constructions, one of which was presumably used to achieve the result.

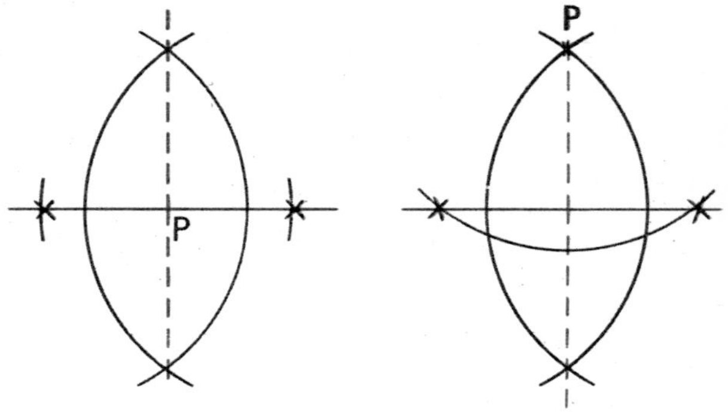

Egyptian architects doubtless knew how to erect a perpendicular from a horizontal line at a given point p, and how to drop a perpendicular from a given point.

recipe of this sort for making a right angle is clear because one table cites 15 such triads of integers. There are several simple recipes for generating them. One is the following. If $h^2 = p^2 + b^2$:

$$b^2 = h^2 - p^2 = (h+p)(h-p)$$

Let us now put $(h-p) = n$, so that $(h+p) = (2p+n)$, whence

$$b^2 = 2pn + n^2 \text{ and } p = \frac{b^2 - n^2}{2n}$$

We may get this into a more useful shape by writing $b (= qn)$ as the product of two integers, so that

$$p = \frac{q^2n^2 - n^2}{2n} = \frac{n}{2}(q^2 - 1)$$

In this expression, p will be an integer if either n is even or q is odd. Hence we can generate triplets as follows by painting in even values of n or odd values of q.

$$n = 1 : q = 3 \text{ so that } b = 3 : p = \frac{(3^2 - 1)}{2} = 4 \text{ and } h = 5$$

$$n = 2 : q = 2 \text{ so that } b = 4 : p = (2^2 - 1) = 3 \text{ and } h = 5$$

In this way we derive (*inter alia*) the following:

3.4.5	9.40.41	13.84.85
5.12.13	11.60.61	16.63.65
7.24.25	12.35.37	17.144.145
8.15.17		19.180.181

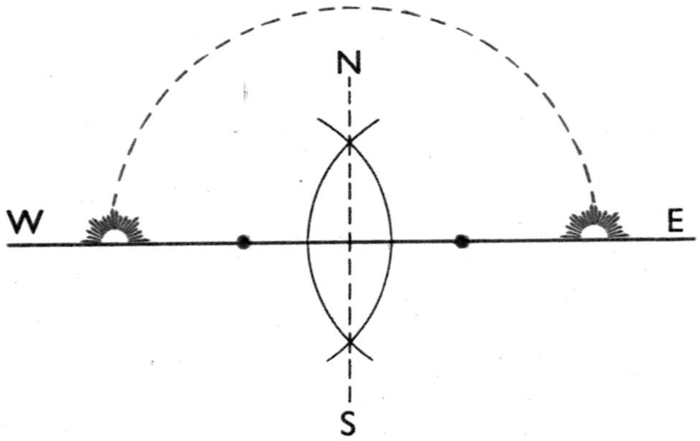

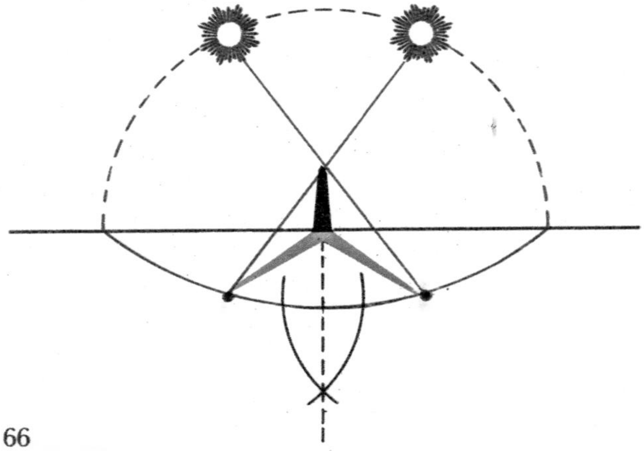

(1) Bisect line joining sun's rising and setting positions at equinoxes. (2) Find position where sun's noon shadow is shortest by bisecting angle between two shadows of equal length before and after noon.

Another formula which generates Pythagorean triads is $\frac{1}{2}\ (n^x \pm n^{-x})$, likewise the triplets:

$$2N\ ;\ N^2-1\ ;\ N^2+1$$

The possibility of finding an indefinitely large number of integers which satisfy the Pythagorean equation has encouraged later mathematicians without success to see whether whole numbers can satisfy an equation such as $h^3 = p^3 + b^3$ or more generally $h^n = p^n + n^n$ $(n > 2)$. In the seventeenth century of our era, Fermat left behind him a note to the effect that he had been able to show that no integers do so. No record remains of Fermat's so-called *last theorem*, and no one else has succeeded in providing a substitute.

Any observation of the elevation of the sun, if based on the length of its shadow, or any use of shadow reckoning to measure the height of a building, presupposes the recognition that: (a) the edge of a beam of light is a straight edge; (b) the ratios of corresponding sides of equiangular triangles are equal. The first assumption, in keeping with experience of people used to seeing dust scintillating in a shaft of light through a narrow slit, is implicit in astronomical interpretation at the most primitive level; and a reference in an extant papyrus to the instructions which stone masons received when facing a pyramid seems to justify our confidence that the Egyptians recognised the second as a rule of procedure. We know more definitely that the temple architects of Mesopotamia did so.

Measures of tribute and merchandise impinge on our theme in connexion with the discovery of rules for assigning the volume of a vessel. The most elementary case is the volume (hb^2) of a cuboid of square base (b) and height (h), whence that of a prism from the rule for the area of a triangle. The Egyptian assessors had a correct rule for the capacity $(\pi r^2 h)$ of a cylindrical granary of radius (r) and height (h). What is most remarkable is that they could correctly prescribe, as cited in the Moscow Papyrus, the volume of a section of a pyramid. Though we do not know how

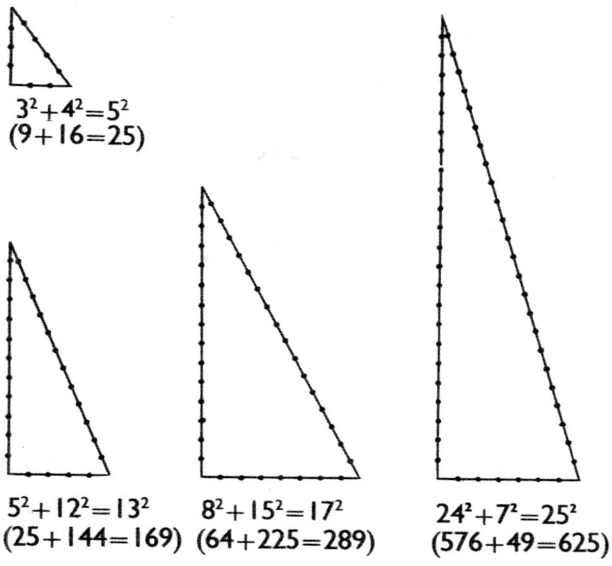

Egyptian geometers were aware of at least one example of the Pythagorean rule that the square on the longest side of a right-angled triangle is equal to the sum of the squares on the two remaining sides: the case of a triangle whose three sides measure 5, 4 and 3 units. Babylonian mathematicians knew of 15 Pythagorean triads of integers of the kind above.

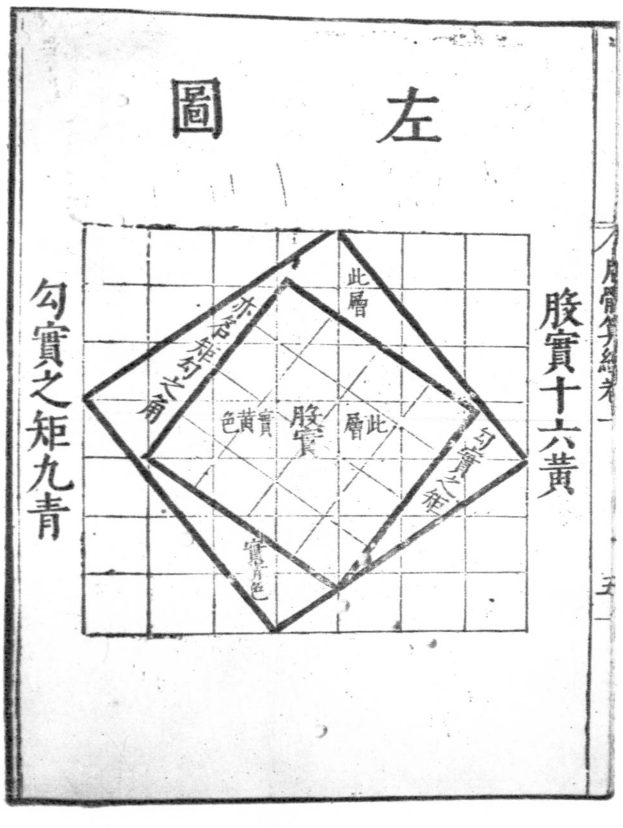

This extremely early example of Chinese block printing was used in vindication of the Pythagorean theorem. Chinese tradition associates it with the mathematician Chou Pei, probably a contemporary of Pythagoras.

Cl.4

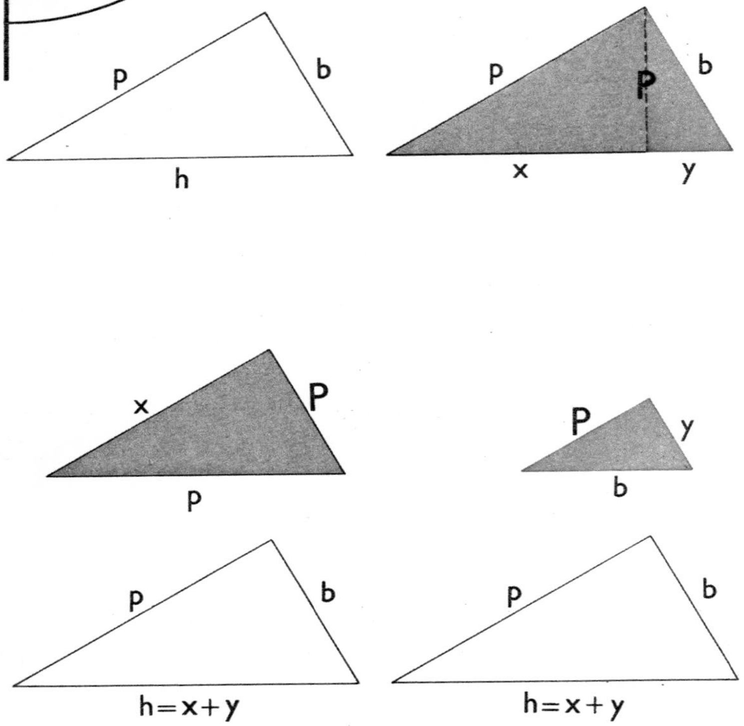

Perpendicular from right angle to hypotenuse divides a right-angled triangle into two triangles, each equiangular to it.

$$\frac{p}{x+y} = \frac{x}{p} \qquad \frac{b}{x+y} = \frac{y}{b}$$

$$\therefore p^2 = x(x+y) = x^2 + xy \qquad \therefore b^2 = y(x+y) = y^2 + xy$$

$$\therefore p^2 + b^2 = x^2 + 2xy + y^2 = (x+y)^2 = h^2$$

In a right-angled triangle of which h is the longest side and the other two are p and b, $p^2 + b^2 = h^2$.

they arrived at the rule, we may plausibly reconstruct their approach by consideration of the simpler issue, *i.e.* the volume of the whole pyramid. Here we do so with the candid admission that we may be in error. When indulging in the luxury of speculating about how people in remote antiquity achieved what they did, one is on comparatively safe ground if one is able to formulate the question: how might they have solved the problem with the means they certainly had at their disposal? Otherwise, one's best guess is to invoke what means they might well have had.

Whether the Egyptian geometers were actually familiar with figurate numbers is as yet unknown. What we certainly know is that the triangular and pyramidal numbers were familiar to the Babylonians of 2000 B.C. In the sixth century B.C. the Greeks were fascinated by them, and regarded them with superstitious veneration. If the Egyptians had some familiarity with them, it is easy to see one way in which they may have reached a correct evaluation for the volume of a pyramid, though it is unlikely that they would have traversed all the steps in the argument which follows.

Since the figurate representation of a pyramid is n layers high on a base whose side is a line of n points, an atomist of the fifth century B.C. might have guessed the correct formula for the volume of a pyramid in terms of its height (h) and base width (B), *viz.* $\frac{1}{3}hB^2$, if he had been sufficiently good at manipulating large numbers to note that the proportionate error entailed in the approximation $(2n+1)(n+1)n.6^{-1} \simeq \frac{1}{3}n^3$ becomes less and less as we increase n. He would then have set about providing the formula with a proof by recourse to geometrical reasoning. Democritus, the father of atomism, records the considerable debt of Greek geometry to his predecessors by boasting of his mastery of all the lessons he had learned from the Egyptian rope-stretchers; and he himself derived in a geometrical way a minor adaptation of the Egyptian formula for regular pyramids on pentagonal, hexagonal, *etc.* bases, all of which have analogues in the domain of figurate numbers.

However, the only geometry we need to rely on, if already equipped with the figurate formulae for the sums of the natural numbers and of their squares, is far below what the Egyptians knew two thousand years before Democritus. We shall lay out the derivation in the sign language of our own algebra.

We recall that the Great Pyramid was first built step-wise like the earliest Egyptian ones and the Toltec monuments of Mexico, the smooth facing being a secondary stage of the construction. In effect, it was therefore *escribed* about the edges of a sequence of cuboids of equal thickness (s) and bases whose sides of length B diminished by a fixed quantity (a). If we conceptualise a pyramid inscribed *within* such a structure, the same rule holds; but the summation involves an additional layer. If the escribed step pyramid has n steps starting with one whose base (B) is the same as that of the pyramid, the lengths of the sides of the layers are:

$$B,\ B-a,\ B-2a,\ B-3a\ldots B-(n-1)a$$

The lengths of the sides of the layers of the inscribed step pyramid include all the above except the first.

Now the area of the rth block is $(B-ra)^2$ and its volume is $s(B-ra)^2$. If we denote the volume of the escribed step pyramid by V_e and that of the inscribed by V_i, the volume (V) of the smooth pyramid is such that $V_e > V > V_i$; and

$$V_e = s\sum_{r=0}^{r=n-1}(B-ra)^2 = s\sum_{r=0}^{r=n-1}(B^2-2raB+r^2a^2)$$

$$V_i = s\sum_{r=1}^{r=n-1}(B-ra)^2 = s\sum_{r=1}^{r=n-1}(B^2-2raB+r^2a^2)$$

Whence we may write:

$$V_e = sB^2\sum_{r=0}^{r=n-1}1 - 2saB\sum_{r=0}^{r=n-1}r + sa^2\sum_{r=0}^{r=n-1}r^2$$

$$V_i = sB^2\sum_{r=1}^{r=n-1}1 - 2saB\sum_{r=1}^{r=n-1}r + sa^2\sum_{r=1}^{r=n-1}r^2$$

Cl.5

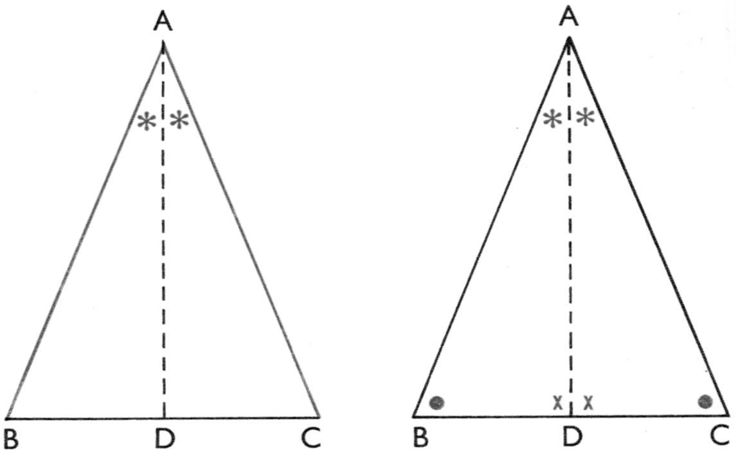

If two sides of a triangle are equal, the angles opposite them are equal; and conversely, if two angles are equal the corresponding opposite sides are equal.
In both cases we dissect the triangle into two parts by a line which bisects the apical angle.
In the first case we know: AB = AC. Since AD which bisects BAC is common to both triangles they have two sides and an enclosed angle equal. They are therefore identical and the corresponding angles ACB and ABC are equal.
In the second case we know: ABC = ACB. The two triangles have by construction two pairs of equal angles, whence by Claim 1 three pairs are equal and both triangles have equal angles at extremities of AD. Hence they are identical and corresponding sides AB and AC are equal.

Cl.6

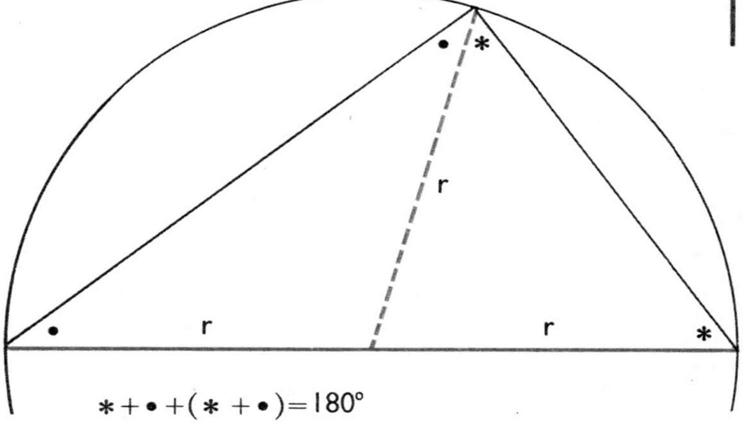

∗ + • + (∗ + •) = 180°
∴ ∗ + • = 90°

In a semi-circle, the lines joining a point on the boundary to the ends of the diameter are at right angles.
A straight line from compass point to apex cuts the triangle into two isosceles triangles, one containing the equal angles • •, the other the equal angles ∗∗.
• + ∗ + (• + ∗) = *180°* ∴ (• + ∗) = *90°*.

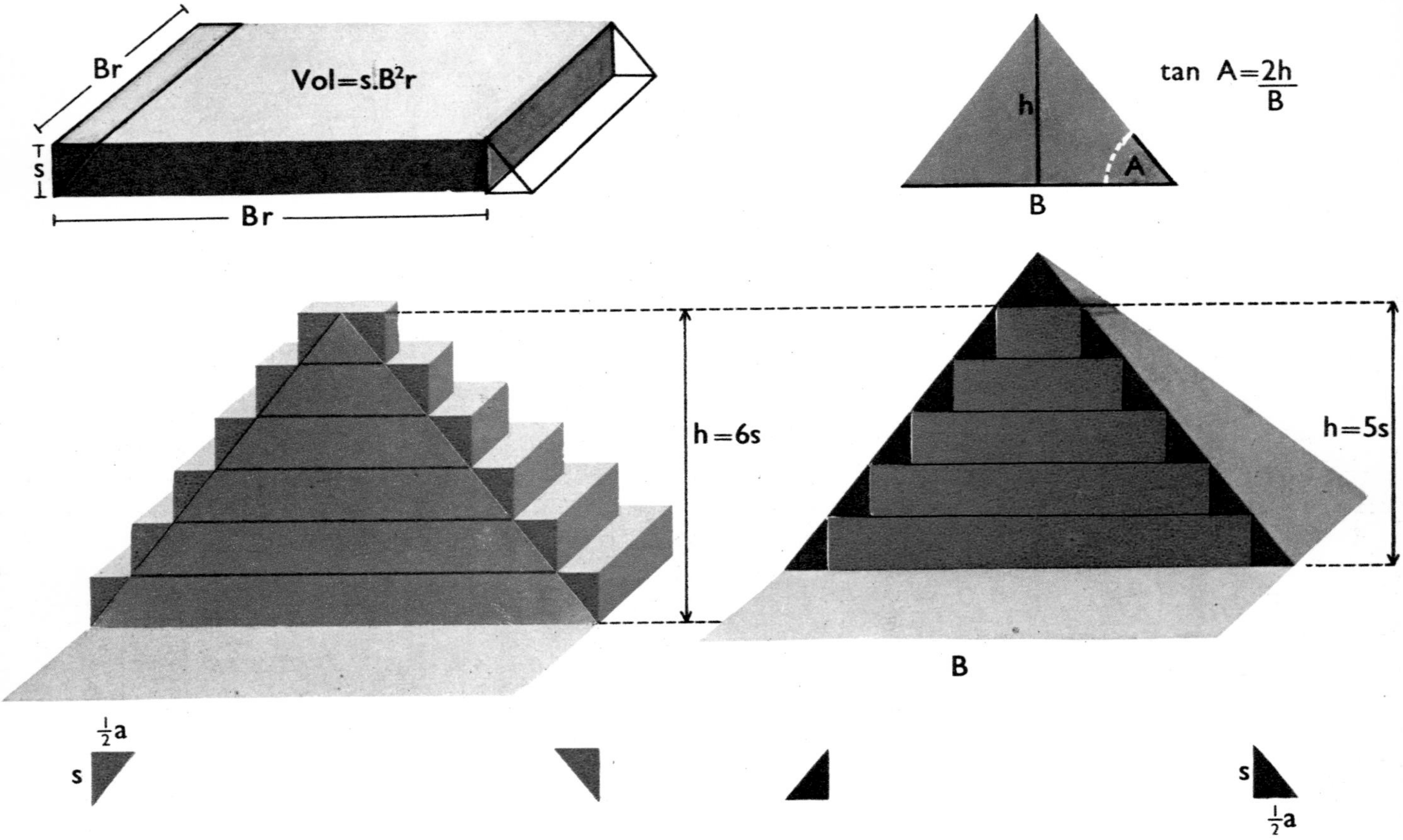

$$\frac{s}{\frac{1}{2}a}=\tan A\ \therefore\ \tfrac{1}{2}a=s\div\tan A\ \therefore\ a=\frac{sB}{h}$$

The Great Pyramid, like the Pyramid of Sakkara below, was built step-wise before masons added the smooth facing. It was, in effect, escribed about the edges of a sequence of cuboids of equal thickness (s) but of regularly-decreasing square base. If B — a represents the length of the lowest layer, the lengths of layers immediately above it are B — 2a, B — 3a, B — 4a, etc. From this, from a knowledge of figurate numbers and from geometry of a kind familiar to Egyptian surveyors, we can derive the formula for the volume of a pyramid.

The first term of V_e signifies adding sB^2 n times and the first term of V_i signifies adding sB^2 $(n-1)$ times. The next two terms respectively correspond to the first $(n-1)$ triangular and the first $(n-1)$ pyramidal numbers, being the same for V_e and V_i. Thus we get

$$V_e=snB^2-sn(n-1)aB+\frac{sn(n-1)(2n-1)}{6}a^2$$

$$V_i=s(n-1)B^2-sn(n-1)aB+\frac{sn(n-1)(2n-1)}{6}a^2$$

Our picture shows that $sn=h$ and $a=sB\div h$, so that

$$V_e=hB^2\left(\frac{1}{3}+\frac{1}{2n}+\frac{1}{6n^2}\right)$$

$$V_i = hB^2\left(\frac{1}{3} - \frac{1}{2n} + \frac{1}{6n^2}\right)$$

If there are only ten layers, we can thus say that the expressions (E) in brackets reduce to $\frac{1}{3}+\frac{31}{600}$ and $\frac{1}{3}-\frac{29}{600}$ or $E \simeq \frac{1}{3} \pm \frac{1}{20}$. For $n=100$, $E \simeq \frac{1}{3} \pm \frac{1}{200}$; for $n=1000$, $E \simeq \frac{1}{3} \pm \frac{1}{2000}$ and so on. As the number of layers (n) becomes larger and larger both expressions thus approach the same limiting value – one third – so that:

$$V = \frac{hB^2}{3}$$

Given the formula (πr^2) for the area of a circle, this reasoning leads us to the corresponding formula $(\frac{1}{3}\pi r^2 h)$ for the volume of a cone. We have no record of it before we reach the Greek period; but it may be significant that the first of the Greeks to disclose it was Democritus. Needless to say, we do not need to credit the Egyptians with carrying out the exploration as thoroughly as above. By computing the volumes of the enclosing and enclosed step pyramids for successively smaller steps, it would not be hard to see that the volume of the smooth pyramid lies between limits becoming progressively nearer to $\frac{1}{3}hB^2$. Even so, a merely empirical approach to the problem anticipates what we may call the *two-sided* method of exhaustion which enabled Archimedes to bound π between definite limits, *viz.*

$$3\tfrac{1}{7} > \pi > 3\tfrac{10}{71}$$

One indication of the possibility that later discoveries will vindicate a higher level of mathematical sophistication than some scholars have been willing to concede to the Egyptians is a type of astronomical text recovered from the tombs of the 20th dynasty Pharaohs (*circa* 1100 B.C.). Neugebauer reproduces a fragment of such from the tomb of Rameses VII. This exhibits a lay-out to determine the time at night throughout the year by depicting the altitude of particular stars located by one co-ordinate against the hour as indicated by its position on a second co-ordinate at right angles. Such a grid lay-out has an astonishing similarity to the familiar graphical devices of our own age.

Before we take leave of antiquity, it is fitting to emphasise once more how considerable was the debt of the Founding Fathers of Greek geometry both to Egypt and to Mesopotamia. During the century of the birth of Thales (*circa* 640–550 B.C.), himself one of its citizens, Miletus, a leading focus of trade among the Greek-speaking communities along the seaboard of Asia Minor, established no less than ninety free colonies along the coast of the Mediterranean and the Black Sea with commercial settlements in Egypt itself. According to tradition, Thales travelled in Egypt and, like Democritus, learned what he could of Egyptian geometry. Seemingly also, he travelled in the East; but we may confidently surmise that Thales and Pythagoras (*circa* 550–500 B.C.) had contacts with Babylonian mathematics through their trade rivals from whom they acquired (*circa* 800 B.C.) the art of alphabetic writing and from whom it is likely that they learned to mint coinage not much earlier than 600 B.C.

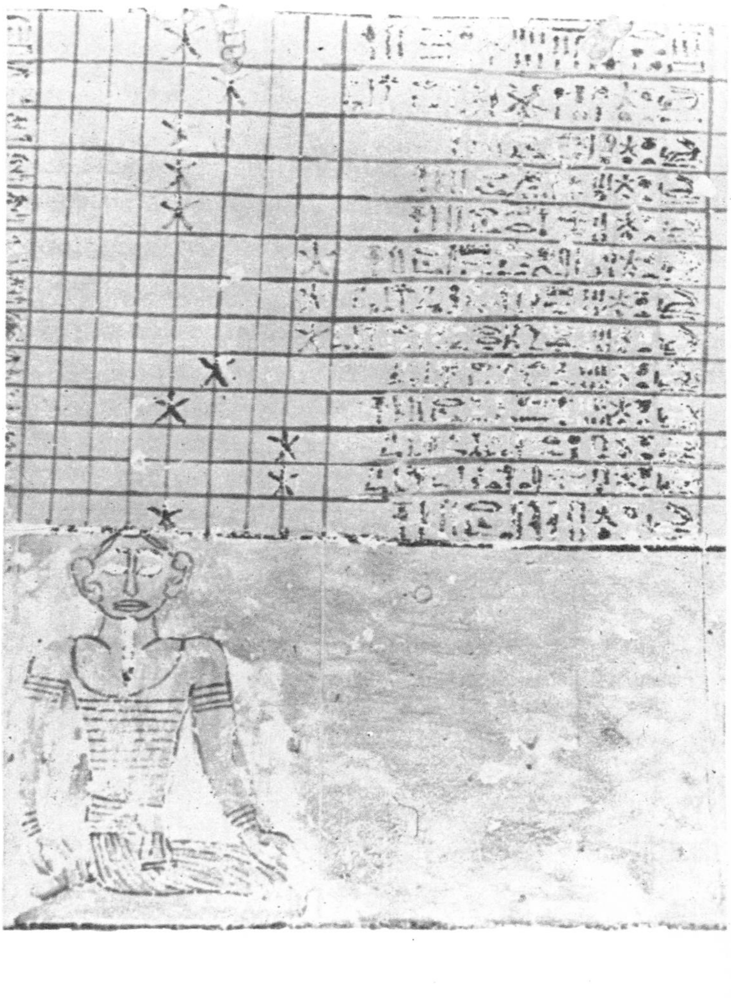

This star map, from the tomb of Rameses VII, locates the altitude of particular stars by one co-ordinate against the hour on a second co-ordinate at right angles to it.

Concerning the level of mathematical knowledge among the Phoenicians we know next to nothing. The earliest communities of these people, who spoke a Semitic language allied to Akkadian, were Tyre and Sidon. About 1600 B.C. they recognised the suzerainty of the Pharaoh Amasis I, but a Hittite invasion of Syria about three centuries later released them from their dependence on Egypt. The parent city states of Tyre, Sidon and Byblos later came under Assyrian (876–605 B.C.) and later still under Chaldean (605–538 B.C.) rule during the lifetime of Thales. From the end of their subservience to Egypt they were thus in lively contact with Mesopotamian civilisation, as attested by their possession of tablets in the language and script of contemporary Babylonia. By 1000 B.C. there were Phoenician colonies along the coast of Spain, and Phoenician ships carrying cargoes of copper, tin and gold were trading with the Scillies, if not with the mainland of Britain.

The Tarshish of the Book of Jonah was a Spanish colony. The greatest of all the Mediterranean colonies was Carthage, founded in 800 B.C. Carthage was far from the original centre of worship of Baal and Astarte, the Bel and Ishtar of the Babylonians. There the social structure at the time of the Punic Wars assumes some of the facies of a secular state, though the royal houses of Phoenicia claimed divine descent, as did the Pharaohs. If it is assuredly a gross exaggeration to call the Greek-speaking communities democracies at any stage of their existence, there is this significant difference between them and their Phoenician instructors. In the Western World, the Phoenicians are the last custodians of a temple culture. The Greeks of the century of Thales, though superstitious, were not priest-ridden; and the several dictators who usurped successive oligarchies in the Greek city states did not base their claims on their sky-born origins.

The latter days of Carthage witness the birth of Roman sea power. The same gangster society butchered Archimedes, destroyed the first library of Alexandria, institutionalised the gladiatorial contest and the public crucifixion of slaves, thereby contributing to our common civilisation those exalted concepts of law and order subsumed by the *Pax Romana* as a model for government by remittance men in Kenya and Rhodesia. Scipio finally destroyed all trace of Carthage (146 B.C.). Otherwise our record of how much the one world of tomorrow owes to people of different colours, creeds and countries might contain another chapter. What little we know about their contribution justifies the belief that the Phoenicians exploited some of the secrets of Babylonian astronomy in the service of scientific navigation and bequeathed to their Greek-speaking rivals a new beginning.

Some of the series mentioned earlier in this chapter have played a very important role in the making of the modern theory of probability; and we may usefully take this opportunity of contrasting the vast economy of the sign language of modern algebra with the verbal formulation of one and the same set of rules over a period which extends from 2000 B.C. or earlier to about A.D 1650. Such economy depends on recognising and labelling for future use patterns which turn up frequently, preferably by a convention which is easily memorisable by analogy with another already familiar one. In the formulae which now convey the formation of figurate numbers, several such characteristic patterns recur, for instance products such as: 1.2; 1.2.3; 1.2.3.4 *etc.*, usually written 2!; 3!; 4! *etc.*, so that

$$\mathbf{n! = 1.2.3.4.5 \ldots (n-2)(n-1)n}$$

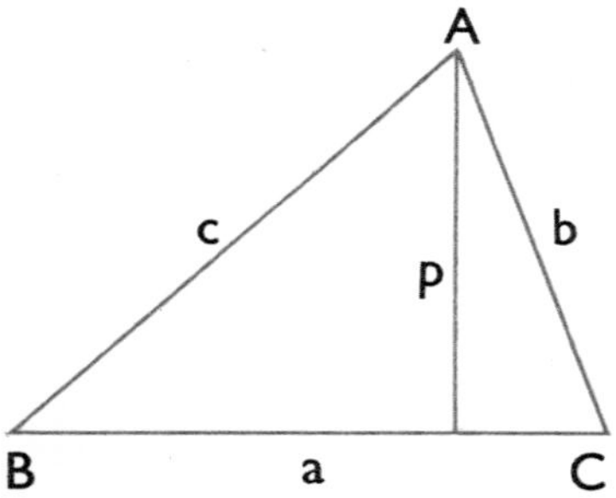

$$\sin B = \frac{p}{c} \text{ and } \sin C = \frac{p}{b}$$

$$\therefore c \sin B = p = b \sin C$$

$$S = \tfrac{1}{2} pa$$

$$\therefore \tfrac{1}{2} ac \sin B = S = \tfrac{1}{2} ab \sin C$$

and

$$\frac{\sin B}{b} = \frac{\sin C}{c}$$

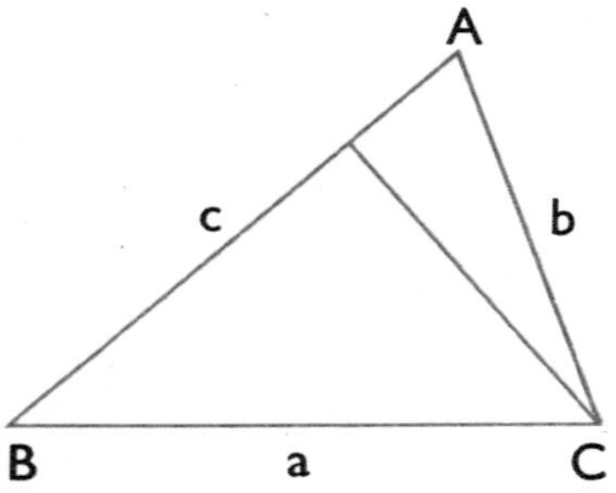

By a similar construction

$$a \sin B = b \sin A$$

$$\therefore S = \tfrac{1}{2} bc \sin A$$

and

$$\frac{\sin A}{a} = \frac{\sin B}{b} = \frac{\sin C}{c}$$

Area (S) of a triangle in the sign language of sines.

In this notation $5!=5(4!)$; $4!=4(3!)$ *etc.* More generally, $n!=n(n-1!)$ Thus $1!=1(0!)=1$ so that $0!=1$. It is therefore better to write:

$$n!=1.1.2.3.4....(n-2).(n-1).n$$

therefore $0!=1$; $1!=1.1$; $2!=1.1.2$; $3!=1.1.2.3$... *etc.*

By analogy with the usual definition of exponents on the left, we may make it both easy to recall and more brief to record another pattern of products as on the right:

$n^0=1$	$n^{(0)}=1$
$n^1=1.n$	$n^{(1)}=1.n$
$n^2=1.n.n$	$n^{(2)}=1.n.(n-1)$
$n^3=1.n.n.n$	$n^{(3)}=1.n.(n-1)(n-2)$
$n^4=1.n.n.n.n$	$n^{(4)}=1.n.(n-1)(n-2)(n-3)$

In this shorthand,

$$\mathbf{n^{(r)}=1.n.(n-1)(n-2)..(n-r+1)}$$

$$\text{and } \mathbf{n^{(n)}=n!}$$

In formulae for the build-up of figurate numbers and the solutions of problems of choice or chance, we frequently meet expressions involving products of the form n^r and $n^{(r)}$ as above, and frequently also the ratio of such products to the product $r!$. Hence it is convenient to write

$$\mathbf{n_{(r)}=\frac{n^{(r)}}{r!}} \quad \text{e.g. } \mathbf{n_{(3)}=\frac{n(n-1)(n-2)}{1.2.3}}$$

$$\mathbf{n_r=\frac{n^r}{r!}} \quad \text{e.g. } \mathbf{n_4=\frac{n.n.n.n}{1.2.3.4}}$$

In this notation the *n*th natural number is expressible in the form:

$$\frac{n}{1!}=n_{(1)}$$

Whence we may write the build-up of the triangular from the natural, and the tetrahedral from the triangular, numbers in the compact form

$$\mathbf{\sum_{r=1}^{r=n} r_{(1)}=\frac{n(n+1)}{1.2}=(n+1)_{(2)}}$$

(triangular numbers)

$$\mathbf{\sum_{r=1}^{r=n}(r+1)_{(2)}=\frac{n(n+1)(n+2)}{1.2.3}=(n+2)_{(3)}}$$

(tetrahedral numbers)

Thus the build-up of the member of rank *n* in the series of dimension $(d+1)$ is expressible for the entire family of such series in terms of the series of lower dimension (d) in the very compact form:

$$\sum_{r=1}^{r=n}(r+d-1)_{(d)}=(n+d)_{(d+1)}$$

In this notation $0!=0^{(0)}$ which we interpret as unity, so that $r_{(0)}=1$ for all r and this exhibits the build-up of the natural numbers consistently from the summation of units, *viz.*

$$\sum_{r=1}^{r=n} r_{(0)}=n_{(1)}=(n+0)_{(0+1)}$$

The verbal translation of the left-hand expression in the last formula but one is as follows:

> to get the *n*th term of the series of dimension $d+1$, add up from rank 1 to rank *n* all the terms of the series of dimension *d*.

The interpretation and manipulation of such examples of planning human communication economically calls for no exercise of unusual subtlety. It involves an act of memorisation on all fours with learning a list of the parts of irregular French verbs. Thereafter the process of translation proceeds with the minimum expenditure of effort.

Chapter 4 The Greek Dilemma

In the Western half of the world too many expensively uneducated people still identify the Glory that was Greece pre-eminently with the names of Plato and of Aristotle. It would indeed be difficult to cite any two men who have done more over so long a period of time to stifle human inventiveness; and if they were truly representative of the intellectual achievements of Greek civilisation, we might reasonably write off the Greek contribution as a prolonged setback. The circumstance that each of them taught at some time in what is now the capital of a sovereign state perpetuates a widespread misunderstanding about what we here mean by it. When one speaks of what the Greeks contributed to modern knowledge and to modern habits of thought, more especially with reference to mathematics, one may thus need to remind oneself that one is not talking either about what was ever a single political unit comparable in any way to the Roman Empire or about people mainly domiciled on the mainland of Greece.

What was common to the people commonly called Greeks, or as they preferred to call themselves *Hellenes*, was the circumstance that they spoke inter-communicable dialects of one Indo-Germanic language. Before 600 B.C. this considerable speech community had independent city states from the entry to the Black Sea along an extensive seaboard of Asia Minor, in many of the off-shore islands, as well as on the mainland of what we now call Greece. It embraced Cyprus, Crete and several city states of South Italy and Sicily, including Crotona, Elea, Tarentum and Syracuse. Before 400 B.C. it had settlements on the Libyan coast at Cyrene and in South Gaul at Massilia (now Marseilles).

During the years 350–323 B.C. the armies of Philip and of Alexander annexed the Middle East and all Egypt. On his death, Alexander's generals founded dynasties, of which only those of the Ptolemies of Egypt and of the Seleucids of Syria and Mesopotamia are relevant to our theme. Thenceforward for six centuries, the city of Alexandria, founded in 332 B.C. to commemorate the conqueror, became the capital of Greek culture and the custodian of its scientific literature. Throughout the whole period of Roman supremacy initiated by the destruction of Carthage at the end of the Third Punic War, the Greek language retained its supremacy as the means of communicating secular knowledge, and the Greek speech community remained intact and scattered over a much more extensive territory than it had occupied in 500 B.C.

Greek tradition, of which no documentary remains antedate a century after his demise, endorses Thales as the Founding Father of Greek mathematics. The putative date of his death at a very advanced age closely coincides with that of the birth of Pythagoras and with the year of the ascent to the throne of the Persian monarch Cyrus who later subdued the Greek settlements in Asia Minor. A contemporary mathematician

Detail from an Attic bowl of about 480 B.C. showing a lesson in progress in a gymnasion. Into mathematics – an essential part of the curriculum – the Greek-speaking peoples first introduced the notion of proof. Their inefficient numeral system and their veneration of compass and rule as the only drawing instruments worthy of mathematical free grace placed them in a serious dilemma in the field of mensuration.

states that Thales himself

> introduced the ideal of establishing by exact reasoning the relations between the different parts of a figure . . . This was a phenomenon quite new in the world, and due, in fact, to the abstract spirit of the Greeks.

If we seek to interpret history with as much concern for the proper use of words as mathematicians aspire to when framing a definition of the infinite, we shall not be content with dismissing the making of Greek mathematics by recourse to a word so diversely exploited in the several contexts of spirits of salts, high spirits, strong spirits and spirits of the deep. The Greek spirit brooded over the Mediterranean basin at least a thousand years before the notion of proof obtrudes into the historic record. During that time, Greek-speaking peoples had colonised Crete and Cyprus, where they had acquired some familiarity with at least two forms of syllabic writing, the Minoan Literal B and the Cypriotic, each ill-adapted to the structural requirements of an Indo-European language. They had not as yet proved themselves as expert as their Phoenician trade rivals in maritime enterprise; and they had made no conspicuously unique contribution to our common civilisation. When they first emerge, a little before 700 B.C., in the penumbra of the written record, they have equipped themselves with the new art of alphabetic writing. This they had learned from their Semitic schoolmasters who had already sailed north and south beyond the Pillars of Hercules.

Henceforth, the picture discloses a combination of circumstances uniquely propitious to curiosity and to disciplined debate. By the time when there was first any written record of Greek mathematics, Greek-speaking master pilots and merchants had learned from their Egyptian customers the use of papyrus as a writing surface vastly more portable and more easily stored than the clay tablets of the Semitic civilisations of antiquity. They had also institutionalised the tribal dance as a dramatic spectacle, the words of which they had begun to put in writing. The practice of recording the spoken word thriftily, and at length, in a script which was no longer a temple secret, was one of the greatest human innovations of all time. The Greek drama set the pattern for the dialogue form in which Plato and others recorded the course of philosophic dispute. It bequeaths to us an understanding of Greek thought far more intimate than we can ever hope to have about the intellectual content of any pre-existing culture.

In the boyhood of Thales, syllabic writing in the Greek tongue still persisted in Cyprus. The Greek alphabet was a comparatively recent (*circa* 750 B.C.) acquisition; and Greek traders were starting to use a minted coinage. Thenceforward, and for no reasons intelligibly ascribable to the abstract spirit of the Greeks, whatever that may mean, we descry the efflorescence of a social pattern hitherto without its like in the human story. In widely dispersed communities which shared a common speech, prosperous merchants and master mariners were becoming literate in a new way, unhampered by a powerful pre-existing priesthood, eager to adapt to their own advantage whatever useful knowledge they could gain from travel, and becoming accustomed to settle their legal disputes by open discussion in accordance with a nascent code for the proper conduct of an argument.

Long before the lifetime of Thales, colonies of Greek-speaking peoples lived in peaceful co-

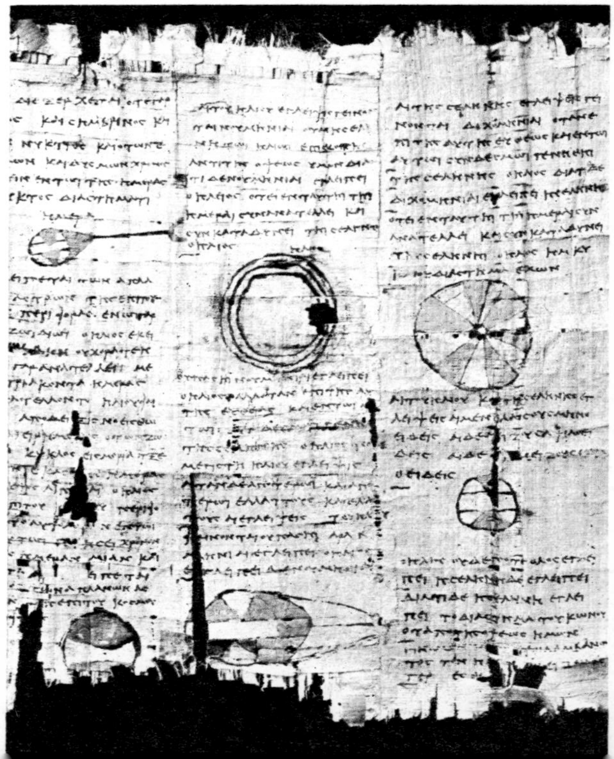

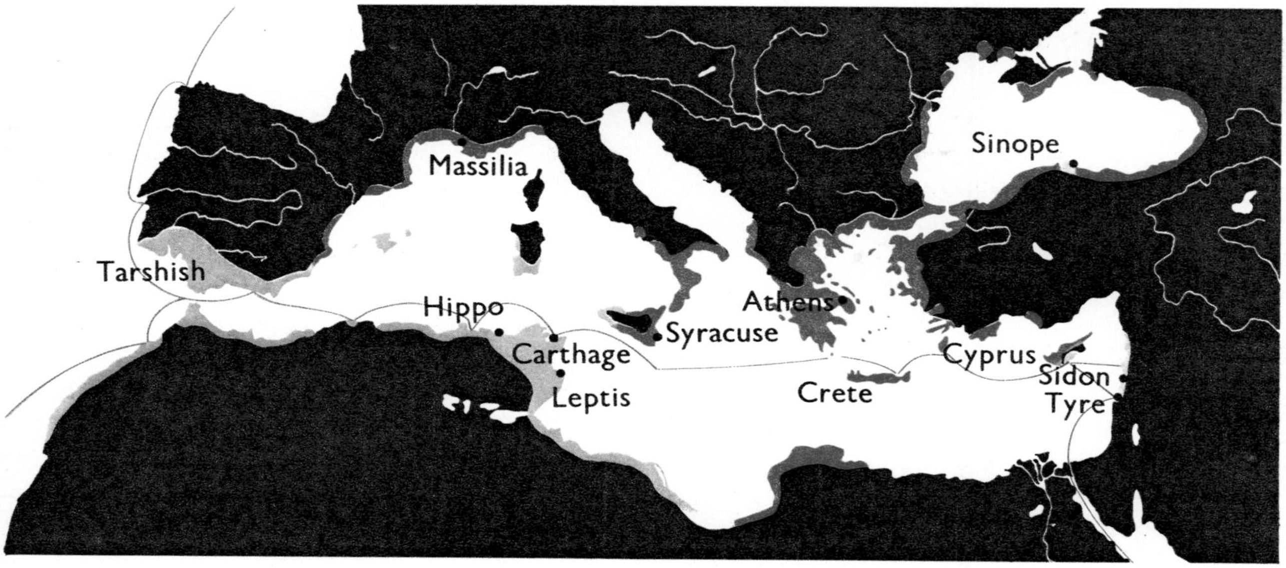

The Mediterranean world of about 1100 – 350 B.C. (Phoenician settlements pink, Greek red.)

Cypriotic vase, contemporary with Thales of Miletus, bearing the name TA-LE-SE in a syllabic script. Only later, when they acquired an alphabetic script, did the Greeks record the spoken word at length.

Part of a Greek astronomical papyrus written in Egypt between the fourth and second centuries B.C. Egyptian papyrus helped the Greeks to codify and store the mathematical lore of earlier civilisations.

Phoenician ship such as King Sennacherib of Assyria might have seen during his last days at Nineveh.

Given the earth's radius (R) and a knowledge of Pythagoras's Theorem, one can calculate distance (d) from observer at sea level to tip of ship's mast of known height (h) about to dip below the horizon. $(R+h)^2 = R^2 + d^2 . 2Rh + h^2 = d^2$. *If R is 4000 miles and h is 52.8 feet* ($= 0.01$ *miles*), $d^2 = 2(4000 \times 0.01) + (0.01)^2$, *and* $d = \sqrt{80 + 0.0001} \simeq 8.95$ *miles.*

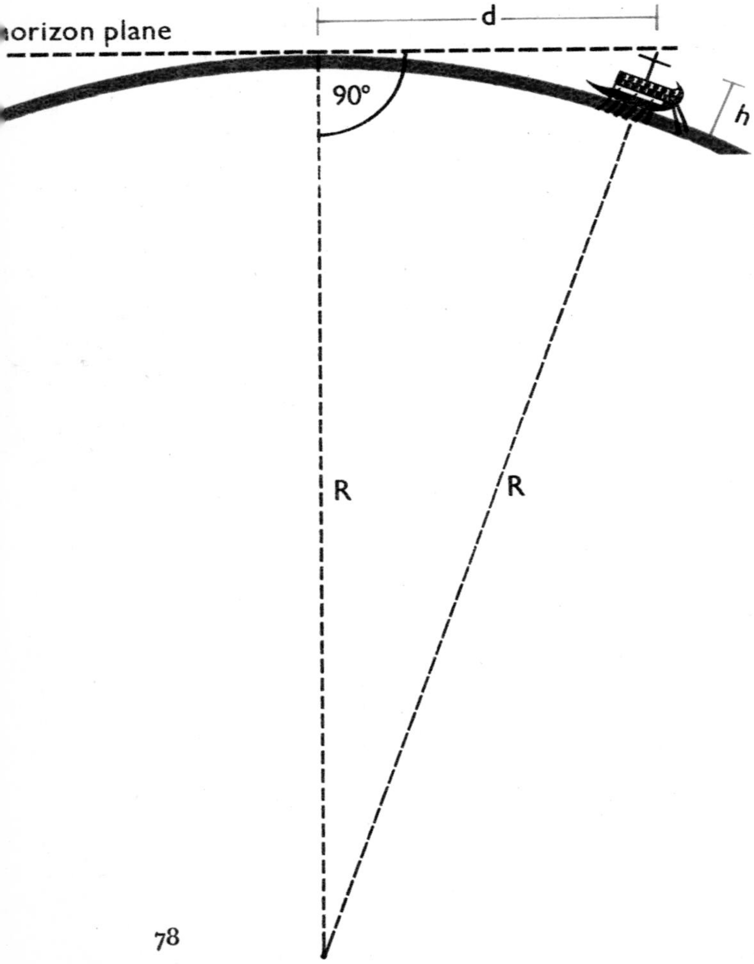

existence with settlements of the Semitic Phoenicians along the coasts of Asia Minor and elsewhere in the Mediterranean basin; and there is ample evidence of inter-marriage between them. Indeed, tradition asserts that Thales himself was of Phoenician descent on one side. By his time, there were teachers from whom pilots could learn the mysteries of Babylonian astronomy in Mediterranean ports. According to Herodotus, Phoenician ships circumnavigated Africa in the century of the birth of Thales, during whose lifetime a Carthaginian sea captain called Hanno took his ships further along the coast of equatorial West Africa than any European commander succeeded in doing till after the death of Henry the Navigator.

It is difficult to exaggerate the importance to Western culture of these early cruises of Phoenician galleys beyond the confines of the land-locked Mediterranean, north and south along the Atlantic seaboard. Hitherto, the dependence of navigation on astronomical lore had been slight, and its contribution to a new picture of the heavens negligible. The custody of the calendar and periodic adjustments by adding leap months or leap years circumscribed any practical value of the work of a priestly caste domiciled in temple sites, usually, and especially in Mesopotamia, remote from the sea. Aside from the attention bestowed on celestial portents, the correct forecasting of which reinforced the authority exerted by the Establishment over the credulous lower castes, the temple observatories disclosed no phenomenon other than the visibly circular edge of the earth's shadow during a lunar eclipse to controvert the naïve impression that the surface of the earth is flat.

Though the mariner's experience of the sight of land rising as a ship approaches land or of a ship mast sinking below the horizon plane when sailing out of port is difficult to accommodate with such a belief, one might say much the same about the viewpoint of those who first navigated at night by the stars visible in the latitudes of the Mediterranean basin; but we may infer how early navigational science attained a new orientation from the fact that a Greek sea captain in 400 B.C. could determine the latitude of Marseilles, then a settlement of the city state of Phocea on the Greek mainland, correct to one tenth of a degree. A wholly new world picture had now forced itself on the attention of a secular craft by the sheer exigencies of tracking a course. The recognition of the earth's rotundity is not, as people unfamiliar with life at sea believe, an exploit of speculative subtlety. It intrudes into every means by which the pilot can locate his port of call when his course has to traverse a wider range of latitude than the dimensions of the Mediterranean circumscribe. To ships which kept close to shore as they sailed northward along the coast of Europe or southward along that of Africa, new skies now unfolded. Every northerly star at its highest point in its course overhead became, and equally so, higher as the ship pushed northward, lower as southward.

As he tracked north to the Tin Isles, the pilot would discover that the length of the midsummer day becomes longer and that of the midwinter day shorter. Voyage by voyage, he learned to locate a port of call by the transit height of a particular star or the length of the sun's shadow on a particular day. Canopus, only less bright than Sirius, would have disappeared beyond recall when he had pushed little further north than the Pillars of Hercules. If he reached the equator, where a star near the pole would just graze the horizon or fail to rise at all, he would see the noon sun directly overhead on the equinoxes. Ere then, he would have seen it directly overhead on the Tropic of Cancer (Lat. $23\frac{1}{2}°$ N) at the noon of midsummer day. (Here, as elsewhere, seasonal and astronomical references are from the viewpoint of an observer in the northern hemisphere.) Perhaps, as the record does indeed disclose, the most challenging novelty greeted him when he passed southward beyond it. Within the tropics, he would see the sun's shadow pointing south at some periods of the year and otherwise north, as at home.

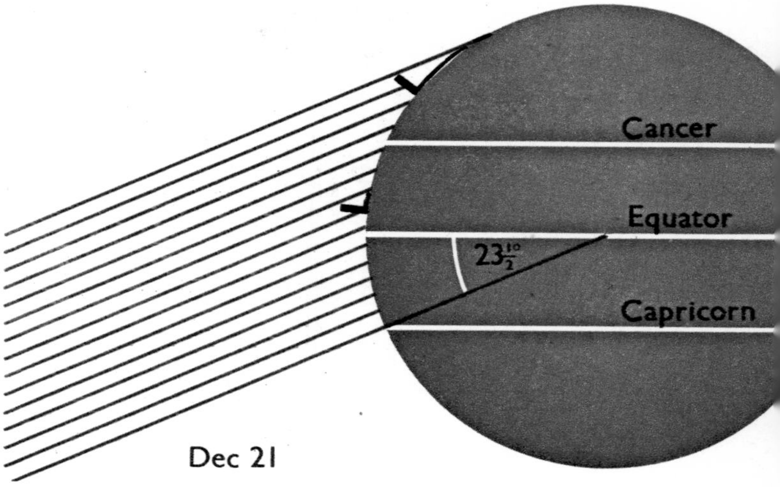

Dec 21

One fact which confronted seamen when they first sailed southward beyond the Tropic of Cancer offers strong evidence of the earth's rotundity. North of the tropics, noon shadows always point north.

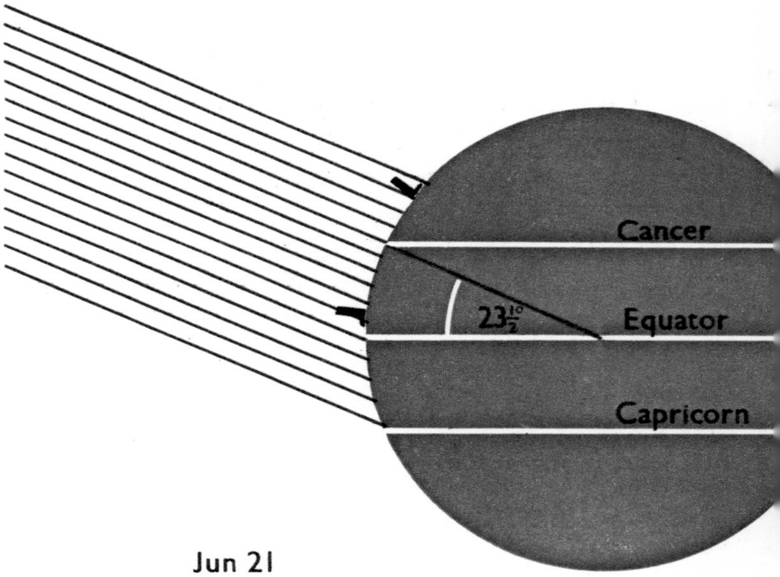

Jun 21

Within the tropics they point north at some seasons, south at others. Top diagram shows noon shadows cast by towers at Athens and Accra on December 21. Lower one shows shadows of same towers on June 21.

Cl.7

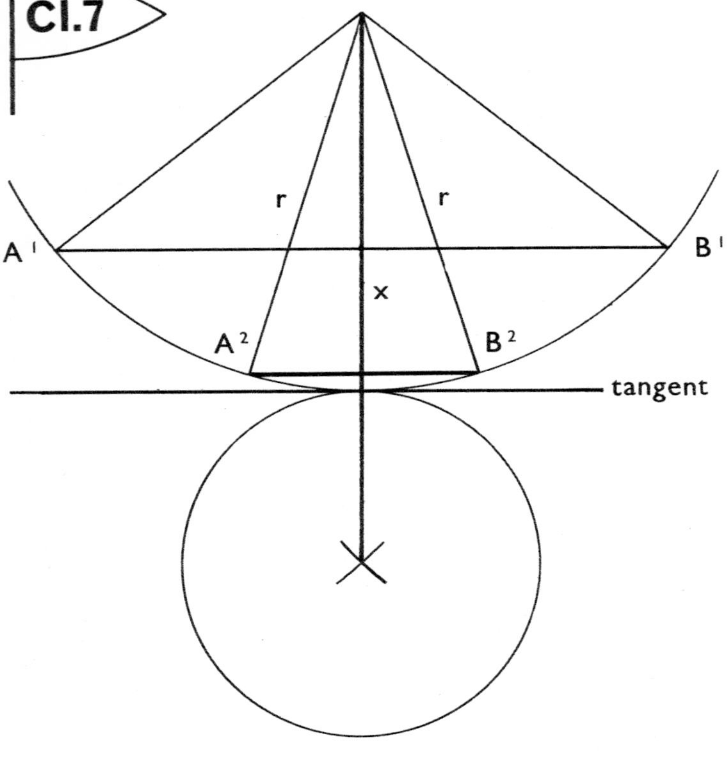

The perpendicular from the centre to a chord of a circle bisects it, and bisects the arc it subtends. When the chord becomes indefinitely small, it grazes the circle at right angles to the line which joins its centre to the centre of the circle.

In learning to navigate by latitude, Mediterranean man also and inescapably learned that the earth is spherical. Astronomy and geography, hitherto non-existent in the modern sense of the term, thus demanded a new geometrical orientation. A geometrical construction relevant to our terrestrial location or position relative to the fixed stars must henceforth exploit two assumptions: (a) the three angles of a triangle make up two right angles; (b) if the horizon plane is tangential to the earth's spherical surface, the plumb-line suffices to define both an axis of reference for the zenith distance of a celestial body and a straight line which joins the observer to the earth's centre.

However, geometry cannot get us anywhere unless we also invoke physical considerations. To use the sun's shadow or what primitive astrolabes the navigator has at his disposal for assessing the altitude of a star also signifies a dual presumption about the propagation of light. The temple astronomers, accustomed to see dust dancing in sunbeams or moonlight penetrating the narrow slit of a hut, a temple door or the mouth of a cave, had already taken for granted what primers of physics still in circulation convey by the assertion that light travels in straight lines. For reasons less obvious, they also assumed that light rays from a sufficiently distant source are parallel.

All of our geographical lore and all astronomical thinking, till the detection of the parallaxes of fixed stars in the time of Gauss, rely on the two preceding assumptions about how light travels, and much contemporary discussion about the credentials of Euclid's geometry is more than necessarily unintelligible because too few writers sufficiently emphasise the need to be clear about whether we do or do not mean by the terms *straight* and *parallel* anything which we can define in terms of human experience other than the propagation of light. If we cannot do so, no experiment can vindicate what is merely a matter of definition, but if we can give them an independent reference to terrestrial experience, how far the terms are applicable to the propagation of light is an experimental issue. One of the least

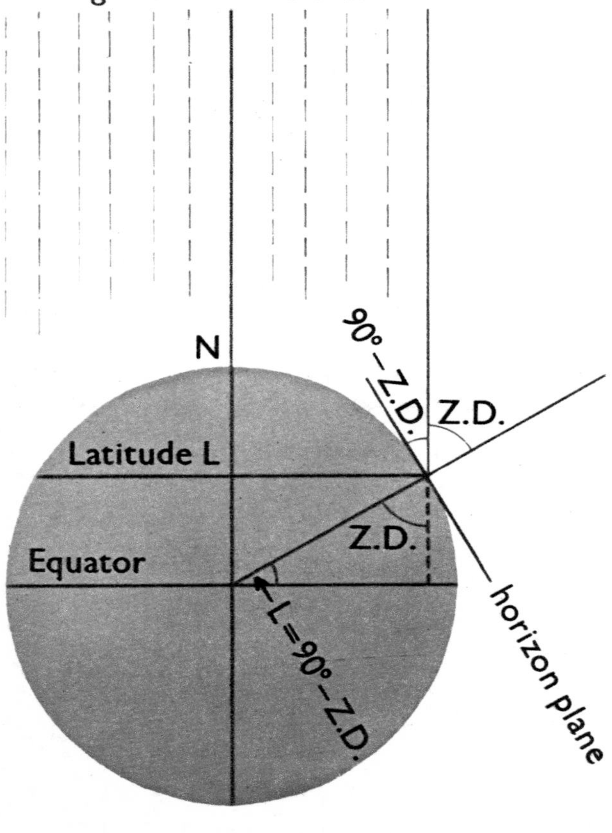

Determining latitude from the Pole Star exploits two assumptions: that the angles of a triangle equal two right angles and that the horizon plane is tangential to the earth's surface. Star's altitude and observer's latitude both equal 90° – Star's z.d.

satisfactory attempts to define a straight line, the one attributed to Archimedes, illustrates one aspect of the dilemma.

If we say that a straight line is the path of shortest distance between two points, we raise several difficulties. One is that we have not disclosed how we measure distance without first defining a straight line. A second is that all our measurements of distance in outer space depend on the path a ray of light traverses. A biological illustration discloses more forcefully the inadequacy of the Archimedean formula. The part of our own internal ear from which we largely derive recognition of spatial relations is an elaborate variant of a sense organ widespread in the animal kingdom. In its simplest form, as in jellyfish, squids or shrimps, it is a sac containing a floating solid body (*statolith*), the inertia of which forces it against a sensitive surface in the direction opposite to the animal's movement. It is experimentally possible to replace the statolith of a shrimp by iron filings, whence also to make the gravitational pull on it negligible in a sufficiently strong magnetic field. This forces the shrimp to orientate its movements with reference to the lines of force of the latter. If the lines of force of the magnetic field are curved, the shrimp's shortest distance between two points will be a curved line.

The foregoing remarks suffice to emphasise that Greek mathematics had its birth in circumstances uniquely propitious to the awakening curiosity of a maritime mercantile community in which there was a relatively high proportion of free men with an appetite for litigation and political debate. The demand for literacy encouraged respect for teachers with fraternities of pupils bound to them at first by ritual vows of secrecy, as were the disciples of Pythagoras (*circa* 550–500 B.C.). Happily, such secrecy could not outlive controversy in open forum between contending schools. The traveller or merchant pilot who announced a new acquisition to a regimen of mensuration and scientific navigation had to justify his credentials in public discussion, and he had to do so in conformity with the emerging

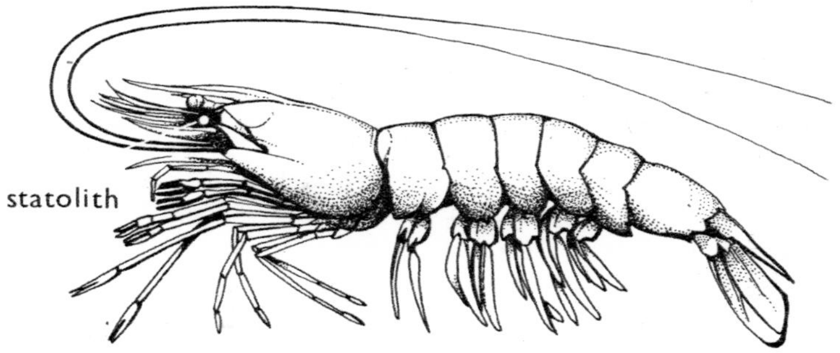

The shrimp derives recognition of spatial relations from a sac containing a solid body the inertia of which forces it against a sensitive surface in a direction opposite to the shrimp's movements.

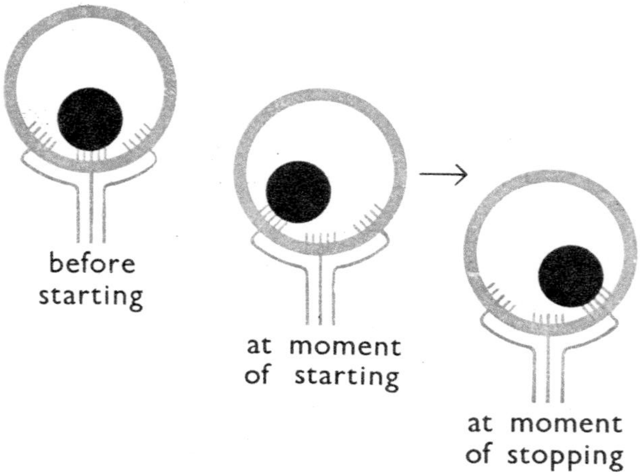

It is possible to replace this body (statolith) by iron filings. In a strong magnetic field the shrimp then orientates its movements with reference to the lines of force. Its shortest distance between two points will then be a markedly curved line.

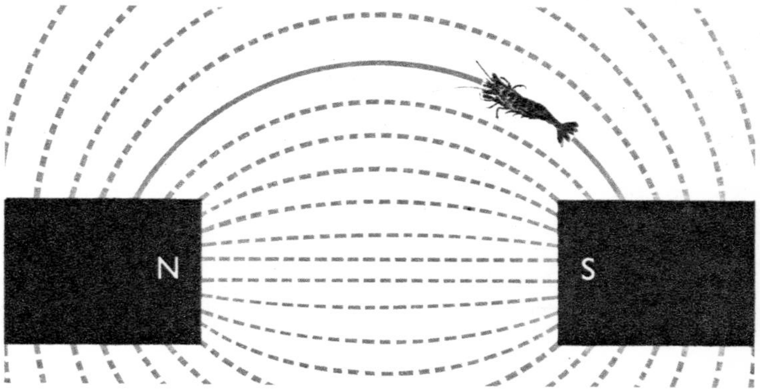

This scene from an Attic red-figure vase reminds us that Greek mathematics came into being among free men with an avid appetite for debate and litigation.

etiquette of an unfolding legal argument. As set forth about 300 B.C. by Euclid, the founder of the Alexandrian school, the principles of Greek geometry are thus the record of a great litigation. If the tempo of his *Elements* is exasperating, a sufficient explanation is the close affinity between the Greek attitude to proof and the leisurely procedure of a law court.

Having read so far, the reader who may have forgotten what he or she learned about geometry at highschool level may soon be in deep water. It will then be well to skip the rest of this chapter, and return to it after reading the next one, which will give a retrospective summary of the Great Litigation. In what remains of this, we shall try to get a glimpse of how much of what puzzled the Greeks has bequeathed to our own century enigmas not as yet fully resolved, but less difficult to recognise as such if we go back to where they first began to puzzle people.

The period during which the Greek-speaking communities of the Mediterranean world made their enduring contribution to the making of mathematics is divisible into three major phases. The first, which has left no documentary remains, covers the period from Thales and Pythagoras to Democritus, roughly 600–400 B.C. The leading writers of this period pursued the quest of proof with an eager curiosity attuned to the practical uses of mathematical enquiry; but disputes between rival schools soon forced geometry into a more academic mould. The keynote of the second phase is the teaching of Plato (430–349 B.C.). It culminates in the system of Euclid who leaned heavily on Eudoxus (408–355 B.C.), a pupil of Plato. Excessive formalism ended when Euclid ceased to teach in Alexandria. His death antedates by a few years the birth (*circa* 287 B.C.) of Archimedes, whose sympathy for invention signalises a new beginning. The third, *i.e.* Alexandrian, phase, characterised by a retreat from formalism, an outburst of inventiveness and a keen appreciation of the practical applications of mathematics, is divisible into two periods less in terms of content and of temper than because of external pressures.

The dynasty of the Ptolemies was on its last legs when Julius Caesar arrived (47 B.C.) in the city. Some say that the first great secular library of the Western World went up in flames; others, that Caesar despatched a large part of its stock with the consent of Cleopatra to the quayside, where fire destroyed it during the retreat of his forces. Thereafter, there is a black-out in Alexandria till we reach the middle of the second century A.D. An event of the reign of the first Augustus may throw some light on the slow recovery of intellectual activity in what remained, till the Christianisation of Rome, the pre-eminent centre of Greek culture. In Euclid's time there had been little or no merchandise of written matter; and throughout the two succeeding centuries the library founded by Ptolemy I was the only considerable secular repository of mathematical lore in the world of antiquity. Meanwhile, we find no trace of a flourishing trade in scrolls before the time of Octavius. It may well be that this was the signal of a steady build-up of a second collection in Alexandria. Certainly, an accessible store of earlier contributions was prerequisite to any considerable opportunities for the exercise of mathematical talent.

In the middle phase of the Greek period, the teaching of Plato had a lastingly banal influence, which we may trace partly to his gentlemanly contempt for manual labour. Otherwise, he might have condoned the use of a potter's wheel or a lathe with as good reason as the use of a compass. To say so, however, is at best a half truth. The quaint trinitarian numerology of the *Timaeus* discloses a mystique which confers on particular figures – triangles, circles, spheres and others (e.g. regular polygons, pyramids, cones) derived therefrom – a peculiar propriety in the providential plan. Thus a curve, other than a circle, can be worthy of mathematical free grace only if traceable on the surface of a figure elected to salvation in accordance with rule and compass ritual. A cone combined in its make-up the

THE QUADRATRIX OF HIPPIAS (circa *425* B.C.)
The line OS = a *advances through 90° in* t *seconds at fixed angular speed, whence through A degrees in* $A.t \div 90$ *seconds. The line ST advances through a horizontal distance* a *in* t *seconds, whence through* $PQ = r \sin A$ *in* $r.\sin A.t \div a$ *seconds while OS rotates through A°. Hence* $(r.\sin A.t \div a) = (A.t \div 90)$; *and if we write* $k = (a \div 90)$:

$$r.\sin A = k.A$$

We speak of r *as the radius vector of the point P.*
To trisect any angle between two lines placed so that they meet at O, draw perpendiculars from OR to the points where they cut the curve. Trisect the segment of OR between them, erect perpendiculars at the two intervening points, and lines from O to the points where these intersect the quadratrix complete the construction.
Squaring the circle depends on the fact (Chapter 5) that $90° = \frac{\pi}{2}$ *in radian measure, so that we may substitute* $K = (2a \div \pi)$ *for* k *in the above if we measure A in radians. As A approaches zero,* $\sin A \div A$ *approaches unity and* r *becomes OV, so that* $OV = K = 2a \div \pi$. *If* $a = 1$

$$\frac{ON}{OV} = \frac{\pi}{2} \quad \text{or } SR = \pi.(OV)$$

We have thus constructed two lines whose ratio is the irrational π.

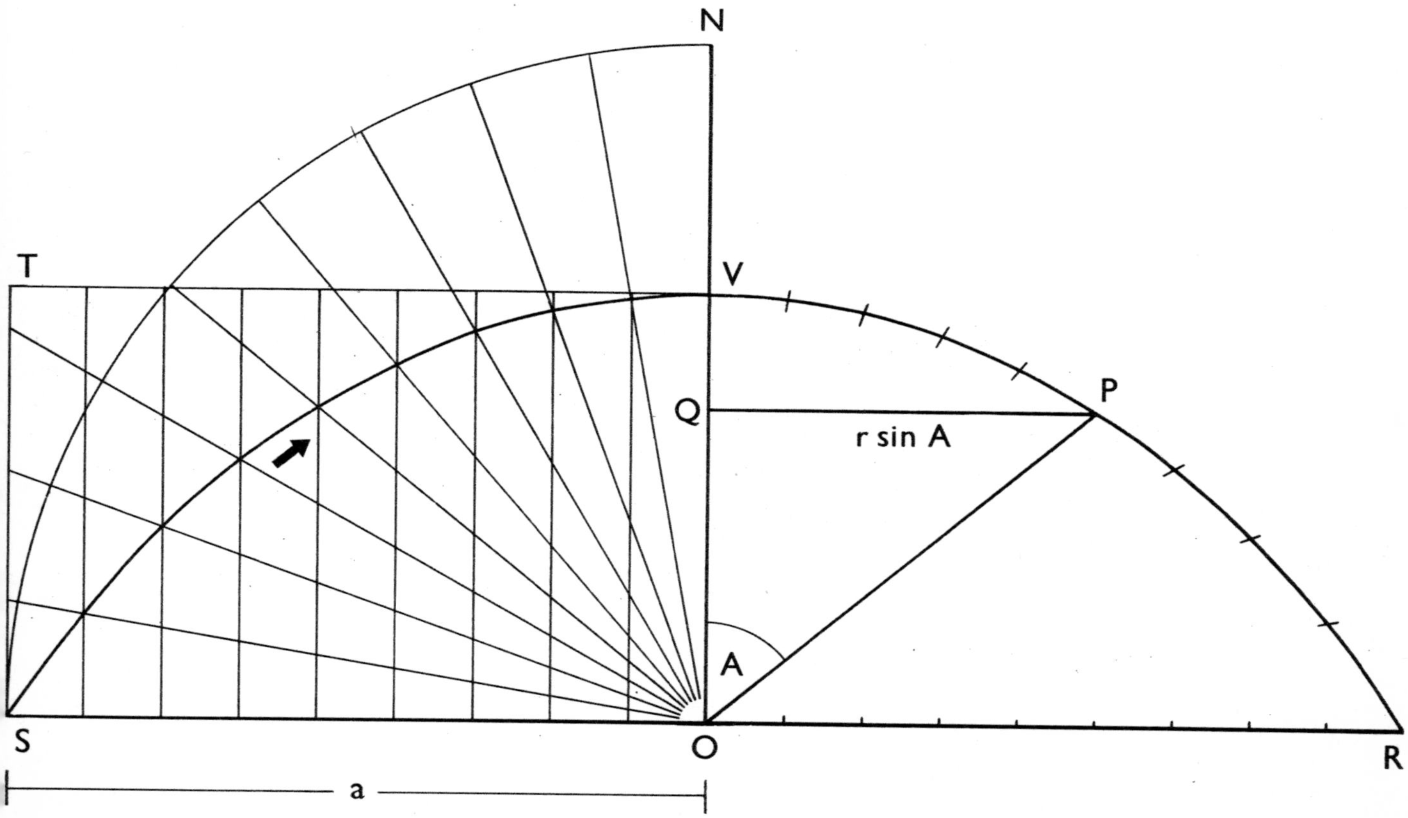

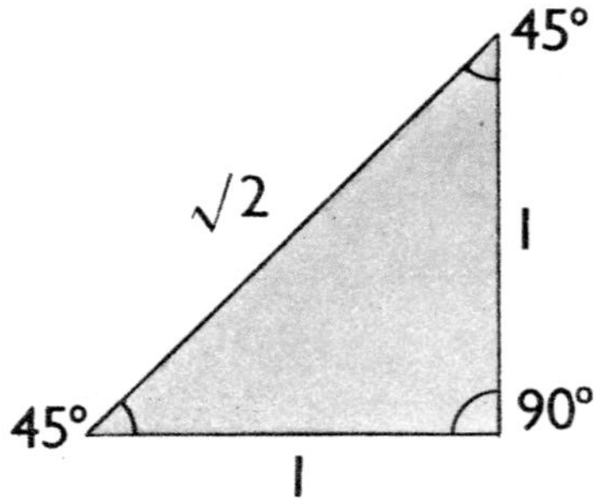

Two sides of this right-angled triangle are of unit length ∴ length of hypotenuse is $\sqrt{2}$.
The doctrine of incommensurability merely means that hypotenuse and another side cannot both match up exactly with scale divisions on same measuring rod.

spiritual perfection of the circle and the triadic mystery of the eternal triangle. Plato's ukase therefore endowed the properties of conic sections with a respectability out of all proportion to their utility at the time, and temporarily stifled curiosity about curves conceptually generated by tracing the pathway (*locus*) of a moving point.

Our earliest record of a temerity so sacrilegious from Plato's viewpoint is the curve known as the quadratrix, discovered by Hippias who flourished in Elea shortly before Plato's birth. Tradition ascribes to Hippias the use or invention of a mechanical device for generating such a curve. Plutarch's remarks on the Maestro's attitude to the use of any mechanical device (other than the compass) as an aid to geometrical reasoning is worthy of citation. In his *Life of Marcellus* he comments on:

> Plato's indignation at it and his invective against it as the mere corruption and annihilation of the one good of geometry thus shamefully turning its back upon the unembodied objects of pure intelligence.

However, Greek mathematics was already bogged down by several enigmas before Plato sounded the retreat into the shadowy domain of his cherished and hitherto unidentifiable universals. The first to emerge was the Pythagorean dilemma. Though it subsumes the whole problem of irrational algebraic numbers such as $10^{\frac{1}{7}}$, the original and somewhat mystifying statement of it is that the side of a square and its diagonal are *incommensurable*. In the domain of mensuration, this means: (a) if the side of a square is of unit length, the diagonal is $\sqrt{(1^2+1^2)}=\sqrt{2}$; (b) if the diagonal of a square is of unit length, the side is $\frac{1}{\sqrt{2}}$. From the Greek viewpoint (*circa* 400 B.C.), the issue was not so simple, if only because the Greeks had no number signs in which to translate explicitly the meaning of $\sqrt{2}$. Translated into contemporary sign language, we may restate the argument, as it has come down to us from Eudoxus, through Euclid, as follows:

(1) $\sqrt{2}$ is not an integer, therefore it is an improper vulgar fraction $(1+f)$ between 1 and 2;
(2) An improper fraction between 1 and 2, if also respectable, must be expressible in terms of integers k and m as:

$$1+\frac{k}{m}=\frac{m+k}{m}=\frac{n}{m}$$

(3) if $\sqrt{2}$ is expressible in this way:

$$2=\frac{n^2}{m^2} \text{ so that } 2m^2=n^2$$

(4) since only squares of even numbers are even, n is therefore even and replaceable by another integer $2r$ so that

$$4r^2=2m^2 \text{ and } m^2=2r^2$$

(5) it thus seems that m is also even; but if m and n must both be even, we can find two integers to express their ratio in two ways only:

(a) if n is greater than m, their ratio must be an integer greater than 2, whence not between 1 and 2;
(b) if m is greater than n, their ratio must be an integer less than 1, whence also not between 1 and 2.

As a slice of arithmetic, this is merely a sideline to the issue we have illustrated in Chapter 2 by considering the particular meaning of $\sqrt[3]{5}$. We have there seen that the naïve interpretation of a ratio as a relation between two numbers reducible to the relation between the numerator and denominator of a vulgar fraction in one sense deprives us of the right to speak of $\sqrt{2}:1$ as a ratio; but we have also anticipated the possibility of representing it as a nonterminating nonperiodic basal fraction, and we have exhibited it as a continued fraction to any order of precision we require for practical purposes. However, this does not clarify the meaning of the term *incommensurable* in the context of side and diagonal.

We can begin to be clear about what the translators of the Greeks did and did not mean by incommensurable magnitudes, if we give a metrical interpretation to the first three approximations for $\sqrt{2}$ already obtained by developing it as a continued fraction:

$$1\tfrac{1}{2}=1\tfrac{6}{12}>\sqrt{2}>1\tfrac{2}{5}\ (>1\tfrac{4}{12})$$
$$1\tfrac{2}{5}=1\tfrac{48}{120}<\sqrt{2}<1\tfrac{5}{12}$$

If we suppose that the side of a square exactly matches one foot on a scale graduated in feet, inches and tenths of an inch, the first two approximations convey that the diagonal will lie between 1 ft. 4 in. and 1 ft. 6 in. The second together with the third conveys that it will lie between 1 ft. $4\frac{8}{10}$ in. and 1 ft. 5 in. Since the difference between the third and fourth ($1\frac{12}{29}$) approximation is much less than 0.1 in., further approximation is irrelevant unless we use a scale with finer intervals. How far it is profitable to continue the process therefore depends on how small we wish to make a *zone of uncertainty*, undetachable from the matching process of scale measurement in real life, by sub-dividing the interval.

Evidently, therefore, the statement that the diagonal of a square is incommensurable with its side does not mean that we cannot measure both on the same scale in the sense in which we can hopefully undertake to match an edge with a scale to a level of precision satisfactory for practical purposes. What the assertion that $\sqrt{2}:1$

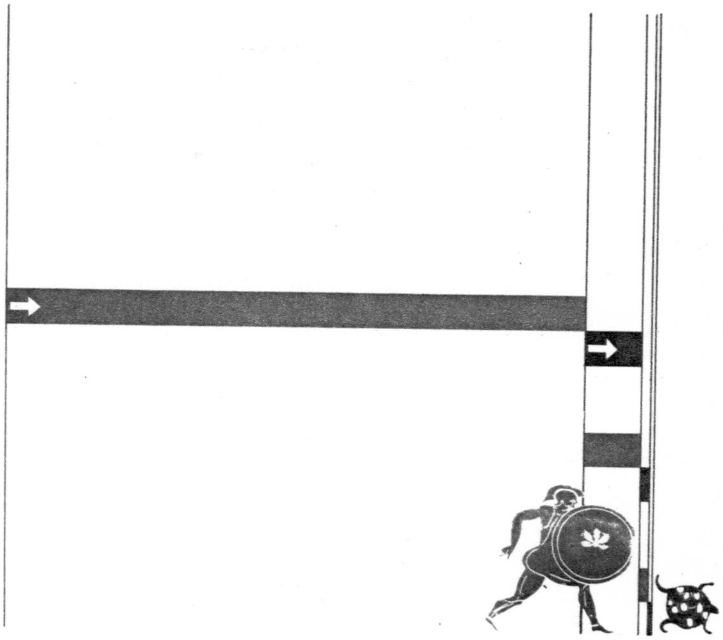

Zeno's puzzle is whether Achilles can win a race on these terms: Tortoise has a start of 1 unit, but Achilles runs ten times as fast. While Achilles runs 1 unit, tortoise runs $\frac{1}{10}$ and is still ahead. While Achilles runs $\frac{1}{10}$, tortoise runs another $\frac{1}{100}$, and so on. A graph makes it clear that the tortoise loses the lead at $1.\dot{1}$ units from the starting point.

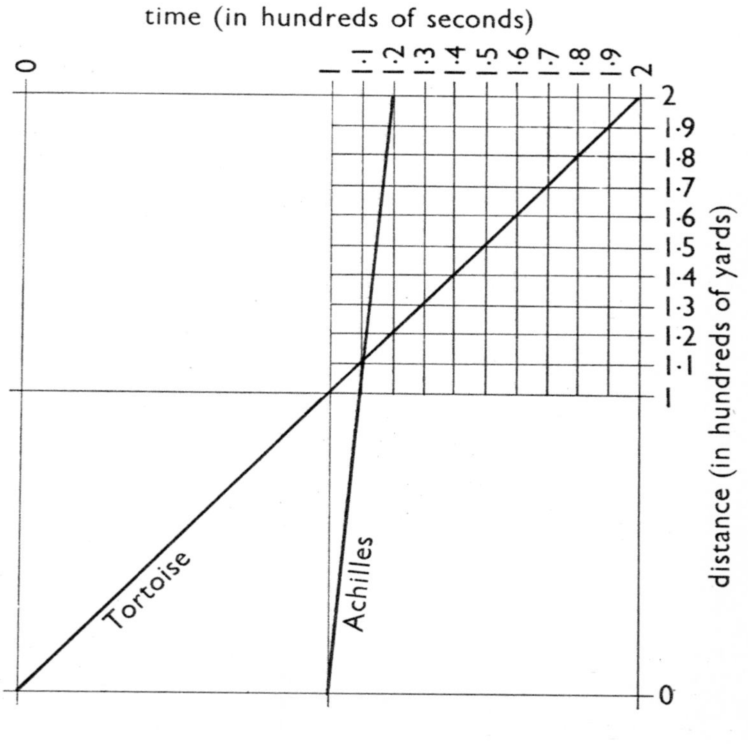

is not expressible as the ratio of two integers $m : n$ correctly signifies is that no scale able to guarantee an exact match of a side of a square at the *m*th division can confer an exact match of the diagonal at any other (*n*th) division – or *vice versa*. If we take the particulate structure of matter seriously, it is difficult to see how we should ever hope to guarantee the truth of an assertion of this type about any two measurements.

The secret brotherhood founded by Pythagoras at the Italian port of Frotona was the first of many schools whose teachers criticised and debated with one another. In short, the Great Litigation was soon a free-for-all. A rival school founded in Elea, also in southern Italy, was among the first to challenge the Pythagorean teaching; and its founder Zeno, born (*circa* 496 B.C.) within a few years of the likely date of the death of Pythagoras, started the case for the prosecution by propounding a sequence of riddles in which some of our contemporaries still see a core of deep thinking. Others regard the paradoxes of the Sophists merely as an illuminating sidelight on how the Greek numeral system obstructed clear thinking about computation.

For instance, one of the paradoxes takes this form. Achilles and a tortoise run a race. The tortoise runs a mile of the course before Achilles, running ten times as fast as the tortoise, starts. When Achilles has covered his first mile the tortoise will therefore be 0.1 of a mile further ahead. When Achilles has completed this distance, the tortoise will be 0.01 of a mile ahead of him ... and so on. Are we to conclude that the tortoise will always be a little ahead? Only if we assume that there is no limit to the following sum of all the distances it puts between itself and Achilles, *viz.* $1+0.1+0.01+0.001 \ldots$ In our notation this sum is $1.\dot{1}=1\frac{1}{9}$, and it is therefore clear that the tortoise can never be more than $1\frac{1}{9}$ of a mile ahead. When Achilles has gone so far, he will pass the reptile.

From one point of view, we may therefore dismiss the paradox, by stating that it is only a paradox if we have not grasped the possibility that the sum of a limitless succession of positive fractions may have a finite – and manageably small – limit. However, Zeno knew well enough that Achilles would win the race, and posterity charitably attributes to him the intention of focusing our attention on the difficulty of accommodating the notion that space is infinitely divisible with the notion that motion at a finite speed is in progress. In short, can we mean anything intelligible when we speak of the speed with which an object is moving in an immeasurably small interval of time? Newton was perhaps the first mathematician to give an emphatic affirmative answer to this question. His contemporary, Bishop Berkeley, the Zeno of the seventeenth century of our own era, restated the case for the prosecution.

In the background of Berkeley's counterattack, we may seemingly discern a second limitation of Greek number lore. In contradistinction to the difficulty of conceiving a limit to the sum of a limitless succession of diminishing fractions, it was equally difficult in 500 B.C., and long after that, to conceive that a fraction may more and more closely approach a finite value as we make the numerator and the denominator each as near to zero as we care. Again, we may say that no difficulty arises in the domain of practical mensuration, and our own numeral system prepares us for the shock by permitting us to write 0.2 as:

$$\frac{1}{5}=\frac{0.1}{0.5}=\frac{0.01}{0.05}=\frac{0.001}{0.005}=\frac{0.0001}{0.0005}$$

$$=\frac{0.00001}{0.00005} \text{ and so on.}$$

As an example of a limiting ratio, we may recall the Babylonian algorithm, written in the form shown on the right:

$$a^2-b^2=(a+b)\,(a-b) \text{ or } \frac{a^2-b^2}{a-b}=a+b$$

If we put $a=b$, both the numerator and the denominator of $(a^2-b^2)\div(a-b)$ vanish; but the identity on the right-hand side signifies that its value does not detectably differ from $2a$ when b

is not detectably less than a. We may make this explicit in another way, if we put $b=a-\triangle$ in which $\triangle(<a)$ is positive, so that

$$\frac{a^2-b^2}{a-b}=\frac{a^2-(a-\triangle)^2}{\triangle}=\frac{2a.\triangle+\triangle^2}{\triangle}=2a+\triangle$$

However small we make $\triangle$, whence however close to a is the value of b, the expression on the right remains finite and therefore as close to the limiting value $(2a)$ as we like to make it. Thus the ratio of two small quantities need never become less than some finite quantity even at the point when they pass out of the realm of measurement.

En passant, we may generalise this type of limiting ratio for future use. First note that:

$$(a^3-b^3)\div(a-b)=a^2+ab+b^2=3a^2$$
(when $a=b$)

$$(a^4-b^4)\div(a-b)=a^3+a^2b+ab^2+b^3=4a^3$$
(when $a=b$)

It is easy to satisfy oneself that a more comprehensive formula is

$$(a^n-b^n)\div(a-b)=a^{n-1}+a^{n-2}b$$
$$+a^{n-3}b^2\ldots ab^{n-2}+b^{n-1}$$

In the expression on the right, there are n terms, each of which we can make as near to a^{n-1} as we like by making the difference between a and b sufficiently close. We now write customarily the result in the form:

$$Lt_{a=b}\frac{a^n-b^n}{a-b}=na^{n-1}$$

We can picture this *limit* for the case $n=2$ by saying that the rectangle $(a-b)(a+b)$ merges into a line of length $(a+b)$ when $(a-b)$ vanishes. Likewise, if $n=3$ the solid figures of total volume $(a-b)(a^2+ab+b^2)$ then merge into a surface of total area a^2+ab+b^2; but rule and compass litigation does not permit us to interpret the general rule last stated. The Platonic embargo on any construction other than rule and compass procedure therefore made the geometer who upheld his discipline as a recipe for practical mensuration vulnerable to attack on another front, and the more so because he could not now provide a recipe for visualising some of the ratios which mensuration of the plane and solid invokes.

One way of interpreting many of Euclid's propositions invoked by the late Alexandrian architects of trigonometry is that they are residues of a period when geometers still hoped to complete such a programme. For instance, Euclid does in fact prescribe (*Book IV, 12*) a construction from which we can derive:

$$\tan 18^\circ=\frac{\sqrt{5}-1}{\sqrt{(10+2\sqrt{5})}}$$

This result emerges in the rationale for inscribing a regular pentagon in a circle and hence constructing an angle of 72° (=360°÷5); and it is therefore interesting to note in passing that Euclid's recipe for making a polygon of 15 equal sides (*Book IV, 16*) tells you how to make an angle of 24° (=360°÷15), whence by successive

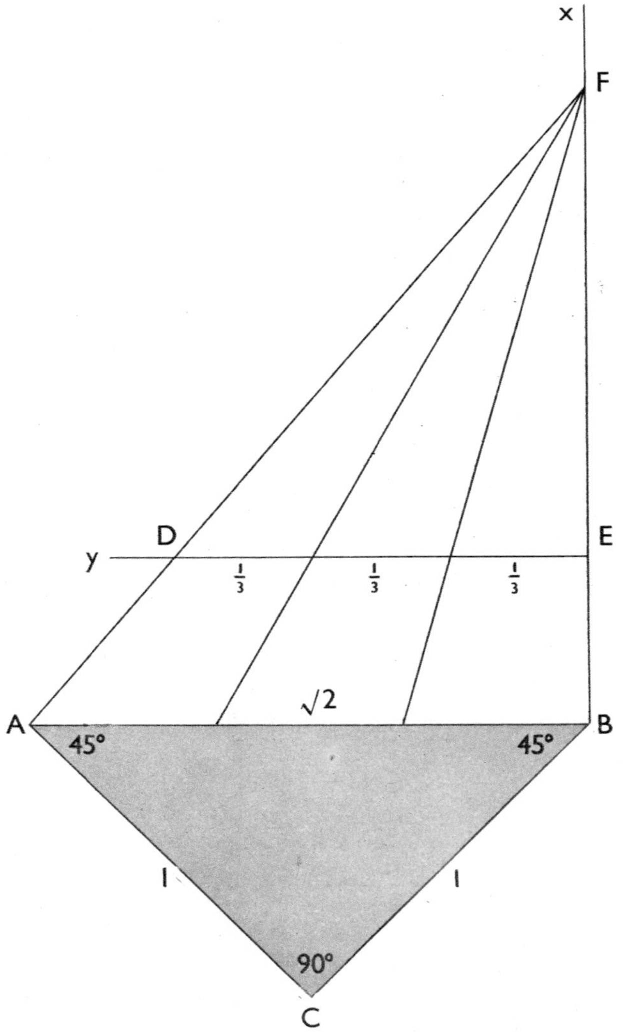

Within the framework of Plato's rule-and-compass procedure, we can make a construction which vindicates the law of division for two incommensurables. Here the line AB, of length $\sqrt{2}$, is divided by 3.

bisection, 12°, 6° and 3°, a result exploited by Ptolemy for the construction of his trigonometrical table. Since $72=3(24)$, the angle 72°, like 90° ($=30°+30°+30°$), is amenable to rule and compass trisection; but no such rule prescribes how to trisect *any* angle, whence no Euclidean recipe for making a trigonometrical table with intervals of 1°. We do not certainly know that Euclid's Greek contemporaries or predecessors used one three-hundred-and-sixtieth of the perimeter of a circle of unit radius as a unit of angular measurement. Hence we cannot be sure that this limitation of Plato's embargo conferred on the impossibility of trisecting any angle its peculiar interest; but it is a likely explanation of the fact.

To face up to the Greek dilemma when the Great Litigation reached its climax, let us be clear about what Plato's prohibition means. We can use a compass to construct a circle or the arc of a circle of equal radius about a fixed point. We can also use a straight-edge to join two points or to extend a line in either direction; but we have no licence to move a straight-edge till it takes up a convenient position as in a recipe for trisecting an angle attributed to Archimedes. Within the framework of this restriction we can make a construction which vindicates the laws of multiplication, subtraction and division for two incommensurables (e.g. $\sqrt{2}$ and 1) in such a situation as we convey by $(a\pm b)^2=a^2\pm 2ab+b^2$. We can likewise represent the ratio of two incommensurable areas, e.g. a square on unit side and a rectangle of $\sqrt{15}$ square units. In Euclid's *Elements* we can also find many constructions which we can use for the solution of quadratics involving irrationals and exploited in the first trigonometrical tables. For instance:

(a) the construction (*Euclid II, 11*) to divide a line into two segments so that the square on one is equal to the rectangle whose adjacent sides are the whole line and the residual segment is equivalent (for a line of unit length) to the assertion:

$$x^2+x-1=0 \text{ if } x=\frac{\sqrt{5}-1}{2}$$

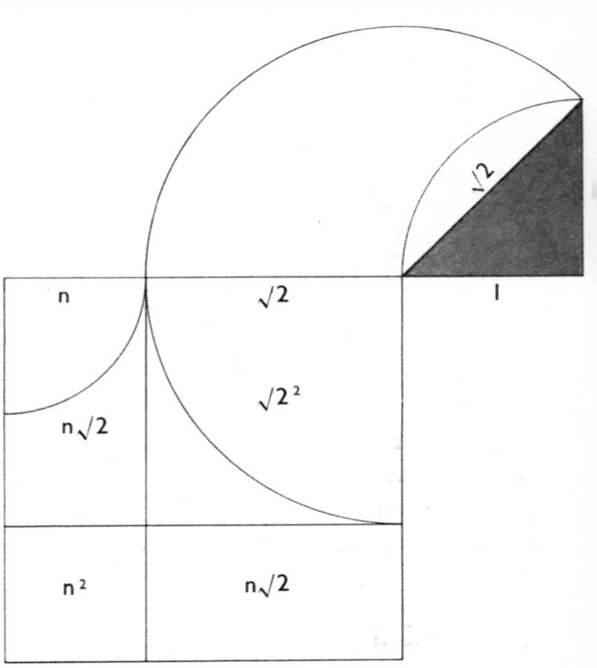

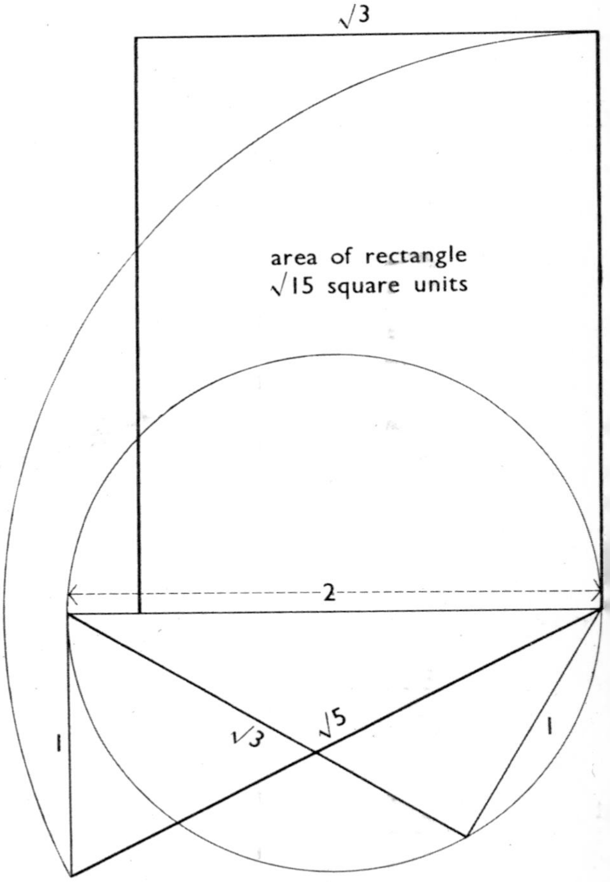

Top left:

Plato's Queensberry Rules of Geometry do not prevent us from making a visual aid to show that $\sqrt{2}$ *exists and behaves like any so-called rational number in the algorithm* $(n + \sqrt{2})^2 = n^2 + 2n\sqrt{2} + (\sqrt{2})^2$.

Top right:

Rule-and-compass construction showing $\sqrt{2} \div \sqrt{3}$. *Produce OA (length 1 unit) to S. From A draw AQ at right angles to OS. Using O as compass-point and with compass opened to 2 units, draw an arc intersecting AQ at B. The sides of triangle OAB then measure 1,* $\sqrt{3}$ *and 2 units respectively.*
At any convenient point (C) on OA produced, make a right angle OCP. Bisect the right angle PCS, and from C mark off 1 unit on the bisector at D. From D, with compass opened to 1 unit, mark off the point E on the line PC. The sides of triangle CDE then measure 1, 1 and $\sqrt{2}$ *units respectively.*
Join BE and produce, to cut OS at T.

By Claim 2, $\frac{TC}{TA} = \frac{EC}{BA} = \frac{\sqrt{2}}{\sqrt{3}}$

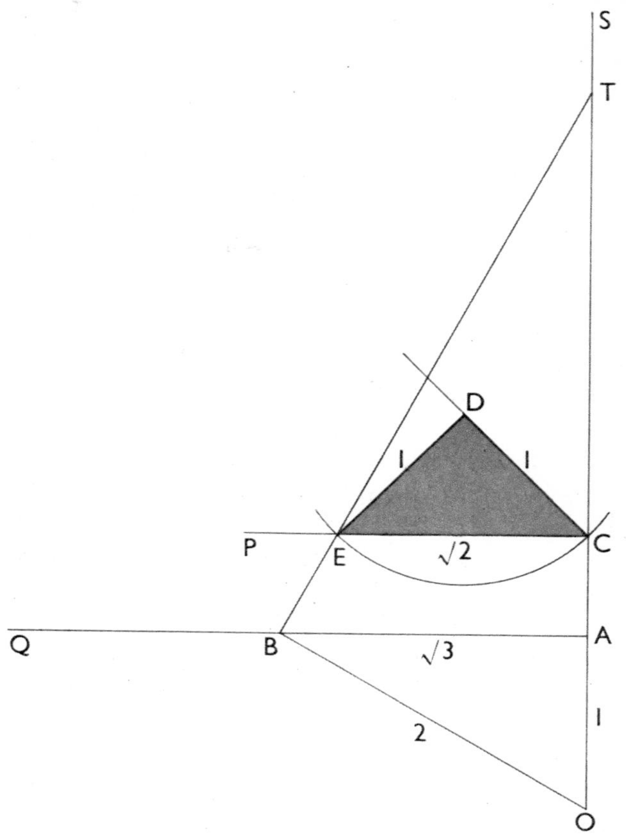

Bottom left:

The sides of the large right-angled triangle measure 1, 2 and $\sqrt{5}$ *units respectively. Those of the smaller one (constructed in accordance with Claim 6) measure 1, 2 and* $\sqrt{3}$ *units. Opening our compass first to* $\sqrt{5}$ *units then to* $\sqrt{3}$ *units, we can construct a rectangle* $\sqrt{5}$ *by* $\sqrt{3}$ *whose area is* $\sqrt{15}$ *square units.*

Bottom right:

When Euclid invites us to divide a line of unit length into two segments so that the square on one segment is equal to the rectangle whose adjacent sides are the whole line and the residual segment, he is in fact inviting us to solve the quadratic $x^2 = 1 - x$, *or* $x^2 + x - 1 = 0$.
In this construction AB, of unit length, is cut into two segments, $AF = x$ *and* $FB = 1 - x$. $AC = \frac{1}{2}$ *unit,* $BC = CD$ *and* $AD = x = AF$.
In the right-angled triangle CAB, $BC^2 = AC^2 + AB^2 = \frac{1}{4} + 1 = \frac{5}{4}$. *Since* $BC = CD = \frac{1}{2} + x$, $BC^2 = \frac{1}{4} + x + x^2 = \frac{5}{4} \therefore x^2 + x = 1$ *and* $x^2 + x - 1 = 0$.

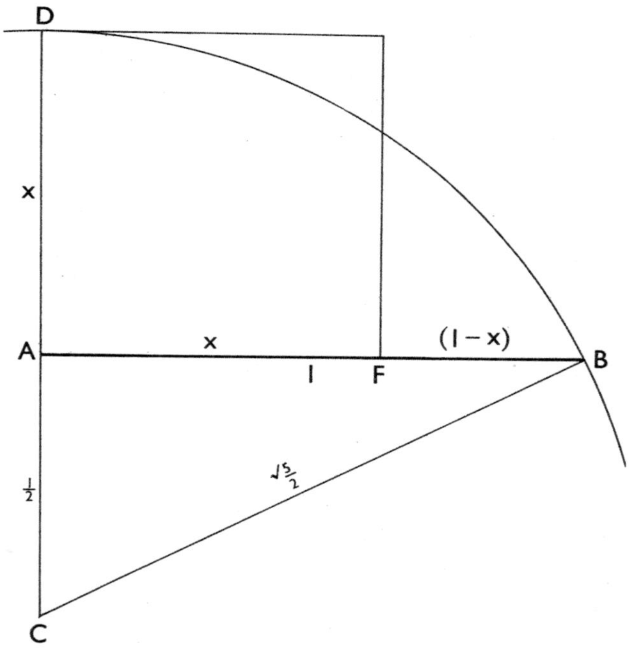

(b) what is implicit (*Euclid VI, 8*) in the proposition that the square of the perpendicular from the right angle to the hypotenuse of a right-angled triangle is equal to the rectangle whose adjacent sides are segments into which it divides, is equivalent (hypotenuse of unit length) to the assertion:

$$x^2-x+p^2=0 \text{ if } x=\frac{1+\sqrt{(1-4p^2)}}{2}$$

In so far as one can easily decipher any intelligible meaning in the tenth book of Euclid's *Elements*, this sort of thing may well be an obituary on earlier attempts to complete the programme of vindicating a tie-up between geometry and computation. Be that as it may, two enigmas defeated all efforts to do so. One was that of doubling or halving the cube. The Greek geometers could find no rule and compass procedure for constructing the side of a cube whose volume (x^3) is twice that of another cube ($x^3=2y^3$). If the length of the side of the latter is unity, the problem signifies $x=\sqrt[3]{2}$; and failure to solve it signifies that it was impossible to vindicate the numerical status of $\sqrt[3]{2}$ in the highest court of appeal. A second setback, usually referred to as the problem of squaring the circle, was the lack of any orthodox rule and compass recipe for drawing the side (s) of a square whose area is exactly equivalent to that of a circle. Here the problem implies $s^2=\pi r^2$, whence the ratio of s^2 to r^2 is the irrational π.

No rule and compass procedure here means no *finite* number of rule and compass operations. If we remove the restriction implied by *finite*, Hippias had long ago solved the problem of squaring the circle and trisecting an angle; but we cannot lay out the whole path of a curve such as the quadratrix without successively making an infinite number of circles and lines to accommodate all points that lie on it. The proof that neither of the objectives last named is achievable did not come till the middle of the nineteenth century. That mathematicians could still be interested in it throws a flood of light on the motivation of a vast mopping-up operation of the last hundred years. There was still a gap to fill if pure mathematics has to be a water-tight logical system. A deductive geometry which proceeds within the framework of Platonic restrictions, as transmitted by Euclid, cannot provide a comprehensive rationale for manipulating all real (*i.e.*, both all irrational and rational) numbers according to the same rules.

By 300 B.C., the rule and compass embargo had made matters more difficult and had driven geometry deeper into a formalism which could offer no lucid aid to the practical tasks of mensuration; but in the Greek setting of bondage to an alphabetic vocabulary of number, the notion of incommensurability imposed on any self-consistent system of mensuration which rejects the notion of limitless divisibility the inescapable obligation to make a choice. Either one must relinquish the notion of absolute equality at no cost to the practical serviceability of the outcome; or one must dispense with the concept of a finite scale interval. If we adopt the second alternative, we have to face the difficulty of justifying the relevance of our calculus to its practical applications; but no such difficulty could disturb those who shared Plato's expressed aristocratic contempt for painstaking and methodical observation of astronomical phenomena.

The fact that no one has explored the first possibility to the bitter end does not justify the belief that the programme is inherently unrealisable; but the outcome would call for a drastic overhaul of the statement of our propositions. For instance, the most we could claim for the relation between the lengths (p, b, h) of the sides of a right-angled triangle on a scale of intervals $2\triangle$ units would be a statement of some such form as:

$$p^2+b^2+\triangle^2-6\triangle < h^2 < p^2+b^2+\triangle^2+6\triangle$$

However, natural science which invokes mathematics does not necessarily, or commonly, proceed from the particular to the general in accordance with the Baconian recipe. More often, we set up a model and paint in its limitations after testing its applicability. In the domain of measurement it is therefore permissible to regard Euclidean lines and figures as fictions which do not exactly correspond to anything in the real world, though none the less useful if we can subsequently define their precise relevance to the use we intend to make of them. When we are able to do so, we shall have accomplished everything which the somewhat misleading current expression *arithmetisation of geometry* can conceivably convey to the physicist with every reason to welcome assurance about the reliability of rules of measurement. Neither the Greeks who followed the Platonic tradition, nor their successors who reverted to a more naïve and, as it turned out, more exhilarating mood, did any such thing.

Plato's contemporaries took a different course; and it is difficult to see how they could have done otherwise. Having at their disposal no number symbolism capable of accommodating the notion of a ratio except in terms of exact divisibility of one integer by another, they tailored a definition of equal ratios to sidestep both the need for invoking the notion of divisibility – whether limitless or not – and the intractability of redefining a ratio in its own right; and the reader may legitimately wonder why later generations of commentators have bestowed so much veneration on the formal definition of equiproportionality attributed to Eudoxus and expanded at great length by Euclid in Books V and X of the *Elements*. We shall say that $p : q = m : n$ if integers A, B, C satisfy the conditions:

(1) $Am > n$ when $Ap > q$
(2) $Bm < n$ when $Bp < q$
(3) $Cm = n$ when $Cp = q$

As laid down in the definitions of Book V of the *Elements*, this criterion is formally consistent with rule and compass regime, because we can multiply lines and add them whether we can or cannot match them exactly with a finite number of intervals of any scale we can construct – or conceive. On the other hand, it is difficult to see how it can throw light on, or provide any justification for, what we are doing when we numerically evaluate a trigonometric ratio in a form such as $(1+\sqrt{5})\div(1-\sqrt{5})$. If it does, Euclid does not disclose the clue; and the Alexandrian architects of the first trigonometrical tables calculated them on the only conceivably relevant assumption that it is possible to divide two irrational numbers by one another to an acceptable precision level.

One can indulge endlessly in unprofitable discussions about what St. Paul, Karl Marx or Charles Darwin really (*sic*) meant, unless one keeps one's feet on the solid ground of what they said and of what meaning their words could conceivably convey in the context of contemporary utterance. A dispassionate reading of the fifth and tenth books of the *Elements*, unbiased by the translator's comments, which would be unnecessary if the translation were itself adequate, at least justifies uncertainty about the author's end in view; and our uncertainty deepens if we seek to interpret it in its own Platonic context. The Platonic view was that geometry was an intellectual exercise for sharpening the controversial wit, and what relevance it might have to measurement or to computation was an unfortunate, if unavoidable, liability. Of itself, the mere fact that the English translators have chosen to use *magnitude* – a Latin, not a Greek, word – to embrace what we now call rational and irrational numbers in a so-called real domain does not force us to conclude that Euclid had any such marriage in mind if, to be sure, the intention was a marriage with computable offspring.

A more reasonable view is that geometry, by now cut adrift from any more secure foothold in the real world than Plato's *unembodied objects of pure intelligence*, had become a highly sophisticated game of chess ending in stalemate. On the evidence before us, the amendment of Eudoxus imposed on his successors the obligation to speak henceforth of lines as to lines, squares as to squares, circles as to circles, spheres as to spheres, etc, in contradistinction to ratios of lengths, of areas or of volumes. Thus there is no sufficient reason to state that Euclid (or Eudoxus) re-

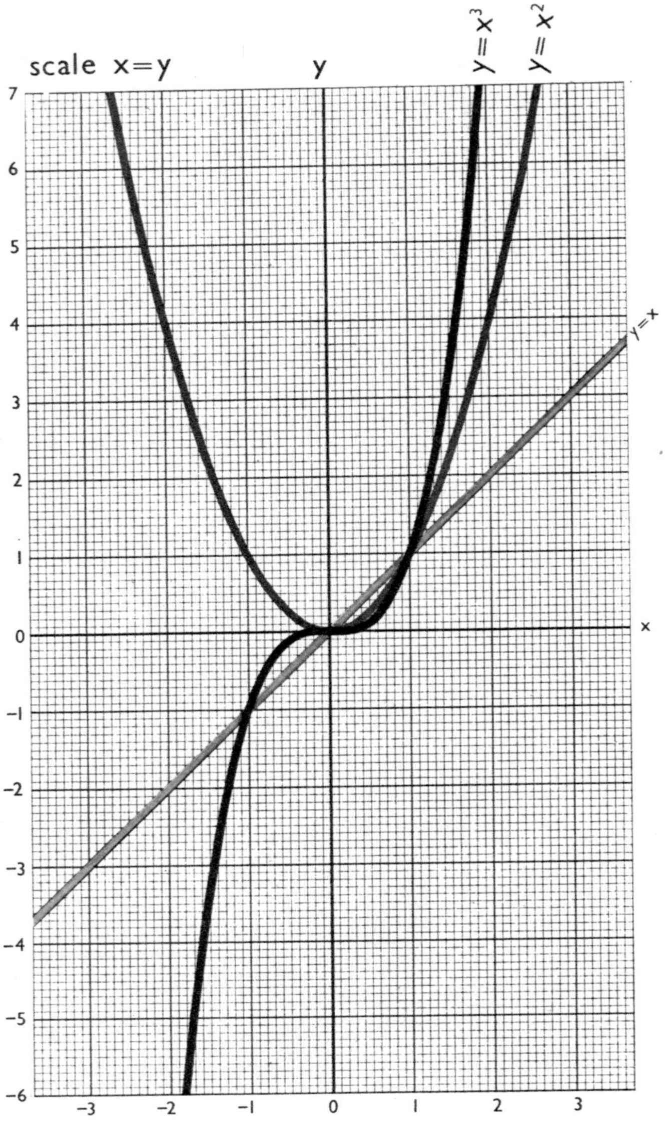

The graph enables us to visualise x^1 *(line),* x^2 *(area) and* x^3 *(solid) for negative as well as for positive values of any real number.*

cognised that π is in any sense a definite *number*. What Euclid (XII, 2) states is that *circles are to circles as the squares of their radii.* The form of the assertion makes no reference to a number which the doctrine of magnitudes dispensed with the need for defining as such; and if Euclid intended to convey the existence of such a number, it is difficult to explain why he failed to anticipate the Alexandrian recognition of the arc-radius ratio as a basis for angular measurement.

A new era of active discovery began when Archimedes gave π a bounded value in a strictly metrical statement which set limits to the zone of uncertainty on both sides ($3\frac{10}{71} < \pi < 3\frac{10}{70}$). By so doing, he reinstated number explicitly in geometry; but he himself gave no formal redefinition of what a ratio is *per se*. At the beginning of the sixteenth century of our own era, Descartes married algebra to a geometry which finally rejected the tyranny of rule and compass, visually accommodated x^4 or x^{10} no less than x^0 (*point*), x^1 (*line*), x^2 (*area*) and x^3 (*solid*), and eventually provided a niche for negative or positive values of any real number; but neither Descartes nor those who exploited the uses of the new geometry in the two succeeding centuries made any attempt to define the precise relevance of the doctrine of incommensurables to the practical tasks of mensuration.

Meanwhile, and till nearly the middle of the nineteenth century, no one questioned Euclid's authority as a geometer; but no one seems to have recognised that his geometry could provide no sufficient foundation for all the uses in which Descartes, Newton and their successors had enlisted numbers. Such is the theme of a large tract of mathematics developed during the last century. Its exploration has led into many bypaths in the search for some concept embracing, but not circumscribed by, number. It is perhaps too early to forecast what will remain of a programme which we may properly call the *arithmetisation of arithmetic*, if it fully accomplishes its target. The quest for finality has never yet justified the end in view; but the quest may be rewarding if it helps us to solve at least one fundamental problem of human communication. Stated simply, the issue is: what do we mean by definition?

From 300 B.C. till the middle of the nineteenth century, mathematicians of the Western World unanimously and explicitly subscribed both to the belief that Euclid's geometry is a logically water-tight system and to the dogma that it is an indispensable basis for interpretation of space relations in general – including those of outer space in particular. Implicitly, they also regarded it as the high court of appeal for what rules we use when manipulating numbers variously classified as negative or positive, rational or irrational and real or (as then called) complex. Though there would have been good enough reason to question this doctrine on its own merits, the main target of early dissent was an assumption often wrongly referred to as the *parallel postulate*. The only intelligible reason why Euclid's nineteenth-century critics picked on this exhibit is that a host of much earlier commentators had bestowed lavish attention on it. If we dig deeper, it might seem strange that the autopsy on Euclid's credentials did not start from a recognition of the need to re-examine the basis of the system against the background of some agreement about what we mean by defining words.

This is a topic on which dictionary makers have not even yet come to terms with common sense. To define a word by recourse to other words without going round in circles, it is necessary to have a minimum battery of words whose meaning we can *either* validate by recourse to some sort of identity parade *or* demonstrate by means of non-linguistic visual aids. It is therefore important to be clear about the distinction between three categories of statement set forth at the beginning of the several books of the *Elements* and translated respectively as *definitions*, *postulates* and *axioms*. In reverse order, we may say:

(a) an *axiom* is what we may take for granted at a verbal level, *e.g.*, the notion of *equality*;

(b) *postulates* specify what procedures are ad-

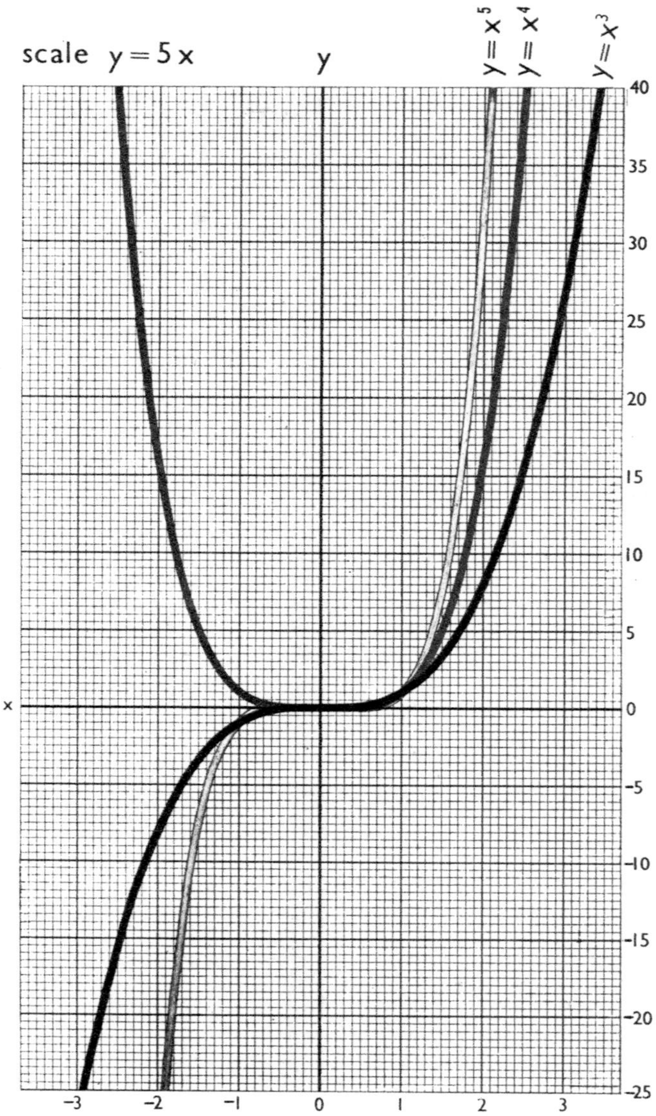

It also enables us to visualise x^4, x^5 and so on. Here $y = x^3$ is plotted again. Note how change of scale alters its shape for range of values opposite.

missible as a device for establishing proof with the endorsed means (*i.e.* rule and compass) at our disposal, *e.g.* the right to draw a circle about a point as centre;

(c) a *definition* is a verbal clarification of a term employed.

The foregoing reversal is intentional, because we cannot realistically clarify one word by recourse to other words unless we can first reach agreement about what it is admissible to assume without recourse to formal definition. For instance, we may define a circle in a meaningful way as the locus of a point which moves so that it is always equidistant from a fixed point, if we concede that: (a) points are merely convenient fictions used in mapping space and, as such, not amenable to factual definition; (b) we all agree about what we mean by the distance between two points. Alternatively, the postulate that we can draw a circle with the means at our disposal suffices to define the circle itself, if we: (a) explicitly state what device we have at our disposal, i.e. compass; (b) illustrate how we use it to draw one. In short, a statement of definition before a statement of axioms and postulates is either thoroughly bad exposition or plain nonsense.

The arbitrariness of what Euclid takes for granted comes into focus most clearly when we get to grips with what we mean by the distinction between the sort of path we represent by a curved line and the sort of path we represent by a straight one. The only such path of which we can have any experience in outer space is that of the propagation of light or other electromagnetic radiation. In the field of outer space, the distinction arises when we speak of the bending of a beam in a gravitational field or by refraction of the earth's atmosphere. If we prefer to use the term *straight* for the path of propagation in empty space where gravitation is negligible, there is then no self-evident tie-up between such a definition and Euclid's system. We have seen why the Archimedean definition of a straight line is doubly defective in the more restricted domain of our terrestrial experience; and it is not easy to frame a more satisfactory one. In the age of

The concept of rectilinearity is not easy to justify by common terrestrial experience. To the Greeks a tight-drawn rope may have seemed a test, but in fact such a rope follows the course of a catenary curve. Here the curvature is by no means negligible.

Democritus, the rope-stretcher's procedure seemed to justify the concept of rectilinearity, but any rope drawn tight between two pegs follows the course of the curve called a *catenary* till it reaches breaking point. To be sure, the curvature may be scarcely detectable on the small scale of a few feet; but it is by no means negligible over a distance of several miles. How much the concept takes for granted becomes clearer if we take the earth's rotundity seriously. A justifiable rule and compass reconstruction of a ship's course projected on a plane presupposes that we have a recipe for making a straight-edge; and the fact that Euclid's system offers no such recipe is perhaps its greatest flaw when we extrapolate the same procedure into the Sputnik domain.

We have seen that Euclid's postulates lay down what we can do with the permissible means at our disposal, *inter alia* to draw a straight line between two points. In its Platonic context we must interpret this to mean that we can use the straight edge of a ruler. If the meaning we attach to a straight line is to have any intelligible

Above: preparing a plane surface. Below: testing it with beam computer equipment which detects irregularities of $\pm$ 0.0005". We can define the straight-edge intelligibly only by saying that it is ground to fit such a surface as closely as possible. Euclid reverses the craftsman's order, 'defining' the plane in terms of an undefined straight-edge.

reference to how we can construct one, we therefore shoulder at the outset the obligation to indicate how we make – as nearly as possible – what we call a straight-edge. The engineer's answer is that we grind it to fit as closely as possible to a plane surface. We then have to press our enquiry a step further, by asking what grinding process generates two flat surfaces. To justify a rule and compass procedure on such a small scale of relevance to terrestrial experience, we must therefore first define a plane as the conceptual limit of a grinding process applied to irregular or uniformly curved surfaces, and subsequently define a line as the conceptual limit of an edge produced by a comparable process. If we do follow this recipe to define what we mean by a straight line in terms of the means at our disposal, we do not commit ourselves to statements referable to an outer space into which the grinding process of the engineer cannot intrude; and we

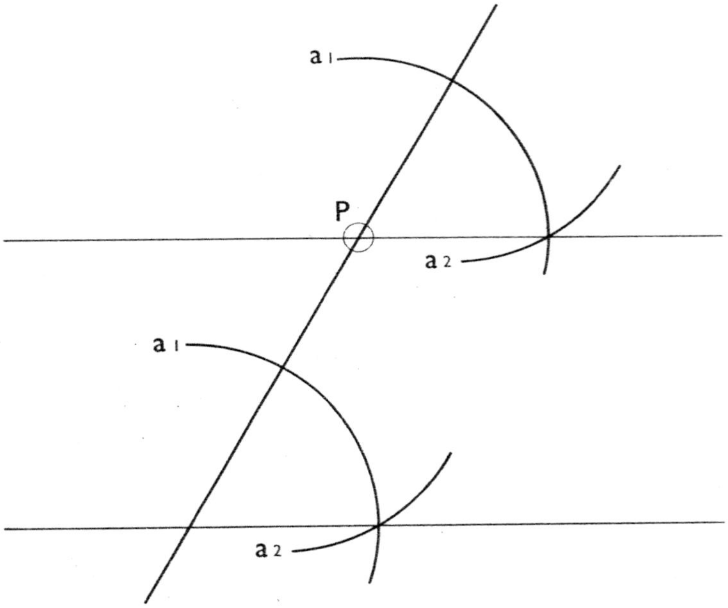

Rule-and-compass construction for drawing a straight line through P, parallel to a pre-drawn straight line in the same plane.

cannot luxuriate in a definition of parallelism tailored, as is Euclid's, to its requirements.

Euclid's definitions of point, straight line and plane at the beginning of Book I of the *Elements* reverse the engineer's order. They have therefore no tie-up with the postulates which disclose what we claim to be able to do. The first seven definitions are as follows (Heath's translation):

1. *A point is that which has no parts, or which has no magnitude.*
2. *A line is length without breadth.*
3. *The extremities of a line are points.*
4. *A straight line is that which lies evenly between its extreme points.*
5. *A surface is that which has only length and breadth.*
6. *The extremities of a surface are lines.*
7. *A plane surface is that in which any two points being taken, the straight line between them lies wholly in that surface.*

The use of the word *evenly* in (4) begs the question, because we require either two plane surfaces or at least another straight line of reference to give it a meaning. As a definition, (4) therefore conveys nothing at all. We now turn to Euclid's definition of what he means by parallel straight lines:

35. *Parallel straight lines are such as are in the same plane and which, being produced ever so far both ways, do not meet.*

This excursion into outer space discloses no clue to how we can draw through a point a straight line parallel to a second one. The fundamental property which enables us to do so is the topic of three formal proofs (*Book I, 27–29*), ostensibly to show: (a) that every straight line cutting two parallel lines in the same plane makes equal corresponding and alternate angles therewith; (b) that two straight lines in the same plane are necessarily parallel if this condition holds. A rule and compass prescription for drawing through a point P a line CD parallel to a line AB is: (a) to draw through P a line crossing AB at Q; (b) make at P to QP produced an angle equal to PQB (or QPA); but this assuredly does not imply that any other line crossing *AB* and *CD* will be equally inclined to both. To pull this out of the bag, Euclid has to introduce the famous twelfth axiom which has been the topic of a prodigious output of portentous discussion; and the simple reason for this is that the 35th definition does not suffice for the job in hand. The axiom is as follows:

12. *If a straight line meets two straight lines so as to make the two interior angles on the same side of it taken together less than two right angles, these straight lines, being continually produced, shall at length meet on that side on which are the angles less than two right angles.*

Since so many very clever people have protested that this is not obvious, it is clearly not an axiom in the sense that it satisfies the identity parade criterion for laying the foundations of a verbal definition. What is equally worthy of comment is that it does not help to clarify a definition which suffers from the same defect, *viz.*, that it carries us into a realm beyond where the means at our disposal end, having therefore no tie-up with any rule and compass recipe for drawing a straight line parallel to a second straight line. If we take the stand that our definitions should have an intelligible relation either to our postulates or to our axiom, we are free to sidestep the so-called proof embodied in the propositions (*I, 27–29*) by replacing the twelfth axiom by another, equally arbitrary but less time-consuming *vis à vis* our future intentions, *viz.*

> *if any straight line is equally inclined to two other straight lines in the same plane, every other straight line which crosses them is equally inclined to both.*

Our definition of parallel straight lines consistent with rule and compass procedure will then be:

> *two straight lines in the same plane are parallel if equally inclined to a third one crossing both.*

Chapter 5 The Great Litigation

Among mathematical treatises of the Greek-speaking world before the Christian era, the most influential in Europe after their assimilation by the Moslem world were the 13 books of Euclid's *Elements*, the versatile writings of Archimedes, a comprehensive work by his contemporary Appolonius on conic sections, a topic on which Euclid had himself elaborated in a lost text, the earlier discoveries of Menaechmus, a tutor of Alexander the Great and a pupil of Plato, also the collected works of Ptolemy known to us by their Arabic name as the *Almagest*. Very little of the contents of the *Elements* – other than the arrangement – was novel when it appeared. All the propositions which have greatly influenced the subsequent history of mathematics had been demonstrated by one or other of a host of predecessors including Democritus, Hippocrates of Cheos, Antiphon, Eudemus, Eudoxus, the Cyrenian Theodorus who taught Plato, Thaetetus, Speusippus and Xenocrates who in turn succeeded the Thomas Arnold of Athens as headmasters of the Old School.

The *Elements* takes precedence over any subsequent composition as the greatest best-seller among manuals of instruction throughout the ages. Its make-up is worthy of comment, if we ask ourselves how much of it we now really need to know. The first four books contain 115 demonstrations and the sixth 33. The 25 demonstrations of the fifth book do not deal with figures. They expound the concept of equiproportionality as taught by Eudoxus by reference to line segments. With two exceptions in Book XII, the foregoing round off the treatment of plane figures. Some 70 demonstrations of Books XI–XIII deal with three-dimensional problems and solid figures. The 33 and 115 demonstrations of Books VI and X respectively discuss what we should now call algebraic problems in the language of lines and areas.

The seventh, eighth and ninth books of Euclid do not deal with geometrical issues. They subsume *arithmetic* in the original sense of the term, *i.e.* the theory of numbers such as primes, the Pythagorean triads and relations between squares and cubes. Some of the propositions are now obvious. Thus VIII, 22 states that c is a perfect square if a is a perfect square when $a:b::b:c$ (*i.e.* $ac=b^2$). We have had occasion to notice already a result given in Book IX, *viz.* that there is no upper limit to the number of primes.

To some extent, the enormity of Euclid's programme is attributable to a very considerable merit. That is to say, Euclid sets forth as a separate proposition the rationale of any construction he uses to disclose the build-up of a figure, albeit his initial disposal of parallels and his definitions of the straight line, plane, square and rectangle fall short of so admirable a promise. To some extent, the score also piles up because of the inclusion of demonstrations which involve merely verbal reference to definitions or axioms without recourse to any dissection. Not a few other demonstrations

Fresco from a villa at Pompeii, showing a slave raising water by treading an Archimedean screw. Throughout the period dominated by Plato and Euclid, mathematicians had little contact with manual work. Proof flourished at the expense of progress. By the time of Archimedes, Alexandria had become not only the leading centre of learning but also a busy centre of the world's work. In this new setting mathematics re-established itself as the powerful ally of astronomy, navigation and engineering.

with reference to areas have little or no relevance to mensuration, and signify to ourselves elementary algebraic identities such as the already-mentioned formulae for expanding $(a \pm b)^2$ or for factorising $a^2 - b^2$. Be that as it may, what now most makes the execution of the programme seem so long-winded is Euclid's reluctance to exploit the notion of a ratio. Reasons for this reluctance in the author's social context should be intelligible. Its consequence was a tortuous detour involving exploits of ingenuity worthy of a better cause. The equiproportionality of equiangular triangles does not make its appearance (VI, 4) till the score is nearly 120 propositions; and the theorem of Pythagoras, which is fool-proof if we make use of this principle, does not occur till the end (I, 47 and 48) of the first book.

If we dismiss as trivial those demonstrations which invoke no dissection of a figure, and take full advantage of the fact that the notion of a ratio need not be to us, as to the Greeks, a nightmare, we can cover the same ground by a vastly less devious path, and the ground we nowadays need to cover is a much smaller territory. The outcome of many constructions useful in their own context is now obtainable with less effort by recourse to trigonometry, to projective and coordinate geometry, or to the Newton-Leibniz calculus. What is essential is what we cannot dispense with as a prerequisite to an understanding of the foregoing. We can then condense the outcome of the Great Litigation in the judgment of ten famous Civil Cases.

Before we do so, let us pay a tribute to another meritorious feature of Euclid's procedure. Throughout the *Elements*, he is meticulous about separately demonstrating a proposition and its converse. For instance, the proposition that the square on the longest side of a right-angled triangle is equal to the sum of the squares on the other two sides does not dispense with the need to prove that a triangle is right-angled if $a^2 = b^2 + c^2$. The distinction anticipates what later mathematicians imply when they distinguish between a *sufficient* condition, a *necessary* condition and a condition which is both necessary and sufficient. Two examples from everyday life suffice to illustrate the use of the two terms. Food is a necessary but not a sufficient condition of happiness. Consumption of more than a specifiable quantity of gin in a specifiable period is a sufficient condition for producing intoxication, but it is not a necessary one, because whisky will produce the same state. As Bertrand Russell remarks, the difference which the terms imply is one we all too frequently dispense with in political argument.

If we now attempt to record the most famous Civil Cases of the Great Litigation, we need not pause to record in the Statute Book definitions of point, line, plane and parallel already discussed

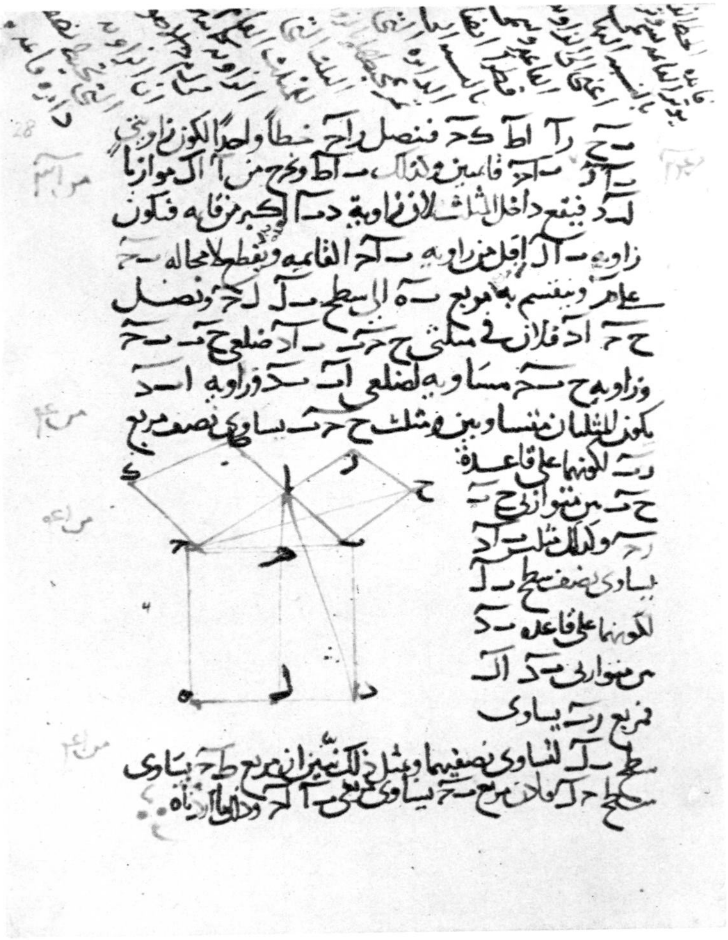

Page from an Arabic commentary on Euclid (c. A.D. 1250). After their assimilation by the Moslems, the thirteen books of Euclid's Elements *exercised a more powerful influence on European mathematics than almost any other treatise of ancient Greece.*

in our last chapter; but we shall need to be explicit about which Statute Book entries tie up with *admissible pleas* (postulates) by citing as *judicial rulings* (axioms) what assertions we may take for granted with the support of the Bench, *viz*:

JUDICIAL RULINGS

1. Every point on the boundary of a circle is equidistant from the centre.
2. The inclination of any segment of a straight line to any other adjacent segment is equivalent to two right angles.
3. Either of two criteria suffices to specify a unique triangle:
   (a) the lengths of two sides and the angle between them;
   (b) the length of one side and the angles at its extremities on one side of it.
4. Equal segments of circles with equal radii subtend equal arcs.
5. Any circular arc is longer than the chord which joins its ends.
6. The unit of area is the square whose side is of unit length.

We may dismiss as trivial judgments, because they invoke no testimony from the pathologist to exhibit a dissection:

7. From 2 above: *opposite angles of intersecting straight lines are equal.*
8. From 1 and 4: *equal segments of circles with equal radii subtend equal angles at their centres.*
9. From 4 and 8: *equal arcs of circles with equal radii subtend equal angles at the centre.*
10. From 2 and 9: *a right angle at the centre of any circle subtends an arc equal to a quarter of its boundary.*
11. From 3: *two triangles are equal if they satisfy either of the specifications (a) or (b).*
12. From 1: *circles are equal if their radii are equal.*
13. From 6: *the area of a rectangle is the product of the length of two adjacent sides.*

What was a Euclidean postulate we shall regard as any procedure which is admissible as a plea to call in evidence. For instance, if one proposes to drop a perpendicular from the apex of a triangle to the opposite side, one must initially satisfy the Court that one can do so with the means at one's disposal. In Euclid's legal code, the Court has ruled that the means at one's disposal are: (a) the straight edge of a ruler to join two points, to extend a straight line or to draw a straight line through a fixed

82 EUCLID'S ELEMENTS.

PROPOSITION 47. THEOREM.

In a right-angled triangle the square described on the hypotenuse is equal to the sum of the squares described on the other two sides.

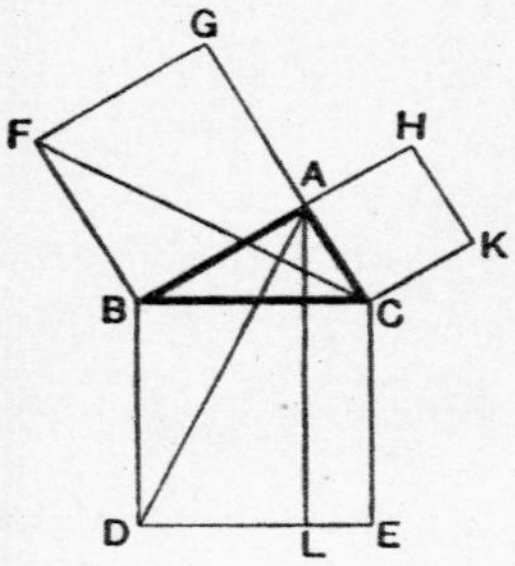

Let ABC be a right-angled triangle, having the angle BAC a right angle:
then shall the square described on the hypotenuse BC be equal to the sum of the squares described on BA, AC.

Construction. **On BC describe the square BDEC;** I. 46.
and on BA, AC describe the squares BAGF, ACKH.
Through A draw AL parallel to BD or CE; I. 31.
and join AD, FC.

Proof. **Then because each of the angles BAC, BAG is a right angle,**
therefore CA and AG are in the same straight line. I. 14.

Now the angle CBD is equal to the angle FBA,
for each of them is a right angle.
Add to each the angle ABC:
then the whole angle ABD is equal to the whole angle FBC.

Certainly Euclid proved an all-time best-seller among writers of manuals of instruction. This page of Hall and Stevens' Text-Book of Euclid's Elements, *first published in 1888, would have seemed strikingly familiar to students at the Moslem universities.*

point; (b) a compass for drawing circles or arcs of circles of equal radii. Accordingly, Counsel for the Defence may submit, in accordance with the proper uses of rule and compass:

ADMISSIBLE PLEAS

14. Make a circle equal to another circle (from 1 and 12).
15. Cut off a segment of a line equal to another line or extend it accordingly (from 1).
16. Make an angle equivalent to another angle (from 4 and 10).
17. Make a triangle equal to another triangle (from 3, 14 and 15).
18. Bisect an angle (from 1 and 11).
19. Bisect a straight line (from 1 and 11).
20. Erect a perpendicular at a fixed point on a straight line (from 1 and 11).
21. Make any rectangle (from 14 and 19).
22. Erect on a line a perpendicular passing through a fixed point outside it (from 1 and 11).
23. Draw through a point a straight line parallel to another straight line (from 16 and the definition of parallels).

One of the above calls for comment, both because it puts the spotlight on a semantic deficiency of Euclid's system and because it draws attention to the pivotal role of the much-disputed twelfth axiom. Euclid defines a rectangle as a four-sided figure, all of whose angles are right angles. This is not an operationally meaningful definition unless we can cite a construction to show that such a figure can exist. With rule and compass we can make a four-sided straight-line figure defined either as: (a) one with two equal sides each at right angles to a third; (b) one with equal opposite sides and with a right angle between one pair of adjacent sides. From (a) it is easy to show that the diagonals are equal; and from (b) it is easy to show that the opposite angles are equal. Without recourse to the properties of parallels or to Claim 1 (p. 21), we cannot show that the two recipes lead to the construction of equivalent figures, whence that a four-sided rectilinear figure, all of whose four angles are right angles, exists. Without recourse to the disputed axiom, it would be easy to prove that the fundamental properties of parallel straight lines follow from the rule and compass recipe for draw-

ing two of them, if we could prescribe a rule and compass recipe for making a figure with the properties with which Euclid's definition gratuitously equips the rectangle.

We are now ready to record all the ten most famous Civil Cases of the Great Litigation, *i.e.*, those we need to quote as precedents in any subsequent appeals to the several High Courts of trigonometry, projective or co-ordinate geometry, *etc.* With each claim presented, we shall record important exhibits for the defence mentioned as awards in a favourable judgment of the court.

THE TEN MOST FAMOUS CIVIL CASES

Claim 1. In a triangle, $A+B+C=180°$.
If we assume by recourse to Euclid's 12th axiom what we can prove about the essential properties of parallel straight lines, or if we adopt as axiom the existence of straight lines having such properties, a customary demonstration of Claim 1 as exhibited earlier is straightforward. Alternatively, we may *assume* that all the four angles of a rectilinear figure constructed in accordance with either rule and compass recipe for making a rectangle are right angles. It is then easy to show that: (a) the sum of the three angles of a right-angled triangle is equal to two right angles; (b) the sum of the three angles of any other triangle is equal to the difference between the sum of the angles of two right-angled triangles and 180°. Thus the verdict in favour of Claim 1 depends on a judicial ruling, and we might with equal propriety adopt Claim 1 as such at the outset. It would then be easy to prove that the Euclidean rectangle and Euclidean parallel straight lines have properties consistent with the uses we make of them in later litigation.

Awards:

(i) If $C=90°$, $B=90°-A$, and $A=90°-B$; whence also $\cos 0=1=\sin 90°$ and $\sin 0=0=\cos 90°$

(ii) right-angled triangles are equal if they have one side and one angle $<90°$ equal (from 3);

(iii) triangles are equal if each side of one is equal to one side of the second.

Greek vase of about 470 B.C., showing citizens engaged in orderly debate in the market place.
It was in this atmosphere that Greek geometry became the subject of a great litigation, with a rigid code of judicial rulings and admissible pleas.

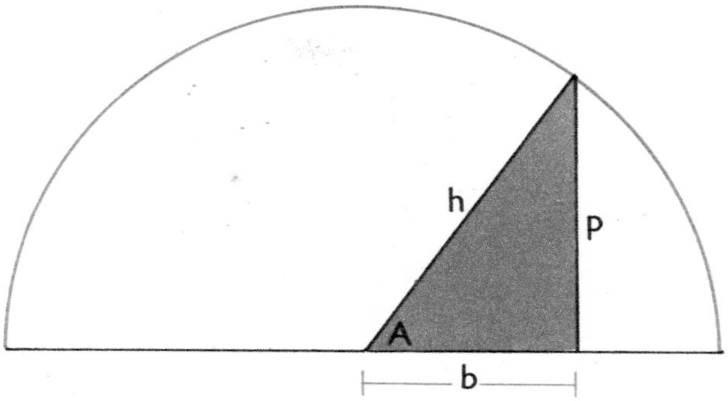

sin A = p ÷ h = cos (90−A)°
cos A = b ÷ h = sin (90−A)°
tan A = p ÷ b = sin A÷cos A

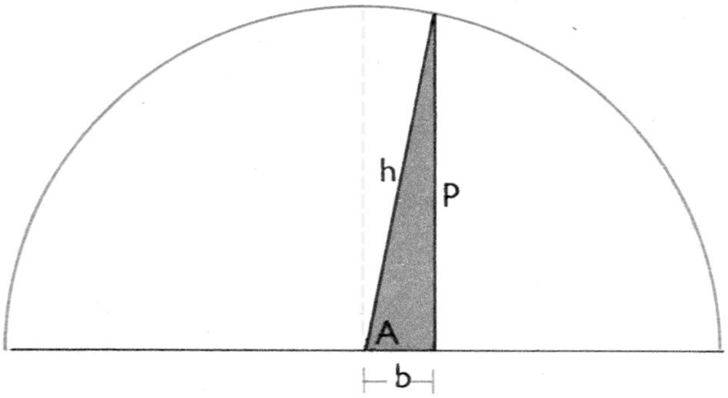

As A tends to 90°, b approximates to 0 and p approximates to h

sin 90° = 1, cos 90° = 0, tan 90° = ∞

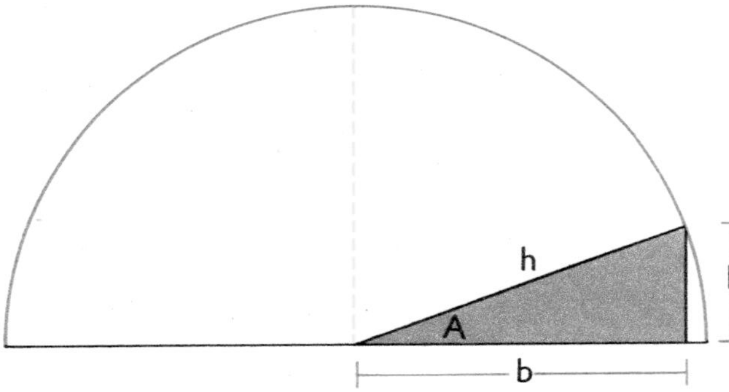

As A tends to 0°, p approximates to 0 and b approximates to h

sin 0° = 0, cos 0° = 1, tan 0° = 0

First step towards a table of trigonometrical ratios. (See Claim 1, Award (i).)

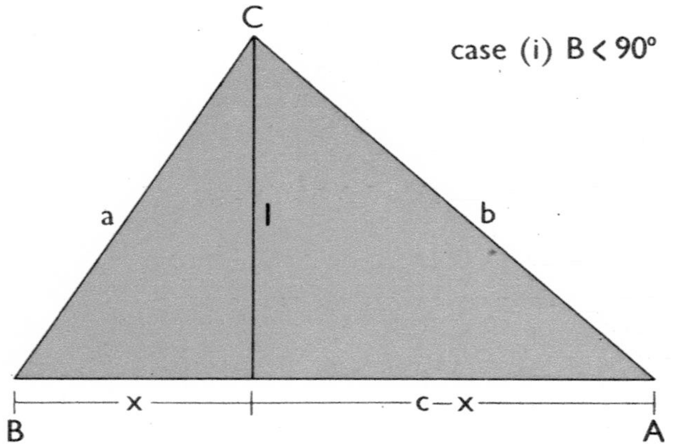

$$x = a \cos B \qquad l + x^2 = a^2$$

$$\begin{aligned} b^2 &= l + (c-x)^2 \\ &= l + c^2 + x^2 - 2cx \\ &= a^2 + c^2 - 2ac.\cos B \end{aligned}$$

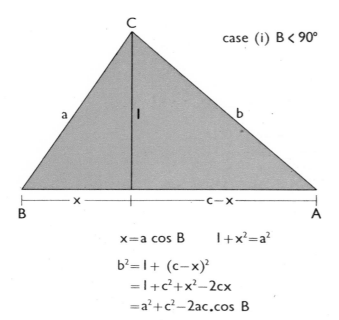

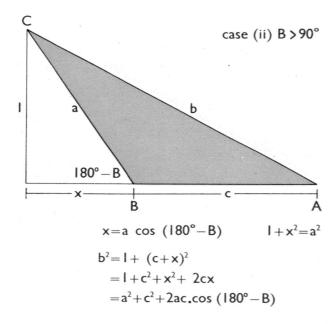

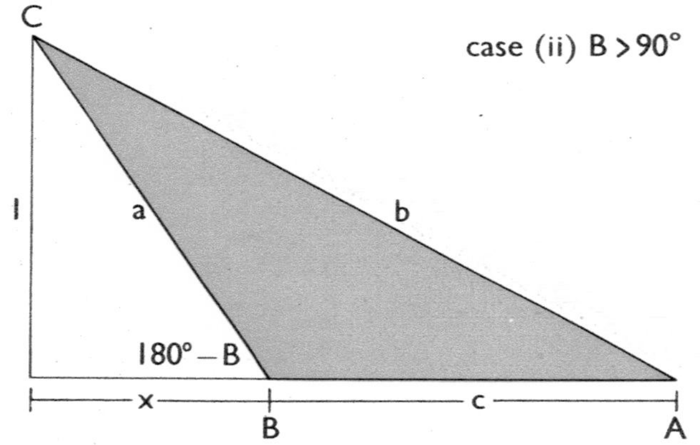

$$x = a \cos (180° - B) \qquad l + x^2 = a^2$$

$$\begin{aligned} b^2 &= l + (c+x)^2 \\ &= l + c^2 + x^2 + 2cx \\ &= a^2 + c^2 + 2ac.\cos (180° - B) \end{aligned}$$

Cosine formula for solution of triangle. (See Claim 4.)

The property last cited is not on all fours with 3(a) and (b) of the foregoing judicial rulings, because the rule and compass recipe for making a triangle, having specified the lengths of its three sides, admits of two possibilities. By joining the apices of the two triangles so constructed on opposite sides of the base line, we can dissect them into equal pairs of right-angled triangles.

Claim 2. In a right-angled triangle, sin A, cos A and tan A do not depend on the length of the sides for fixed A.

Award:

sin A = cos $(90° - A)$ and cos A = sin $(90° - A)$ (from Claim 1)

Claim 3. The area (S) of a triangle is equal to half the product of the length of any one side and that of its perpendicular distance from the opposite apex (*i.e.* $S = \frac{1}{2}pb$).

Award:

(i) in trigonometrical symbols, we may write this in the form
$S = \frac{1}{2}ab.\sin C = \frac{1}{2}ac.\sin B = \frac{1}{2}bc.\ \sin A$

(ii) from the same construction, we derive also

$$\frac{\sin A}{a} = \frac{\sin B}{b} = \frac{\sin C}{c}$$

Claim 4. In a right-angled triangle of which h is the longest side, and the other two are p and b:

$$p^2 + b^2 = h^2$$

This is demonstrable by dissecting the right-angled triangle into two triangles (p. 68) equiangular with it by dropping a perpendicular (P) from the right angle on to the longest side (*hypotenuse*). Since cos 90° = 0 (and sin 90° = 1), the *converse* statement is implicit in an important relation which is true for any triangle. If a is the longest side and:

(i) $A < 90°$ $\quad a^2 = b^2 + c^2 - 2bc.\cos A$

(ii) $A > 90°$ $\quad a^2 = b^2 + c^2 + 2bc.\cos (180° - A)$

Awards:

(i) on dividing throughout by h^2 in the above, it follows that:

$$(\sin A)^2 + (\cos A)^2 = 1$$

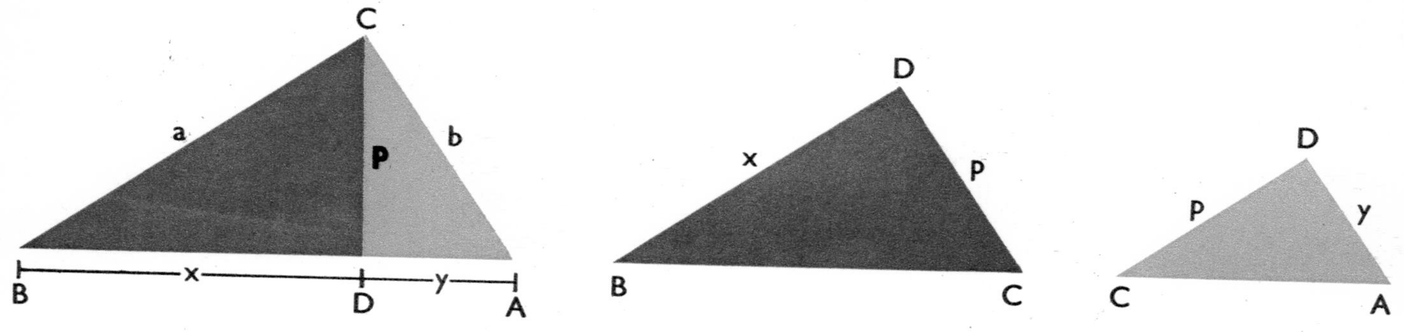

$$\frac{x}{p}=\frac{p}{y} \quad \therefore p^2=xy \quad \therefore p=\sqrt{xy}$$

The length of the perpendicular from right angle to hypotenuse is the geometric mean of the length of the segments into which it divides the hypotenuse. (See Claim 4, Award (ii).)

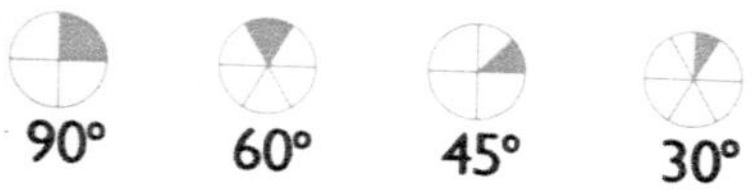

Further steps towards a trigonometrical table. (See Claim 5, Award (ii).)

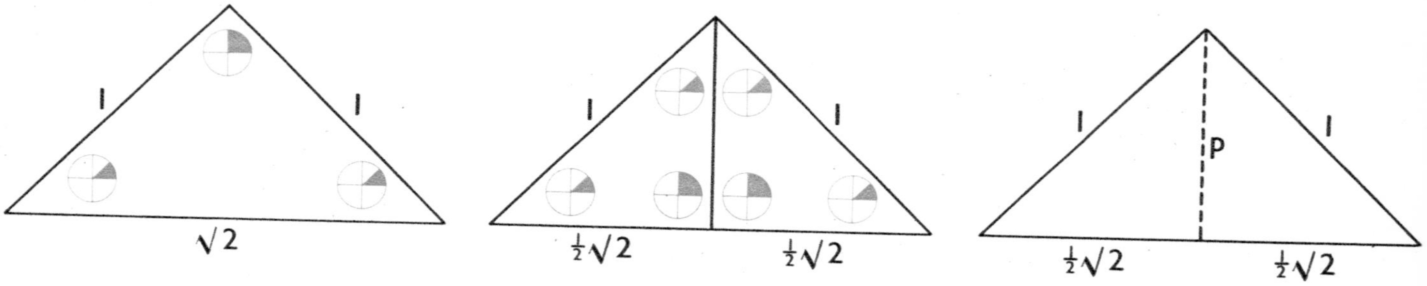

$$p^2=1^2-(\tfrac{1}{2}\sqrt{2})^2=\tfrac{1}{2} \quad \therefore p=\sqrt{\tfrac{1}{2}}$$

$$\sin 45° = \sqrt{\tfrac{1}{2}} \div 1 = \sqrt{\tfrac{1}{2}}; \quad \cos 45° = \tfrac{1}{2}\sqrt{2} \div 1 = \sqrt{\tfrac{1}{2}}; \quad \tan 45° = p \div \tfrac{1}{2}\sqrt{2} = 1$$

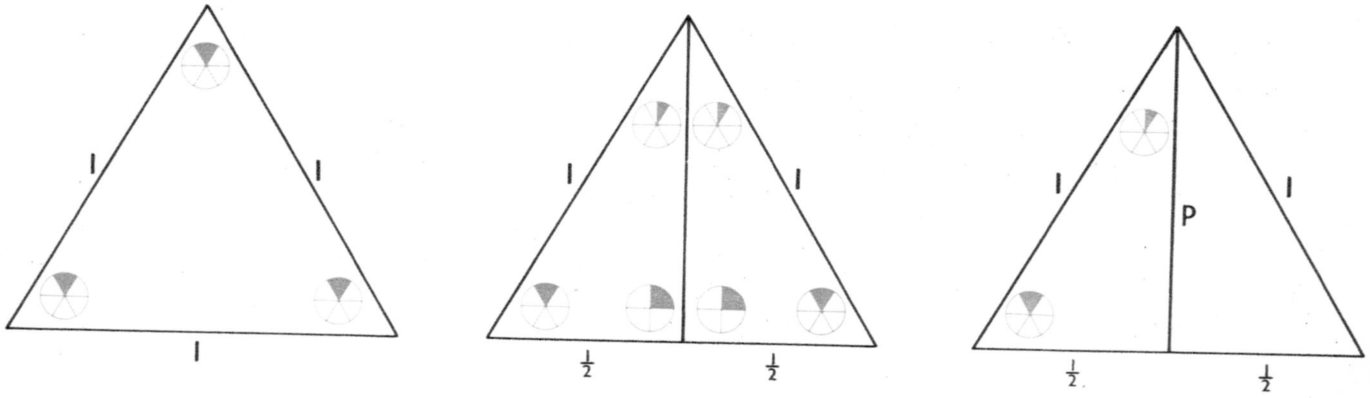

$$p^2=1^2-(\tfrac{1}{2})^2=\tfrac{3}{4} \quad \therefore p=\frac{\sqrt{3}}{2}$$

$$\sin 30° = \tfrac{1}{2} \div 1 = \tfrac{1}{2}; \quad \cos 30° = p \div 1 = \frac{\sqrt{3}}{2}; \quad \tan 30° = \tfrac{1}{2} \div p = \sqrt{\tfrac{1}{3}}$$

$$\sin 60° = p \div 1 = \frac{\sqrt{3}}{2}; \quad \cos 60° = \tfrac{1}{2} \div 1 = \tfrac{1}{2}; \quad \tan 60° = p \div \tfrac{1}{2} = \sqrt{3}$$

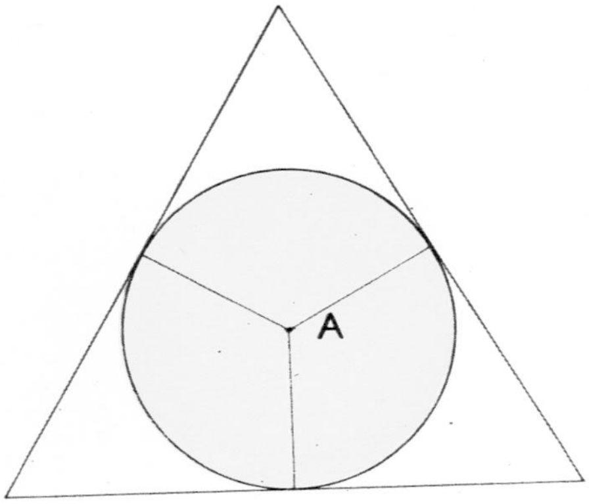

A = 120°
= 360° ÷ 3

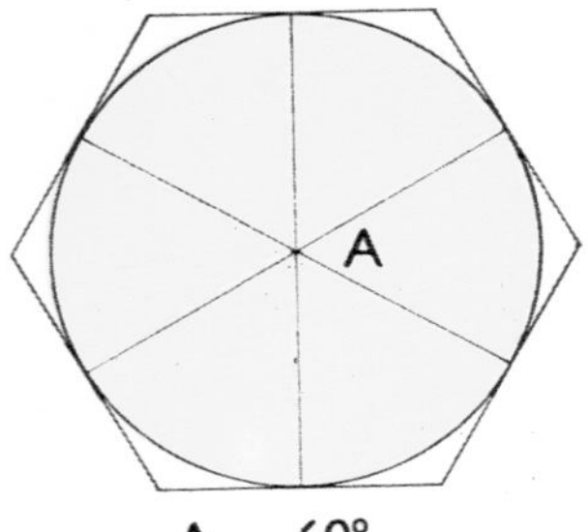

A = 60°
= 360° ÷ 6

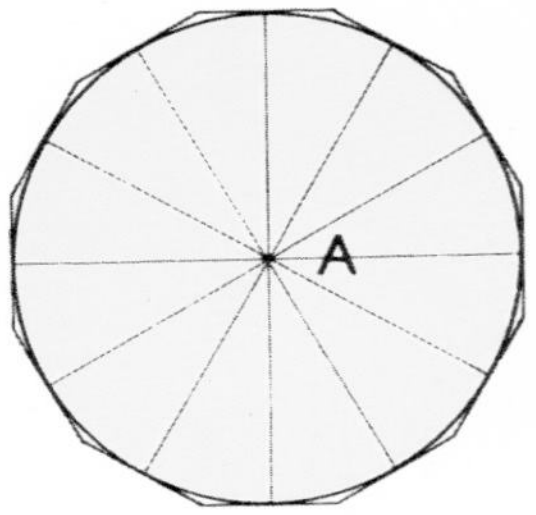

A = 30°
= 360° ÷ 12

whence also from Claim 2:

$$\sin(90^\circ - A) = \sqrt{1-(\sin A)^2}$$

(ii) if P, the perpendicular from the right angle to the hypotenuse, cuts the latter in segments $x+y=h$:

$$P^2 = x.y$$

Claim 5. If two angles of a triangle are equal, the sides opposite them are equal (and conversely).
Awards:
(i) construction for 60°, 45°, and hence by bisection 30°, 15°, $7\frac{1}{2}$°, $3\frac{3}{4}$°, *etc.*, $22\frac{1}{2}$°, $11\frac{1}{4}$°, $5\frac{5}{8}$°, *etc.*, likewise 75° (=30°+45°), $37\frac{1}{2}$°, $18\frac{3}{4}$°, *etc.*
(ii) sin (*etc.*) 60°, 45°, 30° (from Claim 3).
By recourse to (ii) we can henceforth construct a simple trigonometrical table which is easily memorisable in the following form.

	0°	30°	45°	60°	90°
sin	$\left(\frac{0}{4}\right)^{\frac{1}{2}}$	$\left(\frac{1}{4}\right)^{\frac{1}{2}}$	$\left(\frac{2}{4}\right)^{\frac{1}{2}}$	$\left(\frac{3}{4}\right)^{\frac{1}{2}}$	$\left(\frac{4}{4}\right)^{\frac{1}{2}}$
cos	$\left(\frac{4}{4}\right)^{\frac{1}{2}}$	$\left(\frac{3}{4}\right)^{\frac{1}{2}}$	$\left(\frac{2}{4}\right)^{\frac{1}{2}}$	$\left(\frac{1}{4}\right)^{\frac{1}{2}}$	$\left(\frac{0}{4}\right)^{\frac{1}{2}}$
tan	0	$3^{-\frac{1}{2}}$	1	$3^{\frac{1}{2}}$	∞

Here the last entry is the customary symbol meaning limitlessly large (so-called infinity). The surveyor can use this table to measure the height of a mountain or the width of a river by choosing a base at the extremities of which his theodolite sights a selected point at different angles (*e.g.* 30° and 60°).

Claim 6. In a semi-circle, the lines joining a point on the boundary to the ends of the diameter are at right angles.
Award:
We obtain a construction for making a line whose square is equal to the rectangle whose sides are the segments of another. From Award (ii) of Claim 3, P is the perpendicular from the right angle dividing the hypotenuse into segments x and y. If x and y are segments of a line a, bisect the line and describe a semi-circle of which it is the diameter. The required line of length P, such that $P^2 = xy$, is the perpendicular from the point dividing a into x and y to the periphery of the semi-circle. This result is later necessary to show how we can use the curve known as the cissoid to double the cube, and it provided the mathematicians of the eighteenth century of our era with a basis for a geometrical interpretation of $i^2 = -1$.

Claim 7. The perpendicular from the centre to a chord of a circle bisects it, and bisects the arc it subtends.

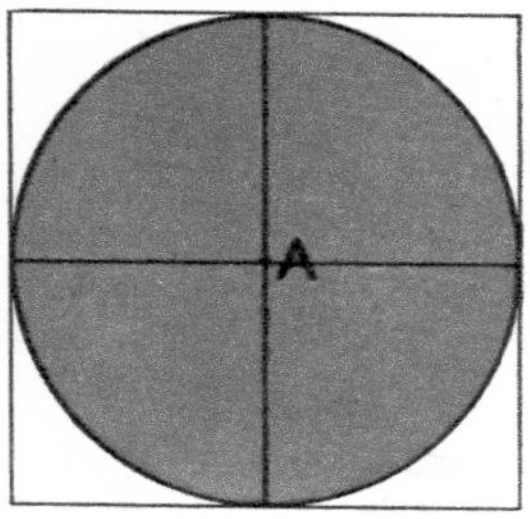

A = 90°
= 360° ÷ 4

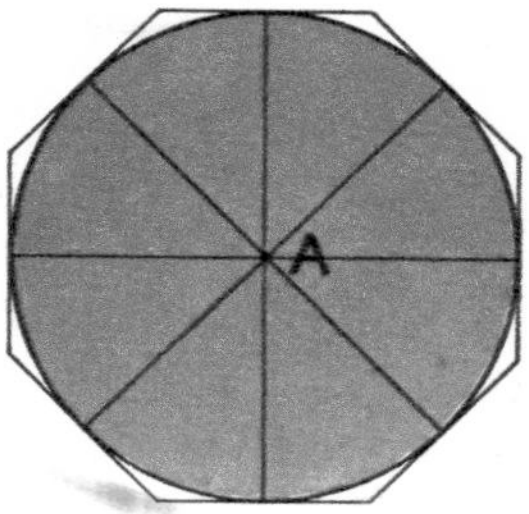

A = 45°
= 360° ÷ 8

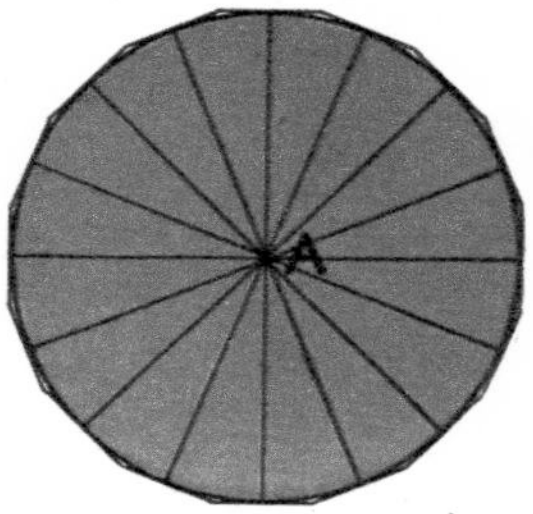

A = $22\frac{1}{2}$°
= 360° ÷ 16

Awards:

(i) the plumb-line points to the centre of the earth, being at right angles to the horizon plane;

(ii) by successively joining the tangents to the points where radii of a circle incline at an angle of (360° ÷ *N*) it is possible to circumscribe a polygon of *N* equal sides about a circle, whence to construct escribed polygons of 3, 6, 12, *etc.*, and 4, 8, 16, *etc.*, equal sides;

(iii) given the earth's radius ($r \simeq 4000$ miles) we can calculate the distance at which a mountain becomes visible at sea level, or conversely the distance of the horizon at sea level from the mountain top.

From (i) and Claim 1, we derive the simplest methods for determining our latitude approximately from the altitude of the Pole Star or that of the Equinoctial noon sun. We can also define the angular position (*declination*) of any star at transit (highest altitude) relative to the plane of the equator, when we have fixed the exact altitude of the celestial pole as the mean of two readings for the Pole Star separated by a 12-hour interval. Having tabulated the declination of stars, or of the sun throughout the seasons, we can use the altitude at transit of any star or the altitude of the noon sun on any day to determine our terrestrial latitude exactly.

Claim 8. The angle subtended by a half chord perpendicular to the diameter is twice as great as

By joining the tangents to the points where radii of a circle incline at an angle of (360° ÷ N), we can circumscribe a polygon of N equal sides about a circle. We can thus construct escribed polygons of 3, 6, 12, etc. and 4, 8, 16, etc., equal sides. (See Claim 7, Award (ii).)

An ornate but simple theodolite of 1586. Of itself the theodolite is merely an instrument for measuring vertical and horizontal angles. Its value in surveying depends on the application of claims and awards established in the Great Litigation.

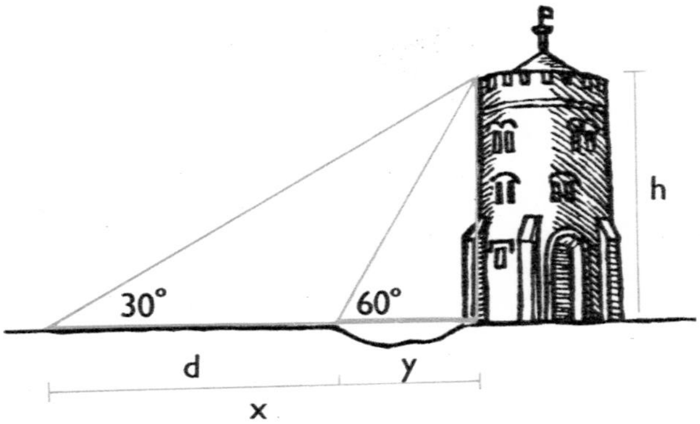

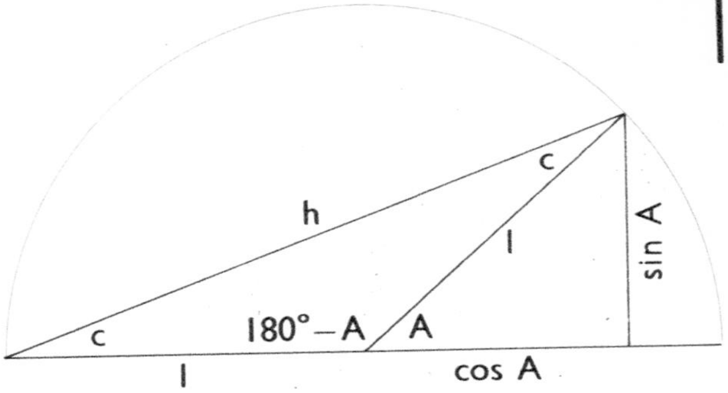

$$2c + 180° - A = 180° \quad \therefore 2c - A = O \quad c = \tfrac{1}{2}A$$

Finding the height of a tower surrounded by a moat.
We can measure $d = (x - y)$, *but not* x *or* y. *From table of tangents we know that* $\frac{h}{x} = \frac{1}{\sqrt{3}}$, *or* $h \cdot \sqrt{3} = x$.

We also know that $\frac{h}{y} = \sqrt{3}$, *or* $y = \frac{h}{\sqrt{3}}$.

Since $x - d = y$, $h \cdot \sqrt{3} - d = \frac{h}{\sqrt{3}}$.

By multiplying both sides of equation by $\sqrt{3}$ *we get*

$$3h - d\sqrt{3} = h$$
$$2h = d\sqrt{3}$$
$$h = \frac{\sqrt{3}}{2} \cdot d$$

Since we can therefore measure d, *we can calculate* h.

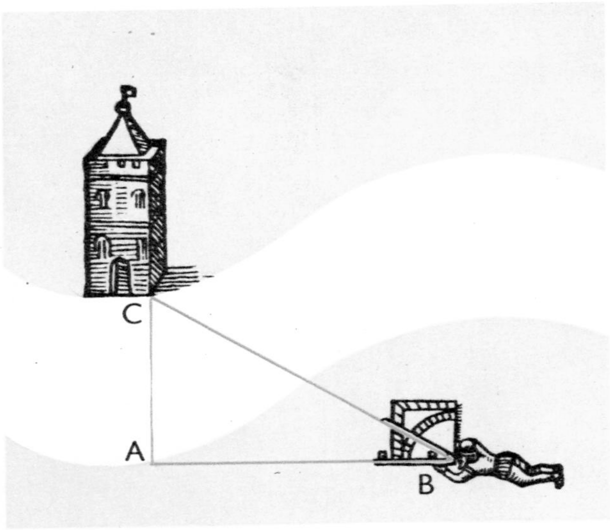

Measuring the width of a stream dry-shod.
Select a point A, *opposite a landmark* C *on far bank.*
Peg out a base-line at right angles to AC.
Find the point B *on this line from which you can sight* C *at exactly* 30° *to* AB. *Measure* AB.
$AC = AB \tan 30° = AB \div \sqrt{3}$.

The angle subtended by a half chord perpendicular to the diameter is twice as great as the angle which the corresponding arc subtends at the diameter's opposite extremity.

AWARD:

$$\cos \tfrac{1}{2}A = \frac{1 + \cos A}{h}$$

$$h^2 = (\sin A)^2 + (\cos A + 1)^2 = (\sin A)^2 + (\cos A)^2 + 2\cos A + 1 = 2 + 2\cos A$$

$$h = \sqrt{2(1 + \cos A)}$$

$$\therefore \cos \tfrac{1}{2}A = \frac{1 + \cos A}{\sqrt{2(1 + \cos A)}} = \sqrt{\frac{1 + (\cos A)^2}{2}}$$

$$\sin \tfrac{1}{2}A = \frac{\sin A}{\sqrt{2(1 + \cos A)}} = \frac{\sqrt{1 - (\cos A)^2}}{\sqrt{2(1 + \cos A)}}$$

$$= \sqrt{\frac{1 - \cos A}{2}}$$

Whence if $\tfrac{1}{2}A = B$

$$2(\cos B)^2 = 1 + \cos 2B$$
$$\cos 2B = 1 - 2(\cos B)^2$$

Right: Decorative frieze from a sixteenth-century German instrument, showing some current surveying problems.

the angle which the corresponding arc subtends at the diameter's opposite extremity.

Awards:

(i) we derive from this

$$2\left(\sin \frac{A}{2}\right)^2 = 1 - \cos A$$

and

$$2\left(\cos \frac{A}{2}\right)^2 = 1 + \cos A$$

(ii) whence also from Award (i) of Claim 3

$$4\left(\sin \frac{A}{2}\right)^2 \left(\cos \frac{A}{2}\right)^2 = 1 - (\cos A)^2 = (\sin A)^2$$

$$\sin A = 2 \sin \frac{A}{2} . \cos \frac{A}{2}$$

By use of the value obtained for 45° and 30° we can now make a table for sines (or cosines) of $22\frac{1}{2}°$, $11\frac{1}{4}°$, $5\frac{5}{8}°$, *etc.*, and 15°, $7\frac{1}{2}°$, $3\frac{3}{4}°$, *etc.* By using the result obtained from Award (i) of Claim 3, we can also obtain values for sin $(90° - 22\frac{1}{2}°) = \sin 67\frac{1}{2}°$, sin $(90° - 15°) = \sin 75°$, *etc.* This, in essence, was the basis of the first table of trigonometrical ratios constructed by Hipparchus (*circa* 150 B.C.).

Claim 9. The areas of polygons with the same numbers (2^n or 3.2^n) of equal sides are proportional to the squares of the radii of the circles which they circumscribe and to the squares of the radii of the circles which circumscribe them. We can make a polygon of N or $2N$ equal sides, if we can divide 360° by N, since we can build up the figure by lines joining the centre to the boundary of a circle from N or $2N$ triangles each having two sides (radii) and an included angle or one side (radius) and the two angles at its extremities equal. The qualification that N must be of the form 2^n (4, 8, 16, 32, 64, 128 . . .) or 3.2^n (3, 6, 12, 24, 48, 96 . . .) is a concession to rule and compass etiquette. From Claim 3, we see that

(i) the area of a polygon of $2N$ equal sides inscribed in a circle of radius R is

$$NR^2 \sin \frac{360°}{2N}$$

(ii) the area of a polygon of N equal sides circumscribing a circle of radius R is

$$NR^2 \tan \frac{360°}{2N}$$

In the foregoing expressions, $N \sin \frac{360°}{2N}$ depends only on N, and the same is true of $N \tan \frac{360°}{2N}$. Hence the areas of two regular polygons having the same number of sides are in the same ratio as the squares of the radii of their inscribed or circumscribed circles.

Awards:

(i) The area (A) of a circle lies between that of the circumscribed polygon of N equal sides and that of the inscribed polygon of $2N$ equal sides, being greater than the latter:

$$NR^2 \tan \frac{360°}{2N} > A > NR^2 \sin \frac{360°}{2N}$$

(ii) as N becomes larger, these boundaries approach one another more and more closely, so that we may regard A as proportional to

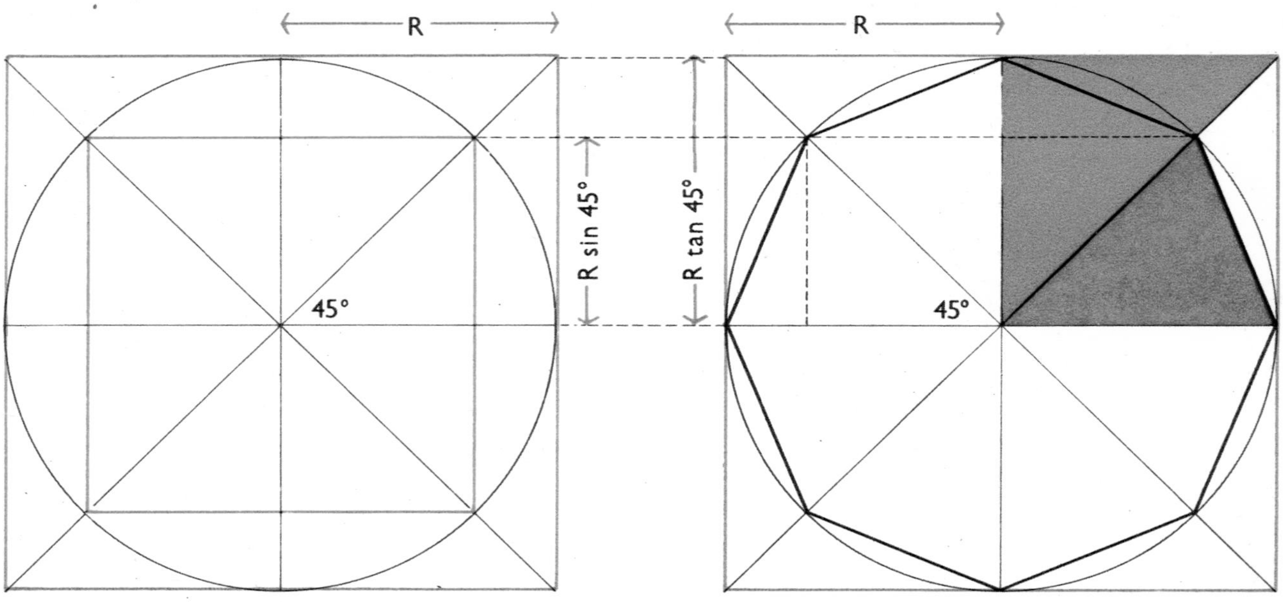

perimeter of escribed square $= 8\ R \tan 45°$

$= 8\ R \tan \frac{360°}{8}$

perimeter of inscribed square $= 8\ R \sin 45°$

$= 8\ R \sin \frac{360°}{8}$

area of escribed square $= 8\ (\frac{1}{2}\ R \times R \tan 45°)$

$= 4\ R^2 \tan \frac{360°}{8}$

area of inscribed octagon $= 8\ (\frac{1}{2}\ R \times R \sin 45°)$

$= 4\ R^2 \sin \frac{360°}{8}$

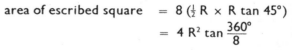

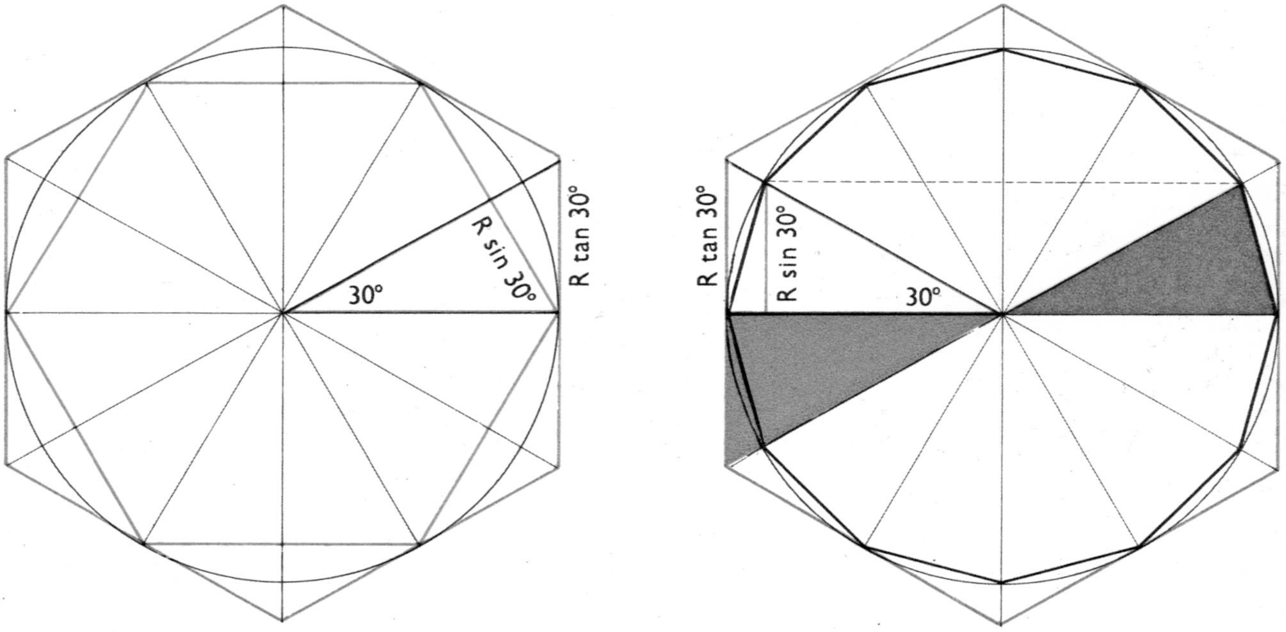

perimeter of escribed hexagon $= 12\ R \tan 30°$

$= 12\ R \tan \frac{360°}{12}$

perimeter of inscribed hexagon $= 12\ R \sin 30°$

$= 12\ R \sin \frac{360°}{12}$

area of escribed hexagon $= 12\ (\frac{1}{2} R \times R \tan 30°)$

$= 6\ R^2 \tan \frac{360°}{12}$

area of inscribed dodecagon $= 12\ (\frac{1}{2}\ R \times R \sin 30°)$

$= 6\ R^2 \sin \frac{360°}{12}$

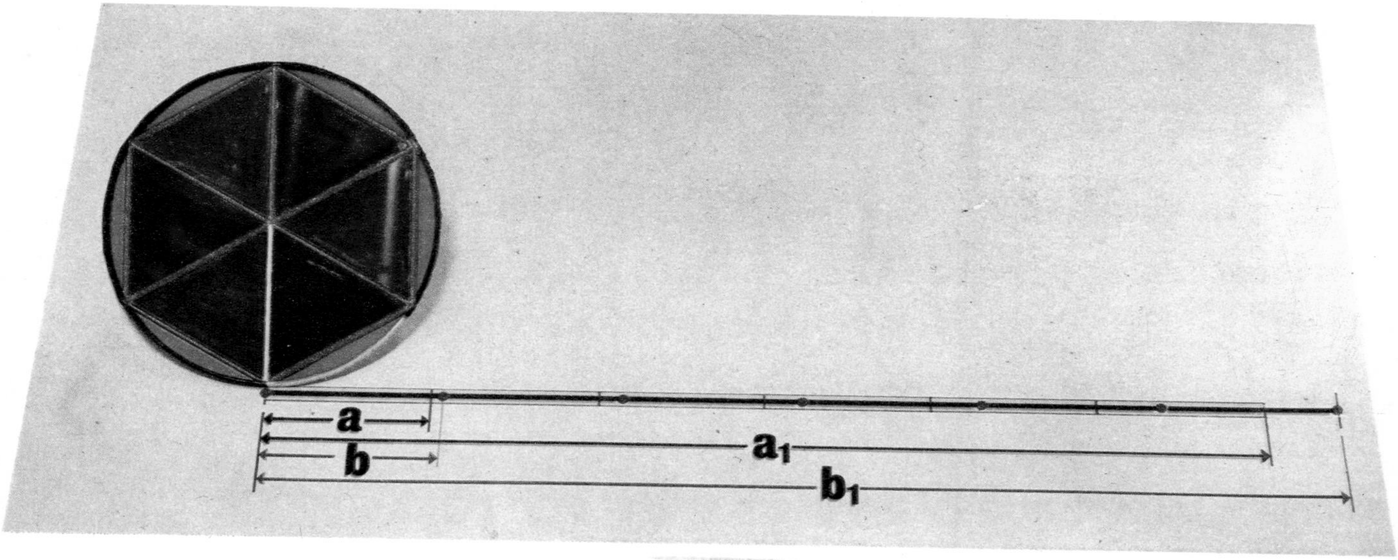

Left: (*Claim 9.*) *Areas of polygons with the same numbers* (2^n *or* 3.2^n) *of equal sides are proportional to the squares of the radii* (R) *of the circles which they circumscribe and to the squares of the radii of the circles which circumscribe them, being:*

inscribed $\frac{1}{2}N \,.\, R^2 \,.\, \sin \frac{360°}{N}$

escribed $N \,.\, R^2 \,.\, \tan \frac{360°}{2N}$

Left: (*Claim 10, p. 112.*) *Perimeters of inscribed and escribed polygons of* N ($= 2^n$ *or* 3.2^n) *equal sides are proportional to the diameters* ($D = 2R$) *of circles respectively enclosed by or enclosing them, being:*

inscribed $N \,.\, D \,.\, \sin \frac{360°}{2N}$

escribed $N \,.\, D \,.\, \tan \frac{360°}{2N}$

The advantage of measuring angles in radians is that the angular measure of a circular arc tells us the length of the arc in linear units. If r *is the radius of the circle and* A *is the angle in radians,* $r.A$ *(the length of the arc which subtends the angle* A *at the centre) is in the same linear units as* r*. In the diagram above,* a = *length of radius,* b = *length of arc subtending 60° at centre;* a_1 = *length of periphery of hexagon,* b_1 = *length of circumference of circle.*

THE TRIAD OF MENAECHMUS

In each case plane cuts cone at right angles to edge.

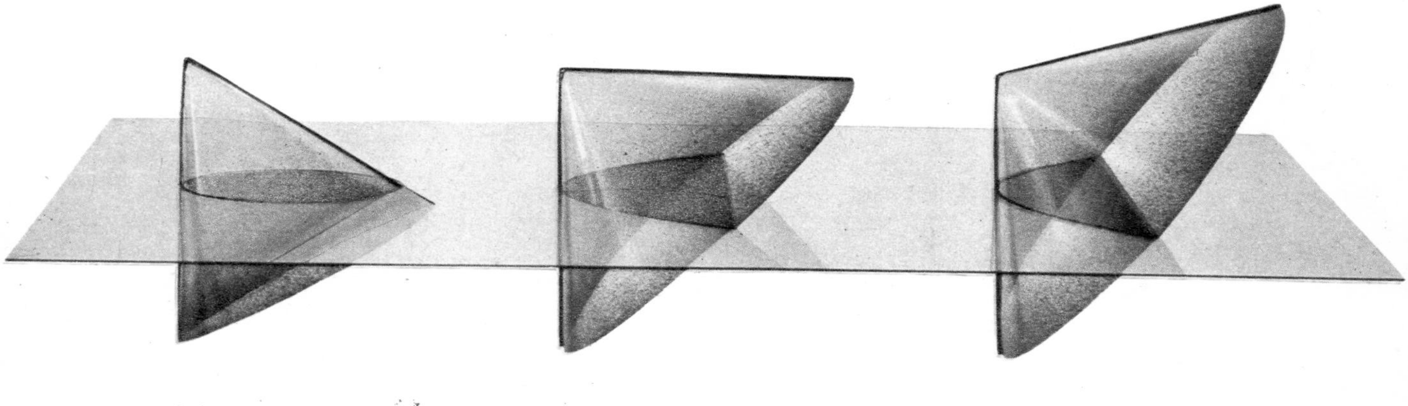

Apical angle $< 90°$
ELLIPSE
(*Orbit of planet, electron, sputnik.*)

Apical angle $= 90°$
PARABOLA
(*Trajectory of cannon ball in empty space.*)

Apical angle $> 90°$
HYPERBOLA
(*Trajectory of alpha particle near atomic nucleus.*)

Alexandrian coin of second century A.D., *depicting the famous Pharos at Alexandria.*

Reconstruction of town-plan of Alexandria, showing grid street-layout and probable sites of main buildings.

R^2 and express the area of a circle as the product of an undetermined fixed number π and the square of the radius (*i.e.* $A = \pi R^2$).

(iii) if we put $R=1$, we therefore derive

$$N \tan \frac{360^\circ}{2N} > \pi > N \sin \frac{360^\circ}{2N}$$

By recourse to the half-angle formulae (Claim 8) we can now refine the boundaries of π successively as below:

N	$N \sin \frac{360^\circ}{2N}$	$N \tan \frac{360^\circ}{2N}$	Mean
3	2.5981	5.1962	3.8972
4	2.8284	4.0000	3.4142
6	3.0000	3.4641	3.2321
8	3.0614	3.3137	3.1876
12	3.1058	3.2154	3.1606
16	3.1214	3.1827	3.1521
24	3.1327	3.1596	3.1462
32	3.1365	3.1517	3.1441
48	3.1393	3.1461	3.1427
64	3.1403	3.1441	3.1422
96	3.1410	3.1427	3.1419
128	3.1413	3.1422	3.1417

In this way, we eventually reach a mean correct to six significant figures $\pi \simeq 3.14159$.

Claim 10. The perimeters of the inscribed and escribed polygons of $N(=2^n \text{ or } 3.2^n)$ equal sides are proportional to the diameters of the circles respectively enclosed by or enclosing them, being:

$$\textit{inscribed } N.D \sin \frac{360^\circ}{2N}$$

$$\textit{escribed } N.D \tan \frac{360^\circ}{2N}$$

Awards:

(i) we may accordingly bound the perimeter (P) of a circle thus:

$$N.D \sin \frac{360^\circ}{2N} < P < N.D \tan \frac{360^\circ}{2N}$$

(ii) from this, we see that the limiting form is $P = \pi.D = 2\pi R$;

Carving showing Roman ships nearing a lighthouse.

(iii) since the lengths of arcs of the same circle are in the same ratio as the angles they subtend at the centre, we can adopt as a convenient unit of angular measurement the angle subtended at the centre of a circle of unit radius by an arc of unit length. Since the perimeter of such a circle is 2π units of length, 2π units (called *radians*) correspond to 360°, whence one radian corresponds to $(360 \div 2\pi)^\circ \simeq 57\frac{1}{3}^\circ$, and 1 degree to $(\pi \div 180) \simeq 0.0175$ radians. It is useful to memorise:

$$\left(\frac{3\pi}{2}\right)^R = 270^\circ \qquad \pi^R = 180^\circ \qquad \left(\frac{\pi}{2}\right)^R = 90^\circ$$

$$\left(\frac{\pi}{3}\right)^R = 60^\circ \qquad \left(\frac{\pi}{4}\right)^R = 45^\circ \qquad \left(\frac{\pi}{6}\right)^R = 30^\circ$$

(iv) in this system of measurement, the angle a is numerically equal to the length a of the arc which subtends it at the centre of a circle of unit radius, so that if $a \leqslant \frac{\pi}{2}$

$$\sin a < a < \tan a$$

$$1 < \frac{a}{\sin a} < \frac{1}{\cos a}$$

Since $\cos a$ approaches 1 as a approaches 0, the ratio $a : \sin a$ becomes nearer to unity as a approaches zero. We customarily write this important limiting ratio in the form

$$\underset{a=0}{Lt} \frac{\sin a}{a} = 1$$

In fact, if $a \leqslant 0.03316$ which corresponds to $A^\circ \leqslant 1\frac{7}{8}$, $a = \sin a$ correct to 5 decimal places. To derive many formulae used in optics, mechanics and electrostatics, one makes the substitution $\sin a = a$ for small values of a. It is therefore important to be clear about what we mean by *small.*

In fact 1 degree is small in the sense that it is less than a fiftieth of a radian. The following figures are therefore instructive.

Note on Radian Measure

Degree	*Sine*	*Radian*	*Tangent*
1	0.01745	0.01745	0.01746
5	0.08716	0.08727	0.08749
10	0.17365	0.17453	0.17633
15	0.25882	0.26180	0.26795
30	0.50000	0.52360	0.57735
45	0.70711	0.78540	1.00000

Claims 10 and 9 lead to identical formulae bounding π for fixed N. Archimedes obtained (*circa* 230 B.C.) a value of π based on the mean figure for polygons of 96 sides. Ptolemy (*circa* 150 A.D.) was content with a value equivalent in our notation to 3.1416. Independently by the same method (*circa* A.D. 250), as Needham tells us, Liu Hui, a Chinese mathematician, obtained the more accurate figure 3.14159 based on the polygon of 3072 sides. About A.D. 480, Tsu Chhung-Chih bounded the limits even more sensitively as $3.1415926 < \pi < 3.1415927$.

The foregoing statement of claims subsumes what is still of relevance to modern needs, though the *Elements* alone do not embrace the whole content of Greek geometry up to 300 B.C. Euclid himself wrote a lost treatise of four books on Conic Sections, the earliest treatment of which seems to be (p. 111) that of Menaechmus (*circa* 350 B.C.). The latter recognised the three curves known as the parabola, the ellipse and the hyperbola as figures traced respectively on the surface of a right-angled, an acute-angled and an obtuse-angled right cone in a plane perpendicular to the edge. Appolonius, born in Perga about

260 B.C. but educated in Alexandria, finalised the Euclidean programme by composing a treatise of eight books on Conic Sections embodying 387 propositions, and was seemingly the first geometer to demonstrate the construction of all the three curves last named as sections of one and the same cone, right or oblique. His work is a *tour de force* of intellectual ingenuity in the Euclidean tradition, and continued to influence the work of mathematicians such as Newton after Cartesian geometry had made it possible to deal with the same topics in a vastly less laborious way.

The completion of the *Elements* signalises the end of an intellectual epoch in the setting for a hitherto unique efflorescence of the human intellect. It seems that Euclid was about thirty-five years of age when Alexander conceived, in the winter of the Egyptian campaign (331 B.C.), the project of founding the city of Alexandria to commemorate his victories, and entrusted its execution to Cleomenes, his Collector of Revenues. Alexander himself died in Babylon (323 B.C.) before he had occasion to inspect the plans eventually completed under the supervision of Demetrios of Phaleron, a pupil of the great Greek botanist Theophrastus, when one of his generals, Ptolemy Soter, became Satrap of Egypt. If Euclid, who died in Alexandria, settled there soon after its foundation, he was well over forty years at the time. Since he did not die long before the abdication of the founder of the Ptolemaic dynasty, and may have died after him, he must have lived to see the foundation of the Museum and Library. We may infer that he was one of its earliest teachers; but we may doubt whether he survived to appreciate the New Look which Greek geometry would assume in a social *milieu* so novel as the city of his adoption.

With Egyptian, Greek and Jewish quarters at its inception, Alexandria was uniquely cosmopolitan. It was in close contact with the ancient temple culture of the Nile and, at least from 250 B.C. till the Roman occupation, in lively contact with the Greek-speaking colonies founded by the Seleucids

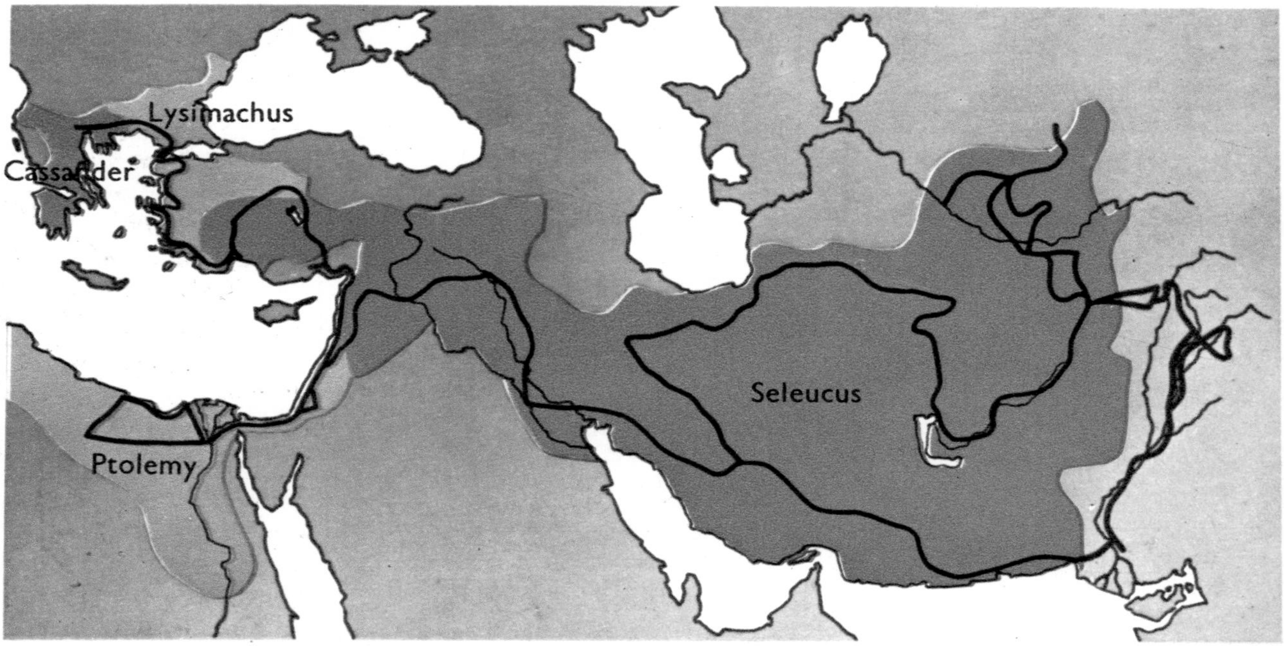

The routes which Alexander followed during his main campaigns and the partition of the Empire he built.

among the great temple sites of Mesopotamia. During the first six hundred years after its foundation, it could claim that every noteworthy man of science in the Western World had studied or taught there. If only because of the unrivalled opportunities for study and publication it could offer to all the literate communities of the West, its foundation was the prelude to hitherto incomparable circumstances for pooling, consolidating and stimulating intellectual endeavour, unhampered by priestcraft or political censorship.

Though he was a soldier by upbringing, it was the ambition of the founder of a dynasty which terminated with the death of Cleopatra to gather in his capital scholars of countries under his ample patronage. The home of the Muses was far more than a museum in the contemporary sense of the term. Its refectories, living quarters and lecture rooms foreshadow the University of the Middle Ages. They provided subsistence, a place of residence and opportunities for teaching to scholars of mathematics, medicine, mechanics, astronomy, geography, history, grammar, philosophy and literary criticism in close propinquity to a library the like of which no previous civilisation had seen. Eventually, the latter seems to have assembled as many as three quarters of a million volumes transcribed on scrolls of parchment or papyrus, presumably for the most part in the Greek language. At a time when Greek was the *lingua franca* of Western culture, the contents harvested much of the intellectual achievement of the temple libraries of remoter antiquity in a script accessible to the scrutiny of scholars of all countries in the Western World.

Thus scholarship in so novel and so challenging a setting had at its command instruments of human communication vastly more efficient than any society had hitherto been able to exploit, either in the West or in the East. It had also day-to-day contact with the world's work in a *milieu* unusually propitious to invention and to the application of theoretical discoveries. To Alexandria, as Morris Kline (*Mathematics in*

Mosaic from Pompeii showing Alexander and Darius at Battle of Issus.

Ptolemaic figurine showing an Archimedean screw being worked treadmill-fashion. The device consists of a long helical flange free to revolve inside a close-fitting open cylinder, one end of which is placed below the surface of the water.

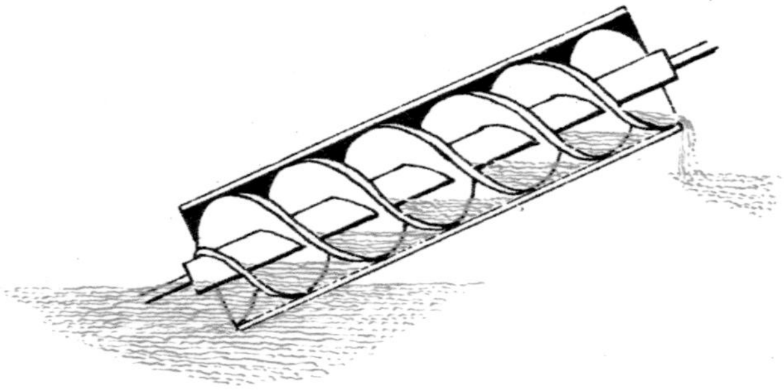

Western Culture) remarks, "came traders and businessmen from all corners of the world. In the harbour were ships that brought wines from Italy, tin from Wales and amber from Sweden. The Alexandrian traders not only spread Greek culture over the world, but brought back to Alexandria knowledge that had been acquired in other countries." In short, means, motive and opportunity conspired to further enquiry, especially in astronomy and geography. Doubtless, the seemingly incredible speed of Alexander's armies whose routes took within their scope the entire Persian Empire embracing Phoenicia, the south shores of the Caspian, Assyria, Iraq, Iran and the province of Sogdiana with its capital Samarkand, now in the U.S.S.R., before they finally crossed the Indus and overran what we now call the Punjab, whetted an appetite for geographical knowledge and opened a window facing Eastward. What top secrets of military surveying contributed to common knowledge we do not know; but some of them would be the talk of the bazaar in the Greek quarter of Alexandria and in the Macedonian colonies of the further East. Accordingly we may be sure that the scholar of Alexandria in 250 B.C. knew far more about the overland trade routes which linked China to the West than did the contemporaries of Euclid's early manhood.

Pari passu with the growing prosperity of the city, the Ptolemies devoted lavish funds to public works of irrigation, harbour construction and architecture, including the erection of the redoubtable Pharos (*lighthouse*). The scale of such undertakings invoked new principles of engineering to accomplish what the Pyramid architects could do with a vast labour force of slaves no longer available in a community predominantly made up of free men. Accounts of some inventions of the period have the flavour of fable, but perhaps only because no illustrated treatises on mechanics were current beyond the confines of China before printing began in the West. Certainly, we have no reason to doubt the construction of water-driven clocks with an ingenuity unthinkable in the context of Plato's time. Such, then, was the social scene at the time when Archimedes, born in the Sicilian Greek kingdom of Syracuse a few years before the putative date of Euclid's death, spent several years of study in the home of the Muses among the scrolls of the Library. In the remainder of this chapter, we shall concentrate on some of the new beginnings in Alexandria during the two centuries following Euclid's death.

Foremost among his successors is Archimedes (287–212 B.C.) himself, the author of the first recorded treatises on mechanical principles. Aside from the formulation of the principle of equilibrium on the inclined plane by Leonardo da Vinci (*circa* A.D. 1500) and Stevinus (*circa* A.D. 1590), his two master principles, that of the lever and that of buoyancy, subsumed all the essentials of mechanics known before the work of Galileo. He applied his knowledge to the design of catapults for the defence of his native Syracuse against the Romans, and is also noteworthy as the inventor both of a water-powered planetarium and of a pump commonly known as the Archimedean screw. Throughout all his work, Archimedes uses geometry as a tool for mensuration, as when he derives the area ($4\pi r^2$) and volume ($\frac{4}{3}\pi r^3$) of a sphere, or when he bounds the value of π ($3\frac{10}{71} < \pi < 3\frac{1}{7}$) by determining the areas of the 96-sided regular polygons enclosed by and enclosing a circle. A break with the Euclidean tradition is most spectacular when, with a profane audacity from the Platonic viewpoint, Archimedes applies the principle of the lever to determine the area enclosed by lines cutting off a segment of a parabola. In the course of the argument which leads to the last-named recipe, he gives the sum of the geometric series

$$\frac{4}{3} = \frac{1}{4^0} + \frac{1}{4^1} + \frac{1}{4^2} + \frac{1}{4^3} \cdots$$

This seems to be the earliest recorded example of a convergent series, and its author's eminently practical attitude to the problems of computation is manifest in a treatise named *ψαμμιτης* (*Sand Reckoner*). Here he shows that he appreciates the rule $10^a . 10^b = 10^{a+b}$; and he makes a plea for a

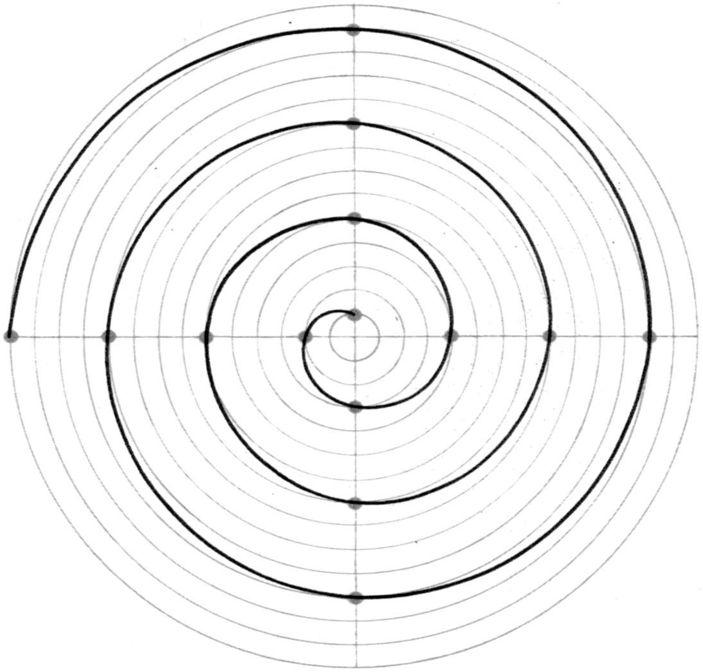

THE ARCHIMEDEAN SPIRAL
Archimedes' geometry took time *into account. Here a line rotates about a fixed centre at fixed angular speed. A point moves forward at fixed linear speed from the centre. The spiral is the trace of its path. To construct it, draw concentric circles with radii increasing by equal increments, mark off with dots equal multiples of the same angle (here 90°) of revolution on successive circles, and join dots evenly.*

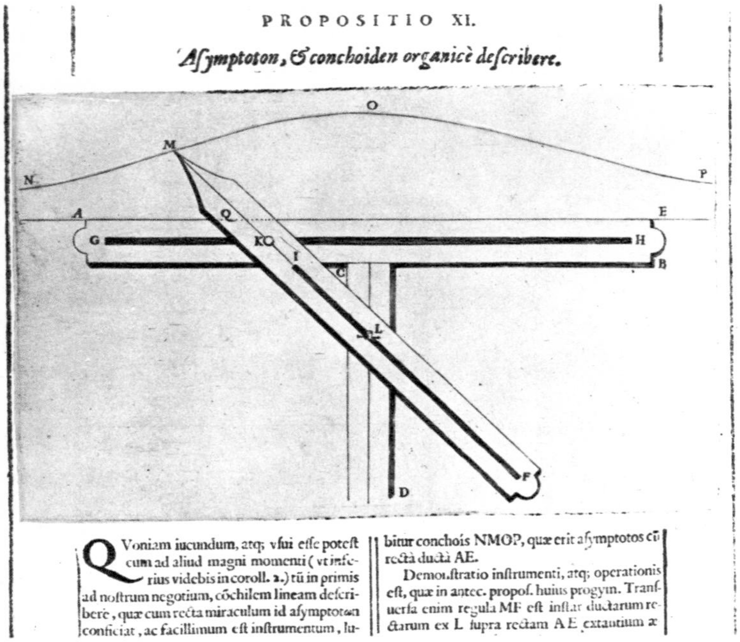
PROPOSITIO XI.

Asymptoton, & conchoiden organicè describere.

QVoniam iucundum, atq; vsui esse potest cum ad aliud magni momenti (vt inferius videbis in coroll. 2.) tũ in primis ad nostrum negotium, cõchilem lineam describere, quæ cum recta miraculum id asymptoton conficiat, ac facillimum est instrumentum, lubitur conchois NMOP, quæ erit asymptotos cũ rectà ductà AE.

Demonstratio instrumenti, atq; operationis est, quæ in antec. propos. huius progym. Transuersa enim regula MF est instar ductarum rectarum ex L supra rectam AE extantium æ

A mechanical device used for generating the curve known as the conchoid, from Bettini's Apiaria Universae Philosophiae Mathematicae *of 1641. In defiance of Platonic tradition, Nicomedes of Alexandria first employed such a device c. 180* B.C.

THE CONCHOID OF NICOMEDES

The line OBN rotates right or left about 0, cutting the line CD. The curve is the trace of all the points on the rotating line at the same distance NB = d from the point where it cuts CD. We call the distance of a point on the curve from O its radius vector (r). From the figure we see:

$r = PQ + QO = d + QO$ *and* $OB = QO \cos A$.

If $OB = c$, *the polar equation which describes the curve is:* $r = d + \dfrac{c}{\cos A}$.

radical reform of the Greek numeral system. Needless to say, the plea fell on deaf ears. Clever people cannot change the means of everyday human communication unless they can carry with them a large body of well-informed public opinion.

That the inventor of the Archimedean screw also left behind him a treatise on the spiral registers a revolt against the Platonic tradition in more than one way. Here we see indisputably a close unity between theory and its applications; and we encounter an idiom entirely foreign to the timeless domain of Euclid. With flagrant disregard for Platonic proprieties, Archimedes expounds his theme in terms of a point moving at fixed linear speed along a line which revolves about a fixed centre at fixed angular speed. Needless to say, no finite number of rule and compass manipulations can trace its course.

Half a century later, another Alexandrian invented a mechanical device for generating a curve the use of which makes it possible to solve each of the three major problems for which the Euclidean rule and compass regimen can offer no recipe. The so-called *conchoid* of Nicomedes (*circa* 180 B.C.), like the quadratrix of Hippias, makes it possible both to trisect an angle and to construct a line the measure of whose length is the irrational π; but it also makes it possible to double the cube. An easier construction for making two lines equivalent to the sides of cubes whose volumes are in the ratio 2:1 is achievable by recourse to the properties of another curve called the *cissoid*, invented by Diocles a little earlier. To outline its course, one draws lines from one extremity (O) of the diameter of a semi-circle to meet the tangent at its opposite extremity (M). If such a line cuts the semi-circle at Q and the

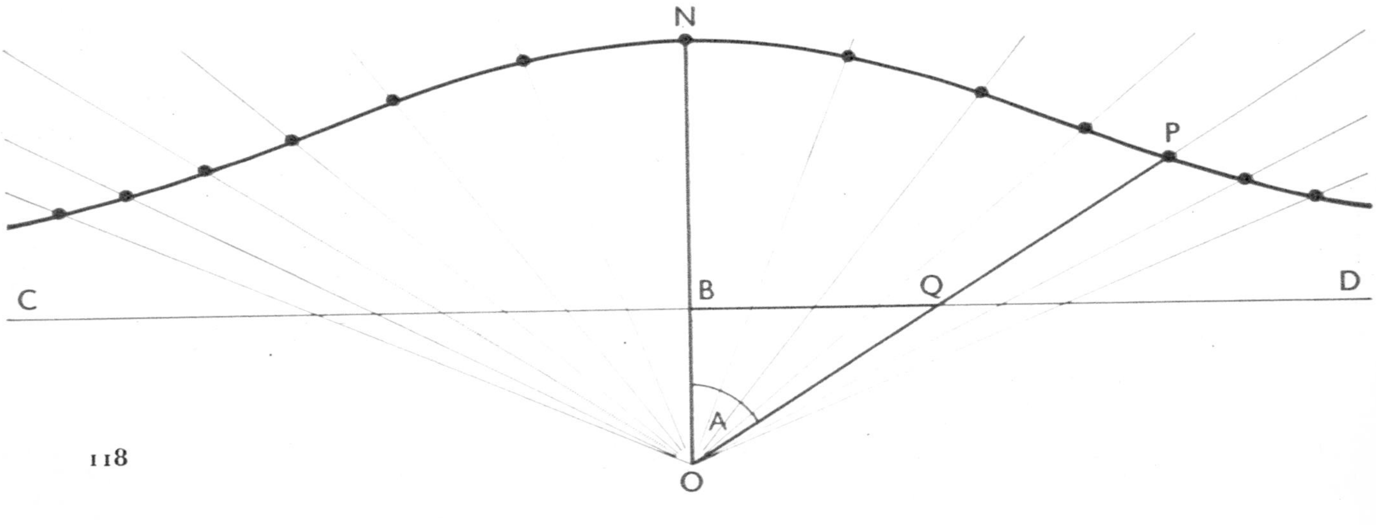

tangent at R, $OP = QR$ is the radius vector of a point P on the curve. This property suffices to define the curve in terms of the length of the radius vector (ϱ) and the angle A which it makes with the base line OM of unit length as $\varrho = 2 \sin A . \tan A$.

The relation of Alexandrian mathematics to the advance of scientific geography and astronomy will occupy another chapter, where we shall see that the impact of the continuing Babylonian tradition was considerable in the time of Hipparchus (*circa* 150 B.C.). Perhaps we may trace to the same Seleucid source the revival of interest in algebraic themes by his contemporary Hypsicles, who wrote extensively on the properties of figurate numbers. What is in one way the most striking example of a break-away from the Euclidean viewpoint is a useful recipe for mensuration due to Heron, an Alexandrian seemingly of Egyptian stock. We do not know with certainty the date of his birth or death. What we know about his remarkable discoveries is at second hand. He may well have made others, of which we have no knowledge.

Almost certainly, Heron lived after the end of the first phase of the Alexandrian culture, likely enough about A.D. 50, before its second efflorescence. However, it is quite possible that he flourished more than a century later in the beginning of the second phase. Like Archimedes, he was also an inventor, one of his achievements being the construction of a rotary steam bellows. Among his contributions to mathematics is a recipe for determining the area of a triangle from the lengths of its sides. We may translate the derivation in the sign language of sines and cosines by recourse to the following relations already cited:

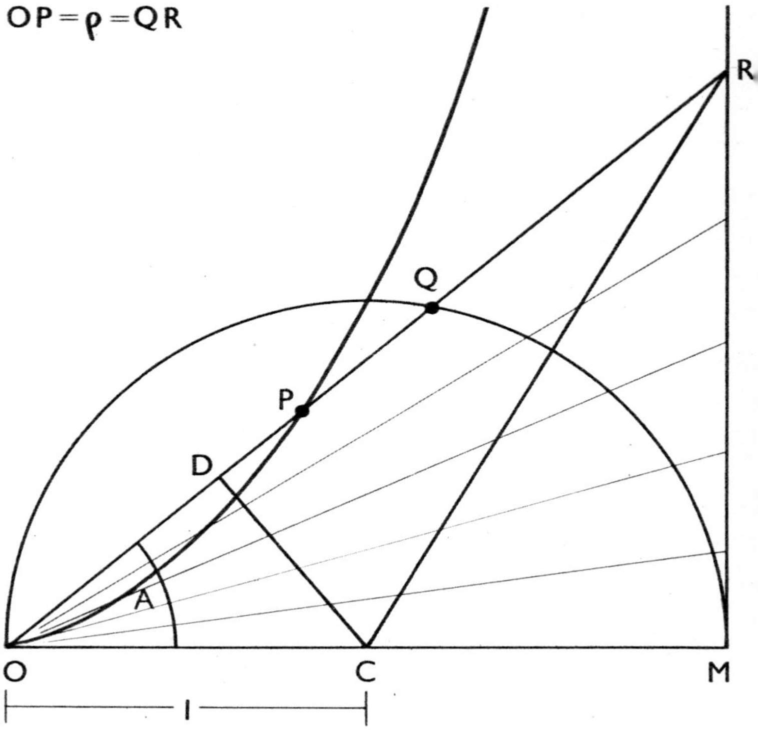

THE CISSOID OF DIOCLES

Here $RM = 2 \tan A$ *and* $RO = \dfrac{2}{\cos A}$

$$RM^2 + 1 = RC^2 = CD^2 + (DQ + \varrho)^2$$
$$= CO^2 + 2\varrho OD + \varrho^2$$
$$\therefore 4(\tan A)^2 + 1 = 1 + \varrho(\varrho + 2OD) = 1 + \varrho . RO$$
$$\therefore 4(\tan A)^2 = \frac{2\varrho}{\cos A} \text{ and } \varrho = 2 \sin A . \tan A$$

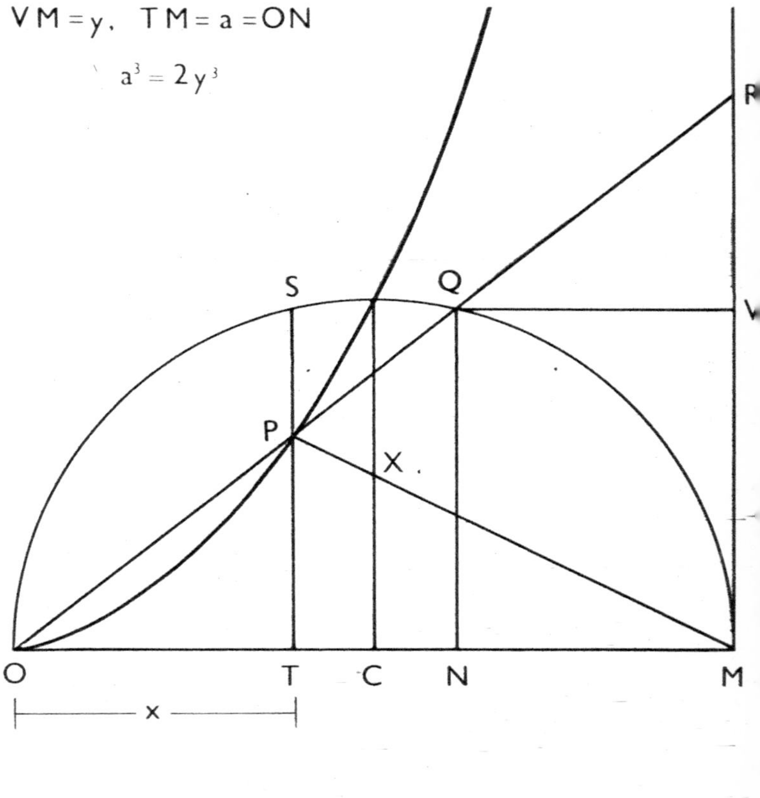

DOUBLING THE CUBE

From the fundamental property of the cissoid
$OP = \varrho = QR$, *whence* $QV = x$ *and* $ST = QN = y$.
The construction makes $CX = \frac{1}{2}CM = \frac{1}{2}r$.
In the right-angled triangle OSM (see Claim 6),
$ST^2 = OT . TM$; *i.e.* $y^2 = ax$

$$\frac{PT}{TM} = \frac{CX}{CM} \text{ so that } \frac{PT}{d} = \frac{1}{2}$$

$$\frac{PT}{OT} = \frac{a}{2x} = \frac{QN}{ON} = \frac{y}{a} \text{ so that}$$

$$a^2 = 2xy \text{ and } a^3 = 2axy = 2y^3$$

$$2bc.\cos A=b^2+c^2-a^2 : 2\left(\sin\frac{A}{2}\right)^2=1-\cos A$$

$$2\left(\cos\frac{A}{2}\right)^2=1+\cos A \text{ and } \sin A=2\sin\frac{A}{2}.\cos\frac{A}{2}$$

If we write s for *half* the sum of the sides of a triangle, so that $2s=(a+b+c)$:

$$2bc+b^2+c^2-a^2=(b+c+a)(b+c-a)=4s(s-a)$$
$$2bc-b^2-c^2+a^2=(a+b-c)(a-b+c)$$
$$=4(s-b)(s-c)$$

We then derive:

$$2\left(\sin\frac{A}{2}\right)^2=\frac{2(s-b)(s-c)}{bc} \text{ and } 2\left(\cos\frac{B}{2}\right)^2=\frac{2s(s-a)}{bc}$$

$$\sin\frac{A}{2}\cos\frac{A}{2}=\frac{\sqrt{s(s-a)(s-b)(s-c)}}{bc} \quad \text{and}$$

$$\sin A=\frac{2\sqrt{s(s-a)(s-b)(s-c)}}{bc}$$

We have seen that we can express the area (S) of a triangle in the form $S=\frac{1}{2}bc.\sin A$, whence

$$S=\sqrt{s(s-a)(s-b)(s-c)}$$

What is most interesting about this result is less the fact that it considerably simplifies the task of measuring the area of a triangular piece of land than its challenge to the traditional attitude to the geometrical rationale of recipes for computation. A formula which expresses an area as the square root of a three-dimensional product has no title to mathematical free grace in Euclidean theology.

When Alexandrian culture revived at the end of the first century of our own era, geography and astronomy dominated the scene; but mathematicians steeped in a tradition which accommodated preoccupations of the Pythagorean *mystique* failed to respond to the challenge of an imperative need for speedier methods of computation to deal with large numbers. Instead, we find Nicomachus (*circa* A.D. 100) absorbed in the quest of perfect and amicable numbers. The former are numbers equal to the sum of all their factors, *e.g.*, $6=3+2+1$, $28=14+7+4+2+1$. The next one is 496. Cajori cites the five following as: 8128, 23550336, 8589869056, 137438691328 and 2305843008139952128. Euclid gave the rule for generating them as follows: if $2^{p-1}-1$ is prime, then $2^{p-1}(2^p-1)$ is perfect, *e.g.* if $p=3$, $2^{p-1}-1=3$ (which is prime) and $2^{p-1}(2^p-1)=4.7=28$.

Two numbers are amicable in the Pythagorean hierarchy if the sum of the factors of each is equal to its fellow. Thus the factors of 284 (being 142, 71, 4, 2, 1) add up to 220, the factors of which (being 110, 55, 44, 22, 20, 11, 10, 5, 4, 2, 1) add up to 284. Nicomachus did not discover a rule for generating such felicitous couples, though there is one, first given by a Moslem mathematician Tabit ibn Korra, *viz.* $2^n.pq$ and $2^n.r$ are amicable if n, p, q, r are integers of which p, q, r are primes expressible in the form $p=3.2^n$; $q=3.2^{n-1}$; $r=3^2.2^{2n-1}$. The examples cited are obtainable when $n=2$. A seemingly original discovery of Nicomachus himself hints at the role of figurate numbers in the background of the arithmetic of antiquity. He disclosed the possibility of expressing the cubes of the integers as sums of successive odd numbers (U_n) in accordance with the following pattern:

$1^3=1$	U_1
$2^3=3+5$	U_2+U_3
$3^3=7+9+11$	$U_4+U_5+U_6$
$4^3=13+15+17+19$	$U_7+U_8+U_9+U_{10}$

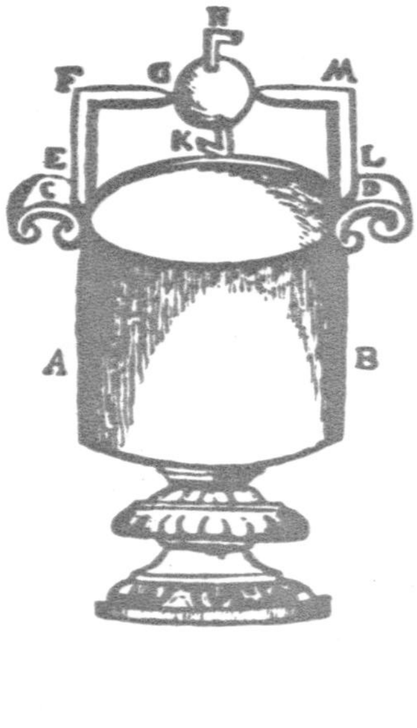

Left: Sixteenth-century reconstruction of rotary steam bellows invented by Heron of Alexandria, one of many mathematicians and inventors who followed the Archimedean revolt against Platonic tradition.
In the device on the right, attributed to Heron, fire caused air in the altar to expand and force water out of a sealed container into a bucket. The increasing weight of water in the bucket acted on ropes wound round spindles and opened the temple doors.

Each sequence corresponding to n^3 here has n terms each differing by 2 from its predecessor and/or successor. The clue to the build-up is that the rank of the last term is the nth triangular number, *e.g.* the last term in the sequence for 4^3 is the tenth odd number and 10 is the 4th triangular number. If we substitute the formula $\frac{1}{2}n(n+1)$ for the nth triangular number for r in the expression $(2r-1)$ for the odd number of rank r, we derive the last term in the sequence as n^2+n-1. If we represent the sequence as $x+1$, $x+3$, $x+5$ . . ., the last is $x+2n-1=n^2+n-1$ only if $x=n^2-n$. Thus the first is n^2-n+1. Whence a general expression for the terms of the nth sequence is $n^2-n+2r-1$ in which $r=1, 2, 3 \ldots n$ and the sum is

$$\sum_{r=1}^{r=n} n^2-n+2r-1=(n^2-n-1)\sum_{r=1}^{r=n}1+2\sum_{r=1}^{r=n}r$$

$$=n(n^2-n-1)+n(n+1)=n^3$$

After Nicomachus, only Diophantos is noteworthy as an innovator in the Alexandrian scene, albeit less in terms of his enduring contribution to knowledge than because his failure to accomplish a daring project puts the spotlight on an obstacle which the Greek-speaking peoples of the Western World never surmounted of their own initiative. To the last, Greek mathematics remained mainly preoccupied with geometric themes and ill at ease in the use of any sign language but that of geometric figures. The Alexandrians never surpassed – or even caught up with – the Chaldean temple culture in the art of computation; and it is doubtful whether they outstripped their predecessors in manipulation of problems nowadays labelled as algebraic.

At the end of the sixteenth and at the beginning of the seventeenth century of the Christian era, the transition to a shorthand based partly on verbal abbreviation, till only the initial letter remained, and partly on conventional signs such as + and − employed in the newly-printed commercial arithmetics was effortless, if none the less gradual, for Europeans already equipped with the Hindu numerals. It would not have been surprising if their immediate predecessors in India and in the Moslem world had done the same at an earlier date; but the character of their scripts had made it well-nigh impossible for the Egyptians,

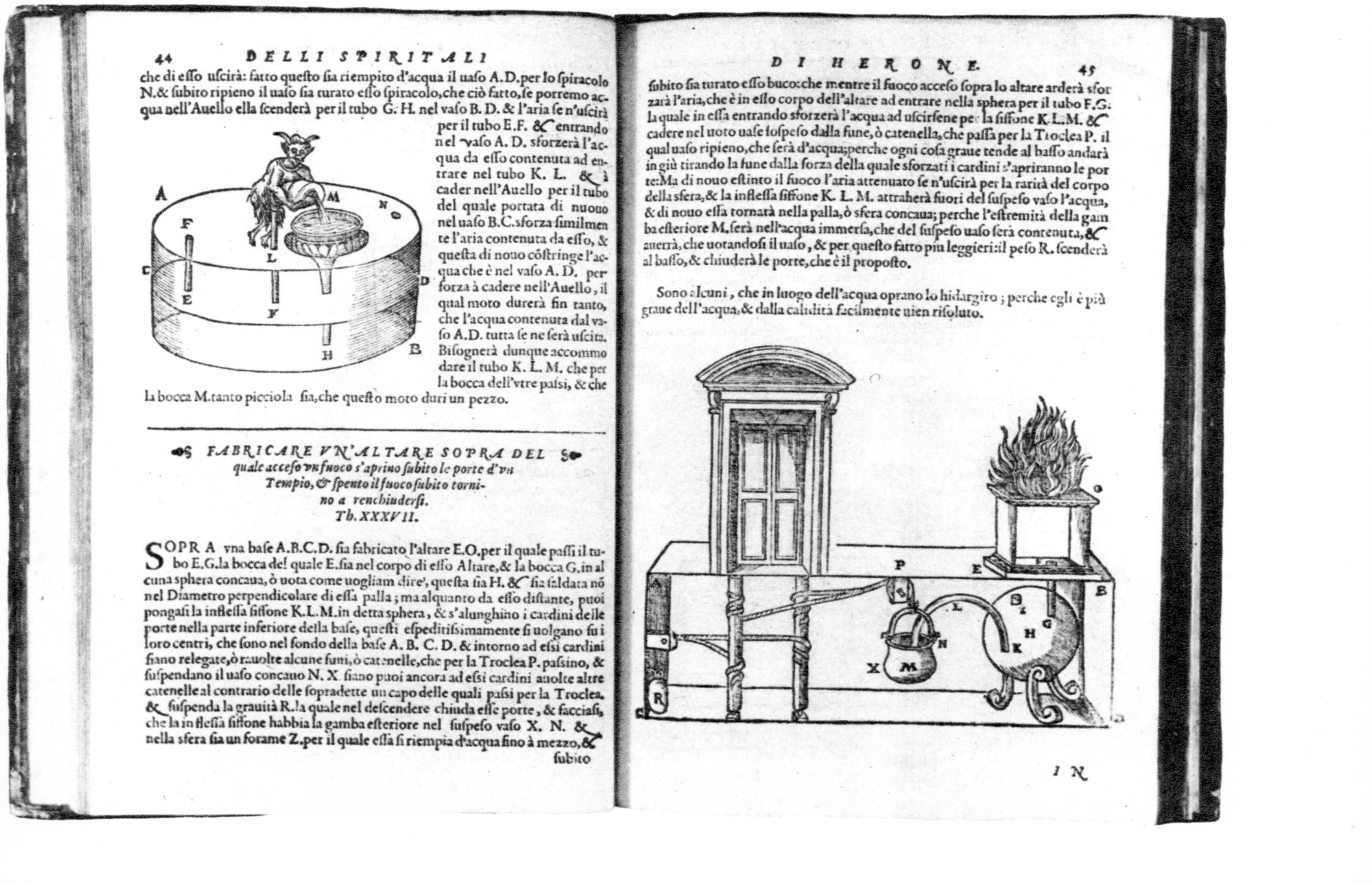
44 DELLI SPIRITALI

che di esso uscirà: fatto questo sia riempito d'acqua il uaso A.D. per lo spiracolo N. & subito ripieno il uaso sia turato esso spiracolo, che ciò fatto, se porremo acqua nell'Auello ella scenderà per il tubo G. H. nel vaso B. D. & l'aria se n'uscirà per il tubo E.F. & entrando nel vaso A. D. sforzerà l'acqua da esso contenuta ad entrare nel tubo K. L. & à cader nell'Auello per il tubo del quale portata di nuouo nel uaso B.C. sforza similmente l'aria contenuta da esso, & questa di nouo cõstringe l'acqua che è nel vaso A. D. per forza à cadere nell'Auello, il qual moto durerà fin tanto, che l'acqua contenuta dal vaso A.D. tutta se ne serà uscita. Bisognerà dunque accommodare il tubo K. L. M. che per la bocca dell'vtre passi, & che la bocca M. tanto picciola sia, che questo moto duri un pezzo.

FABRICARE VN'ALTARE SOPRA DEL quale acceso vn fuoco s'aprino subito le porte d'vn Tempio, & spento il fuoco subito tornino a renchiudersi.
Th. XXXVII.

SOPRA vna base A.B.C.D. sia fabricato l'altare E.O. per il quale passi il tubo E.G. la bocca del quale E. sia nel corpo di esso Altare, & la bocca G. in alcuna sphera concaua, ò uota come uogliam dire, questa sia H. & sia saldata nõ nel Diametro perpendicolare di essa palla; ma alquanto da esso distante, puoi pongasi la inflessa siffone K.L.M. in detta sphera, & s'alunghino i cardini delle porte nella parte inferiore della base, questi espeditissimamente si uolgano su i loro centri, che sono nel fondo della base A. B. C. D. & intorno ad essi cardini siano relegate, ò rauolte alcune funi, ò catenelle, che per la Troclea P. passino, & suspendano il uaso concauo N. X. siano puoi ancora ad essi cardini auolte altre catenelle al contrario delle sopradette un capo delle quali passi per la Troclea, & suspenda la grauità R. la quale nel descendere chiuda esse porte, & facciasi, che la inflessa siffone habbia la gamba esteriore nel suspeso vaso X. N. & nella sfera sia un forame Z. per il quale essa si riempia d'acqua fino à mezzo, & subito

DI HERONE. 45

subito sia turato esso buco: che mentre il fuoco acceso sopra lo altare arderà sforzarà l'aria, che è in esso corpo dell'altare ad entrare nella sphera per il tubo F.G. la quale in essa entrando sforzerà l'acqua ad uscirsene per la siffone K.L.M. & cadere nel uoto uase sospeso dalla fune, ò catenella, che passa per la Troclea P. il qual uaso ripieno, che serà d'acqua; perche ogni cosa graue tende al basso andarà in giù tirando la fune dalla forza della quale sforzati i cardini s'apriranno le porte: Ma di nouo estinto il fuoco l'aria attenuato se n'uscirà per la rarità del corpo della sfera, & la inflessa siffone K. L. M. attraherà fuori del suspeso vaso l'acqua, & di nouo essa tornarà nella palla, ò sfera concaua; perche l'estremità della gamba esteriore M. serà nell'acqua immersa, che del suspeso uaso serà contenuta, & auerrà, che uotandosi il uaso, & per questo fatto piu leggieri: il peso R. scenderà al basso, & chiuderà le porte, che è il proposto.

Sono alcuni, che in luogo dell'acqua oprano lo hidargiro; perche egli è più graue dell'acqua, & dalla calidità facilmente uien risoluto.

I N

Babylonians or Chinese to do so. What bogged down any such aspirations which the Alexandrians might have entertained was a circumstance which made calculation intolerably burdensome. Having conscripted the entire alphabet to do service as signs for the integers, any system of literal abbreviations was inescapably beset with ambiguities.

Either for his own use or – as some claim – with the deliberate intention of making a new sign language for exposition, Diophantos (*circa* A.D. 200) made a hardy attempt to devise one. That all authorities do not yet agree about the interpretation of some of his symbols testifies to the meagre success of the project, if the author's intention transgressed beyond the confines of his own convenience. Since many readers of this book are unfamiliar with the order of the letters of the Greek alphabet, we may best convey the salient features of Diophantine symbolism by using Roman signs. The sign values of the numerals would then have this aspect:

1	2	3	4	...	9
a	*b*	*c*	*d*		*i*
10	20	30	40	...	90
j	*k*	*l*	*m*		*r*
100	200	300	400		
s	*t*	*u*	...		

In this system, we should write 253 in accordance with Alexandrian practice as *tnc*, whence it is clear that no letter in any number sign sequence could precede a letter of later rank in the alphabet itself. Accordingly, a contraction such as *cu* for *cube* could not stand for an integer as must *uc* (=303) or *c* (=3) on its own. On the other hand, *sq* (=180) could not do service for *square*, though *qu* would do so. Presumably for greater clarity when exploiting these conventions, Diophantos wrote the second letter as a superscript, *e.g.* in Romanised form q^{u} and c^{u} for the square and cube of any undetermined quantity which we might represent as x^2 and x^3. The actual forms (δ^{υ} and $\varkappa^{\upsilon}$) of these derive from the corresponding Greek words *δυναμις* and *κυβος*. For first powers of an undetermined quantity, which we might write as x, Diophantos used a contraction probably based on the first two letters of *ἀριθμος* (*number*), when the coefficient was 1, but otherwise ςς, which we may take to be a signpost of grammatical plurality. We may Romanise these respectively as *ar* and *ss* below. To make clear where necessary that other numbers were *monads* (*i.e.* coefficients of $x^0=1$), Diophantos used the symbol μ^0 which is self-explanatory.

In accordance with Greek syntax, number signs for coefficients followed ($q^{u}.kb=22x^2$) the abbreviations for undetermined quantities, and in accordance with the number sign conventions ($kb=20+2$) Diophantos conveyed addition by juxtaposition as we convey multiplication (*e.g.* $xy=x\times y$). Thus we may write as below first in our own way, Romanised as explained above, and below as in the original:

$$x^3+13x^2+5x+2$$
$$c^{u}\ ar\ d^{u}\ jc\ ss\ e\ ss\ m^{o}\ b$$
$$\varkappa^{\upsilon}\ \alpha\varrho\ \delta^{\upsilon}\ \iota\gamma\ \varsigma\varsigma\ \varepsilon\ \varsigma\varsigma\ \mu^{o}\ \beta$$

For fractions, Diophantos used a convention which we can Romanise by representing $27\div 12$ as kg^{jb}. For *minus*, he employed a sign (⋔) like the Greek *ps* written upside down, *e.g.*:

$$\varkappa^{\upsilon}\ \alpha\varrho\ \varsigma\varsigma\ \theta\ ⋔\ \delta^{\upsilon}\ \varepsilon\ \mu^{o}\ \alpha$$
$$x^3-5x^2+8x-1=(x^3+8x)-(5x^2+1)$$

The form of the expression set out as above is an eloquent comment on the claim sometimes made on behalf of Diophantos to be the father of algebraic symbolism. The fact that he always collects positive and negative terms in separate groups is consistent with the circumstance that he nowhere explicitly formulated our familiar rules for manipulating the *plus* and *minus* signs, and no symbolism which can accomplish economy of effort is possible until we take this step. Just as the numeral signs which the Attic Greeks constructed from the letters of the alphabet economised space without any economy of effort expended in computation, the symbolism of Diophantos economised space without regularising any procedure to sidestep reliance on verbal reasoning. Its symbols recorded the results of such reasoning as the numeral signs recorded work done on the abacus; but it provided no short-cut to the solution of a mathematical problem.

Had the system, if indeed the intention of Diophantos was to devise one, led to results which simplified sexagesimal computations for astronomers and surveyors, others might have improved on the workmanship of its inventor by increasing the battery of signs like ⋔ no longer recognisable as letters. Had its author recognised that an equation can have negative and positive roots – or several positive roots – his work might have exercised a powerful influence on the mathematical progress of the Hindu and Moslem inheritors of the Alexandrian tradition, though it is difficult to believe that the attempt to interpret the private sign language of Diophantos in Arabic would have encouraged the translator to accept the challenge to do better in a different medium. What remains other than a treatise on figurate numbers in the same vein as that of Hypsicles is an *ad hoc* treatment of equations in the Babylonian manner.

He pays much attention to those which admit of solution in terms of integers, as does the Pythagorean relation $h^2=p^2+b^2$ for $h=5$, $p=3$, $b=4$, *etc.* Such so-called Diophantine equations have subsequently had an understandable fascination for pure mathematicians; but they have had no pay-off in the domain of measurement. One of their few applications at the periphery of practical use is in connexion with the application of the theory of probability in card or urn model situations which admit of no specification of cards or balls other than simple enumeration. The following is an example.

Let us suppose that $6x-4y=8$. By trial and error, we discover without great difficulty that the integers $x=2$ and $y=1$ satisfy the equation; and Diophantos gives no recipe for finding *all* others which do so. If A, B, C are also integers, we may write the equation in the form $Ax-By=C$. Its Diophantine properties (solubility for integer values of x and y) are then as follows:

(i) it admits of no solution, unless C is *either* zero *or* an exact multiple of H, the highest common factor ($=2$ in the above) of A and B;

(ii) it admits of 2 solutions if $C=0$, *viz.* either $x=B$ and $y=A$ or $A=0=B$;

(iii) otherwise, we may generate from the first pair of values (*e.g.* $x=2$, $y=1$ above) one of the following in which n may successively have the values 0, 1, 2, 3;

$$x+\frac{nB}{H} \qquad y+\frac{nA}{H}$$

By painting in successive values of n in the above we get:

$(x=4, y=4)$; $(x=6, y=7)$; $(x=8, y=10)$, *etc.*

Diophantos is perhaps chiefly memorable because his failure to sire algebra in the modern sense of the term illustrates that any isolated culture contains in itself the seeds of its decay. No enlightened person would wish to condone the brutal murder of Hypatia, herself a mathematician, by the monks of St. Cyril; but we may question whether the adoption of Christianity as the official creed of the Roman Empire has much to do with the decadence of Alexandrian learning. When twilight falls on the Alexandrian scene, the mathematics of the Western World had already exhausted its capacity for further growth without a dramatic reform of human communications through contact with the less sophisticated cultures of the Far East.

Chapter 6 The Ptolemaic Synthesis

Some say that Menaechmus replied to the complaints of his pupil Alexander that there is no royal road to geometry; others, that Euclid said the same thing to Ptolemy Soter. If there is any truth in either story, it seems that professional mathematicians of ages other than ours have not always been over-anxious to part with secrets of Temple Lore to their contemporaries. To say as much prompts the question: why were the soldier-builders of the Alexandrian set-up so eager to make Greek mathematics their most enduring monument? A plausible answer is that Platonic mathematics could none the less provide recipes, albeit submerged beneath an avalanche of gamesmanship, for solving practical problems in a social *milieu* ready to exploit human inventiveness. To be sure, we do not, and may never, know the top-secret military surveying of the Alexandrian route marches or indeed those of the less amazing exploits of Hannibal a century later; but we can scarcely doubt that the Greek quarter of Alexandria in the declining years of Euclid's domicile harboured many veterans familiar with some of the data. Their Greek-speaking comrades-in-arms who settled in the Seleucid colonies of Mesopotamia also shared them.

Such is the social background of the situation in which Eratosthenes, now remembered as a mathematician only for a fool-proof recipe for finding prime numbers, laid the foundations of scientific geography with little prescience of its implications for the making of mathematics in the world of today. Eratosthenes (*circa* 250 B.C.) was the fifth Librarian of the Alexandrian Library. There seems to be no convincing evidence that he ever made any original astronomical observations. What makes him noteworthy is that he used his opportunity to exploit to advantage a new repository of human communications. In his day, it was common knowledge that the noon sun's zenith distance ($90° - \textit{altitude}$) at Alexandria on the summer solstice was in round figures, as determined by the sun's noon shadow, $7\frac{1}{2}°$. It was also on record that its image was then visible at noon on the water surface of a deep well in Syene, near the Assuan dam of today and just on the Tropic of Cancer, whence its midsummer noon z.d. was zero. Since Alexandria and Syene lie nearly on the same meridian, it thus appears that a great circle arc $\simeq 7\frac{1}{2}°$ separates them. Either from records of the mean time of route marches over level desert country or from crude maps made by the Egyptian temple surveyors for tax allocation, Eratosthenes knew that the intervening distance is in Anglo-American linear measure $\simeq 520$ miles. A simple diagram (page 126) completes the story, *viz.* that the earth's circumference is approximately 25,000 miles.

From two sources, the fifth librarian of Alexandria would have at his disposal estimates of the meridional distance between two places whose latitudes were assignable with a precision not exceeded before Newton's generation; and had he given due consideration to its merits, one

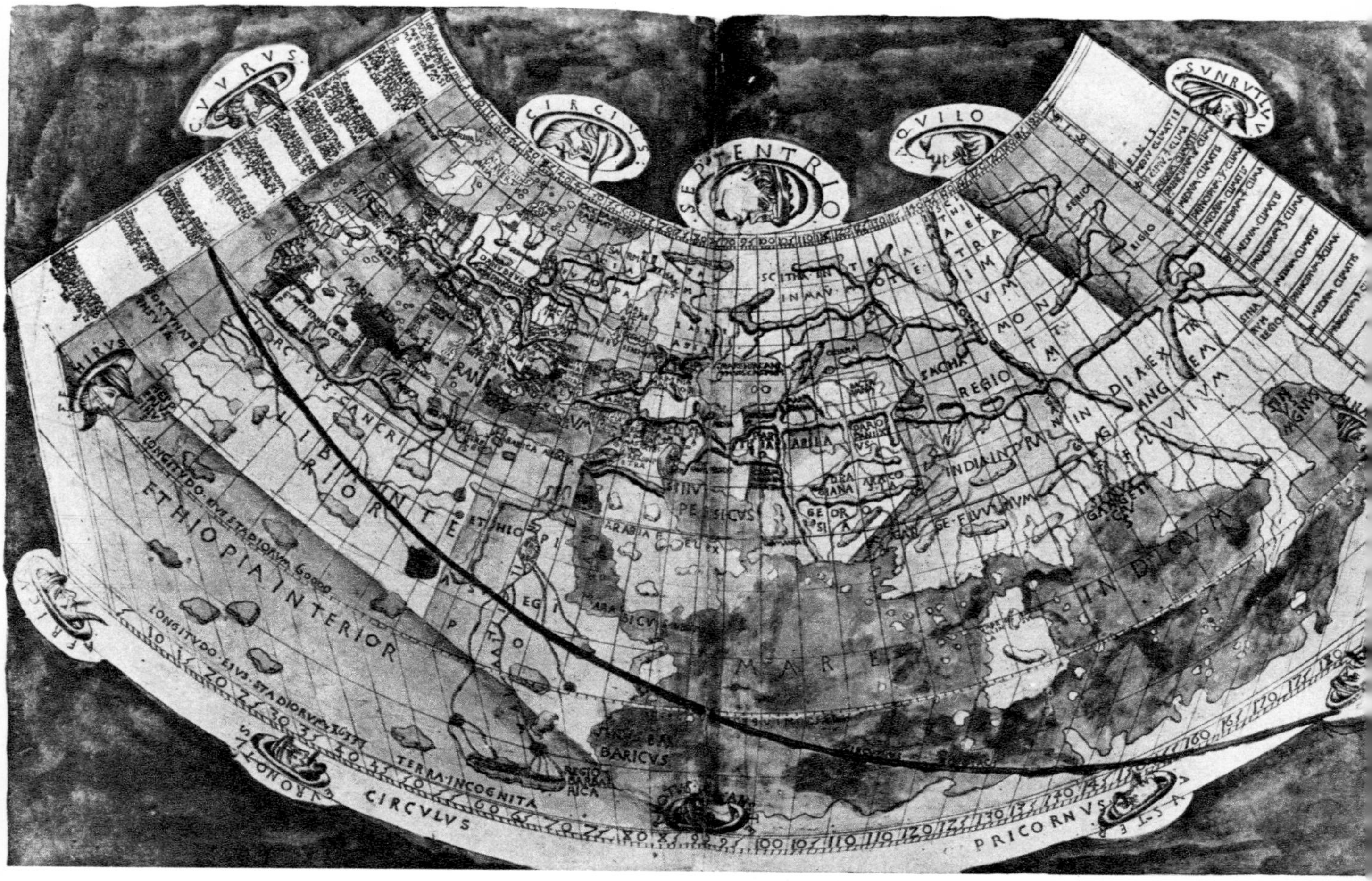

Both these fifteenth-century maps are based on projections devised by Ptolemy c. A.D. 150. That Ptolemy used three different projections shows that he was aware of the problem of invariance and knew that invariance for every useful property of the sphere cannot be achieved by any one method.

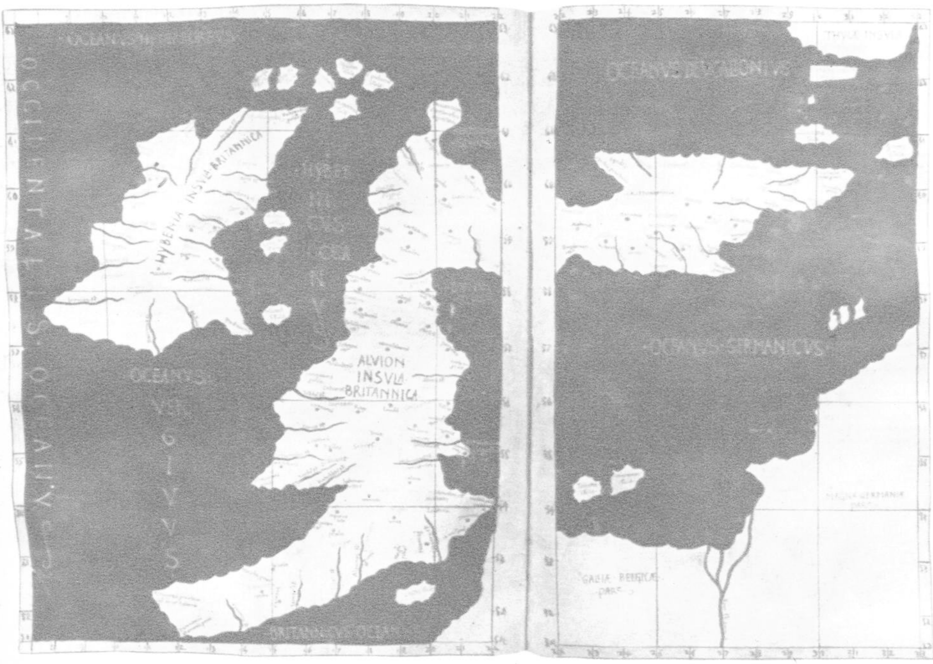

of the greatest of the Alexandrians might have suffered a different fate at the hands of history. During the period A.D. 800–1500, Ptolemy's *Almagest* (*circa* A.D. 150), as commonly known by its Arabic title, enjoyed a prestige on all fours with Euclid's *Elements* in the marriage festival of East and West; but the work of few men who have greatly advanced learning has had less favourable recognition from posterity. Though Greek-speaking, Ptolemy seems to have come from the Coptic quarter, being like Heron in the Seleucid-Mesopotamian tradition which took the art of computation seriously. For that reason alone, he could not enjoy the recognition due to him as a pioneer of pure mathematics before the Gaussian revolt against Euclid's self-evident principles; but his reputation in modern times has been the victim of other circumstances.

Though the outcome encouraged Columbus to believe that there was a short westerly course to India and hence played its part in the colonisation of America, Ptolemy made a grave error of judgment by adopting an estimate of the earth's circumference made a century after that of Eratosthenes by Poseidonius. The latter made use of the difference between the transit z.d. of the star Canopus at Alexandria and at Rhodes. Unless one already had at one's disposal a correct estimate of the earth's circumference, a reliable figure for linear distance at sea was then unobtainable. Hence it is less surprising that his own was grossly deficient than that later geographers preferred the estimate of Poseidonius to that of Eratosthenes.

With better reason, Ptolemy, like his predecessor Hipparchus, also rejected the heliocentric view elaborated by Aristarchus, a contemporary of Eratosthenes and of Archimedes. In its own setting, this had nothing to commend it. Before A.D. 1837, no available instruments could detect the annual parallax of a fixed star; and a society without either wheel-driven clocks or telescopes could not anticipate either of the confirmatory data announced first in Newton's lifetime, *i.e.* the retardation of the pendulum by latitude owing to the earth's spin and the optical phenomenon

This tablet, believed to be a primitive planisphere of the second millennium B.C., reminds us that star-catalogues were not new at the time of Hipparchus. But the discovery of the precession of the equinoxes now set a new problem: how to locate stars in such a way that tables would not need constant revision.

At noon on the summer solstice the sun is vertically overhead at Syene. At Alexandria its zenith distance is $\simeq 7\frac{1}{2}°$. *Since both lie nearly on the same meridian, a great circle arc of* $\simeq 7\frac{1}{2}°$ *separates them. Knowing the distance was* $\simeq$ *520 miles, Eratosthenes estimated the earth's circumference to be* $\simeq 520(360 \div 7\frac{1}{2})$ *miles.*

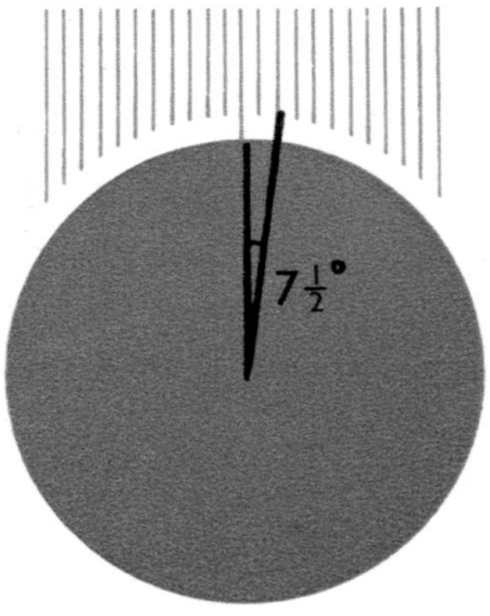

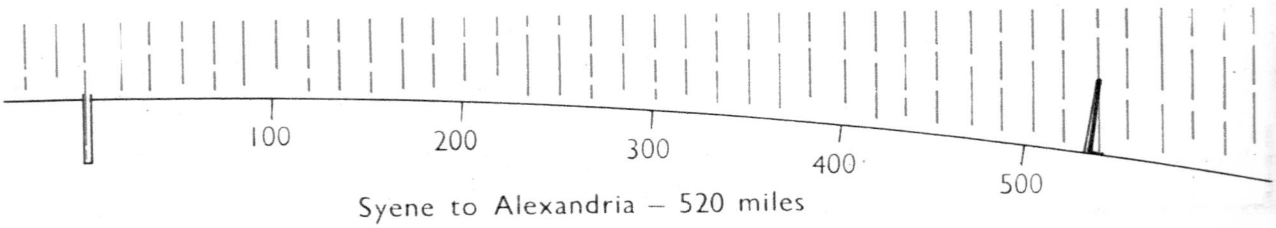

Syene to Alexandria – 520 miles

called stellar aberration. Be it said that Ptolemy, who himself published the first recorded experimental data on the refraction of light, was well aware of the absence of any support for a world picture which is much less mathematically manageable from the viewpoint of the navigator. If observations made nearly two thousand years later showed it to be mathematically preferable in connexion with the vagaries of planetary motion, data available to Ptolemy did not justify the conclusion.

As the pivotal port of contemporary Mediterranean trade with access to the lore of the temple astronomers of the Middle East, Alexandria was a setting uniquely favourable to the advancement of astronomy; and Aristarchus is notable in the forefront of a distinguished line of astronomers who contributed to the Ptolemaic synthesis, if only because he made the first recorded estimates both of the diameters of the sun and the moon and of the distance of the latter from the earth. The method was sound enough; but the figures were grossly deficient, partly because the instrumental errors involved were considerable and partly because nearly all estimates of long distances on the earth's surface available at that time were highly unreliable. The attempt is none the less memorable as an anticipation of the birth of trigonometry.

We have seen that Euclid's *Elements* contained all the essential ingredients for the development of plane trigonometry to the level attained before the co-ordinate method made it possible to deal with direction in contradistinction to static magnitude. We owe the conventions we now use to Hindu mathematicians in the social context (*circa* A.D. 500) of vast irrigation projects, which entailed formidable problems of surveying several centuries after trigonometry had reached a much higher level in Alexandria. There, plane trigonometry

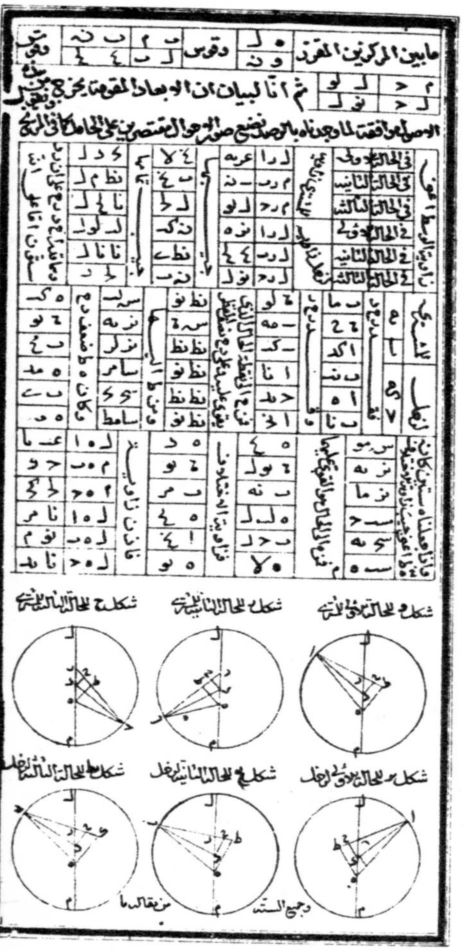

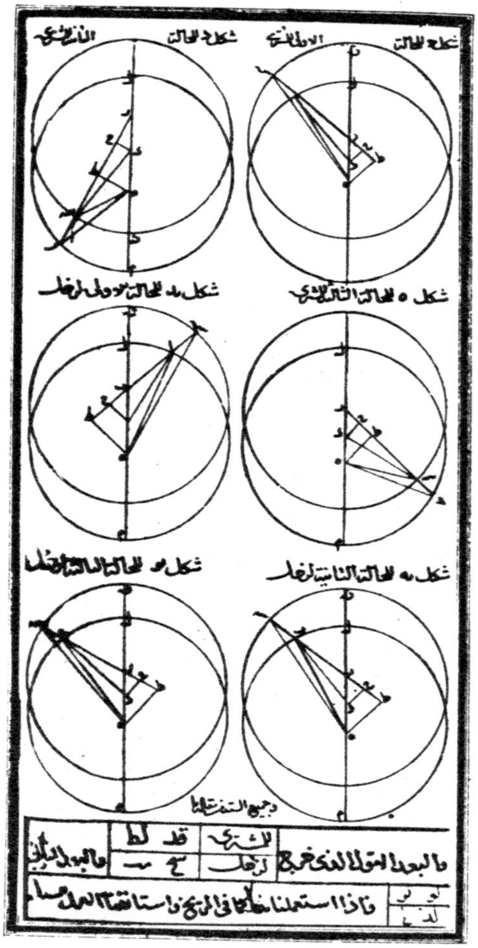

A thirteenth-century Arabic edition of Ptolemy's Almagest. *From A.D. 800 to 1500 this work enjoyed a prestige on all fours with that of Euclid's* Elements. *Visual aids had not yet reached today's standards.*

CELESTIAL SPHERE AND HORIZON PLANE

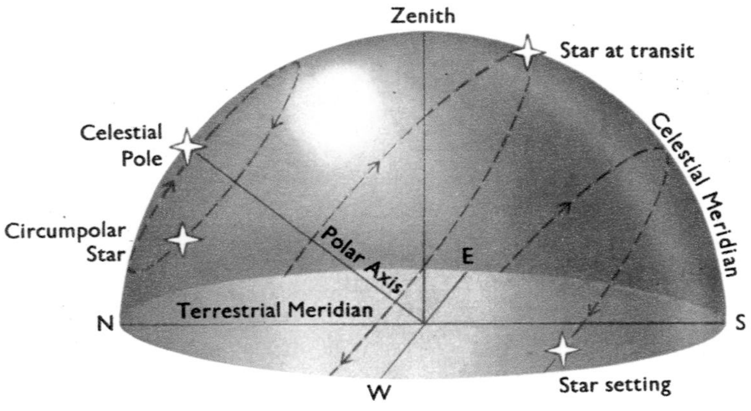

DECLINATION AND TERRESTRIAL LATITUDE

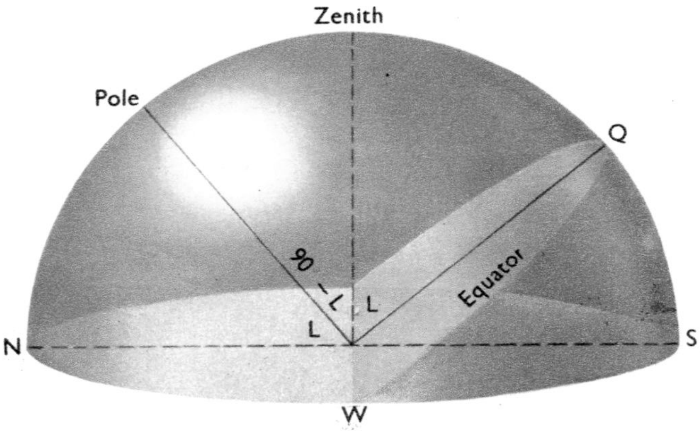

The observer's ***latitude*** *is both the inclination of the plumb-line to the plane of the equator and that of the polar axis to the horizon plane.*
We label as positive (otherwise negative):
(i) terrestrial latitudes north of the equator;
(ii) zenith distances at transit in the northern sector of the plane of the celestial and terrestrial meridians; (iii) declinations of stars in the northern celestial hemisphere. Then Lat. = declination – z.d.

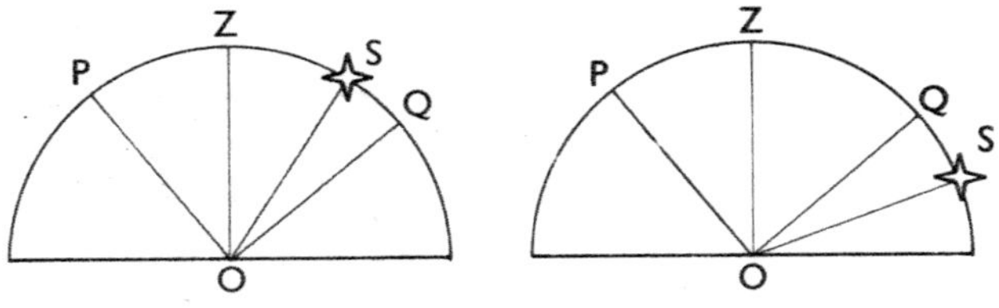

ZOQ = Latitude of the observer
ZOS = Zenith distance of star at transit
SOQ = Star's declination

seemingly emerged without any concern for the needs of the surveyor in response to a different challenge. When available to the Alexandrian astronomer Hipparchus (*circa* 150 B.C.), the Chaldean discovery of the Precession of the Equinoxes faced the maker of a star catalogue with a problem which was then more imperative than it could be after Europe had taken over printing from the Chinese.

To appreciate the nature of the challenge, the reader will need to be familiar with a few elementary data, well established when Aristarchus comes into our picture:

(i) the fixed stars appear to revolve in concentric circles or circular arcs above the horizon plane about a point called the *celestial pole* whose inclination (*altitude*) to the plane itself is the mean of the maximum and minimum altitudes of any one circumpolar (non-setting) star;

(ii) the position of the sun's noon shadow suffices to define the observer's terrestrial meridian and the celestial pole lies on a plane cutting the horizon along this line at right angles thereto;

(iii) we speak of the *zenith* as a point on this plane (the *celestial meridian*) vertically above the observer, the plumb-line being collinear with zenith, observer and the earth's centre;

(iv) any celestial body is highest above the horizon when it crosses the celestial meridian; and we speak of its inclination to the extended plumb-line at transit as its zenith distance (*z.d.*), this being the complement of its altitude;

(v) the observer's latitude is the altitude of the celestial pole; and the z.d. at transit of any celestial body (*e.g.* the equinoctial sun) in the plane of the equator, where the altitude of both poles is zero, is equivalent to the observer's terrestrial latitude;

(vi) the so-called *ecliptic*, *i.e.* plane of the sun's apparent retreat below the eastern horizon at the rate of roughly 1° per day among the fixed stars, inclines to the plane of the equator at about $23\frac{1}{2}°$ and cuts the earth's surface at the two tropics;

(vii) the rising and setting positions of a star with reference to the horizon plane depend only on the observer's latitude and the equatorial inclination (so-called *declination*) of an imaginary line joining the star to the earth's centre where the celestial (and terrestrial) polar axis cuts the equatorial plane;

(viii) the rising and setting times of a star at one and the same latitude on one and the same night depend only on its declination and the so-called *hour-angle* which separates its own imaginary meridian, *i.e.* half-circle passing through it from pole to pole, from the imaginary meridian on which the sun itself then lies;

(ix) on each day of the year in which the sun completes a cycle in the ecliptic, it is convenient to locate its position by its declination and the hour-angle (*Right Ascension*) between its own meridian and a fixed meridian conveniently chosen as the one which contains the point of intersection of its ecliptic path in the northerly direction with the equatorial, *i.e.* the approximate position of the sun on the vernal equinox when its noon z.d. is the observer's latitude, being zero on the equator itself;

(x) the zodiacal constellations are noteworthy solely because they lie on the great circle of the ecliptic; and we speak of the point of intersection of the latter with the equator on the vernal equinox as the *First Point of Aries*, customarily represented by the astrological

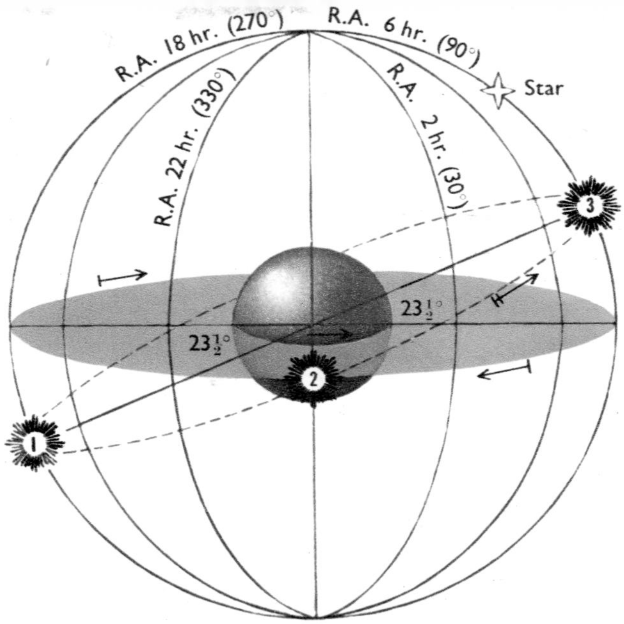

① Sun on Dec 21 R.A. 18 hr. (270°) declination $23\frac{1}{2}°$S

② Sun on Vernal Equinox R.A. 0 hr. declination 0°

③ Sun on Jun 21 R.A. 6 hr. (90°) declination $23\frac{1}{2}°$N

⟶ actual direction of earth's rotation

↦ apparent diurnal motion of celestial sphere (westward)

⇸ apparent annual retreat of sun (eastward)

The ecliptic, i.e. the plane of the sun's annual apparent eastward retreat among the fixed stars, inclines to the equator at $23\frac{1}{2}°$ and cuts the earth's surface at the tropics. The star shown (R.A. 6 hours) makes its transit above the meridian at midday on the summer solstice, at midnight on the winter solstice.

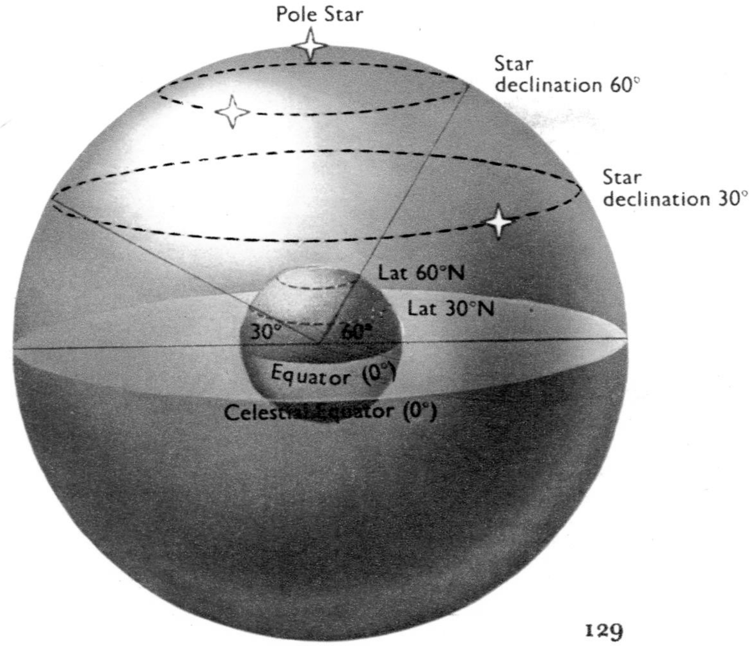

North of the equator, the latitude of a given point on the earth's surface is the angle which a straight line from that point would make with the plane of the equator at the earth's centre. Declination is the angle which a straight line from a point on the celestial sphere would make with the plane of the celestial equator at its centre. South of the equator we label latitudes and declinations as negative.

symbol (♈) for the constellation of the Ram, because the sun's position at the spring equinox corresponded with that of the latter in the time of Hipparchus.

If we pay no regard to the very slow ($\simeq$26,000 year) cycle of changes with respect to the rising and setting times and positions of a star as subsumed by the term *precession*, the most convenient way of mapping the celestial sphere is to define the position of each fixed star by two co-ordinates: its declination and its Right Ascension defined as for the sun. The former suffices to locate the observer's terrestrial latitude from a single determination of the z.d. at transit. The latter suffices to date its time of transit, if there is available a table citing the sun's R.A. on the same day. This method was certainly in use before Hipparchus; and the Chinese employed it, almost certainly uninfluenced by Greek astronomy, early in the Christian era.

The so-called Precession of the Equinoxes signifies that ♈ moves westward along the ecliptic (clockwise from the viewpoint of an observer north of the Tropic of Cancer) at a steady rate ($\simeq$1° in 72 years). This causes both the R.A. and

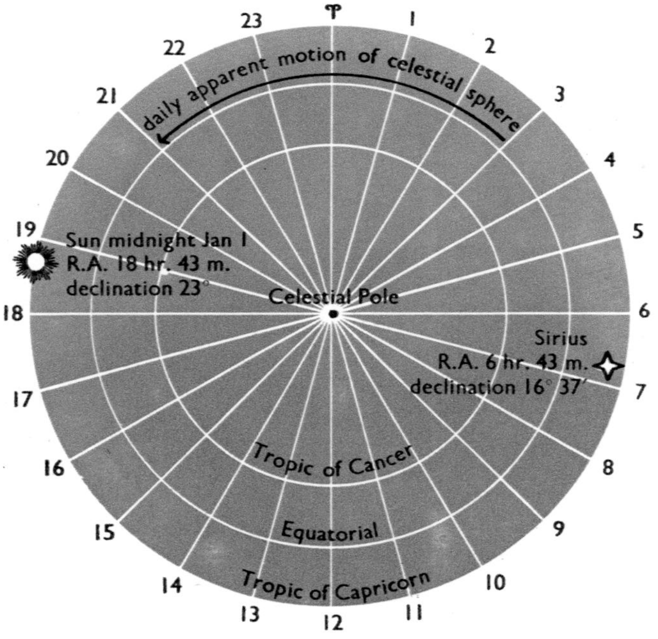

The R.A. of Sirius is 6h. 43m. At midnight on Jan. 1 that of the sun is 18h. 43m., being then 24 – (18h. 43m.) west of ♈. Thus the hour angle of Sirius on that date is almost exactly 12 hours with reference to the sun, and its time of transit will be $\simeq$ 12 hours after noon, i.e. at midnight.

the declination of a fixed star to change appreciably in the course of a century; but if we define its position with reference to corresponding co-ordinates referable to the plane of the ecliptic and to an axis through the earth's centre at right angles thereto, the ecliptic declination (so-called *celestial latitude*) remains fixed for all calculable time. Since also the zero meridian through ♈ changes at a fixed rate, the so-called celestial longitude of a star corresponding to R.A. in equatorial co-ordinates is easily calculable from its value cited at a known date. The current terms longitude and latitude in this context are misleading. Declination and R.A. respectively have a direct relation to terrestrial latitude and to terrestrial longitude; and their determination calls for no trigonometrical *expertise*. Contrariwise, it is not possible to calculate our terrestrial location from ecliptic co-ordinates without recourse to spherical trigonometry.

It should thus be clear why the discovery of precession confronted astronomers with a new challenge. It is not possible to correct tables of declination and R.A. referable to a particular date for use at a much later one without recourse to the formulae we require to derive the ecliptic co-ordinates of a celestial body from equatorial co-ordinates referable to direct observation and far more convenient for every-day use. Hipparchus may therefore have set out to construct what appears to have been the first crude trigonometrical table in order to test the validity of precession from past records or to prescribe a procedure for revising his 850-star catalogue at later dates. Since the Ptolemaic star map locates each of 1008 fixed stars in ecliptic co-ordinates, Ptolemy himself appears to have intended to produce a star catalogue which would need no correction; but another consideration which may have influenced his decision could have been the fact that the orbits of the moon and of the planets lie very close to the ecliptic. Since he compiled a table of declinations for each degree of so-called celestial latitude, the outcome sufficed for the needs of navigation at a time when there was no neans of, nor pressing need for, determining ongitude at sea.

We may assume that the conceptualisation of he dome of the heavens as a hemisphere emerged arly in the astral lore of the temple civilisations, nd therewith a convenient way of representing he orientation of a celestial body towards the arth's centre when the recognition of the earth's phericity made it possible to visualise the apparent motion of the fixed stars in terms of the iurnal motion of the celestial sphere about an xis collinear with the geographical polar axis. uch a visual schema is fully consistent with the act that some celestial bodies are further away rom us than are others. From the viewpoint of n imaginary observer at the earth's centre, the atitude and longitude of an object do not disclose whether the latter is at the bottom of the deepest ocean or at the top of the highest mountain. They merely define the line of vision on which it lies. f we specify a star's location by two angular o-ordinates analogous to terrestrial latitude and ongitude, we may therefore speak as if the elestial bodies lie on it without implying that hey are all equidistant from the earth's centre.

When we make use exclusively of angular co-ordinates referable to inclination to a fixed plane (*e.g.* the equator) and rotation about a fixed axis (*e.g.* the polar), measurement does not necessarily raise problems with which Platonic geometry is unable to deal. In terms of the earth's radius R, plane geometry suffices to describe the track of a ship: (a) along a parallel (L) of latitude through l degrees of longitude as $2\pi R.l.\cos L \div 360$; (b) between L degrees of latitude along a meridian as $2\pi R.L \div 360$. However, plane geometry cannot assess the shortest path between two points which lie both on different meridians and on different parallels. This path is the arc of least curvature, *i.e.* that of a so-called *great circle* of radius R like a meridian or the equatorial; and when problems of mensuration on the sphere involve great circles which intersect obliquely with the equatorial and meridians, they are soluble only in terms of the properties of a triangle whose three sides are intersecting arcs of great circles, being as such specifiable for the circle of unit radius in angular units only.

Each pair of the three sides of such a spherical triangle meet at the intersection of two planes.

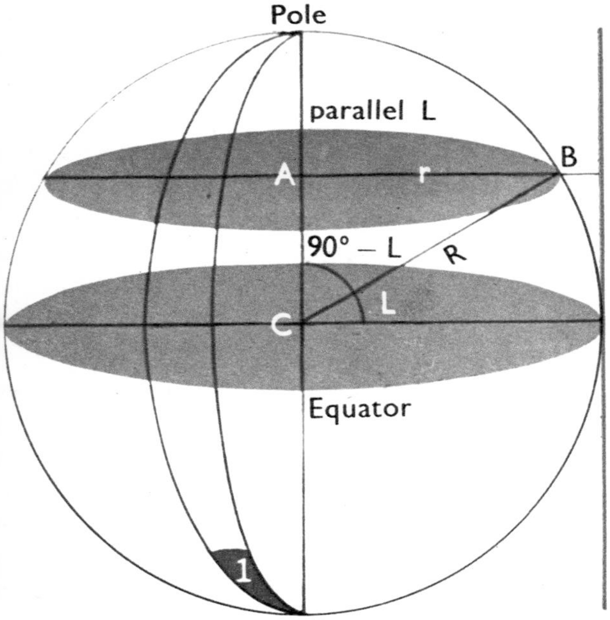

BC = R (earth's radius)

AB = r (radius of the plane of the parallel L)

$$\frac{AB}{BC} = \sin(90° - L) = \cos L$$

$\therefore$ r = R cos L

CALCULATING LENGTH OF l DEGREES OF LONGITUDE AND L DEGREES OF LATITUDE

The circumference of the circle of which r is radius at latitude L is $2\pi r$, and the length of one degree of arc is $2\pi r \div 360$. Thus the distance along an arc of l degrees of longitude is

$$2\pi r.l \div 360 = 2\pi R.l.\cos L \div 360$$

Since a meridian is half a great circle of circumference $2\pi R$, one degree of latitude along a meridian is equivalent to $(2\pi R \div 360)$ linear units and L degrees of latitude $= L.2\pi R \div 360$.

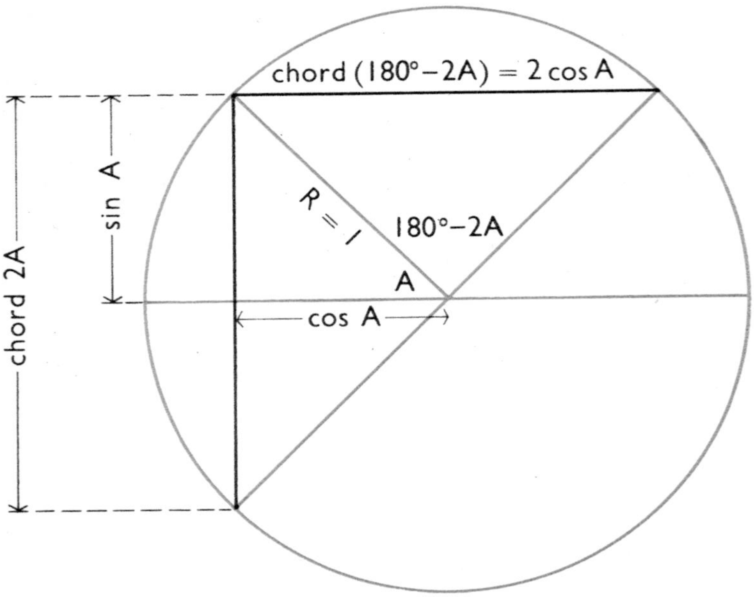

PTOLEMAIC AND HINDU TRIGONOMETRY
Above: in a circle of unit radius
Chord 2A=2 sin A. Chord (180°−2A)=2 cos A.
Below: sin (90°+A)=cos A.

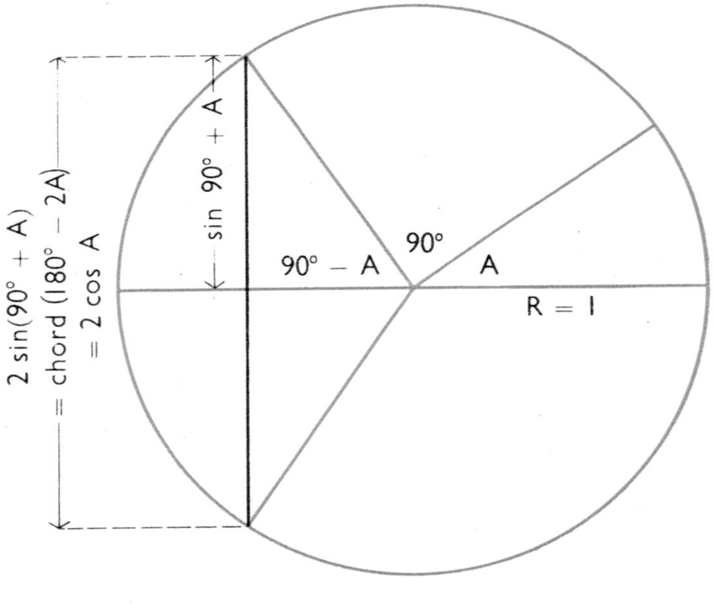

Accordingly, we define its angles by the inclination of the planes in which the appropriate arcs of the three great circles lie, *i.e.* in terms of the inclination of their polar axes. The problem of transforming into ecliptic co-ordinates the equatorial co-ordinates of a celestial body as inferred from local time and altitude at transit by recourse to elementary geometrical considerations is essentially like that of measuring the arcs of terrestrial great circles and the angles at which they intersect.

The accuracy of calculations of this sort depends on that of available tabulated values of the trigonometrical ratios invoked no less than on the attainable precision of the raw data; and the enormous influence of Ptolemy during the Middle Ages is in no small measure due to the fact that he set a new pattern of compiling mathematical tables with a high degree of accuracy (expressed in sexagesimal fractions) and regularity of intervals. His table of chords provided all the essentials for the tables we use today. In a circle of unit radius, the length of a chord which subtends an angle A at the centre is $2\sin \frac{1}{2}A$. Ptolemy's recorded values are expressed as fractions of the *diameter* of a circle of radius 60 units. If the length of a chord is $2a$, we may therefore write $\sin A = a \div 60$ and chord $2A = 2a \div 120 = a \div 60$. Hence the tabular values cited by Ptolemy are equivalent in our notation to writing chord $A = \sin \frac{1}{2}A$ and chord $(180° - A) = \cos \frac{1}{2}A$. To appreciate the procedure of making a table of chords entailed in the Alexandrian setting, we may therefore use our own terms without misunderstanding.

Since $\cos A = \sin (90° - A)$, a modern table of sines suffices for use as a table of cosines if labelled in reverse order, whence a table of tangents is easily derived from the relation $\tan A = \sin A \div \cos A = \sin A \div \sin (90° - A)$. The raw materials at the disposal of Hipparchus were the sines of 90°, 60°, 45°, 30° (and 0°) derived from the theorem of Pythagoras and those of 72° and 24° from Euclid's construction of regular polygons respectively with 5 and 15 sides. We can derive any binary submultiple of these by recourse to one award to

Claim 8, *i.e.* that $(\sin \frac{1}{2}A)^2 = \frac{1}{2}(1-\cos A)$. This, however, leaves large gaps at the upper end of the table, *e.g.* between 90° and 72°. We can fill these in to some extent by using (*see* Claim 4) the rule $(\sin A)^2+(\cos A)^2=1$, whence $\sin 75° = \sqrt{1-(\sin 15°)^2}$ *etc.* However, such a table will always be more closely packed at the extremities than in the middle; and successive items will not correspond to any fixed interval.

To make a tidier table, it is first necessary to be able to find the sine of the sum of two angles. It would have encouraged schoolboys mystified in an earlier generation than ours by the inclusion of Ptolemy's theorem cheek by jowl with the Euclidean sequence to have disclosed that this is in fact equivalent in our symbolism to the formula:

$$\sin A \pm B = \sin A . \cos B \pm \cos A . \sin B$$

By recourse to the formula $(\sin A)^2+(\cos A)^2=1$ we can obtain from this the corresponding formula for cosines:

$$\cos A \pm B = \cos A . \cos B \mp \sin A . \sin B$$

For instance, $\sin 4° = \sin 1° . \cos 3° + \sin 3° . \cos 1°$.

The first of the two foregoing formulae does not provide a recipe for equalising the intervals of the table, unless we invoke an important limit already mentioned. Ptolemy, whose table of chords in $\frac{1}{2}°$ intervals does the same job as a table of sines in $\frac{1}{4}°$ intervals, recognised in effect that $\sin A$ approaches A very closely if measured in radians (*see* award to Claim 10) for angles smaller than $5° = (\pi \div 36)^R$, and it is easy to check that there is no error in the sixth decimal place if we take $\sin \frac{1}{2}° = (\pi \div 360)^R$, whence also easy to level out a table of unequally-spaced values by the addition formula. For instance, we have $60 \div 2^5 = 1\frac{7}{8}$, so that $\sin 1\frac{3}{4}° = \sin 1\frac{7}{8}° \cos \frac{1}{8}° - \sin \frac{1}{8}° \cos 1\frac{7}{8}°$.

$$\text{Sin A} = \frac{QR}{OQ}$$

$$\text{Cos A} = \frac{OR}{OQ} = \frac{PT}{PQ} \quad (\because ♣ = A)$$

$$\text{Sin B} = \frac{PQ}{OP}$$

$$\text{Cos B} = \frac{OQ}{OP}$$

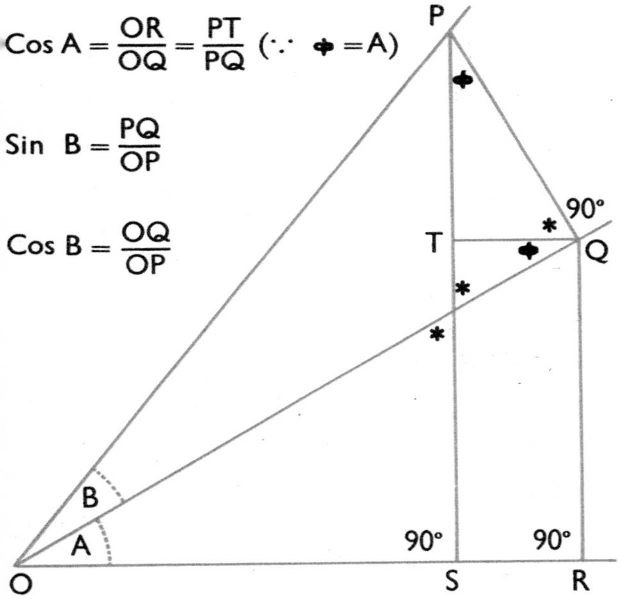

$$\text{Sin (A+B)} = \frac{PS}{OP} = \frac{TS+PT}{OP} = \frac{QR+PT}{OP}$$

$$= \frac{QR}{OP} + \frac{PT}{OP}$$

$$= \frac{QR}{OQ} \cdot \frac{OQ}{OP} + \frac{PQ}{OP} \cdot \frac{PT}{PQ}$$

$$\therefore \text{Sin (A + B)} = \text{Sin A Cos B} + \text{Cos A Sin B}$$

Since $(\cos A)^2 = 1-(\sin A)^2$
$\cos (A+B) = \cos A . \cos B - \sin A . \sin B$.

$$\text{Sin A} = \frac{PR}{OP} \qquad \text{Sin B} = \frac{PQ}{OQ} \qquad \text{Cos B} = \frac{OP}{OQ}$$

Since $90° - A + C = 90°$, $A = C$

$$\therefore \text{Cos C} = \text{Cos A} = \frac{TQ}{PQ}$$

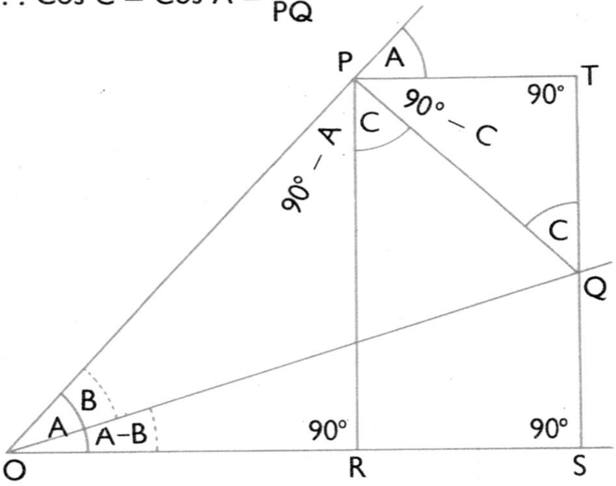

$$\text{Sin (A−B)} = \frac{QS}{OQ} = \frac{TS-TQ}{OQ} = \frac{PR-TQ}{OQ}$$

$$= \frac{PR}{OQ} - \frac{TQ}{OQ}$$

$$= \frac{PR}{OP} \cdot \frac{OP}{OQ} - \frac{PQ}{OQ} \cdot \frac{TQ}{PQ}$$

$$\therefore \text{Sin (A − B)} = \text{Sin A Cos B} - \text{Sin B Cos A}$$

Since $(\cos A)^2 = 1-(\sin A)^2$
$\cos (A-B) = \cos A . \cos B + \sin A . \sin B$.

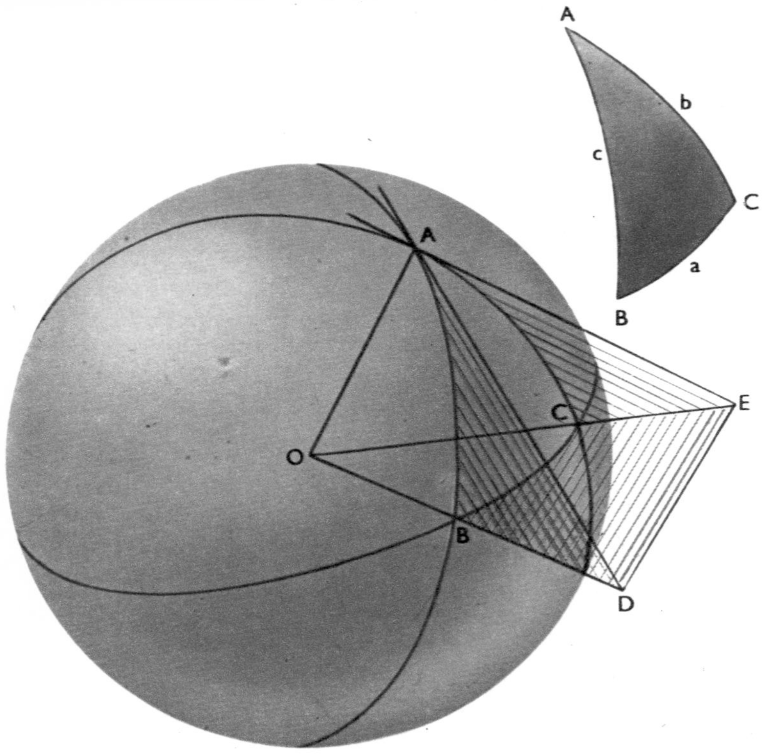

In a sphere of unit radius ($R=1$) *the lengths of the arcs a, b, c are equivalent to their angular measure in radians. AOD, AOE and EOD are planes through the centre along the arcs AB, AC and BC respectively. D and E are so located that the plane AED is tangential to the sphere at A. Here:*

$$OA=R=1 \text{ and } OE^2-AE^2=1=OD^2-AD^2$$

$$OE^2+OD^2-2OE.OD\cos a=AE^2+AD^2-2AE.AD\cos A$$

$$\therefore 1+AE.AD\cos A=OE.OD\cos a$$

$$\therefore \frac{1}{OE.OD}+\frac{AE}{OE}\cdot\frac{AD}{OD}\cos A=\cos a$$

$$\therefore \cos b.\cos c+\sin b.\sin c.\cos A=\cos a$$

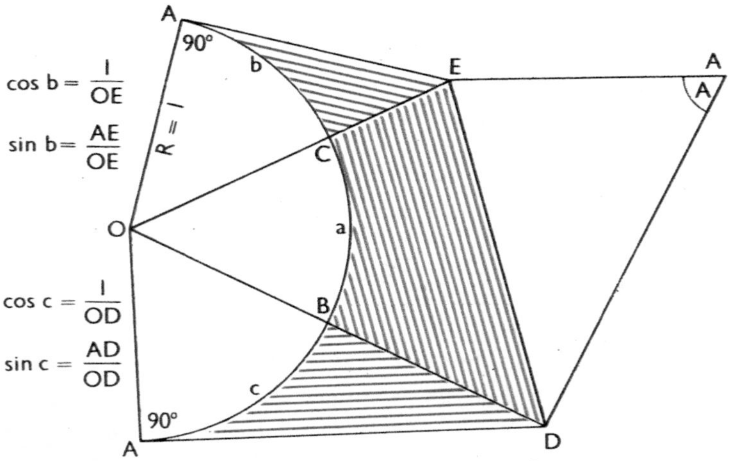

Having got so far, Ptolemy did not take what might now seem to be an obvious step forward by scrapping the degree as a unit of measurement in favour of the radian. One reason for this may be that the radian, though mathematically more meaningful, is an impracticable unit of calibration; but that does not explain why he, and others before him, preferred to divide the circle into 360 primary units in preference to say 384, which admits of successive bisection into 256 intervals of $1\frac{1}{2}$. In early Greek mathematics we meet no trace of the degree, as a unit of angular measurement; and the only unit referred to in Euclid's treatise is the right angle. However, we may be sure that either Hipparchus or his predecessors took over the division of the circle into 360 parts from Mesopotamian sources, though we have no clue to the method of calibration of whatever sort of quadrant Hipparchus used.

Some historians have surmised that the Mesopotamian astronomers hit on the degree by discovering how to inscribe a regular hexagon in a circle. If we divide the circumference into 360 divisions, this admittedly makes 6 centre angles whose measure accords with the base 60 of their system; but any such explanation overlooks the social challenge from which the need of angular measurement first emerges in human experience. It is at least more likely that the Sumerians chose 60 as their base because they first divided the circle in terms of a well-nigh world-wide compromise between solar and lunar time-keeping, *i.e.* the delimitation of the civil year as 12 months of 30 days. Several circumstances point to this conclusion. The Semitic peoples and the Hindus clung to the 360-day convention more than a millennium after they knew better, making the best of a bad job by intercalating leap months at appropriate intervals; and we have already seen that 360 is also the weak link in the chain of the Mayan numeral system. What is perhaps more eloquent is the fact that the Chinese, by the beginning of our era, divided the circle into $365\frac{1}{4}$ divisions, thereby adjusting their unit of angular measurement to a more sophisticated calendar.

Given a table of sines (whence of cosines also), the complete specification of a spherical triangle is possible in terms of two sides and the included angle, of three sides and any single angle or of three angles and any one side. This supplies a recipe for Ptolemy's major preoccupation, *i.e.* changing equatorial into ecliptic co-ordinates. If we use a, b, c, for the sides (in angular measure) and A, B, C, for the corresponding opposite angles, the indispensable formulae in current notation are:

1. $\cos \mathbf{a} = \cos \mathbf{b} \cdot \cos \mathbf{c} + \sin \mathbf{b} \cdot \sin \mathbf{c} \cdot \cos \mathbf{A}$

2. $\sin \mathbf{C} = \dfrac{\sin \mathbf{c} \cdot \sin \mathbf{A}}{\sin \mathbf{a}}$

To derive the first formula we construct: (a) two planes tangential to the arcs b and c inclined at the same included angle A; (b) a third plane including the arc a meeting both the other two at the centre of the sphere. We then conceive it to be possible to

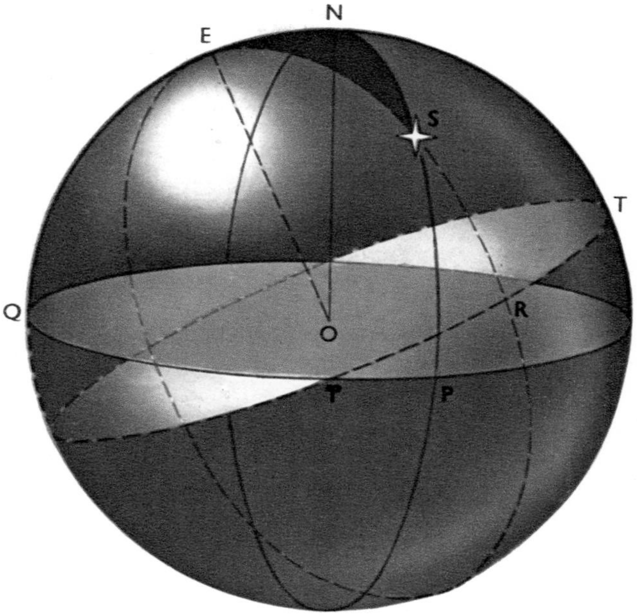

N — Celestial Pole
E — Pole of the Ecliptic
∠QN♈ — 90° = ∠TE♈
♈P — Star's R.A.
PS — Star's declination
♈R — Star's celestial longitude
RS — Star's celestial latitude
EON — Obliquity of the Ecliptic ($\simeq 23\frac{1}{2}°$)

TOP: CHANGING FROM EQUATORIAL TO ECLIPTIC CO-ORDINATES

In the triangle ENS, NS = 90° − star's declination, ES = 90° − star's latitude, EN ≃ 23½°. The angle NES = 90° − star's longitude, angle ENS = 90° + star's R.A.

$\cos NS = \cos ES \cdot \cos EN + \sin ES \cdot \sin EN \cdot \cos NES.$

$\therefore \sin(\text{declin.}) = \sin(\text{Lat.}) \cdot \cos 23\tfrac{1}{2}° + \cos(\text{Lat.}) \cdot \sin 23\tfrac{1}{2}° \cdot \sin(\text{Long.})$

$\sin ENS \div \sin ES = \sin NES \div \sin NS.$

$\therefore \sin(90° + \text{R.A.}) = \cos(\text{Lat.}) \cdot \cos(\text{Long.}) \div \cos(\text{declin.})$

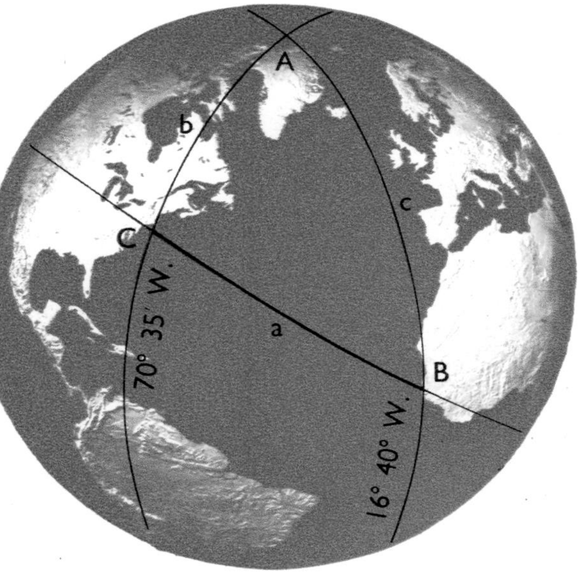

Latitude of B (Bathurst) ≃ 13° 28′ N.
Latitude of C (Nantucket Sound) ≃ 41° 20′ N.

BOTTOM: GREAT CIRCLE SAILING

In the spherical triangle ABC: $A = 70° 35' - 16° 40' = 53° 55'$; $b = 90° - 41° 20' = 48° 40'$; $c = 90° - 13° 28' = 76° 32'$.

$\therefore \cos a = \cos 48° 40' \cdot \cos 76° 32' + \sin 48° 40' \cdot \sin 76° 32' \cos . 53° 55'$

Whence from the tables $a = 54° 16\tfrac{1}{2}'$. *If we take 69½ miles (≃ 25,000 ÷ 360) as the linear value of a degree of a great circle, the length of the course from Bathurst to Nantucket Sound is ≃ 3772 miles. The reader can work out the compass bearing of the great circle route from either point by means of the second formula given in the text.*

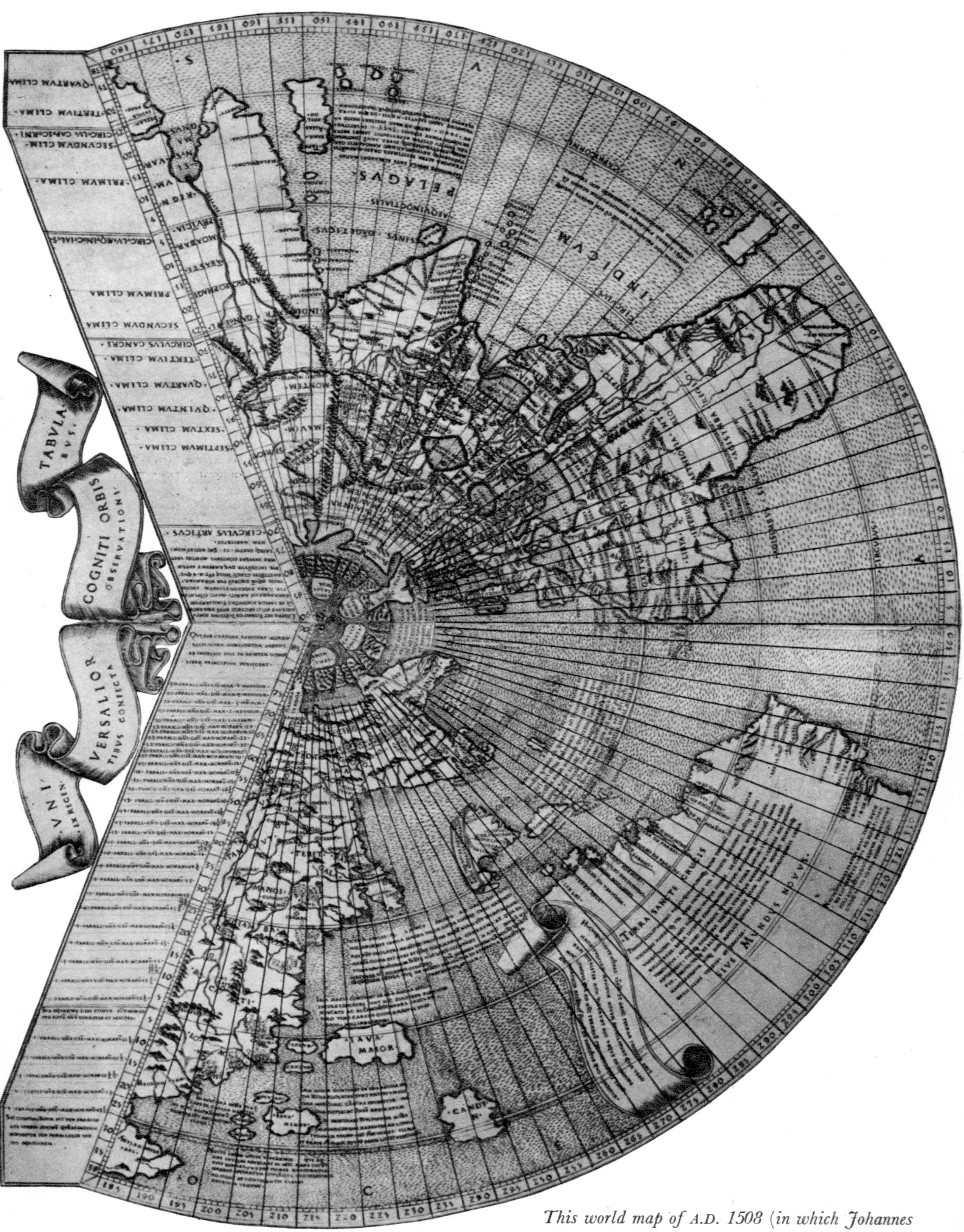

This world map of A.D. *1508 (in which Johannes Ruysch identifies part of the New World with Marco Polo's Asia) is drawn on a conical projection.*

open out the tetrahedron defined by their 6 edges, so that we derive 4 triangles in one and the same plane. The second formula follows from the first by putting $\cos a - \cos b . \cos c = \sin b . \sin c . \cos A$. We then square both sides, substituting $(\cos b)^2 = 1 - (\sin b)^2$ *etc.* and obtain:

$$(\sin A)^2 . (\sin b)^2 . (\sin c)^2 = (\sin a)^2 + (\sin b)^2 + (\sin c)^2 + 2 \cos a . \cos b . \cos c - 2$$

Similarly, by solving for cos b as above, we derive:

$$(\sin B)^2 . (\sin a)^2 . (\sin c)^2 = (\sin a)^2 + (\sin b)^2 + (\sin c)^2 + 2 \cos a . \cos b . \cos c - 2$$

$$\therefore (\sin A)^2 . (\sin b)^2 . (\sin c)^2 = (\sin B)^2 . (\sin a)^2 . (\sin c)^2$$

$$\therefore \frac{\sin A}{\sin a} = \frac{\sin B}{\sin b}$$

For the same reason $\sin A . \sin c = \sin C . \sin a$.

If we take the angle between the ecliptic and polar axes, *i.e.* the latitude of the Tropic of Cancer, as $23\frac{1}{2}°$, we thus have two equations for the unknown celestial latitudes and longitudes of a celestial body in terms of its known declination and R.A., *viz.*:

$$\sin (\textit{declin.}) = \cos 23\tfrac{1}{2}° . \sin (\text{Lat.}) + \sin 23\tfrac{1}{2}° \cos (\text{Lat.}) . \sin (\text{Long.})$$

$$\sin (90° + \text{R.A.}) = \cos (\text{Lat.}) . \cos (\text{Long.}) \div \cos (\textit{declin.})$$

No angle of the right-angled triangle can exceed 90°. Hence one term in the above is meaningless, if we define sines and cosines in terms of its sides; but their interpretation in terms of the radius of a circle and the line which joins the centre to the mid-point of a chord (*see* Claim 7) is both consistent with the alternative definition when $A \leqslant 90°$ and of wider reference. In Ptolemy's notation (p. 132), $2 \cos A = \text{chord } (180° - 2A)$, whence $\sin (90° + A) = \cos A$. Since $\sin 90° = 1$ and $\cos 90° = 0$, this result is consistent with the addition formula cited above, *viz.*:

$$\sin (90° + A) = \sin 90° . \cos A + \cos 90° . \sin A$$

Though less convenient from the viewpoint of the surveyor, the Alexandrian *chord* was therefore

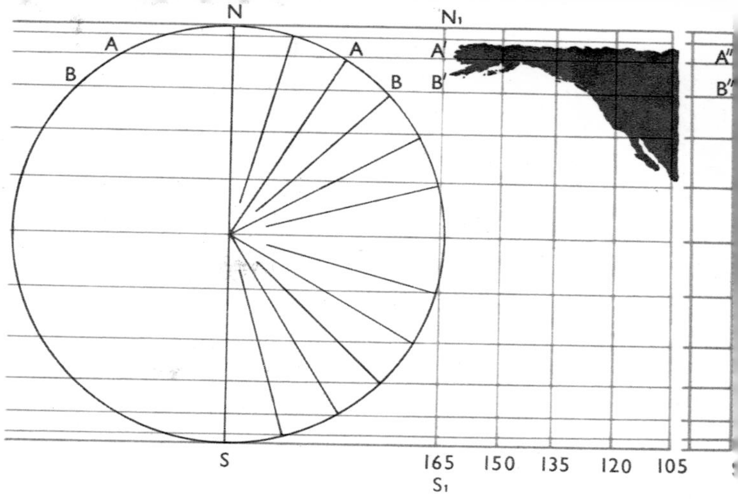

This diagram from T. W. Birch's Maps *(O.U.P.) shows the graphical construction of a cylindrical equal-area projection. This projection achieves invariance for area but grossly distorts shape in high latitudes. For comparison, the same part of North America which appears on the right of the diagram is shown below on the more familiar polyconic projection.*

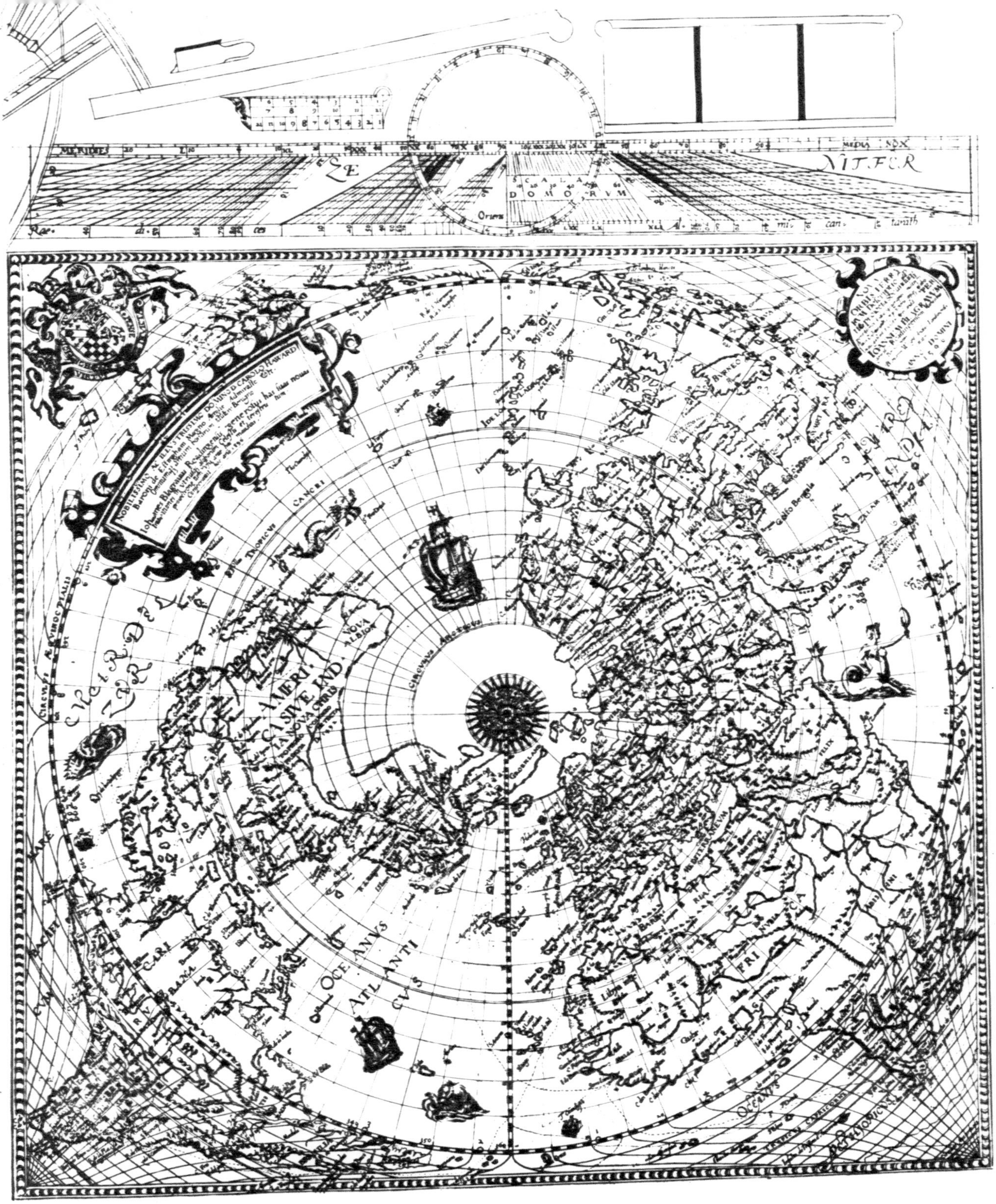

The projection for this late sixteenth-century map was devised by an Oxford mathematician, John Blagrave. The area within the circle is plotted on a polar stereographic projection. The changeover to sharply-converging meridians in the lower corners allows the cartographer to lend a touch of verisimilitude to what would otherwise be a most bald and unconvincing pictorial statement about the southern land-masses.

more explicitly adaptable than the Hindu *sine* to the end in view. On the other hand, it dispensed with the necessity to make a decisive step forward by equipping the sign of a trigonometrical ratio with a meaning referable to the direction in which the radius of a circle can rotate. It may be good for the morale of some of us to realise that it took mankind about 1500 years to take this step confidently, as we shall see in Chapter 9.

There we shall reinterpret the trigonometrical ratios in terms which can accommodate angles of any size. Here it suffices to do so in terms consistent with the addition formulae, *e.g.*:

$$\cos(180°) = \cos(90°+90°)$$
$$= \cos 90°.\cos 90° - \sin 90°.\sin 90° = -1$$
$$\sin(180°) = \sin 90°.\cos 90°. + \cos 90°.\sin 90° = 0$$
$$\sin(90°+A) = \sin 90°.\cos A + \cos 90°.\sin A$$
$$= +\cos A$$
$$\cos(90°+A) = \cos 90°.\cos A - \sin 90°.\sin A$$
$$= -\sin A$$
$$\sin(180°-A) = \sin 180°.\cos A - \cos 180°.\sin A$$
$$= +\sin A$$
$$\cos(180°-A) = \cos 180°.\cos A + \sin 180°.\sin A$$
$$= -\cos A$$

The last expression disposes of an anomaly in an award to Claim 4 cited in Chapter 5, *viz.* the rule for deriving the length of the remaining side of a triangle, if we know the lengths of two sides and the angle between them. If $\cos(180°-A) = -\cos A$, one formula embraces the two rules:

$$A < 90° : a^2 = b^2 + c^2 - 2bc.\cos A$$
$$A > 90° : a^2 = b^2 + c^2 + 2bc.\cos(180° - A)$$

By recourse to the foregoing sign conventions, we can write the two last-cited formulae involving spherical triangles in the more convenient form:

$$\sin(\text{Lat.}) = \cos 23\tfrac{1}{2}°.\sin(\textit{declin.}) -$$
$$\sin 23\tfrac{1}{2}°.\cos(\textit{declin.}).\sin(\text{R.A.})$$
$$\cos(\text{Long.}) = \cos(\text{R.A.}).\cos(\textit{declin.}) \div \cos(\text{Lat.})$$

Given the information that the polar axis rotates at $23\frac{1}{2}°$ around the pole of the ecliptic in 25,800 years, we can use the same formulae to find what was the declination of Sirius at the date of the building of the Great Pyramid. Hence from the inclination of the tunnel then pointing thereto, we can fix the date itself. The two fundamental formulae (1 and 2 above) have other eminently practical uses. If we know its altitude and azimuth (bearing on the horizon plane) at a particular local time, we can calculate what the time and z.d. at transit of a star will be, whence also: (a) its R.A. and declination; (b) the bearings and rising or setting times of a celestial body. We have also at our disposal a recipe for calibrating the Chinese-Moslem sundial in hours of equal length, *i.e.* in units of time which tally with those of a clepsydra (water clock) or a mechanical clock.

Among many others, two milestones of the Alexandrian saga pinpoint progress achieved

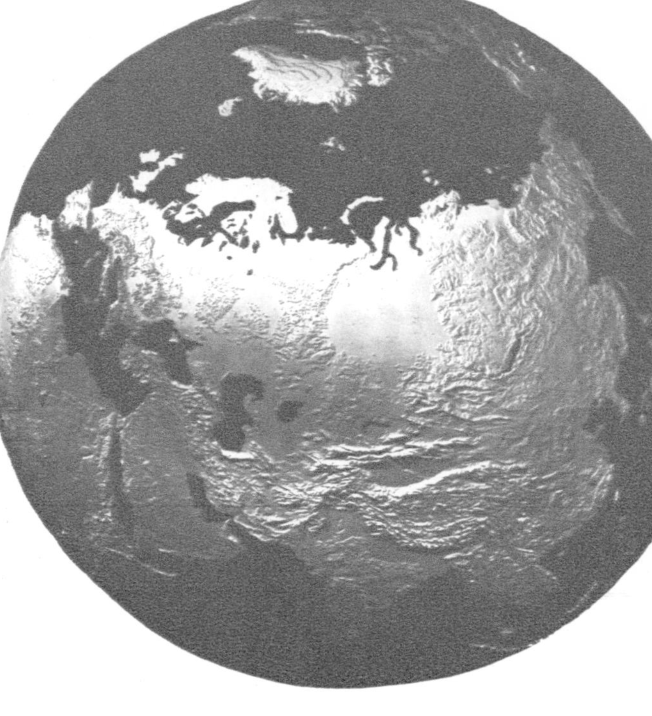

This map is a photograph of a globe some 30 inches in diameter, on which each thousand-foot contour is represented by the addition of a layer of material 0·012 in. thick. Photographing from such a globe by-passes many projection problems but it does not achieve invariance for all properties of the sphere. Proportionate area diminishes towards the periphery.

in the period between Aristarchus and Ptolemy. Hipparchus cites the length of the tropical year as 365 d. 5 h. 55 m. 12 s. (correct value 365 d. 5 h. 48 m. 46 s.) and that of the synodic month as 29 d. 12 h. 44 m. $3\frac{1}{3}$ s. (roughly half a second too much). Ptolemy's estimate of the moon's distance from the earth was $\simeq 61$ times the earth's radius. If we take the estimate of the earth's radius as approximately 4000 miles in close accordance with the determination of Eratosthenes, this makes the distance 244,000 as against the most accurate modern value of 238,862. Ptolemy's own linear estimate of the distance itself was deficient, being based on the figure cited by Poseidonius for the earth's radius.

Of the great synthesists in history, Ptolemy (*circa* A.D. 150) is unique. The geometrical books of Euclid's *Elements* contain no reference to the many contributors elsewhere named. Newton carries the burden of credit for what Omar Khayyam (and his Chinese predecessors), Fermat, Wallis, Isaac Barrow among his elders and even more what Huygens, Flamsteed and Leibniz among his own contemporaries, contributed to the common effort. Contrariwise, most of what we know of the contributions of Hipparchus, Marinus of Tyre (*circa* A.D. 100) and Menelaus (*circa* A.D. 160), we know only because Ptolemy had the uncommon honesty to acknowledge his sources.

Among the foregoing last-named, it seems that Marinus of Tyre first brought the co-ordinate geometry of the celestial sphere down to earth by the introduction of terrestrial longitude into the art of map-making. This innovation was to be of cardinal importance when ships were able to track due west or east across the Atlantic; but the master pilot of antiquity could not as yet make full use of it. Nor need we suppose that he felt the need for it. Before the Columbian voyages, ships kept close to shore by recognition of sea fowl, and latitude sufficed to locate a port on a northerly or southerly track beyond the confines of the Mediterranean. On the other hand, scientific geography could not advance beyond the level it had reached in the time of Eratosthenes while latitude was the only precise criterion of location between places widely separated.

The sole criterion of terrestrial longitude is the time which intervenes between a celestial event as observed on two meridians. This was a simple enough matter in an age already equipped with reliable mechanical clocks. We can then state what time on the zero meridian corresponds to our own noon or to the transit of any celestial body of known R.A. The Alexandrians made great advances in the construction of water clocks; but such devices were not portable. It was therefore impossible to determine longitude at sea. On land, however, local times of the beginnings of eclipses or of lunar occultations of planets at different places might be a matter of sufficient interest to record as portents if coincident with a major battle or other noteworthy event. A few such data were available to Ptolemy, as points of reference.

To fill in the gaps, he had to rely on linear measurements which were liable to gross error magnified by the linear value of the degree assigned by the Poseidonius estimate of the earth's radius. That Ptolemy's world map is so recognisable as such may be partly due to the possibility that errors of the first and errors of the second kind could cancel. However, our theme is not the making of geography. What is relevant to it is that Ptolemy the geographer bequeathed to posterity a challenge to mathematical ingenuity at two different levels. If clever people commonly solved problems without the invitation of an external stimulus, it would be remarkable that no one before the nineteenth century of our own era clearly saw the possibility of putting a pipeline between them.

Except in so far as it gave a powerful impetus to the advance of trigonometry in the Hindu-Moslem world, Ptolemy's pioneer work as a cartographer had little effect on the making of modern mathematics during the millennium after his death; but its subsequent influence was considerable. From the use of a map to chart a ship's course, analytic methods for tracing the genesis of a curve developed by imperceptible steps in the context of the great navigations of the Columbian era; and the representation of a spherical contour on a flat surface explicitly raised for the first time a problem of *invariance*, destined to become a dominant theme of mathematics during the latter half of the nineteenth century. To do justice to Ptolemy's contribution, it is therefore fitting to indicate very briefly what issues arise when we attempt to picture a three-dimensional object in a space of two dimensions.

Two-dimensional representations of outlines on a spherical surface are broadly classifiable as *circumscribed* (*e.g.* cylindrical equal-area), *perspective* and *metrical* (*e.g.* Mercator). Before distinguishing between them, let us return* (p. 131) to some elementary properties of the sphere. If a straight line parallel to the polar axis grazes the equator, a line drawn parallel to the latter at latitude L^R (measured in radians) will intercept a segment $R \cos\left(\frac{\pi}{2}-L^R\right) = R \sin L^R$ on the grazing line, so that the ratio of the length of the vertical projection of the arc between the parallel and the equator to its true length will be $R \sin L^R : R.L^R$. If a straight line inclined to the polar axis at an angle A^R in the same plane as a meridian grazes the sphere at latitude L^R, it follows that $L^R = A^R$. The ratio of the length of the grazing line intercepted between the plane of the equator and the plane of the parallel L^R to the arc measured along the corresponding meridian is $R \tan L^R : R.L^R$. These data suffice to define what, if any, properties are invariant under projection, when we flatten out the trace of a figure projected from the sphere when planes of the meridians and parallels of latitude cut the surface of an escribed grazing cylinder or cone coaxial with its polar axis.

If we unroll a cylinder of flexible plastic lined up in this way, we get a flat surface crossed by equally-spaced rectilinear meridians and rectilinear parallels of latitude becoming more closely packed towards the polar extremities. If we unroll a cone likewise placed with its vertical axis in line with the poles of the sphere, the meridians on the flattened surface will be straight lines diverging from a point (the vertex of the cone) north (or south) of the habitable globe and the trace of the parallels of latitude will be arcs of concentric circles. For the cylindrical projection the rectilineal value of a degree of latitude will be proportionally smaller as we go further from the equator; but the rectilinear distance between two meridians being constant will exceed their correct values on the same scale of measurement. It follows that there is a gross distortion of shape in the far north or far south, and distances measured along either parallels or meridians will not correspond to their correct values.

However, it happens that errors of the two sorts compensate, so that the cylindrical projection will faithfully represent areas of land masses or seas. We may express this by saying that the cylindrical recipe ensures that *area remains invariant under projection*. In side view, the projection of either eastern or western hemisphere on the cone has no such invariant property of use to the explorer, navigator or surveyor; but it does convey a less erroneous impression of the shape of land masses at the polar extremity nearest the vertex of the cone. The escribed cone cannot graze at the equator, unless its vertical angle is zero, in which event it is equivalent to an escribed cylinder; but if it grazes at the Tropic of Cancer it will accommodate without very gross distortion as much of the southern hemisphere as was known in Ptolemy's time. It also suggests recipes for the sort of projections we referred to above as *metrical*.

Whereas the conical projection *sensu stricto* preserves a correct scale of neither area nor distance in longitude or latitude, it is possible to readjust in different ways the spacing of the concentric arcs of latitude on the original projection in which straight lines diverging at equal angles represent meridians diverging equally at either pole. One such adjustment correctly preserves distances in latitude, another preserves distances in longitude; but it is not possible to achieve both results in one figure. If E is the radius of the arc which represents the equator, we can keep the scale invariant with respect to

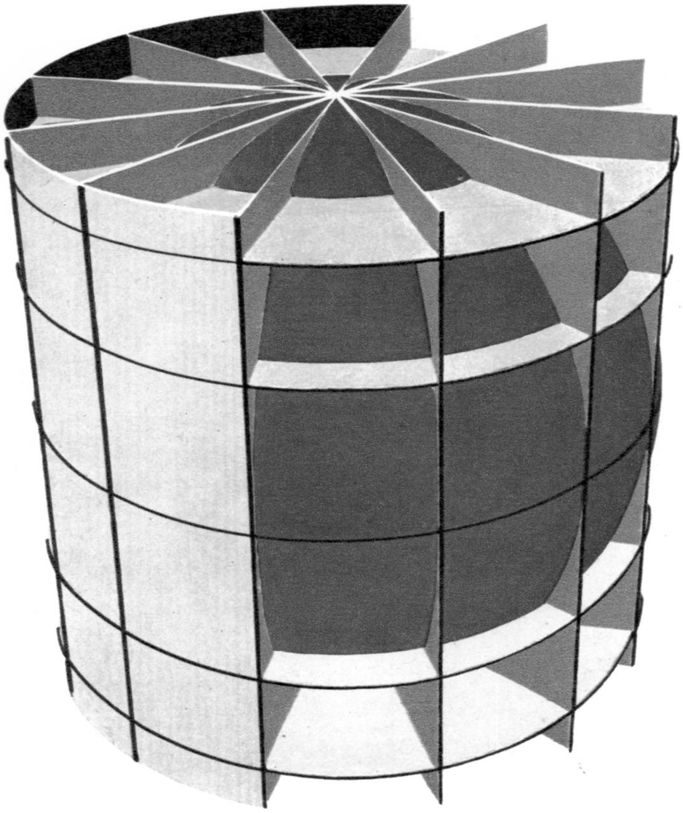

THE INVARIANT PROPERTY OF THE CYLINDRICAL PROJECTION

Here R is the radius of the sphere. For a small four-sided element on the sphere bordered by parallels of latitude L_1 and $L_2 = L_1 + \triangle L$ and by meridians which subtend an arc A in a plane parallel to the equatorial, the area is approximately the product of the height $(R.\triangle L)$ and the mean value of the width $\frac{1}{2}A.R(\cos L_1 + \cos L_2)$, in which $\cos L_2 = \cos(L_1 + \triangle L) = \cos L_1 . \cos \triangle L - \sin L_1 . \sin \triangle L$. As $\triangle L$ approaches its limiting value, $\cos \triangle L \simeq 1$ and $\sin \triangle L \simeq 0$, so that $\frac{1}{2}A.R(\cos L_1 + \cos L_2) \simeq A.R \cos L$, i.e. the product of $R.\triangle L$ and the mean width approaches $AR^2 . \cos L_1 . \triangle L$, the limiting value of which we write in the notation of the Newton-Leibniz calculus as $AR^2 . \cos L . dL$. We shall see in Chapter 10 that this is the limiting value of $AR^2 . \triangle \sin L$ which is the area of the corresponding four-sided element of the cylindrical projection.

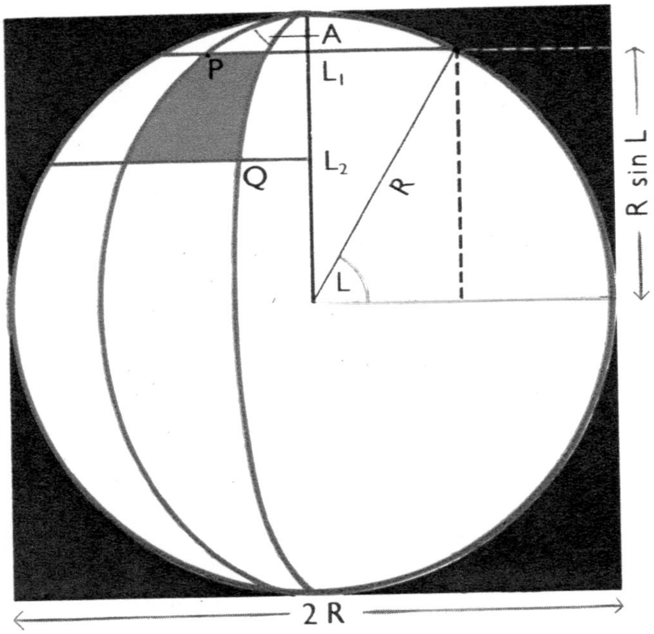

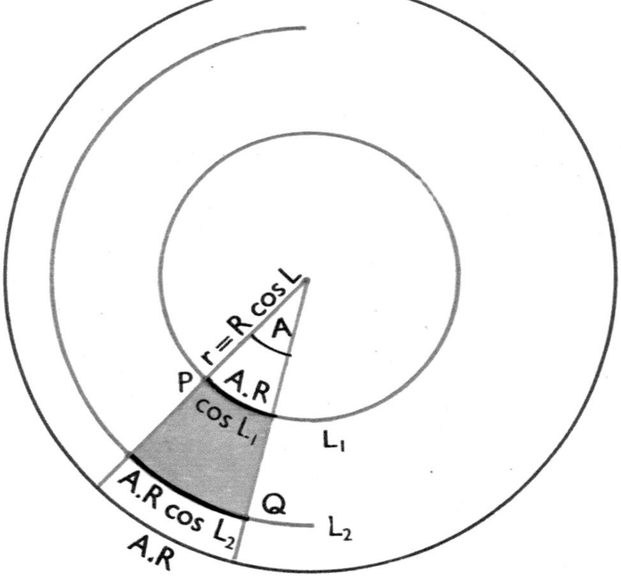

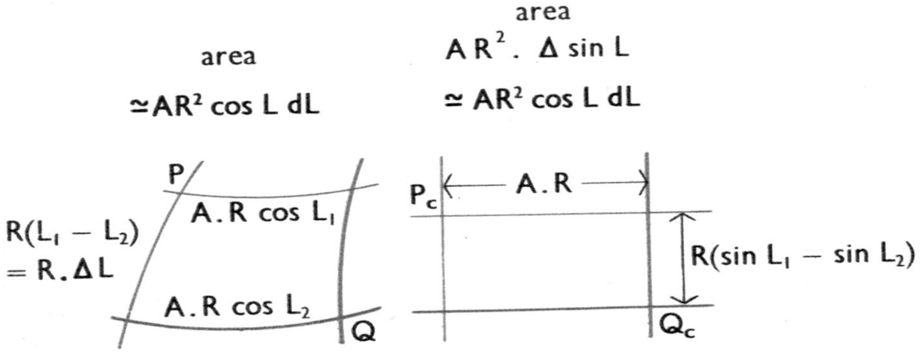

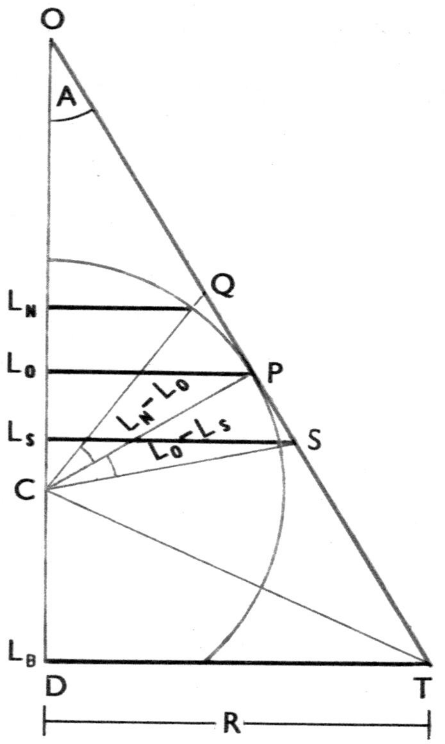

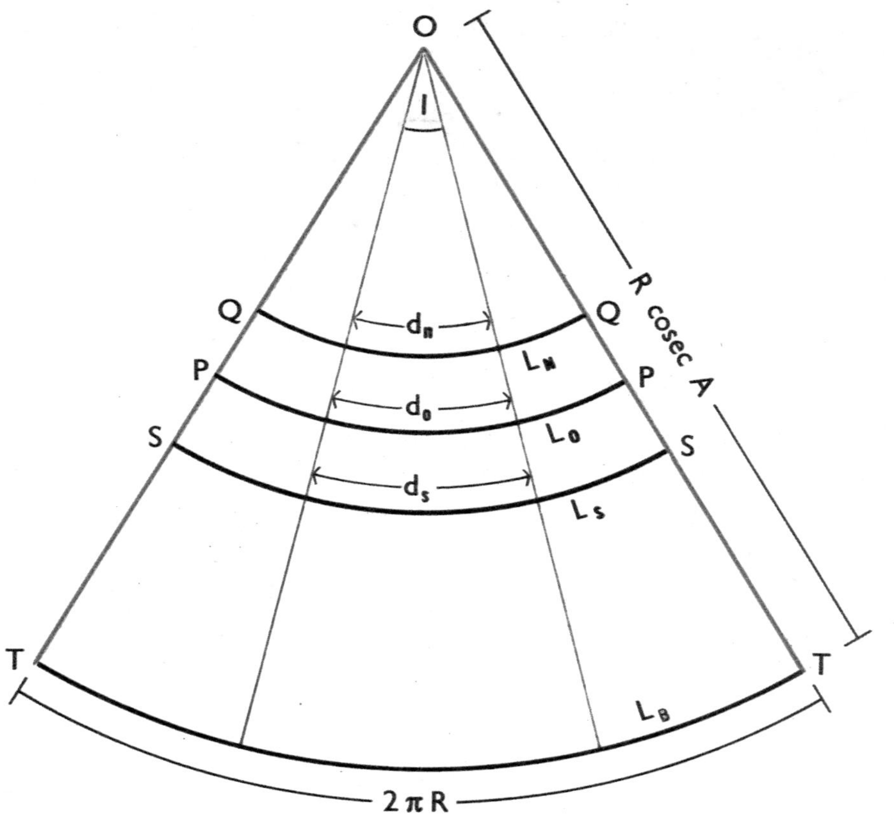

CONICAL PROJECTION

This shows the relative spacing of latitude along the same meridian in the conical projection. Here L_o is the standard parallel, i.e. latitude at which the cone is tangent to the sphere of radius $CP=r$. On the cone the projection of $QCP=L_n-L_o$ is $QP=r\tan(L_n-L_o)$, and the projection of $SCP=L_o-L_s$ is $SP=r\tan(L_o-L_s)$. The projection of $TCP=L_o-L_b$ is $PT=r\tan(L_o-L_b)$. The radius of the base of the cone is $DT=R$. For use in the next figure note the following relations:

$$OP=\frac{r}{\tan A}$$

$$OQ=\frac{r}{\tan A}-r\tan(L_n-L_o)$$

$$OS=\frac{r}{\tan A}+r\tan(L_o-L_s)$$

$$OT=\frac{R}{\sin A}$$

It is common to write the reciprocals of sin A, cos A and tan A respectively as cosec A, sec A and cot A, so that $\cot A=(\tan A)^{-1}$ etc. In this notation:

$$OP=r\cot A;\ OT=R\operatorname{cosec} A$$
$$OQ=r\cot A-r\tan(L_n-L_o)$$
$$OS=r\cot A+r\tan(L_o-L_s).$$

Since $OT=OP+PT$:
$$R\operatorname{cosec} A=r\cot A+r\tan(L_o-L_b).$$

This shows relative lengths of parallels intercepted on the conical projection by the same meridians. We first note that the projection of any point on the same parallel is equidistant from the vertex. The parallels appear on the flat projection as circular arcs, and the meridians as straight lines radiating from the vertex (O). Thus the length of the arc which subtends a longitude difference l (in radian measure) on the standard belt of latitude is $d_o=OP.l$. If d_n and d_s are respectively the lengths of the arcs on the belts at latitudes L_n and L_s:

$$\frac{d_n}{d_o}=\frac{OQ}{OP}=\frac{OP-QO}{OP}=\frac{\cot A-\tan(L_n-L_o)}{\cot A}$$
$$\frac{d_s}{d_o}=\frac{OS}{OP}=\frac{OP+PS}{OP}=\frac{\cot A+\tan(L_o-L_s)}{\cot A}.$$

On the projection A of the previous figure is expressible in terms of the angle $TOT=a$, the radius (r) of the sphere and that (R) of the base of the cone, because:

$$R\operatorname{cosec} A(a)\ =2\pi R$$
$$\therefore\ a=2\pi\sin A$$

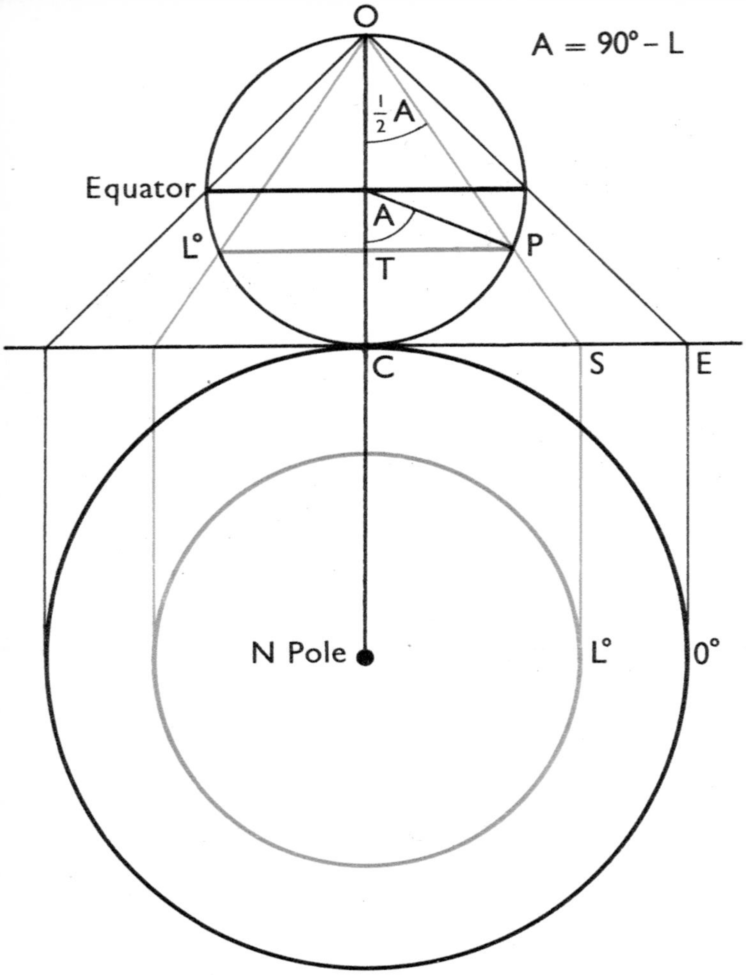

STEREOGRAPHIC PROJECTION OF THE NORTHERN HEMISPHERE

The observer O is at the South Pole. A is the co-latitude (90° — Lat.) of the point P, and S is its projection on the plane of contact at C (the North Pole). PT is the radius of the terrestrial latitude in the plane at right angles to the polar axis (OC) and SC is that of its projection on the plane.

If R is the earth's radius:

$SC = 2R . \tan \tfrac{1}{2}A$ *and* $PT = R \sin A = 2R . \sin \tfrac{1}{2}A . \cos \tfrac{1}{2}A$

$$\therefore \frac{SC}{PT} = \frac{1}{(\cos \frac{1}{2}A)^2} = \frac{2}{1 + \cos A}$$

$$\text{and } SC = \frac{2R \sin A}{1 + \cos A} = \frac{2R \cos(Lat.)}{1 + \sin(Lat.)}$$

At the equator (Lat. 0°), the radius (CE) of the Northern hemisphere projection will be twice that of the equator in orthographic projection.

distance between equally-inclined meridians along different parallels if we mark off concentric arcs from the same vertex but of radius *E*.cos *Lat.* Contrariwise, equally-spaced concentric arcs corresponding to the parallels will keep the scale invariant with respect to distance between the same parallels along any meridian. Though called *conical*, such projections belong to the class better called *metrical*, since the layout corresponds neither to projection of the sphere on a circumscribed figure nor to a third procedure invoked in perspective art work.

The perspective method represents points on the sphere by the interception on a flat surface of lines converging to the eye of a hypothetical observer. If the observer is indefinitely remote from the sphere, the lines are parallel; and the projection (called *orthographic*) corresponds to a moon's eye view of the earth; but it has no useful metrical properties. For most purposes of mechanical draughtsmanship or anatomical exposition, orthographic projection in planes at right angles suffices for the end in view; but it has little to commend it from the viewpoint of the cartographer except as a descriptive device for familiarising children with the layout of the great land-masses.

A second perspective method is of greater interest because it has an invariant property not mentioned above. For the *stereographic* or conformal projection, the focal point of an imaginary observer's vision is on the surface of the sphere, which makes contact with the flat plane at a point on the other side of the sphere exactly opposite the observer. Thus the contact plane may touch at one pole, being parallel to the plane of the equator. In that event, the observer will be at the opposite pole. Alternatively the contact plane may be at right angles to the plane including the polar axis, the Greenwich meridian and the date line. In that event, the co-ordinates of the contact point and of the observer will respectively be Lat. 0°, Long. 0° and Lat. 0°, Long. 180° – or *vice versa*.

How far Ptolemy clearly realised the several

merits and disadvantages of different projections, and thus anticipated the notion of invariance common to so many fields of nineteenth-century mathematics, is enigmatic. That he made three different maps, based on different methods, justifies the verdict that he realised the impossibility of achieving invariance with respect to every useful geometrical property of the sphere by any one method. Of the three he describes, one was a projection of the conical type. A second map aimed at a better compromise by drawing curved lines for meridians as well as parallels. A third one, recently clarified by Neugebauer, is a compromise between the perspective and the metrical approach.

It would not be true to dismiss as negligible the influence of Ptolemy on his immediate successors. Pappus (*circa* A.D. 250), the last of the great geometers of antiquity, seems to have been the first to have recognised a fundamental invariant property of the projection of 4 points in one and the same straight line. If the four points a, b, c, d, on one straight line correspond to A, B, C, D, on another and lines connecting corresponding pairs (Aa, *etc.*) meet at a common vertex V, the cross-ratio rule of Pappus states that:

$$\frac{DB.CA}{DA.CB} = \frac{db.ca}{da.cb}$$

A proof depends on formulae for expressing the area of a triangle alternatively in terms of its height (h) and base or two sides and the sine of the included angle, *e.g.* the area of the triangle $DVA = \frac{1}{2}h.DA = \frac{1}{2}DV.AV.\sin DVA$. Since the four triangles whose bases are DA, DB, CA and CB have the same vertex, their heights (h) are the same; and we may write:

$$\frac{DB.CA}{DA.CB} = \frac{\frac{1}{2}h.DB.\frac{1}{2}hCA}{\frac{1}{2}h.DA.\frac{1}{2}hCB}$$

$$= \frac{\frac{1}{4}DV.BV.\sin DVB.CV.AV.\sin CVA}{\frac{1}{4}DV.AV.\sin DVA.CV.BV.\sin CVB}$$

$$\therefore \frac{DB.CA}{DA.CB} = \frac{\sin DVB.\sin CVA}{\sin DVA.\sin CVB}$$

$$= \frac{\sin dVb.\sin cVa}{\sin dVa.\sin cVb} = \frac{db.ca}{da.cb}$$

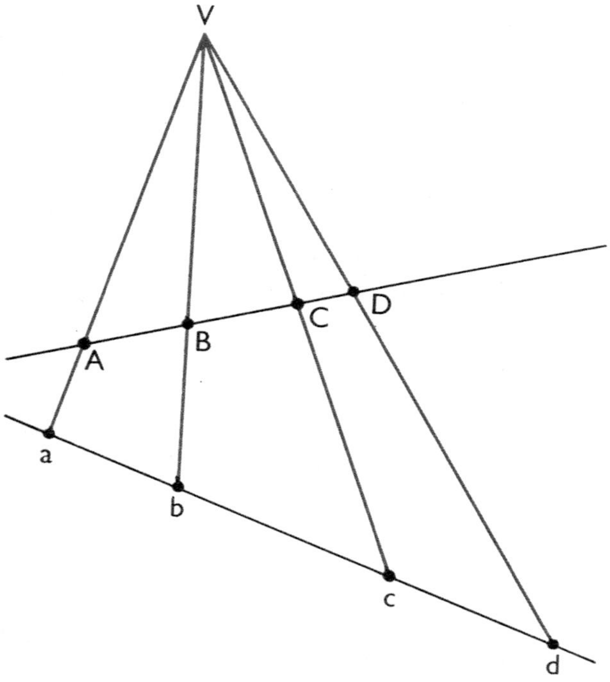

THE INVARIANT CROSS-RATIO

If four points a, b, c and d on one straight line correspond with four points A, B, C and D on another, and if straight lines joining corresponding pairs of points meet at a common vertex, V:

$$\frac{DB.CA}{DA.CB} = \frac{db.ca}{da.cb}$$

This example of invariance appears to have been first recognised by Pappus (c. A.D. 250).

Chapter 7 The Oriental Contribution

Through the labours of two scholars, we have advanced during the last two decades towards a more global view of the making of mathematics. We know much more than heretofore about the contributions of the temple cultures of the West through the erudite studies of O. Neugebauer as summarised in *The Exact Sciences in Antiquity*. Thanks to Joseph Needham (*Science and Civilisation in China*), we also know far more than formerly about the level of mathematical attainment in the Far East during the period which antedates intimate linguistic contacts between East and West through Jesuit missionary enterprise (A.D. 1550–1800). Scholars with different prejudices hold diverse views both about how far the efflorescence of mathematical knowledge in China, in India and in the Moslem world developed independently of the Alexandrian synthesis during the period which antedates the rise of European mathematics in the sixteenth century of our era and about how far they contributed unique ingredients to a new synthesis. It will therefore be wise to be clear at the outset concerning what we certainly can and certainly cannot conclusively assert.

Inter alia, the West owes to the Chinese silk and sugar, paper and printing, gunpowder and lock gates for canals. Before the age of modern industrialism, it is therefore fair to say that the West took from China many more material amenities than it conferred in return. Such flow of inventive skills is consistent with a comparatively primitive level of linguistic communication at an oral level; but it does not justify the belief that there could have been any links through the written record between the alphabetic Alexandrian and the *literati* of a territory where the only medium of scholarship was (and is) an ideographic script comparable to those of the temple cultures of antiquity. Even with India, the iron curtain of an alien speech and a still more alien script proscribed any considerable measure of sophisticated culture contact till the introduction of Buddhism.

Seemingly the first Buddhist monks arrived in China about A.D. 75, but the translation of Buddhist literature by co-operation between monks from India and Chinese scholars did not get under way till about A.D. 150. There is thus a *prima facie* case that any features of Chinese mathematics traceable to a date appreciably earlier than A.D. 200 are indigenous and that Hindu mathematics, such as it was at that time, did not borrow from China. It is therefore worthy of comment that extant Chinese star catalogues in declination and R.A. antedate the time of Hipparchus. Links through Buddhist scholarship between India and the West existed before his time. The convert Hindu king Asoka (*circa* 250 B.C.) dispatched missions of Buddhist monks to both the Seleucid and the Ptolemaic courts. Not improbably some of them, said to have settled in the Sinai peninsula and further north, laid the foundations of the Christian monastic tradition traceable from the date of the Dead Sea Scrolls

Top: Fifth-century cave painting from Tun-Huang, early meeting-place of Chinese and Indian cultures.

Left: The Buddhist monk Hsüan Tsang who journeyed from India to China, c. A.D. *643. By that date Buddhist monks from India had been in contact with China for some six centuries, and scholars of both lands had been co-operating on translations for some five centuries. Only culture contact at that level enabled sophisticated mathematical concepts to cross the frontiers. At the older level of trade contact, the barrier of speech and script had prevented the interchange of any ideas depending largely on verbal expression.*

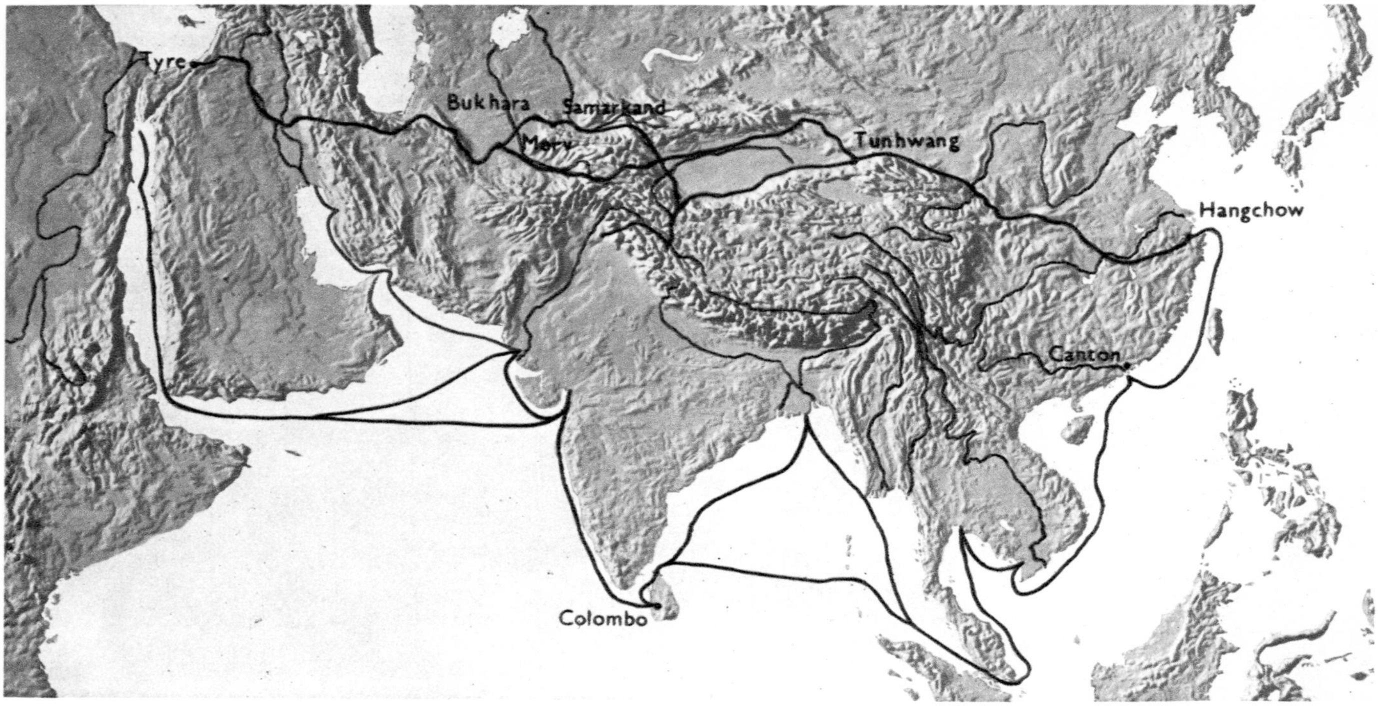

Map of main Indo-Chinese trade routes in use before the Christian era. In the flow of inventive skills along these routes it seems that the traffic from east to west exceeded that from west to east.

This sculpture of about A.D. 700 from the caves of Yun-Kang shows a marked Indian-Buddhist influence.

(possibly 150 B.C.) through the Essenes of the first century A.D. to the Thebaiad.

However, direct evidence of the transmission to the Far East of mathematical knowledge from the Hellenistic civilisations before the translation of Greek texts into Arabic by Persian scholars (A.D. 750–820) is lacking; and we may date the efflorescence of Hindu mathematics by the composition (*circa* A.D. 400) of the *Surya Siddhanta*, an astronomical work which records the moon's distance from the earth as $\simeq 60$ times the radius of the latter, cites a figure for the precessional motion of the ecliptic relative to the equatorial and shows familiarity with the use of terrestrial longitude. All this is consistent with the eastward diffusion of Alexandrian science during the period between that of Hipparchus and later astronomers of the West; but there is no direct evidence for the surmise. What we certainly know is that Hindu trigonometry, in close partnership with surveying for irrigation and with architecture by shadow

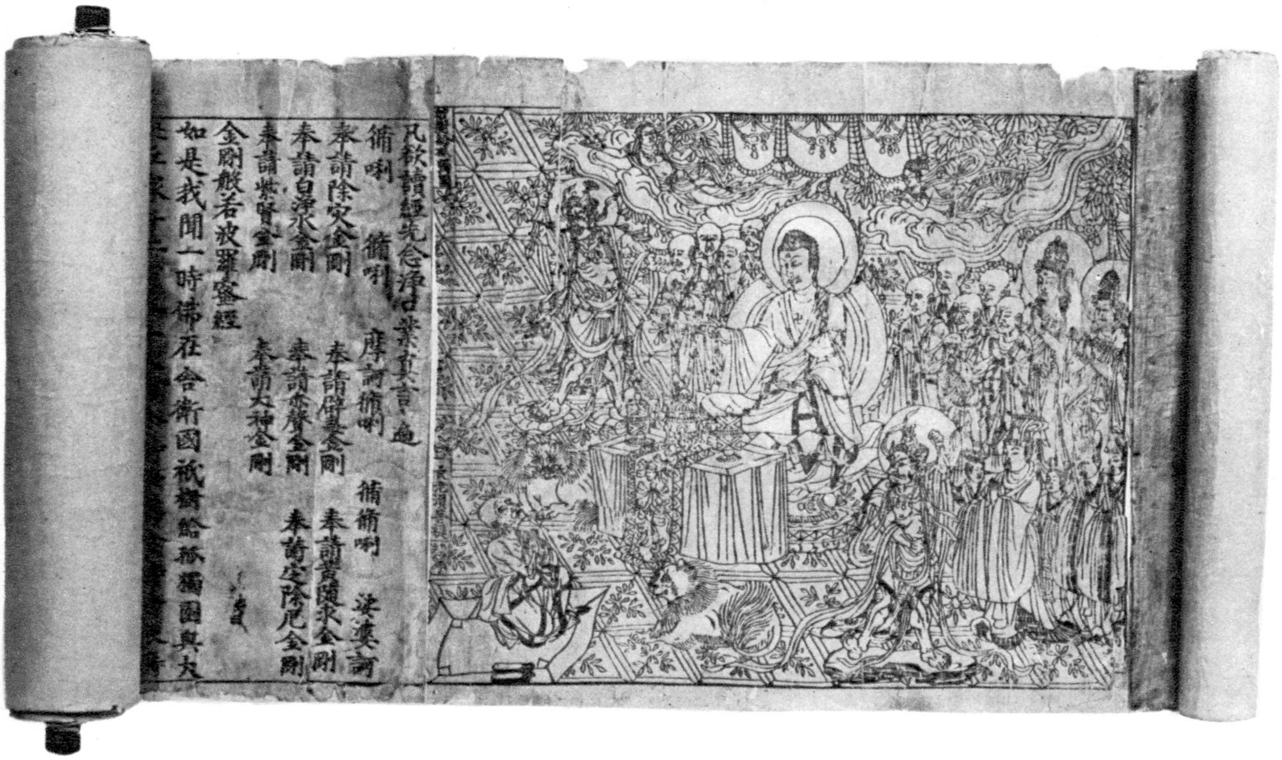

Part of the Diamond Sutra, the earliest extant Chinese block-print of certain date, A.D. *868. The great antiquity of printing in China makes it easier to speak with certainty of mathematical attainment there than in most other regions during the period* A.D. *850–1450.*

reckoning, had assumed the form which has come down to us through Moslem sources in spite of the powerful subsequent influence of the Ptolemaic tables of chords.

Some six centuries elapse between the putative date of the composition of this treatise and the transcription of the earliest extant copy, which may therefore contain amendments and addenda inserted after India had access to the *Almagest*. However, there is little reason to doubt that the Hindus were familiar with the use of *sine*, *sine complement* (cosine), *versed sine* (vers. sin $A = 1 - \cos A$), tangent and its reciprocal (cotangent) by A.D. 500. By that date they had crude tables based on the half-angle formula (Award to Claim 8). Brahmagupta (*circa* A.D. 630), says Winter (*Eastern Science*), "knew the rules for arithmetical operations involving zero, appreciated the use of negative quantities and of negative terms in algebra, studied quadratic equations and had considerable success in the solution of indeterminate (so-called Diophantine) equations." The last item more likely points to Chinese than to Western influence; but the grasp of the zero concept (in contradistinction to the zero sign) reinforces the claim that the latter is indigenously a Hindu creation.

The facts are as follows. The Moslems indisputably transmitted to Europe the numerals we now use, at a time (*circa* A.D. 1100) when Sicily was under Moslem rule and the Moslem universities of Spain were a beacon of learning in an otherwise dark age of Western civilisation. There is indeed no reason to doubt the testimony of Moslem

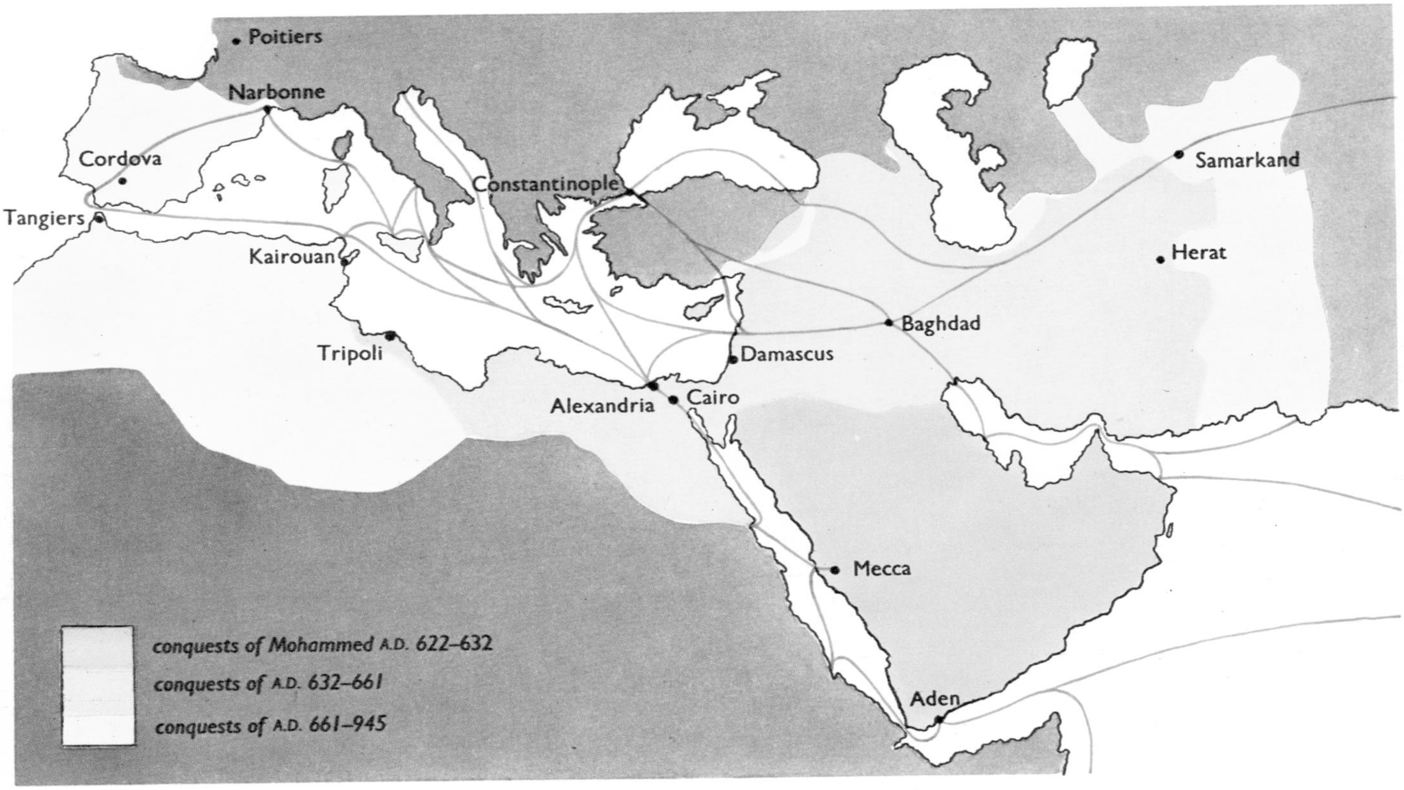

Map showing the growth of the Moslem Empire between A.D. 622 and 945, and some main trade routes within and beyond it. In this great stretch of Europe and Asia, Moslem scholars brought about a fruitful marriage between the mathematics of East and West.

Below is a fourteenth-century miniature depicting Mohammed at the siege of Banu Nadir. It is interesting that such early representations of Moslem conquests were painted only after the Mongols had established dominion over much of Asia.

writers that the use of the nine cyphers with a zero sign came from India. However, extant documentary evidence does not antedate the rise of Moslem civilisation. Our first relics in which zero appears come from the Indo-Chinese border (*circa* A.D. 700); the date of the earliest available inscriptions in which we meet it in continental India is about A.D. 850, somewhat later than its first known appearance, by which time China had block-print books. On epigraphic evidence alone, it is therefore possible to make a plausible case for the view that the zero sign as used by the world of today originated in China; but the efflorescence of the art of calculation in India, three centuries before recovered inscriptions entitle us to say with certainty that the Hindu-Arabic numerals as we know them were already in use, is difficult to reconcile with the belief that they were not. The precise contributions of Hindu and Moslem civilisations to the rules of calculation which the latter transmitted to us are difficult to disentangle, because their elaboration proceeded along parallel lines during a period when there was lively culture contact between India and Persia.

Within less than a century after the flight (A.D. 622) of the Prophet to Medina, his followers had overrun Syria (by A.D. 638), Egypt and Iraq (by A.D. 650), all North Africa and Spain (A.D. 714). Their contribution to the cultures of East and West alike begins with the rule of the Abassid caliphs (*circa* A.D. 750) whose territory embraced Iraq, Persia and the part of Turkestan whose most notable city was Samarkand. To Baghdad, their capital, came scholars with Syriac translations of Alexandrian science by Nestorian monks and doubtless with original texts from the surviving relics of the Greek-speaking colonies of the Near East. The work of translation into Arabic began under the second Caliph, whose finance minister was the son of Barmak, a Moslem convert, previously the hereditary keeper of a Buddhist temple. It is here appropriate to cite Winter (*Eastern Science*):

> the second half of the eighth century A.D. was a period of transmission of knowledge to the

This thirteenth-century picture shows scholars in the Abode of Wisdom at Baghdad, "the most effective academy of science since the museum at Alexandria."

A page from an Arabic translation of Ptolemy's Almagest, *A.D. 1294.*

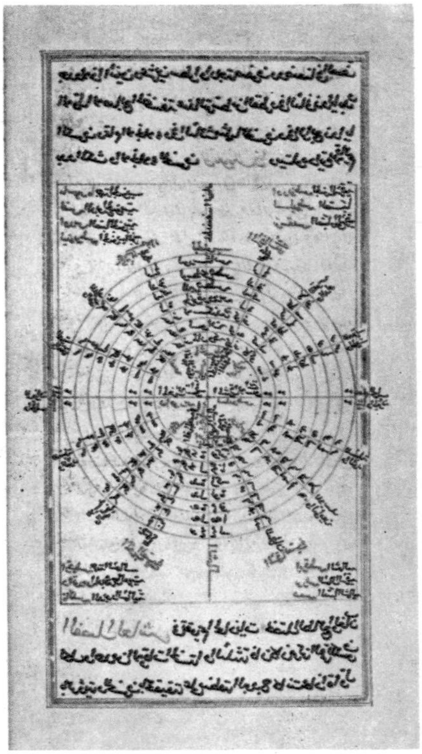

Arabs and scholars were mainly concerned in translation and elaboration of texts from the Syriac, Pahlawi, Greek and Sanskrit, especially under the second and fifth Abassid Khalifs Al-Mansur (754–75) and Harun al-Rashid (786–809). Under the former the city of Baghdad was planned by the Jewish astronomer and astrologer Mashalluh and the Persian astronomer and engineer Al-Nanbakht, and the translation of the Hindu astronomy defined by the term *siddhanta* was made by Muhammad ibn Ibrahim al-Fazari . . .

. . . In the ninth century the work of assimilation and translation went on. The seventh Abassid Khalif, Al-Mamun (813–33), a great patron of science, sent a mission to the Byzantine emperor, Leon the Armenian, to collect Greek manuscripts and established the Abode of Wisdom (Bayt al-hikma) at Baghdad which was the most effective academy of science since the museum of Alexandria.

Till Baghdad itself became the prey (A.D. 1050) of the Seljuk Turks, circumstances were henceforth peculiarly propitious for the intermingling of Alexandrian science with what was indigenously Hindu and what of Chinese culture had penetrated into India through Buddhist contacts. By that date, Moslem scholarship had established itself in the universities of Seville, Cordova and Toledo in Moorish Spain. It would be rash to dogmatise about how far the new synthesis influenced China directly during the three centuries of Baghdad's intellectual glory, the peak period (A.D. 813–833) of which was the reign of the seventh caliph Al-Mamun. Close contact between Chinese and Moslem culture developed when the successors of the Mongol conqueror Jenghiz Khan occupied Baghdad in A.D. 1258 and subsequently adopted the faith of Islam (*circa* A.D. 1300). During the next two centuries, Moslem overlords held dominion over a territory which extended to the borders of North China, into which Mongol armies penetrated deeply (A.D. 1213–23). By the time Tamerlane (Timur) destroyed Baghdad (A.D. 1383), Samarkand was emerging as a new focus of Moslem culture. There it was that Ulugh Beg (A.D. 1436) completed the construction of astronomical tables of which Flamsteed (the first English Astronomer Royal) made extensive use.

Since the particular contribution, directly or indirectly, made to our common civilisation by the Hindu and Moslem cultures intrudes into the great awakening of Europe in the *milieu* of the exploration of the New World, the main theme of this chapter will be what we may plausibly regard as ingredients contributed by China via India, through the medium of the Buddhist connexion, or through Moslem territory under Mongol overlords. First it is necessary to be more explicit than hitherto about the Chinese number script. Side by side with the numeral signs in the characteristic flowing ideographical style (shown in Chapter 2), the Chinese used at the beginning of the Christian era the *rod* forms which consistently preserve the primitive repetitive principle, written vertically (10^0, 10^2, 10^4 . . .) or horizontally (10^1, 10^3 *etc.*) as below:

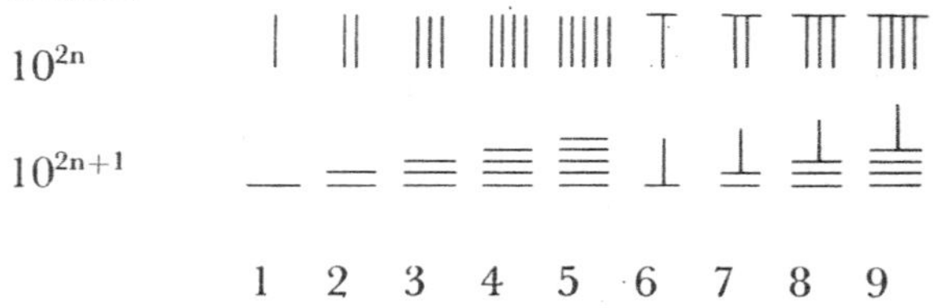

As is true of all the early repetitive numeral scripts, the Chinese followed a place value system, *e.g.* 3 as |||, 30 as ≡, 33 as ≡|||, 330 as |||≡ and 333 as |||≡|||. None the less, ambiguities arise, *e.g.*

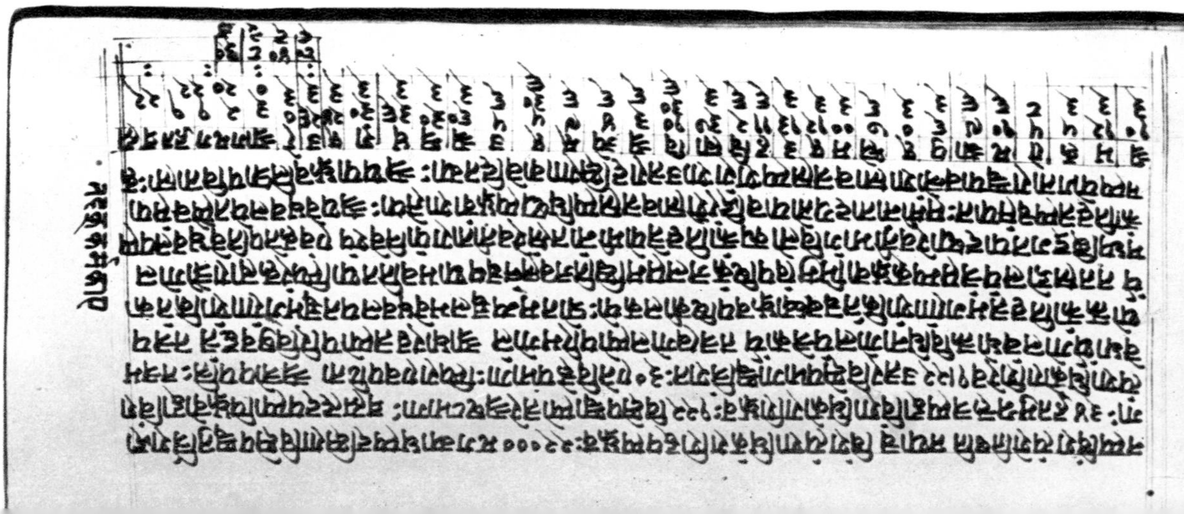

whether to interpret ||| as 3 or 300, ≡ as 30 or 3000, when we transfer these to paper without recourse to the zero symbol, which came into use not later than the eighth century A.D. Thereafter it was possible to represent the subtraction $1{,}470{,}000-1{,}405{,}536=64{,}464$ as follows:

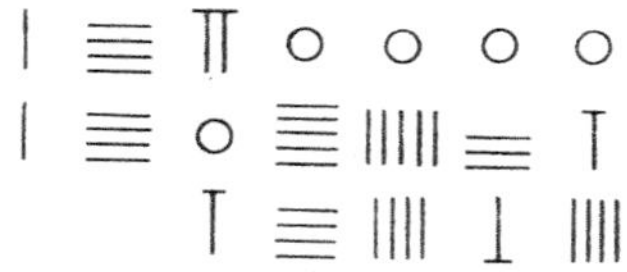

One feature of Chinese mathematics is of special interest in connexion with the use of the rod numerals as a *replica* of computation on the earliest form of the Chinese abacus. This was a flat piece of wood divided into squares like a chess-board, and the computer carried out his task by placing sticks of equal lengths on appropriate squares, leaving vacant a space corresponding to the empty column of the later counting-frame. The English word *exchequer* (counting on the chequer-board with counters) survives from the use of a similar device in medieval Europe.

The grid layout of the counting-board anticipates several peculiar features of Chinese mathematics. One is an early preoccupation with the magic square, and later with the type of algebraic problems now subsumed by the term *combinatory analysis*. Another is a systematic routine for solving linear equations by a chess-board arrangement of detached coefficients on all fours with the use of *determinants* as developed independently but much later in the West. In short, a grid-like layout of the numerical components of algebraic equations, wherewith it was possible to formulate rule-of-thumb methods of solution by the beginning of the Christian era, anticipates by more than a millennium the *matrix* notation which did not take shape in the West till the mid-nineteenth century. Nearly three centuries before then, the Japanese mathematician Seki Kowa had fully anticipated all the more important matrix operations involved in the solution of simultaneous linear equations involving any number of variables. Though it is unlikely that this development directly influenced European mathematics, it is instructive to outline the procedure in modern terms, starting with the simplest type of solution.

We shall first consider the pair of equations: $7x-8y=-21$ and $3x+2y=29$. Instead of proceeding by elimination as in our schooldays, we shall now arrange the terms of the above in our own symbols as on the left below, exhibiting on the right with rods of different thickness to identify their signs (as the Chinese used rods of different colours) the numerical components:

$$7x-8y+21=0$$
$$3x+2y-29=0$$

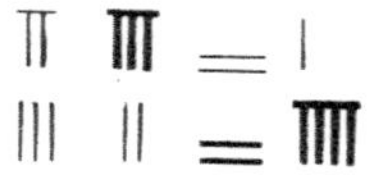

The counting-rod layout constitutes what we now call a 2×3 *matrix*, written as

$$\begin{bmatrix} 7 & -8 & 21 \\ 3 & 2 & -29 \end{bmatrix}$$

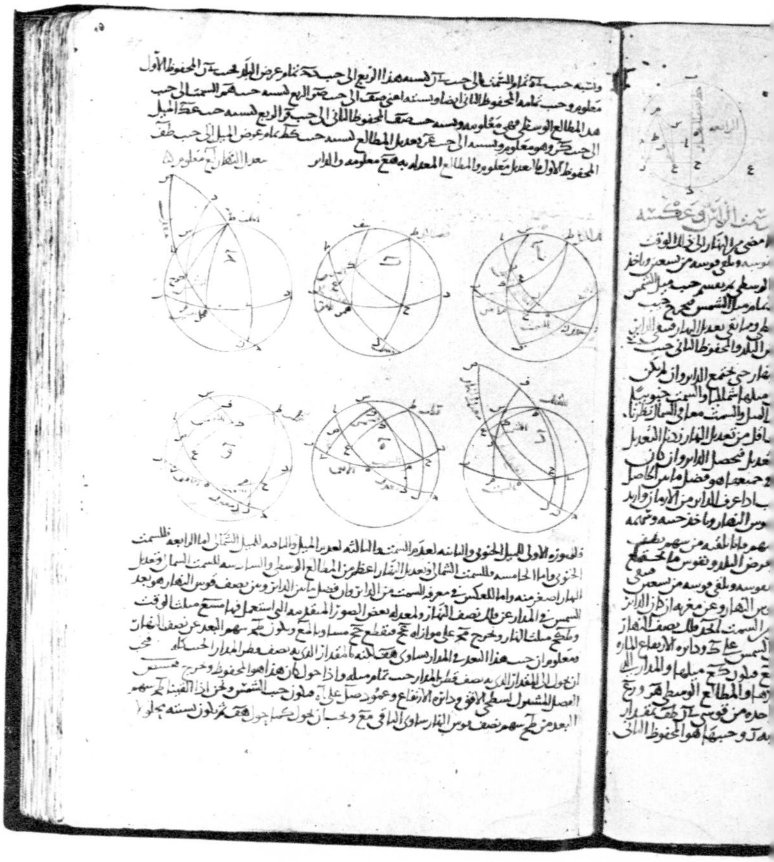

Left: Part of a sixteenth-century transcription of the Surya Siddhanta, *an astronomical treatise composed* c. A.D. 400. *At or soon after the date of composition Hindu mathematicians were doubtless familiar with the use of sine, cosine, tangent and cotangent.*

Right: Al-Beruni's treatise on astronomy (c. A.D. 1044) *based on the* Surya Siddhanta *and other Indian astronomical texts.*

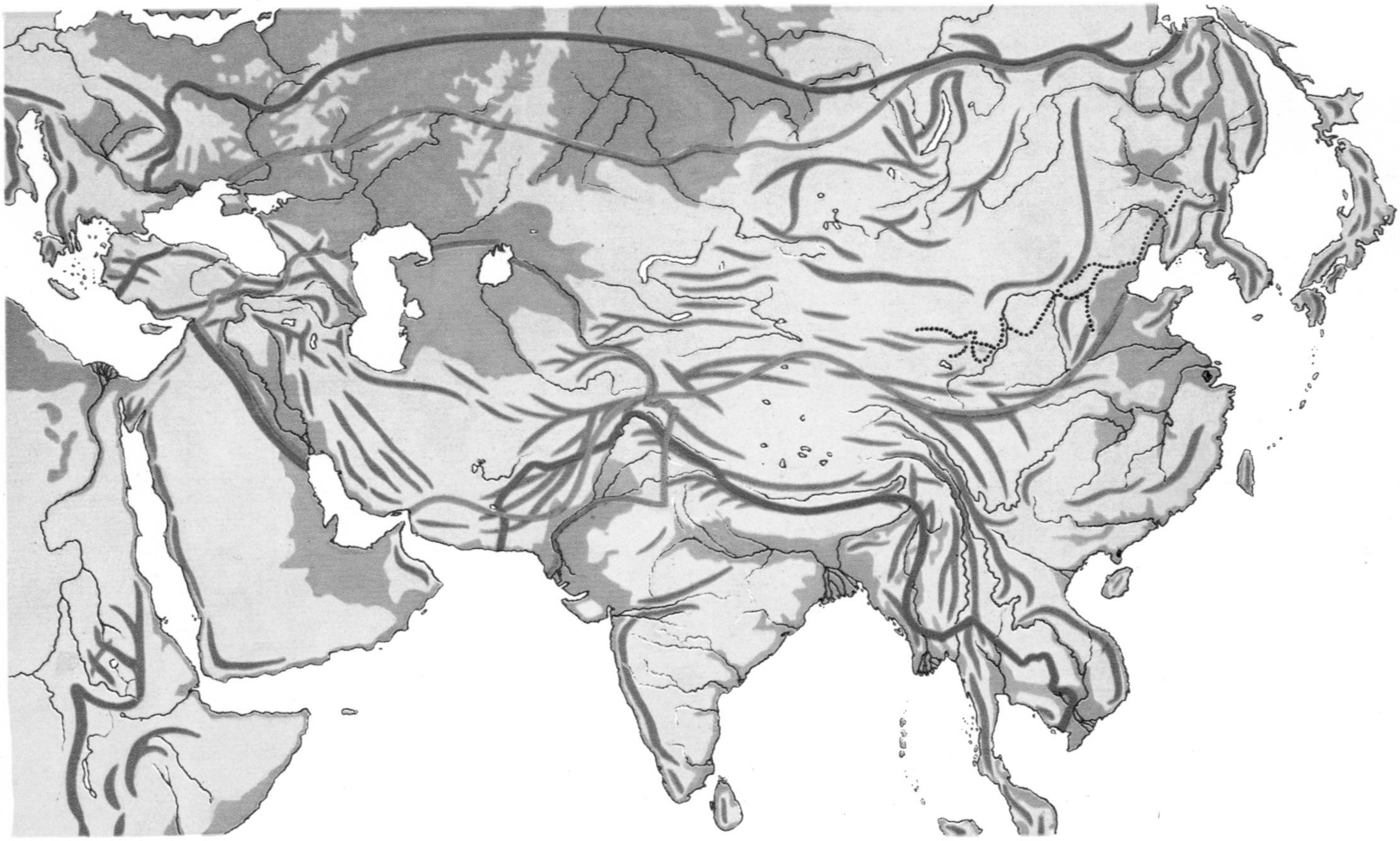

Map showing stages in the spread of Mongol dominion in Asia. One cannot speak with confidence about the effect of the Moslem synthesis on Chinese mathematics until the thirteenth century, when Mongol overlords held sway from Mesopotamia to North China. Such was the impetus of the Mongol thrust eastwards that even Japan felt unsafe. The painting (right) shows Samurai waiting to repel an expected attack in 1281.

A Turkish miniature depicting the siege of Samarkand, the city which became the capital of Tamerlane's domains and an important focus of Moslem culture.

By eliminating one or other column, we can obtain from this three 2×2 matrices. If we use $D(x)$ to signify the result of eliminating the numerical coefficients of the x-column in the grid layout of the two equations cited above, we may briefly write them in the form below:

$$D(x) \quad \begin{vmatrix} -8 & 21 \\ 2 & -29 \end{vmatrix} \qquad D(y) \quad \begin{vmatrix} 7 & 21 \\ 3 & -29 \end{vmatrix} \qquad (Dn) \quad \begin{vmatrix} 7 & -8 \\ 3 & 2 \end{vmatrix}$$

Nowadays, we speak of such an arrangement as a *second order determinant* when we choose to associate it with a numerical value in accordance with the pattern:

$$\begin{vmatrix} a_1 & b_1 \\ a_2 & b_2 \end{vmatrix} \equiv a_1b_2 - a_2b_1$$

If we do interpret a second order determinant in this way, we obtain from the above:

$$D(x) = 190 \qquad D(y) = -266 \qquad D(n) = 38$$

The reader will see that we reach the correct solution ($x=5$, $y=7$) of our equations if we programme the operation in the form:

$$\frac{x}{D(x)} \equiv \frac{-y}{D(y)} \equiv \frac{1}{D(n)}$$

Up to this point, the counting-board layout has admittedly accomplished for us nothing more than the customary schoolbook procedure of *elimination*; but elimination is much more laborious if we have to deal with three variables (x, y, z), and progressively more so when the number of unknowns is greater than three. Even a fool-proof programme is then advantageous, though the determinant procedure is also labour-saving for a more important reason which will emerge later. A clue to a pattern which is adaptable to the solution of linear equations involving any number of unknowns comes to light if we compare the solution of an equation involving only one unknown (*e.g.* $3x+15=0$) with the result already

Greenwich Royal Observatory at the time of Flamsteed. Here as elsewhere Western men of science still leaned heavily on predecessors who worked centuries earlier in the centres of learning of the Mongol Empire.

In fifteenth-century Samarkand, Ulugh Beg prepared a table of co-ordinates of fixed stars. In seventeenth-century England, Flamsteed used this version of it before preparing his own revised star tables.

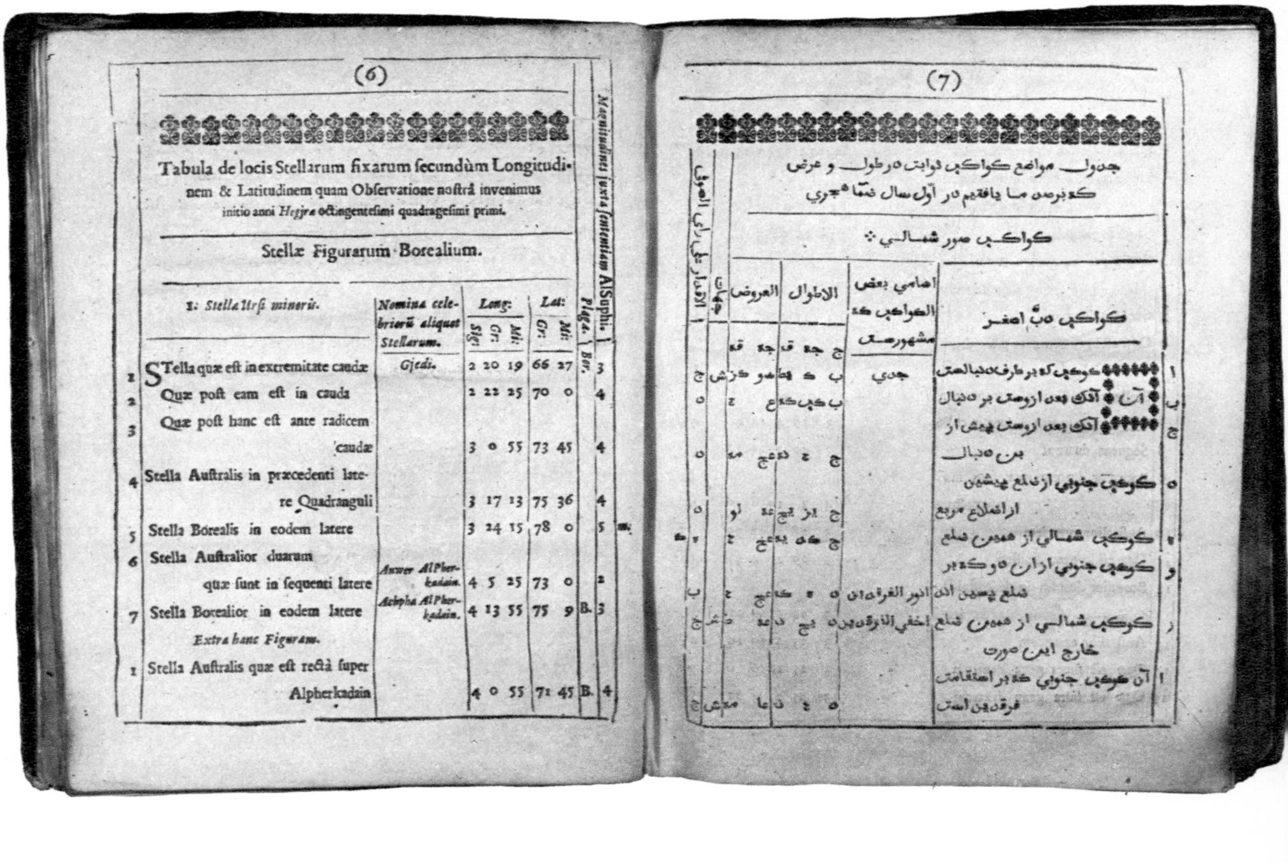

obtained. In this case, the matrix is:

$$[3 \quad 15]$$

From this, we can derive the two *first order* determinants:

$$D(x) = |15| \qquad D(n) = |3|$$

If we interpret such so-called first order (1 row and 1 col.) determinants as numerically equal to the single symbol enclosed by the bars, it follows that:

$$\frac{x}{D(x)} = \frac{-1}{D(n)}$$

We see also that the second order determinant is reducible to first order determinants by:

(a) multiplying each term in the first column with the residual determinant obtained by eliminating all the elements of the same row and column as itself;

(b) summation of the products with alternating signs:

$$\begin{vmatrix} a_1 & b_1 \\ a_2 & b_2 \end{vmatrix} \equiv a_1 |b_2| - a_2 |b_1|$$

The first of the last two results cited suggests a general pattern of solution, *e.g.* for six variables u, v, w, x, y, z:

$$\frac{u}{D(u)} \equiv \frac{-v}{D(v)} \equiv \frac{w}{D(w)} \equiv \frac{-x}{D(x)} \equiv \frac{y}{D(y)} \equiv \frac{-z}{D(z)} \equiv \frac{1}{D(n)}$$

Our second result also suggests a recipe by which we can reduce a determinant of n rows and n columns to a sequence involving determinants of $(n-1)$ rows and $(n-1)$ columns, whence eventually to determinants of order 2. Thus

$$\begin{vmatrix} a_1 & b_1 & c_1 & d_1 \\ a_2 & b_2 & c_2 & d_2 \\ a_3 & b_3 & c_3 & d_3 \\ a_4 & b_4 & c_4 & d_4 \end{vmatrix} \equiv a_1 \begin{vmatrix} b_2 & c_2 & d_2 \\ b_3 & c_3 & d_3 \\ b_4 & c_4 & d_4 \end{vmatrix} - a_2 \begin{vmatrix} b_1 & c_1 & d_1 \\ b_3 & c_3 & d_3 \\ b_4 & c_4 & d_4 \end{vmatrix}$$

$$+ a_3 \begin{vmatrix} b_1 & c_1 & d_1 \\ b_2 & c_2 & d_2 \\ b_4 & c_4 & d_4 \end{vmatrix} - a_4 \begin{vmatrix} b_1 & c_1 & d_1 \\ b_2 & c_2 & d_2 \\ b_3 & c_3 & d_3 \end{vmatrix}$$

The reader can easily satisfy himself or herself that we have now discovered a rule-of-thumb recipe for the solution of simultaneous equations involving three or more variables, for instance the following:

$$5x+3y-5z+21=0$$
$$8x+6y+2z-4=0$$
$$2x-4y-3z+2=0$$

$$\underset{D(x)}{\begin{vmatrix} 3 & -5 & 21 \\ 6 & 2 & -4 \\ -4 & -3 & 2 \end{vmatrix}} \quad \underset{D(y)}{\begin{vmatrix} 5 & -5 & 21 \\ 8 & 2 & -4 \\ 2 & -3 & 2 \end{vmatrix}} \quad \underset{D(z)}{\begin{vmatrix} 5 & 3 & 21 \\ 8 & 6 & -4 \\ 2 & -4 & 2 \end{vmatrix}} \quad \underset{D(n)}{\begin{vmatrix} 5 & 3 & -5 \\ 8 & 6 & 2 \\ 2 & -4 & -3 \end{vmatrix}}$$

$$=-254 \qquad =-508 \qquad =-1016 \qquad =254$$

$$\frac{x}{-254} = \frac{-y}{-508} = \frac{z}{-1016} = \frac{-1}{254}$$

Whence, as one can check by substitution in each equation, $x=1, y=-2$ and $z=4$.

Our calculation will here involve evaluation of four third-order determinants, each reducible by definition into three terms each involving a second-order determinant. The supreme economy achieved by the grid procedure is that a few simple rules, some of which the Chinese had discovered very early in the Christian era, suffice to eliminate one or more of the terms. The most important of such rules are as follows:

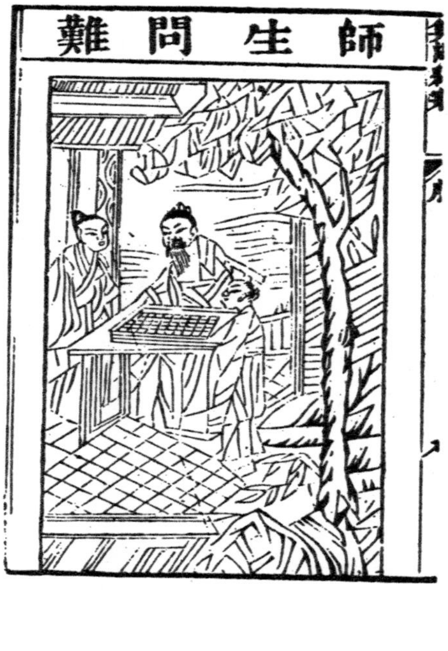

The chequer-board pattern of the Chinese counting-board shown here explains why the Chinese were first to evolve a system of solving algebraic equations by rule-of-thumb methods after placing their numerical components in a chequer-board layout.

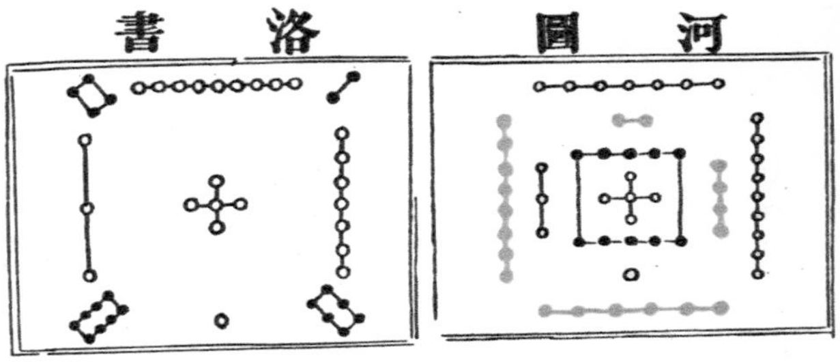

The counting-board may also go far to explain early Chinese preoccupation with magic squares of the kind shown here. (Key in modern numerals below.)

4	9	2
3	5	7
8	1	6

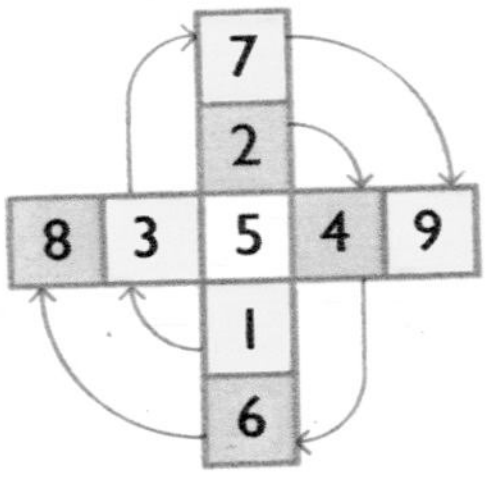

MNEMONIC FOR A SECOND-ORDER DETERMINANT
The use of determinants involves no big intellectual effort. It does involve considerable memorising. The reader may find it worthwhile to work out other aids to the memory of the kind given here.

$$7x - 8y + 21 = 0$$
$$3x + 2y - 29 = 0$$

complete grid of coefficients for both equations reads:

7	−8	21
3	2	−29

From this we can produce three grids of four squares:

D(x)
(eliminating x coefficients)

	a	b
1	−8	21
2	2	−29

D(y)
(eliminating y coefficients)

	a	b
1	7	21
2	3	−29

D(n)
(eliminating n coefficients)

	a	b
1	7	−8
2	3	2

To each grid we may assign the value:

$$a_1 b_2 - a_2 b_1 \equiv \left[\blacksquare \times \blacksquare \right] - \left[\square \times \square \right]$$

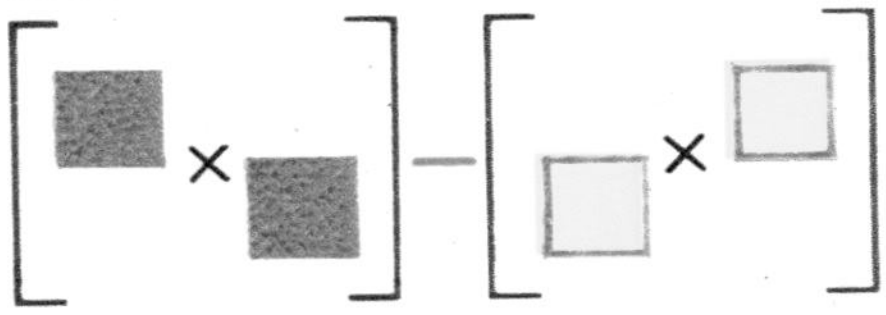

Their respective values are then:

D(x)

$$\left[-8 \times -29 \right] - \left[2 \times 21 \right]$$
$$= 232 - 42$$
$$= 190$$

D(y)

$$\left[7 \times -29 \right] - \left[3 \times 21 \right]$$
$$= -203 - 63$$
$$= -266$$

D(n)

$$\left[7 \times 2 \right] - \left[3 \times -8 \right]$$
$$= 14 - (-24)$$
$$= 38$$

$$\frac{x}{D(x)} = \frac{-y}{(Dy)} = \frac{1}{D(n)}$$

$$\frac{x}{190} = \frac{1}{38} \quad \therefore x = \frac{190}{38} = 5$$

$$\frac{-y}{-266} = \frac{1}{38} \quad \therefore y = \frac{-266}{-38} = 7$$

1. The numerical value of a determinant remains unchanged, if:
   (a) we rotate the arrays through 90° so that rows become columns and *vice versa*;
   (b) add to (or subtract from) any array (row or column) a fixed multiple of each corresponding element in a parallel array.
2. The numerical value of a determinant is zero, if:
   (a) any two parallel arrays are identical;
   (b) all members of any one array are zeros.
3. N-fold multiplication (or division) of every element of single array is equivalent to N-fold multiplication (or division) of the numerical value of the determinant itself.
4. Interchange of adjacent arrays is equivalent to multiplication by -1.

The proof of any of these rules is easy to establish, and it will suffice to give the reader a clue by citing that of 1(b) for the simple case of a second-order determinant:

$$\begin{vmatrix} (a_1 \pm Kb_1) & b_1 \\ (a_2 \pm Kb_2) & b_2 \end{vmatrix} = (a_1b_2 + Kb_1b_2) - (a_2b_1 + Kb_1b_2)$$

$$= a_1b_2 - a_2b_1 = \begin{vmatrix} a_1 & b_1 \\ a_2 & b_2 \end{vmatrix}$$

Needless to say, nimble manipulation of this type comes only with practice; but the reader who

Above: Han Dynasty (206 B.C.–A.D. 220) figurine of men playing a game of chance. Below: nineteenth-century cards used in the game Hwahaw. The Chinese, doubtless due to their addiction to gaming, early realised the significance of figurate numbers in dealing with problems involving chance and choice.

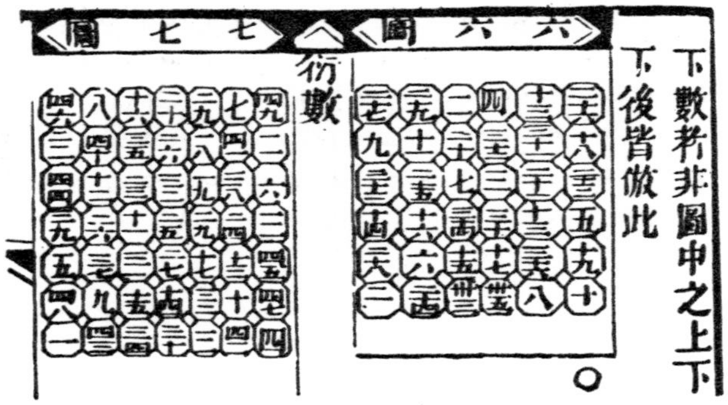

27	29	2	4	13	36
9	11	20	22	31	18
32	25	7	3	21	23
14	16	34	30	12	5
28	6	15	17	26	19
1	24	33	35	8	10

Above: A Chinese magic square of about A.D. 1590. Below: A Japanese magic circle from the works of Seki Kowa who, in the seventeenth century, stated all the more important matrix operations involved in solving simultaneous linear equations with many variables.

五周徑之圖

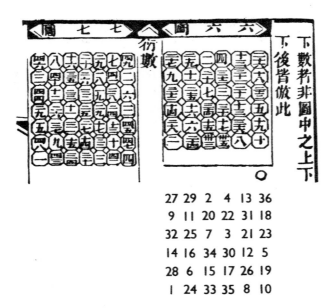

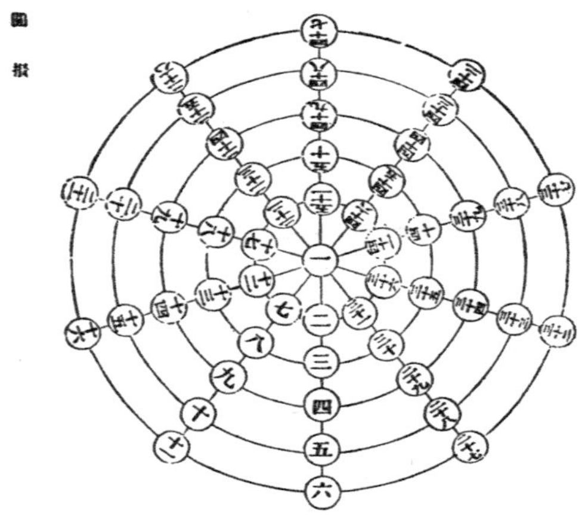

makes the attempt to use the determinant method to solve equations involving five variables and then checks the result by elimination will need no persuasion that it effects a vast economy of time. As an example of such simplification, we may treat as if it were in fact a determinant of the third order the archetypal Chinese magic square, of which rows, columns and diagonals add up to 15:

$$\begin{vmatrix} 4 & 9 & 2 \\ 3 & 5 & 7 \\ 8 & 1 & 6 \end{vmatrix} = \begin{vmatrix} 4 & 9 & 2 \\ 3 & 5 & 7 \\ 8-2(4) & 1-2(9) & 6-2(2) \end{vmatrix} = \begin{vmatrix} 4 & 9 & 2 \\ 3 & 5 & 7 \\ 0 & -17 & 2 \end{vmatrix}$$

$$= \begin{vmatrix} 1 & 4 & -5 \\ 3 & 5 & 7 \\ 0 & -17 & 2 \end{vmatrix} = \begin{vmatrix} 1 & 4 & -5 \\ 3-3(1) & 5-3(4) & 7-3(-5) \\ 0 & -17 & 2 \end{vmatrix}$$

$$= \begin{vmatrix} 1 & 4 & -5 \\ 0 & -7 & 22 \\ 0 & -17 & 2 \end{vmatrix} = \begin{vmatrix} -7 & 22 \\ -17 & 2 \end{vmatrix} = -2\begin{vmatrix} 7 & 11 \\ 17 & 1 \end{vmatrix} = 360$$

Check:

$4(5.6-7.1)-3(9.6-2.1)+8(9.7-2.5)$
$=4(30-7)-3(54-2)+8(63-10)$
$=4(23)-3(52)+8(53)=92-156+424=360$

It is possible to discern the counting-board in the background of another important milestone in Chinese mathematics, undoubtedly transmitted to the West several centuries after the event. From more than one viewpoint, the Chinese had a keen interest in figurate series; and one important outcome of this emerges when we lay out grid-wise the super-family whose 0-, 1-, 2-, 3-dimensional sets are respectively the units, integers, triangular numbers and tetrahedral numbers. The diagonal entries then reveal sequences which are of interest for more than one reason.

By direct multiplication, the reader will easily see that:

$(a+b)^0=1$
$(a+b)^1=a+b$
$(a+b)^2=a^2+2ab+b^2$
$(a+b)^3=a^3+3a^2b+3ab^2+b^3$
$(a+b)^4=a^4+4a^3b+6a^2b^2+4ab^3+b^4$
$(a+b)^5=a^5+5a^4b+10a^3b^2+10a^2b^3+5ab^4+b^5$

If we use successive squares of the counting-board to locate the products involving successive powers of a and b, our numerical entries will follow the pattern shown on the left below. On the right we have laid out our super-family to which the triangular (*etc.*) numbers belong, so that the row (d) on the left corresponds to the first term of the *diagonal* sequence of the dimension d:

1						
1	1					
1	2	1				
1	3	3	1			
1	4	6	4	1		
1	5	10	10	5	1	
1	6	15	20	15	6	1

1	1	1	1	1	1	1	...
1	2	3	4	5	6	7	...
1	3	6	10	15	21	28	...
1	4	10	20	35	56	84	...
1	5	15	35	70	126	210	...
1	6	21	56	126	252	462	...
1	7	28	84	210	462	924	...

In the expanded expression $(a+b)^d$ there are $(d+1)$ terms. So we may tailor our sign language most explicitly to the job in hand if we label the rank in step with the index of b. To do this we must label the *initial* one as the term of rank 0. If we use $C_{r.d}$ to signify the numerical coefficient of the term of rank r, we shall then write:

$$(a+b)^d = C_{0.d}a^d b^0 + C_{1.d}a^{d-1}b^1 + C_{2.d}a^{d-2}b^2 \ldots \ldots C_{d.d}a^0 b^d$$

To discover a general expression for $C_{r.d}$ in terms of integers r and d, we may now set out the triangle side by side with the figurate shadow grid in which $F_{r.d}$ means the figurate of rank r (starting from $r=1$) and dimension d (starting from $d=0$) as in the accompanying layout.

$C_{0.0}$					
$C_{0.1}$	$C_{1.1}$				
$C_{0.2}$	$C_{1.2}$	$C_{2.2}$			
$C_{0.3}$	$C_{1.3}$	$C_{2.3}$	$C_{3.3}$		
$C_{0.4}$	$C_{1.4}$	$C_{2.4}$	$C_{3.4}$	$C_{4.4}$	
$C_{0.5}$	$C_{1.5}$	$C_{2.5}$	$C_{3.5}$	$C_{4.5}$	$C_{5.5}$

$F_{1.0}$	$F_{2.0}$	$F_{3.0}$	$F_{4.0}$	$F_{5.0}$	$F_{6.0}$	...
$F_{1.1}$	$F_{2.1}$	$F_{3.1}$	$F_{4.1}$	$F_{5.1}$	...	
$F_{1.2}$	$F_{2.2}$	$F_{3.2}$	$F_{4.2}$	...		
$F_{1.3}$	$F_{2.3}$	$F_{3.3}$	$F_{4.3}$	...		
$F_{1.4}$	$F_{2.4}$	$F_{3.4}$	$F_{4.4}$	...		
$F_{1.5}$	$F_{2.5}$	$F_{3.5}$	$F_{4.5}$	...		

We now see that the row of coefficients specified on the left by d corresponds to the diagonal starting with rank 1 of dimension d and proceeding upwards from the left. In short:

$$C_{0.d} = F_{1.d} \quad C_{1.d} = F_{2.d-1} \quad C_{2.d} = F_{3.d-2}$$
$$C_{r.d} = F_{(r+1)\,(d-r)}$$

Now we have seen (end of Chapter 3) that

$$F_{r.d} = \frac{(r+d-1)^{(d)}}{d!}$$
$$= \frac{(r+d-1)\,(r+d-2)\ldots(r+1)r}{d!}$$
$$= \frac{(r+d-1)!}{d!(r-1)!}$$

Whence we have

$$F_{r+1.d-r} = \frac{(r+1+d-r-1)!}{(d-r)!\,(r+1-1)!} = \frac{d!}{(d-r)!\,r!}$$

In the last expression $d! = d^{(r)}\,(d-r)!$, so that in the notation of Chapter 3:

$$F_{r+1.d-r} = \frac{d^{(r)}}{r!} = d_{(r)}$$
$$C_{r.d} = d_{(r)}$$

We may therefore re-write the above expanded expression in the form:

$$(a+b)^d = d_{(0)}\,a^d + d_{(1)}a^{d-1}b + d_{(2)}a^{d-2}b^2 \text{ etc.}$$

More compactly, we can express the above – usually referred to as the *binomial theorem* – for an index which is an integer (d) in the form:

$$(a+b)^d = \sum_{x=0}^{x=d} d_{(x)} \cdot a^{d-x} \cdot b^x$$

The formal proof by induction is simple. If $C_{r.d}$ in $(a+b)^d$ is equivalent to $d_{(r)}$, the value of $C_{r.(d+1)}$ in $(a+b)^{d+1}$ is $(d+1)_{(r)}$, *i.e.*

$$C_{r.(d+1)} = \frac{(d+1)!}{r!(d+1-r)!}$$

Now $(a+b)^{d+1} = (a+b)\,(a+b)^d$ and the build-up of $C_{r.(d+1)}$ is evident from inspection, if we examine the genesis of any such product, *e.g.*:

$$\begin{array}{l} a^3+3a^2b+3ab^2+b^3 \\ a\;+\;b \\ \hline a^4+3a^3b+3a^2b^2+\;ab^3 \\ \qquad a^3b+3a^2b^2+3ab^3+b^4 \\ \hline a^4+4a^3b+6a^2b^2+4ab^3+b^4 \end{array}$$

Thus we may write:

$$C_{r.(d+1)}=C_{r.d}+C_{(r-1).d}$$

$$=\frac{d!}{r!(d-r)!}+\frac{d!}{(r-1)!\,(d+1-r)!}$$

$$=\frac{d!}{(r-1)!\,(d-r)!}\left[\frac{1}{r}+\frac{1}{(d+1-r)}\right]$$

$$=\frac{d!}{(r-1)!\,(d-r)!}\cdot\frac{d+1}{r.(d+1-r)}=\frac{(d+1)!}{r!(d+1-r)!}$$

This is the result already anticipated; and it proves that our discovery is true in the domain of positive integers. We shall later see that it has a much wider domain of validity. There is some evidence that the Moslem astronomer, poet and mathematician known to us as Omar Khayyam knew the Chinese triangle in the eleventh century of the Christian era. This is indeed earlier than its appearance in the earliest printed Chinese treatises; but the latter refer to it as if well known over a long period. On the whole, it is more likely that it made its way from China through the Moslem world to the West than that it made its way East and West from Baghdad. Be that as it may, the Binomial Theorem for integer values of n in $(a+b)^n$ reached Europe through Moslem sources, in or before the fifteenth century.

The Chinese (or Omar Khayyam's) layout of binomial coefficients, often referred to as Pascal's triangle on account of its treatment in that author's posthumous treatise (A.D. 1665) on figurate numbers, makes its first recorded appearance in the West on the title page of an early printed European arithmetic by Appianus (A.D. 1527). Towards the end of the seventeenth century it became the focal point for the development of three branches of mathematics: the study of infinite series, the calculus of finite differences and the theory of probability. Neither the Chinese nor the Moslems had any prevision of the first two of these; but the former, at all times (and lucklessly) fascinated by games of chance, recognised at an early date in the Christian era the significance of figurate numbers in the evaluation of problems involving choice, *i.e.* specification of the number of classes one can select if distinguished (*combination*) only by the occurrence of recognisably different constituents and also (*permutation*) by their arrangement *inter se*. They already knew that each of the expressions we have written respectively as $n^{(r)}$ and $n_{(r)}$ has a special significance in connexion with choice.

The first of the last-mentioned answers the question: in how many ways (*linear permutations*) can we lay out in a straight line any r of them from a pool of n distinguishable objects? This will be clear if we consider in how many ways we can lay out 3 of the 5 letters A, B, C, D, E, each on one line of the counting-board, as below. Clearly, n arrangements are distinguishable with reference to the first place in the sequence. We can then

First Place	Second Place	Third Place		
A	A B	A B C	A B D	A B E
	A C	A C B	A C D	A C E
	A D	A D B	A D C	A D E
	A E	A E B	A E C	A E D
B	B A	B A C	B A D	B A E
	B C	B C A	B C D	B C E
	B D	B D A	B D C	B D E
	B E	B E A	B E C	B E D
C	C A	C A B	C A D	C A E
	C B	C B A	C B D	C B E
	C D	C D A	C D B	C D E
	C E	C E A	C E B	C E D
D	D A	D A B	D A C	D A E
	D B	D B A	D B C	D B E
	D C	D C A	D C B	D C E
	D E	D E A	D E B	D E C
E	E A	E A B	E A C	E A D
	E B	E B A	E B C	E B D
	E C	E C A	E C B	E C D
	E D	E D A	E D B	E D C

The three steps by which we ascertain every possible line of three letters chosen from five (*A, B, C, D, E*).

choose any one of the remaining $(n-1)$ to fill the second, having then completed $n(n-1)=n^{(2)}$ ways of placing the first two. We now have $(n-2)$ remaining letters with which to complete the third place in each of the $n^{(2)}$ sequences of two. We have therefore now $n^{(3)}$ ways of filling the first three, and so on. It follows that the number of linear permutations of all the n objects in a pool is $n^{(n)}=n!$.

The expression $n_{(r)}$ for the coefficients of the Chinese triangle has a double significance. First it answers the question: how many $(C_{r.n})$ different classes can we make from r out of n distinguishable objects, if we disregard order? Since each of such classes contains r members which we can lay out in $r!$ different ways in a line as above, the total number of linear arrangements $(n^{(r)})$ is $r!C_{r.n}$, so that $C_{r.n}=n^{(r)}\div r!=n_{(r)}$. It is instructive to lay this out in an orderly way for $n=5$ as follows:

$r=1$ $A\ B\ C\ D\ E$
$C_{1.5}=5$
$r=2$ $[AB.AC.AD.AE]:[BC.BD.BE]:[CD.CE]:[DE]$
$C_{2.5}=10$
$r=3$ $[ABC.ABD.ABE.ACD.ACE.ADE]:[BCD.BCE.BDE]:[CDE]$
$C_{3.5}=10$
$r=4$ $[ABCD.ABCE.ABDE.ACDE]:BCDE$
$C_{4.5}=5$
$r=5$ $[ABCDE]$
$C_{5.5}=1$

If we interpret $C_{0.5}=1$ as the statement that there is only one way of taking *no* object from a class of 5, we see that the total number of ways of choosing 0, 1, 2, 3, 4 or 5 out of 5 is: $1+5+10+10+5+1$. This is equivalent to expanding the binomial expression $(1+1)^5=2^5$. In general, the number of ways of choosing 0, 1, 2 . . . n out of n objects is therefore 2^n and the number of ways of choosing *at least one* is 2^n-1, a result which is useful for coding medical documents for analysis by mechanical devices for sorting, *etc.*

In the domain of choice, the symbol $n_{(r)}$ has a meaning different from the one already cited; and

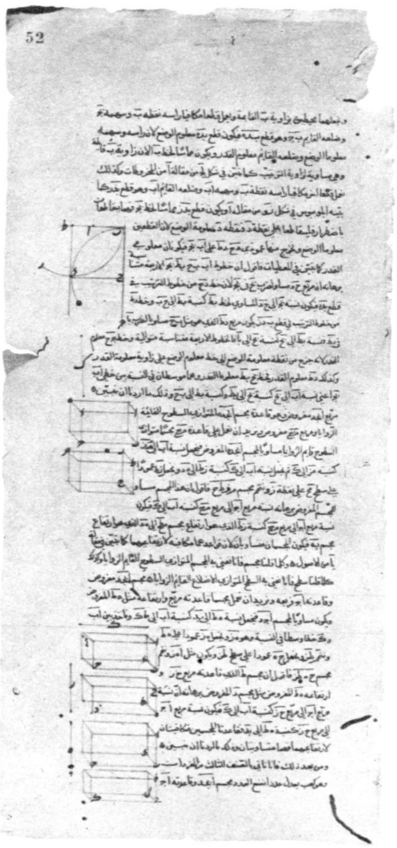

Sixteenth-century manuscript of a treatise by Omar Khayyam. There is some evidence that he knew the Chinese triangle—a visual aid to the binomial theorem—in the eleventh century of our era.

The Chinese triangle (so-called Pascal's triangle) from a work printed about A.D. *1303.*

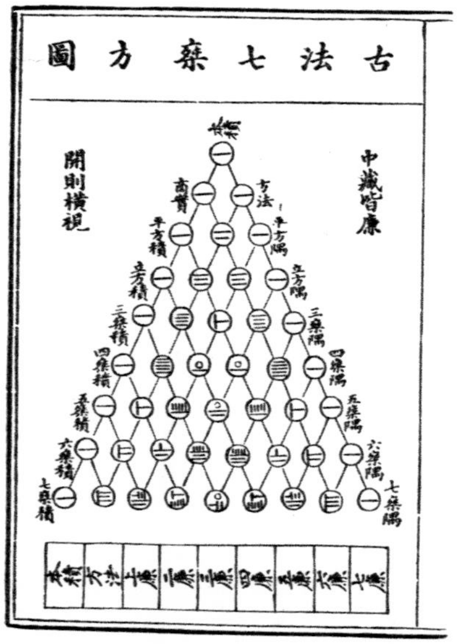

some confusion therefore arises from using such symbols as ${}^{n}C_r$ or $C_{r.n}$ unless they refer to combinations *sensu stricto*. The symbol $n_{(r)}$ also answers the question: in how many recognisably different arrangements ($P_{r.n-r}$) can we lay out on a row of the counting-board *all* of n things when r of them are indistinguishably alike *inter se* and the remaining $n-r$ are likewise indistinguishably alike *inter se* though different from the other set? To answer this conundrum, let us first re-interpret $n_{(r)}$ as follows:

$$n_{(r)} = \frac{n^{(r)}}{r!} = \frac{n(n-1)\ (n-2)\,..\,(n-r+1)}{r!}$$

$$= \frac{n(n-1)\,...\,(n-r+1)\ (n-r)\ (n-r-1)\,..\,3.2.1}{r!(n-r)!}$$

$$\therefore n_{(r)} = \frac{n!}{r!(n-r)!}$$

If all the n objects were distinguishable, the number of linear arrangements of the whole set would be $n!$. If r of them are indistinguishable, $r!$ out of these $n!$ will also be indistinguishable, and the total number if all are distinguishable will be $r!$ as great as otherwise. By the same token, if $(n-r)$ are distinguishable, the number will be $(n-r)!$ as great as otherwise. Thus we may write the number specified as $P_{r.n-r}$ above as the product:

$$r!(n-r)!\ P_{r.n-r} = n!$$

$$\therefore P_{r.n-r} = \frac{n!}{r!(n-r)!} = n_{(r)}$$

When speaking of choice so far, we have assumed that no object can be more than once in a class, as when we draw 5 cards simultaneously from a pack. If we draw them one at a time, *replacing* each before drawing the next, the situation is different. We leave it to the reader to see that the number of linear arrangements of n distinguishable things taken r at a time is then n^r instead of $n^{(r)}$.

How familiarity with figurate numbers gives us a master clue to problems of choice, another problem illustrates forcibly, *viz.* what is the number of different classes of r from a pool of n members, if chosen *with replacement*? We shall leave it to the reader as an opportunity for exercise in discoveries one can make by roaming on the number landscape if one already has a little familiarity with figurate series. Having found a rule, it is less difficult to work backwards to a proof. As an illustration of this truism it is instructive to consider a result well known to the Hindu mathematicians before the Norman Conquest of England, *viz.* a formula for the sum of the cubes of the first n integers, *i.e.*

$$1 = 1^2 \quad 1+8 = 9 = 3^2 \quad 1+8+27 = 36 = 6^2$$
$$1+8+27+64 = 100 = 10^2$$

Anyone familiar with figurates will at once recognise the above as the squares for the first four triangular numbers, whence one suspects that the formula is

$$S_n = \left(\frac{n(n+1)}{2}\right)^2 = \frac{n^4+2n^3+n^2}{4}$$

The proof by induction follows effortlessly:

$$S_{n+1} = \frac{n^4+2n^3+n^2}{4} + (n+1)^3$$

$$= \frac{n^4+2n^3+n^2}{4} + n^3 + 3n^2 + 3n + 1$$

$$= \frac{n^4+6n^3+13n^2+12n+4}{4}$$

$$= \frac{(n+1)^2\ (n+2)^2}{4} = \left[\frac{(n+1)\ (n+1+1)}{2}\right]^2$$

We have not seen the last of figurate series, and the symbol $F_{r.d}$ will provide us with a general definition for later use, *viz.* if K is a constant:

$$F_{r.d} = F_{(r-1).d} + F_{r.(d-1)} \text{ and } F_{r.0} = K$$

In words, this means:

(a) to get the member of rank r in any row after the initial one ($d=0$), add to its predecessor the term above itself;

(*b*) all terms in the initial row are identical.

For the family whose diagonal terms are the rows of Pascal's triangle $F_{1.d} = 1$ for all d. If we set

$F_{r.0}=2$ and $F_{1.d}=1$ for $d>0$, we get the family whose 1-, 2- and 3-dimensional sequences are the odd numbers, the squares and pyramids. By setting $F_{r.0}=3$, 4, *etc.*, keeping $F_{1.d}=1$ for $d>0$, we get figurates whose 2-dimensional forms are pentagons, hexagons, *etc.*, and whose 3-dimensional forms are pyramids with 5, 6, *etc.*, faces excluding the base. For the super-family last mentioned, the sufficient prescription is $F_{1.d}=1$ for $d>0$ and $F_{r.0}=1$, 2, 3, *etc.*, for different sub-families. The number of sides of the 2-dimensional members is $F_{r.0}+2$. The recipe for another super-family is $F_{1.1}=0$ and $F_{1.d}=1$ for $d>1$. The number of sides of the 2-dimensional members of the sub-families is now $F_{r.0}$. When $F_{r.0}=8$, the octagonal 2-dimensional terms correspond to the squares of the odd numbers and the 3-dimensional octagonal pyramids correspond to the sums of the squares of the odd numbers.

Constructing series of different types provides endless opportunities for entertainment on the number landscape, especially if one builds up crystalline forms; and it is one of the calamities of mathematical instruction that the last attempt to exploit the opportunities for exploration on so picturesque a terrain for even the very young was in the first systematic treatise on probability (*Artis Conjectandi*), published by J. Bernoulli two hundred and fifty years ago. One long-forgotten discovery therein will turn up later in our story; but it is relevant to its theme if we close this chapter by referring to an item in the *Liber Abaci*, a book in which Leonardo of Pisa, perhaps better known as Leonardo Fibonacci, a wealthy Italian merchant of the thirteenth century, familiarised Europe with the first-fruits of the Hindu-Moslem (and hence indirectly Chinese also) contribution to the making of mathematics.

The reader may have noticed the use of the term *nonrecurrent* in an earlier chapter. Somewhere in his travels Leonardo picked up the simplest type of a recurrent series, *viz.*

1	1	2	3	5	8	13	21	...
..	..	(1+1)	(2+1)	(3+2)	(5+3)	(8+5)	(13+8)	...

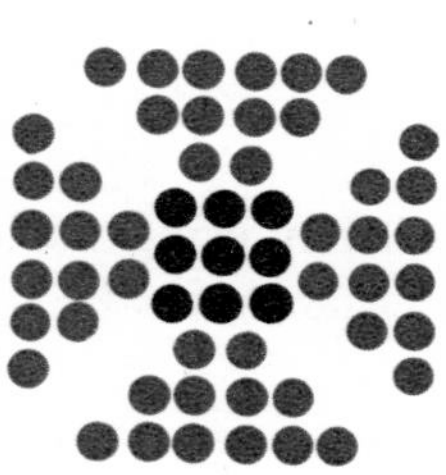

(*a*)

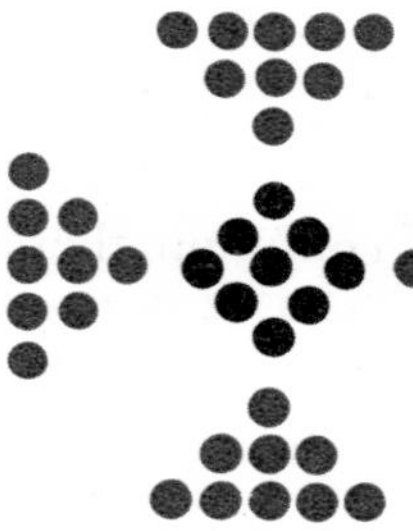

(*b*)

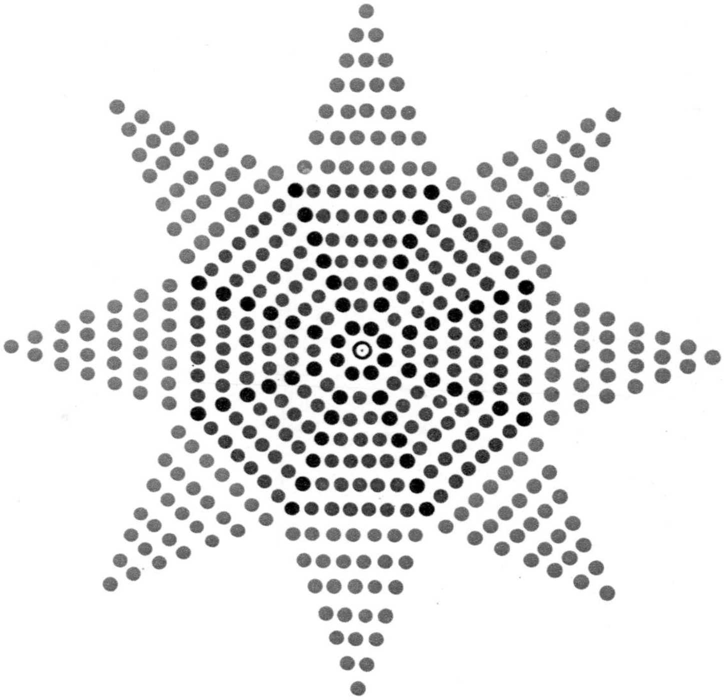

(*c*)

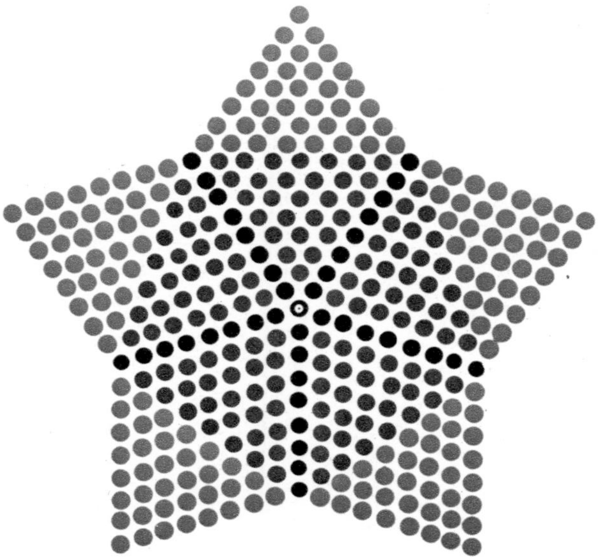

(*d*)

Visualisations of four number series. The reader may wish to supply the formulae. (Answers near end of book.)

To form each term of rank n, we here need to know its relations to its *two* predecessors, *i.e.*:

$$u_n = u_{n-1} + u_{n-2}$$

More than seven hundred years elapsed before this provided an illustration of the pay-off for permitting people seriously concerned with intellectual problems to equate their hobbies with remunerative employment, unless as rich as Leonardo himself. It happens that such series turn up in almost every facet of the part of genetics concerned with how the proportions of individuals of different genotypes change in successive generations under prescribed conditions of mating (*e.g.* brother by sister in conformity with Leonardo's own series expressed in fractional form). This is because we cannot (as Mendel first discovered) know the hereditary make-up of a generation without knowing what is relevant about their grandparents as well as about their parents.

As it happens, one may guess with some plausibility where Leonardo Fibonacci (the son of Bonaccio) did pick this up in his travels, if we lay out the Chinese (or Omar Khayyam's) triangle as below and sum the terms diagonally from the left upwards:

							Diagonal sum
1							1
1	1						1
1	2	1					2
1	3	3	1				3
1	4	6	4	1			5
1	5	10	10	5	1		8
1	6	15	20	15	6	1	13 *etc.*

The Fibonacci series is one of a type of recurrent series also referred to as linear difference series of the second order. In terms of fixed numbers (*constants*) A and B definitive of a particular series of the sort, the general pattern of this type is for the term (U_n) of rank n:

$$u_n = A . u_{n-1} + B . u_{n-2}$$

If $A = \frac{1}{2} = B$, the following are examples of considerable biological interest:

0	1	$\frac{1}{2}$	$\frac{3}{4}$	$\frac{5}{8}$	$\frac{11}{16}$	$\frac{21}{32}$	$\frac{43}{64}$	.....
1	1	1	1	1	1	1	1	.....
$\frac{1}{2}$	$\frac{1}{4}$	$\frac{3}{8}$	$\frac{5}{16}$	$\frac{11}{32}$	$\frac{21}{64}$	$\frac{43}{128}$		.....

A Norman Bishop, Oresme (*circa* A.D. 1360), who was the first person on record to grasp the significance of a fractional index and the relevance of mapping to the notion of an algebraic fraction, studied one of the series for which $A = 1$ and $B = -\frac{1}{4}$. Thereafter recurrent series disappear from sight till the end of the eighteenth century. The series for which Oresme gave the sum of the first n terms is:

$$\frac{1}{2} \quad \frac{2}{4} \quad \frac{3}{8} \quad \frac{4}{16} \quad \frac{5}{32} \quad \frac{6}{64} \quad \ldots\ldots$$

It is possible to derive a summation formula, which the reader may be able to discover, by recourse to the use of figurate numbers; but what is of importance to the biologist is an answer to the question: if we know the first two terms, *i.e.* the proportion of grandparents and parents of different genotypes, how do we calculate the proportions in any later generations? The answer, though elementary, is not so simple as one might suppose. Since few (if any) current college textbooks disclose a clue to it, it may give the reader some entertainment if we do so here. Clearly, we require two consecutive, *e.g.* the two initial, terms to build up a second-order recurrent series step by step, and we shall label these u_0 and u_1. Given the values of these, it is possible to find an expression for u_n without going through the process the hard way, if we take advantage of the fact that we can rig up any geometrical series ($u_n = u_0 r^n$) in the foregoing form in more than one way. For instance, if $A = (r-1)$ and $B = r$, or $A = \frac{1}{2}r$ and $B = \frac{1}{2}r^2$:

$$u_n = A . u_0 r^{n-1} + B . u_0 r^{n-2} = A . u_{n-1} + B . u_{n-2}$$

It goes without saying that the sum of two second-order recurrent series with the same constants A and B is another of the same class, and since we want to make use of two initial terms, we next explore

Page from Jacob Bernoulli's Artis Conjectandi, *published in 1713. Besides being the first systematic treatise on probability it also exploits the picturesque possibilities of constructing various number series.*

$$u = p.r_1^n + q.r_2^n$$

In this, $u_0 = p + q$ and $u_1 = pr_1 + qr_2$, so that by schoolbook elimination (or as an exercise in the use of determinants):

$$p = \frac{u_0 r_2 - u_1}{r_2 - r_1} \text{ and } q = \frac{u_0 r_1 - u_1}{r_1 - r_2}$$

If the foregoing formula satisfies the recurrence relation:

$$pr_1^n + qr_2^n = A(pr_1^{n-1} + qr_2^{n-1}) + B(pr_1^{n-2} + qr_2^{n-2})$$

$$\therefore pr_1^n - Apr_1^{n-1} - Bpr_1^{n-2} = 0 = qr_2^n - Aqr_2^{n-1} - Bqr_2^{n-2}$$

We have thus two quadratics which reduce to:

$$r_1^2 - Ar_1 - B = 0 \text{ and } r_2^2 - Ar_2 - B = 0$$

Two different (of four possible) solutions are:

$$r_1 = \frac{A + \sqrt{(A^2 + 4B)}}{2} \text{ and } r_2 = \frac{A - \sqrt{(A^2 + 4B)}}{2}$$

If $A = \frac{1}{2} = B$, these reduce to $r_1 = 1$ and $r_2 = \frac{-1}{2}$. If we paint these in the foregoing expressions for p and q, we get:

$$p = \frac{u_0 + 2u_1}{3} \text{ and } q = \frac{2(u_0 - u_1)}{3}$$

$$u = \frac{u_0 + 2u_1}{3} + \frac{2(u_0 - u_1)}{3}\left(\frac{-1}{2}\right)^n$$

As is true of the family for which $A = \frac{1}{2} = B$, a recurrent series may oscillate about a limit the terms approach as n becomes larger. This is clear if we write out the first nine terms of the series $A = \frac{1}{2} = B$, $u_0 = \frac{1}{2}$, $u_1 = \frac{1}{4}$, as shown below:

$$\frac{1}{2} \quad \frac{1}{4} \quad \frac{3}{8} \quad \frac{5}{16} \quad \frac{11}{32} \quad \frac{21}{64} \quad \frac{43}{128} \quad \frac{85}{256} \quad \frac{171}{512}$$

$$\frac{256}{512} \quad \frac{128}{512} \quad \frac{192}{512} \quad \frac{160}{512} \quad \frac{176}{512} \quad \frac{168}{512} \quad \frac{172}{512} \quad \frac{170}{512} \quad \frac{171}{512} \quad \cdots$$

In the foregoing formula, $(\frac{-1}{2})^n$ approaches the limit of zero. Thus the limiting value of u_n is $\frac{1}{3}$ as we might surmise from the above. As an exercise, the reader can profitably retrace the foregoing argument for the series disclosed by Fibonacci himself. The formula is:

$$u_n = \frac{3 - \sqrt{5}}{2\sqrt{5}}\left(\frac{1 + \sqrt{5}}{4}\right)^n + \frac{\sqrt{5} - 3}{2\sqrt{5}}\left(\frac{1 - \sqrt{5}}{4}\right)^n$$

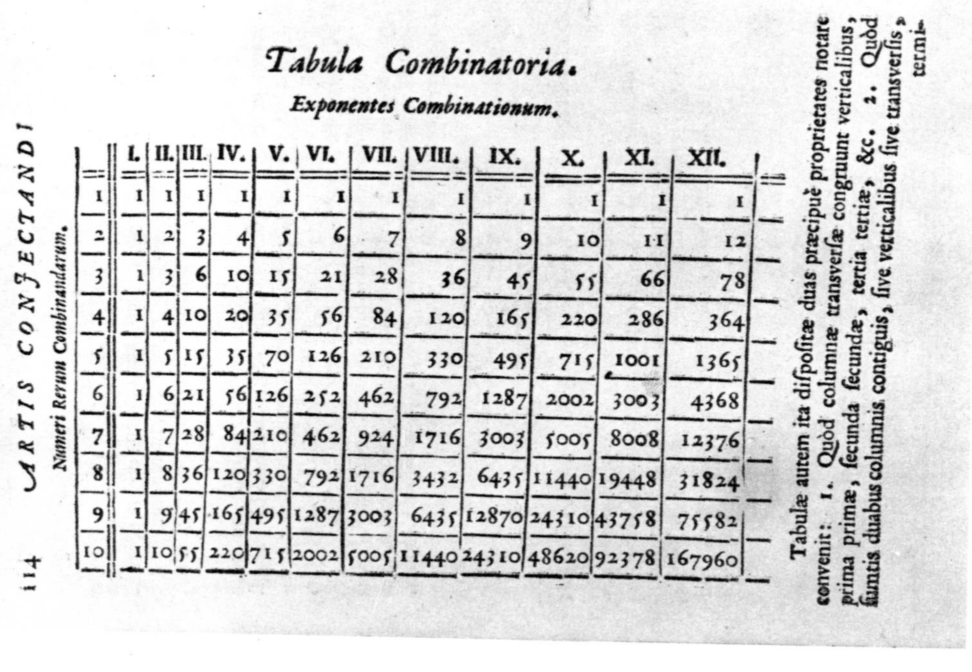

114 ARTIS CONJECTANDI

Tabula Combinatoria.

Exponentes Combinationum.

Numeri Rerum Combinandarum.	I.	II.	III.	IV.	V.	VI.	VII.	VIII.	IX.	X.	XI.	XII.
1	1	1	1	1	1	1	1	1	1	1	1	1
2	1	2	3	4	5	6	7	8	9	10	11	12
3	1	3	6	10	15	21	28	36	45	55	66	78
4	1	4	10	20	35	56	84	120	165	220	286	364
5	1	5	15	35	70	126	210	330	495	715	1001	1365
6	1	6	21	56	126	252	462	792	1287	2002	3003	4368
7	1	7	28	84	210	462	924	1716	3003	5005	8008	12376
8	1	8	36	120	330	792	1716	3432	6435	11440	19448	31824
9	1	9	45	165	495	1287	3003	6435	12870	24310	43758	75582
10	1	10	55	220	715	2002	5005	11440	24310	48620	92378	167960

Tabulæ autem ita diſpoſitæ duas præcipuè proprietates notare convenit: 1. Quòd columnæ transverſæ congruunt verticalibus, prima primæ, ſecunda ſecundæ, tertia tertiæ, &c. 2. Quòd ſumtis duabus columnis contiguis, ſive verticalibus ſive transverſis, termi-

Chapter 8 The European Awakening

A shift of the centre of gravity of Moslem culture during the tenth century of the Christian era from Baghdad to the institutions of higher learning in Spain, more especially that of Cordova, had momentous consequences for an intellectually dormant Europe. Their site was now close to a frontier across which the fruits of Hindu and Chinese inventiveness could make their way into Western Christendom and they first made accessible to Western Europe the major contributions of Greek antiquity through Arab translations. The work of translating the Arabic texts into Latin, undertaken by monks, Jewish physicians or merchants who studied in the Moorish universities of Spain, began in the twelfth century and continued throughout the thirteenth when the knowledge of paper making, block printing and gunpowder trickled beyond the Pyrenees. By A.D. 1200, Latin translations of the works of Euclid, Ptolemy and other Greek authors were as widely available as could be possible before printing from movable type began. There were also Latin versions of the writings of the more prominent Moslem scholars.

Foremost among the last-named was the algebra of al-Kwarismi, who flourished in Baghdad during the seventh Abassid caliphate (A.D. 813–833). The medieval word *algorithm* for the arithmetic of our own schooldays is a corruption of his name; and the Arabic word *algebra* signifies his rule for transposition when solving an equation, *e.g.* $x^2-2x=5x+6$, whence $x^2=7x+6$. The treatise of al-Kwarismi is especially memorable because it sets forth a geometric demonstration to justify our familiar rule for solving a quadratic by completing the square. Written in our symbols, his actual paradigm was

$$x^2+10x=39$$

$$\therefore x^2+10x+25=39+25=64 \text{ and } (x+5)^2=64=8^2$$

Today we customarily boil this down to a formula written thus:

$$x^2+bx+c=0 \text{ if } x^2+bx+\frac{b^2}{4}=\frac{b^2}{4}-c$$

$$\left(x+\frac{b}{2}\right)^2=\frac{b^2}{4}-c \text{ and } x=\frac{-b\pm\sqrt{(b^2-4c)}}{2}$$

From the viewpoint of the pure mathematician, Moslem mathematics was of importance to Europe less because of its indigenous content than because of what it transmitted from the East and preserved from the past. On the other hand, its contribution to the applications of mathematics, in particular to spherical trigonometry, to scientific map making and to astronomy considerably surpassed the level attained in the *Almagest*. It was also very certainly the source of the revival of nautical astronomy in Western Europe, where nautical astronomy throughout the two centuries following the great navigations continued to provide the main avenue of full employment for men with mathematical talent. The Moors still occupied a sizeable territory in the Spanish Peninsula when Prince Henry of Portugal founded (*circa* A.D. 1420) an

Sixteenth-century engraving of the Port of Lisbon, from T. de Bry's Americae. *Throughout the fifteenth, sixteenth and seventeenth centuries the ports of Western Europe were building up a new world-wide network of ocean routes. Nautical astronomy gave full employment to many men with mathematical talent.*

observatory and a school for training pilots at Sagres on one of the promontories of Cape St. Vincent. Both its cartographical materials and its staff were products of the Moorish schools in which the Jewish master navigators who later piloted the Columbian voyages had received their training.

The half century which begins when the ships of Henry the Navigator set out to round the coast of West Africa and ends with the so-called discovery of America was momentous for mathematics because of the revolution in human communications brought about by the introduction of movable type. During the fourteenth century, Western Christendom had become increasingly familiar with the Hindu-Arabic algorithms as an aid to merchant accountancy through hand-copied manuals of commercial arithmetic. There had also been an awakening interest in trigonometry as an aid both to military surveying and to nautical astronomy. It is therefore significant that nautical almanacs, commercial arithmetics and treatises on military surveying together make up a large proportion of the reading matter which left the new hand presses during the first century of their existence.

By the encouragement it gave to the engraver's art, printing indirectly stimulated mathematical curiosity in two ways. In so favourable a setting as the century which witnessed the circumnavigation of the globe, cartography received an immense impetus from the mass production and saleability of maps. To meet the challenge of a new need, its practitioners, among whom Gerardus Mercator (1512–1594) was pre-eminent, had to tackle new technical problems of projective geometry. Contemporaneously, interest in projective geometry had a different focus through the work of the great engraver Albrecht Dürer (1471–1528), himself a geometer. Dürer's own contribution had nothing to do with navigation, being the outcome of a new style of European painting, largely developed in the century before printing from movable type. In the forefront of the movement, Leonardo da Vinci (1452–1519)

The mosque at Cordova. During the tenth century A.D. Spain became the new centre of gravity of Moslem culture. The event was the signal for an awakening of scientific endeavour in Western Europe.

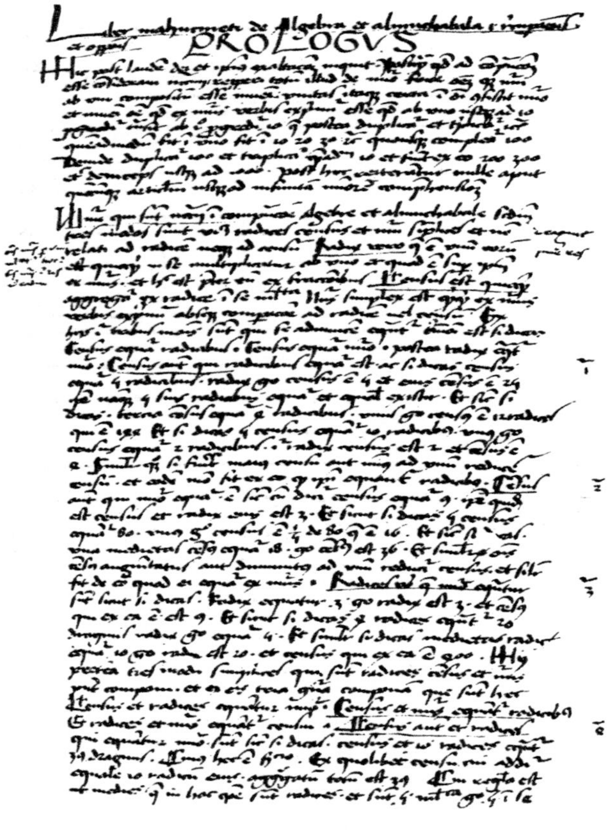
PROLOGVS

Latin translation of a work by al-Kwarismi. Scholars of the universities of Spain began translating Arabic texts into Latin in the twelfth century A.D.

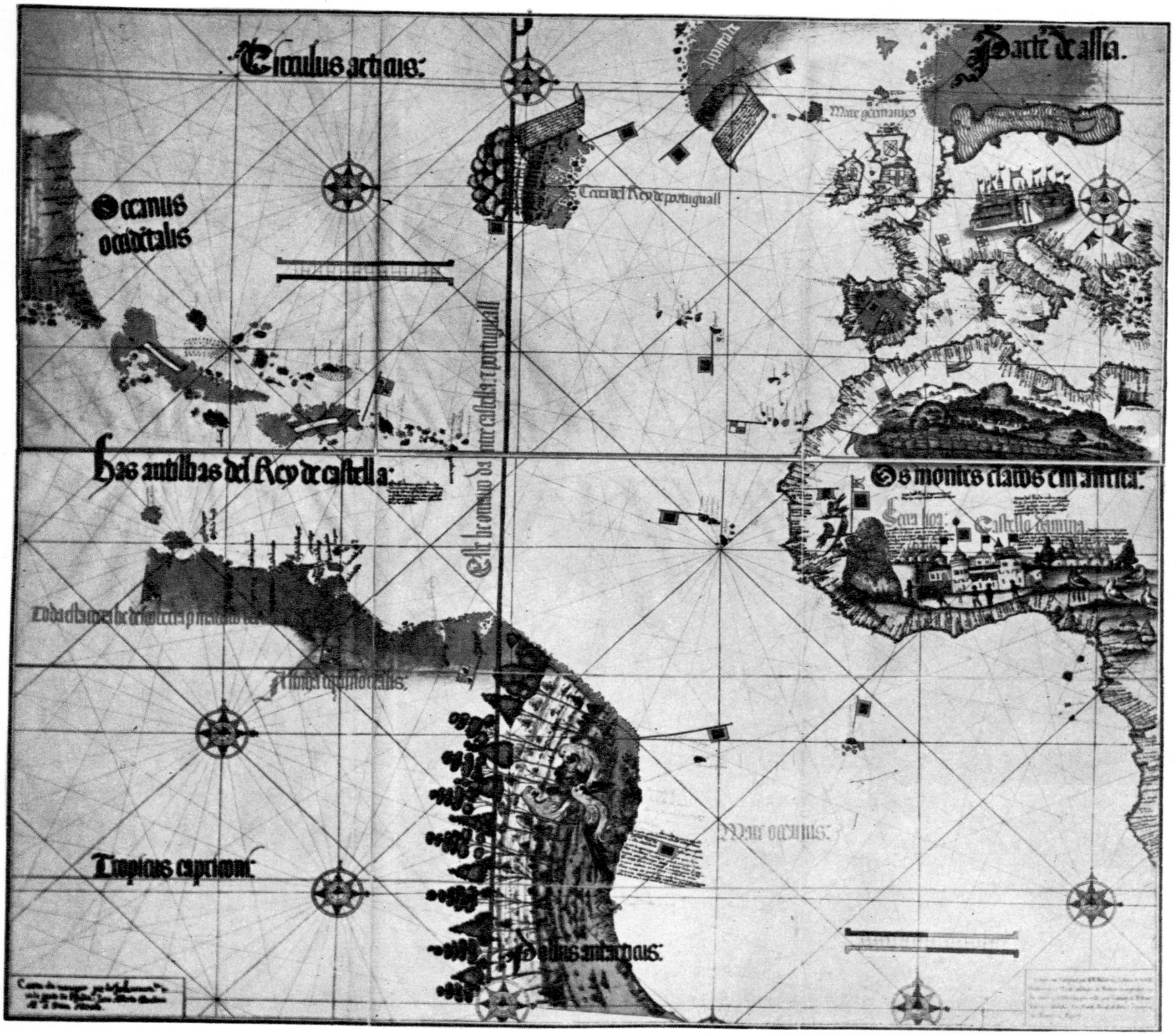

Map of 1502 showing the world as Portuguese mariners then knew it. It marks the West African settlement and fort of S. Jorge da Mina, now Elmina.

Site of the famous school of navigation at Sagres, founded by Prince Henry of Portugal. It was staffed by men trained in the Moorish universities.

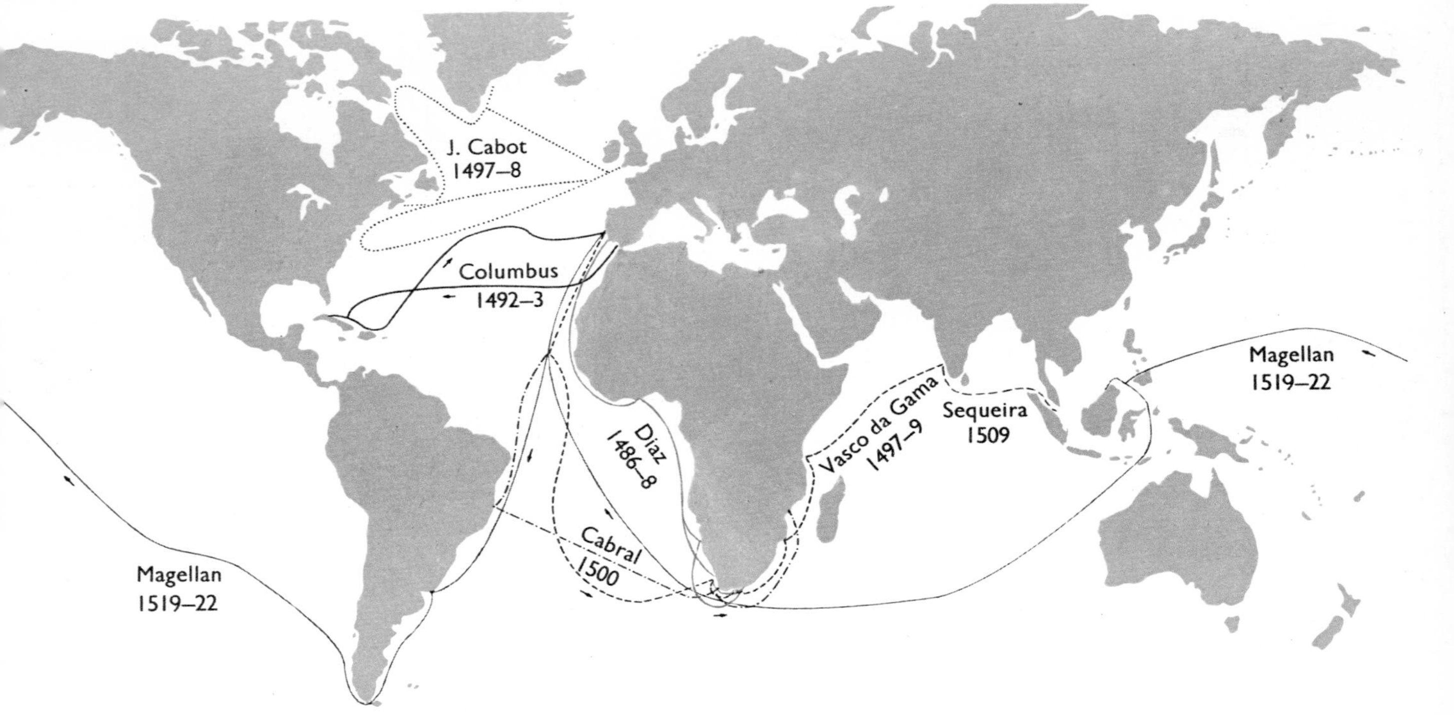

Voyages of discovery shortly before and after 1490 (before brown, after black). The post-1490 east-west voyages first made the longitude problem urgent.

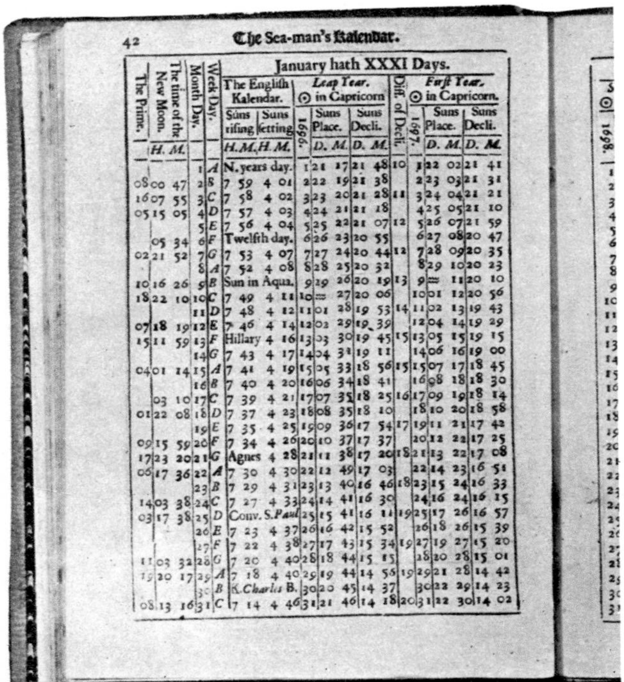

42 The Sea-man's Kalendar.

January hath XXXI Days.

The Prime.	The time of the New Moon. H. M.	Month Day.	Week Day.	The English Kalendar. Suns rising H. M.	Suns setting H. M.	1696.	Leap Year. ☉ in Capricorn. Suns Place. D. M.	Suns Decli. D. M.	Diff. of Decli.	1697.	First Year. ☉ in Capricorn. Suns Place. D. M.	Suns Decli. D. M.
		1	A	N. years day.		1	21 17	21 48	10	1	22 02	21 41
08	00 47	2	B	7 59	4 01	2	22 19	21 38		2	23 03	21 31
16	07 55	3	C	7 58	4 02	3	23 20	21 28	11	3	24 04	21 21
05	15 05	4	D	7 57	4 03	4	24 21	21 18		4	25 05	21 10
		5	E	7 56	4 04	5	25 22	21 07	12	5	26 07	21 59
	05 34	6	F	Twelfth day.		6	26 23	20 55		6	27 08	20 47
02	21 52	7	G	7 53	4 07	7	27 24	20 44	12	7	28 09	20 35
		8	A	7 52	4 08	8	28 25	20 32		8	29 10	20 23
10	16 26	9	B	Sun in Aqua.		9	29 26	20 19	13	9	♒ 11	20 10
18	22 10	10	C	7 49	4 11	10	♒ 27	20 06		10	01 12	20 56
		11	D	7 48	4 12	11	01 28	19 53	14	11	02 13	19 43
07	18 19	12	E	7 46	4 14	12	02 29	19 39		12	04 14	19 29
15	11 59	13	F	Hillary	4 16	13	03 30	19 45	15	13	05 15	19 15
		14	G	7 43	4 17	14	04 31	19 11		14	06 16	19 00
04	01 14	15	A	7 41	4 19	15	05 33	18 56	15	15	07 17	18 45
		16	B	7 40	4 20	16	06 34	18 41		16	08 18	18 30
	03 10	17	C	7 39	4 21	17	07 35	18 25	16	17	09 19	18 14
01	22 08	18	D	7 37	4 23	18	08 35	18 10		18	10 20	18 58
		19	E	7 35	4 25	19	09 36	17 54	17	19	11 21	17 42
09	15 59	20	F	7 34	4 26	20	10 37	17 37		20	12 22	17 25
17	23 20	21	G	Agnes	4 28	21	11 38	17 20	18	21	13 22	17 08
06	17 36	22	A	7 30	4 30	22	12 49	17 03		22	14 23	16 51
		23	B	7 29	4 31	23	13 40	16 46	18	23	15 24	16 33
14	03 38	24	C	7 27	4 33	24	14 41	16 30		24	16 24	16 15
03	17 38	25	D	Conv. S. *Paul*		25	15 41	16 11	19	25	17 26	16 57
		26	E	7 23	4 37	26	16 42	15 52		26	18 26	15 39
		27	F	7 22	4 38	27	17 43	15 34	19	27	19 27	15 20
11	03 32	28	G	7 20	4 40	28	18 44	15 15		28	20 28	15 01
19	20 17	29	A	7 18	4 40	29	19 44	14 56	19	29	21 28	14 42
		30	B	K. *Charles* B.		30	20 45	14 37		30	22 29	14 23
08	13 16	31	C	7 14	4 46	31	21 46	14 18	20	31	22 30	14 02

Nautical almanacs forecast times of celestial events at some fixed point. Only by observing local times of the same events could seamen find longitude at sea.

had realised the need for an optical geometry to equip the practice of perspective art with precise principles; but it is pre-eminently to Dürer that we owe their exploration and exposition. Such was the background of the revival of projective geometry in the writings of the French engineer and architect Desargues (1593–1662).

Mercator's projection impinges on the issue of *invariance* already mentioned for a reason directly related to long-distance westerly navigation, *i.e.* the need to chart a ship's course on a flat surface as nearly as possible along a great circle. No mathematical formula describes the so-called *loxodromic* curve which is the plane projection of a terrestrial great circle; but Mercator devised a projection which makes the best of a bad job from the pilot's viewpoint. Mercator's projection has the merit that straight (so-called *rhumb*) lines equally inclined to the meridians closely follow the true curve of least distance. However, the impetus westerly navigation gave to cartography is by no means its most important indirect contribution to the great outburst of mathematical activity in the sixteenth century. To appreciate the significance of another, we need to bring into the picture the military manuals which familiarised

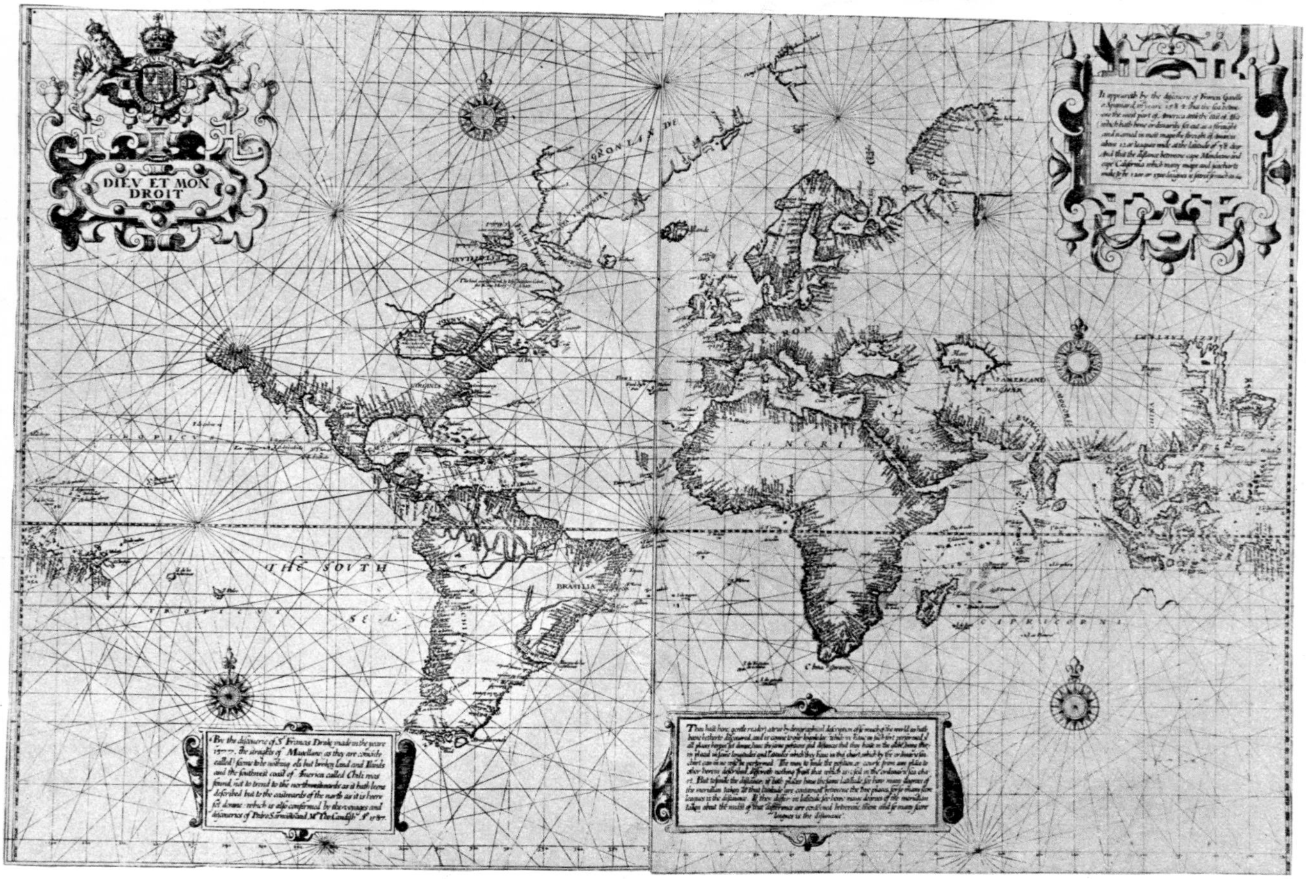

Sixteenth-century Mercator world map. The saleability of printed maps was then challenging cartographers to tackle new problems of projective geometry.

mathematicians of the sixteenth century with how little the professional soldiery understood the mechanics of marksmanship.

The mechanics of antiquity had been exclusively concerned with static equilibrium; but the introduction of gunpowder into warfare had lately created a situation which offered mathematical skill for the first time in history the challenge of solving a purely dynamical problem. How Galileo (1564–1642) successfully met the challenge by formulating the trajectory of the cannon ball in terms of his own experimental discoveries about terrestrial gravitation is a well-thumbed brief. His success led on to another problem, and one directly related to the needs of seamanship.

Westerly navigation insistently called for a method of determining longitude without reliance on comparatively uncommon celestial signals. Within a century after Columbus sighted the

During the same period Albrecht Dürer, engraver and geometer, investigated other problems of projective geometry in the interests of perspective art.

West Indies, the needs of navigation had indeed put the spotlight on the possibility of exploiting an invention unknown to the civilisations of antiquity or to the Oriental world as a means of doing so at any time of the night or at noon on any cloudless day when far from land. By the middle of the sixteenth century, clock making was already a flourishing industry, and geographers realised that a reliable seaworthy clock set in unison with local time at a fixed meridian would dispose of all such difficulties at sea. Though the discovery of the retardation of the pendulum by latitude later proved the unsuitability of a pendulum clock as a chronometer, Galileo's interpretation of the motion of the pendulum in terms of gravitational acceleration assuredly reinforced the hope that ingenuity could indeed design one. Thenceforth, and speedily, dynamical concepts and problems emerging from the needs of clock technology were destined to revolutionise mathematical thought.

The output of commercial arithmetics had an ulterior significance for the progress of mathematics in the two centuries which followed the introduction of printing from movable type. Here for the first time we meet familiar symbols such as +, − and = enlisted to expound the use of an

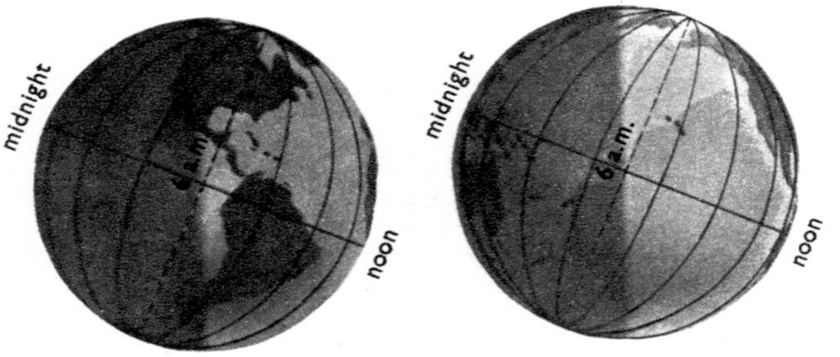

Because of the earth's daily west-to-east spin, the sun (like other heavenly bodies) transits different meridians at ascertainably different times. Longitude differences are thus measurable as time differences.

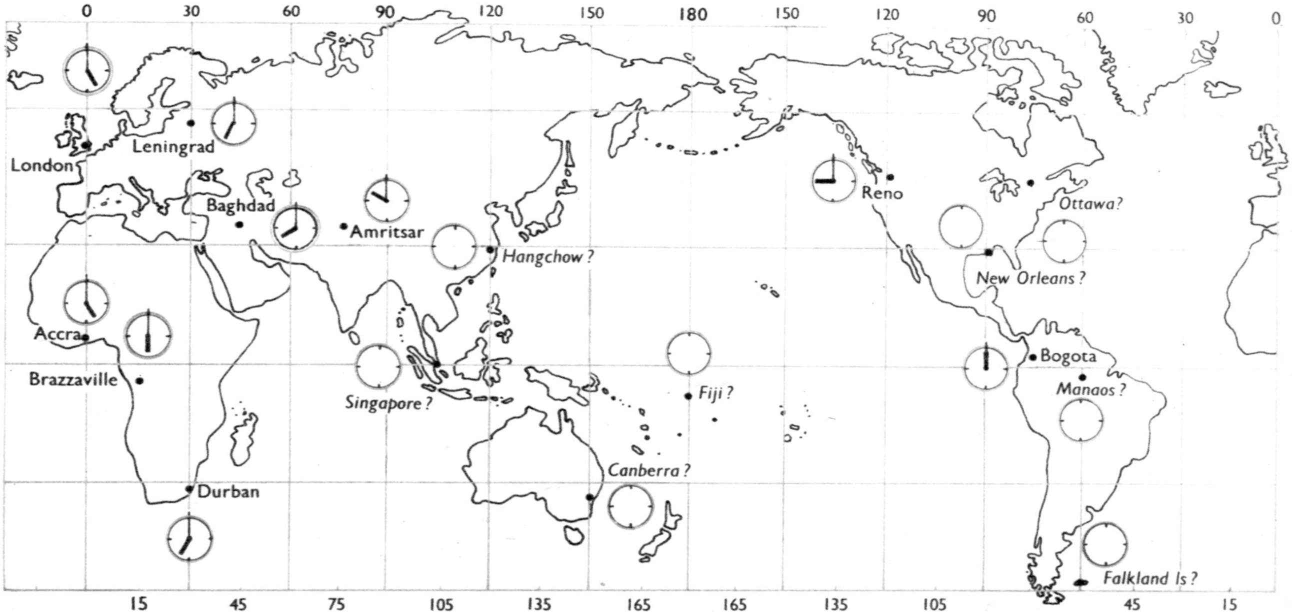

Some of the clock-faces above register local times on different meridians when Greenwich Mean Time is 5 p.m. The reader may care to work out correct times for the blank dials. (Answers near end of book.)

operation. To that extent, printing, which also spread the use of a common staff notation among musicians, contributed to the creation of the common shorthand which we now call algebra. Maybe a sufficient explanation for the popularity of such signs is that they were linguistically neutral, at least by the time they got into use, even if, as some surmise, their origin was literal. Progress towards the adoption of a common algebraic notation for all countries might well have been swifter if one could say the same about the other ingredients, in particular the use of alphabetic signs.

Contractions of words by stages to bare initials was not new in A.D. 1500. The practice gained ground everywhere during the ensuing century and a half; and if Latin had remained the *lingua franca* of scholarship, standardisation of symbols might have been speedier. It was inevitably slow in the sixteenth and seventeenth centuries for the same reason that *p*, which might suggest *pomme* to a Frenchman, would not suggest *apple* to an Englishman. In short, the *milieu* of the cannon ball and the clock was one which witnessed the emergence of aggressive nationalism and therewith the use of the vernacular. The table on page 176 discloses several examples of the multiplicity of conventions brought about by abbreviating vernacular words.

Standardisation mainly came about through the great influence exerted first by Descartes (1596–1650) and then by Wallis (1616–1703) and Newton (1642–1727). Before them and throughout the two centuries A.D. 1500–1700, the printing press, which could now turn out hundreds of copies of a treatise in a small fraction of the time previously taken by a scrivener to make a single one, was in fact enlisting the attention of an ever-expanding audience for any work of scientific or mathematical novelty, and in a social *milieu* which enlisted trained mathematicians in the solution of uniquely novel tasks in a rapidly-

Regulated by a pendulum swinging through a cycloidal arc, Huygens' clock of 1673 (left) augured a solution of the longitude problem by mechanical time-measurement. Some century later, Harrison's second marine chronometer (below) proved capable of showing home-port time correct to within a few minutes over a voyage of several years.

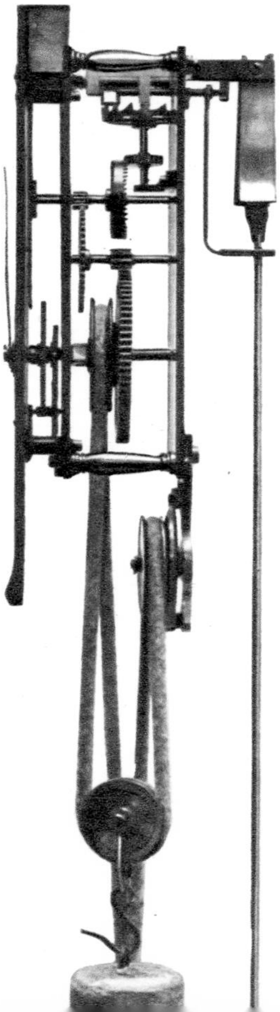

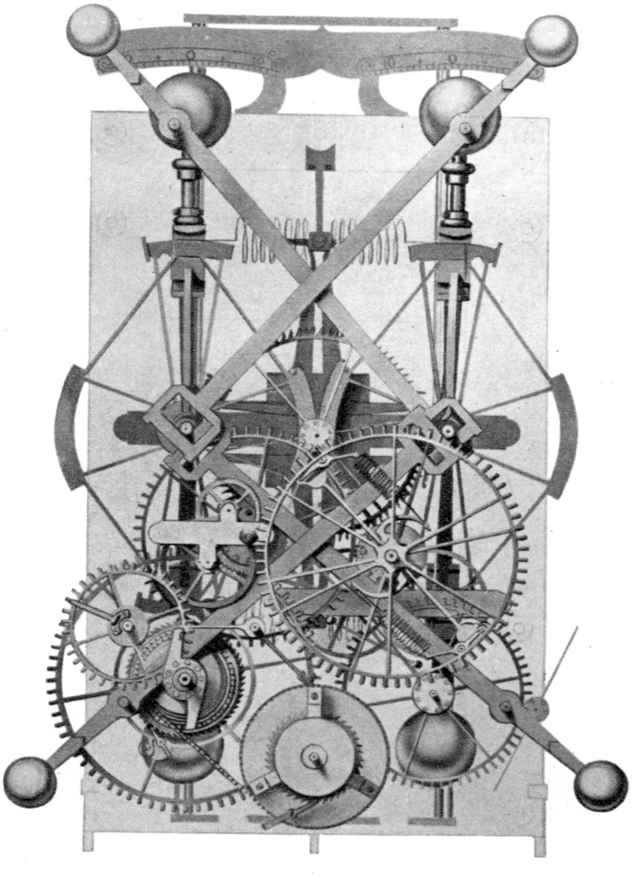

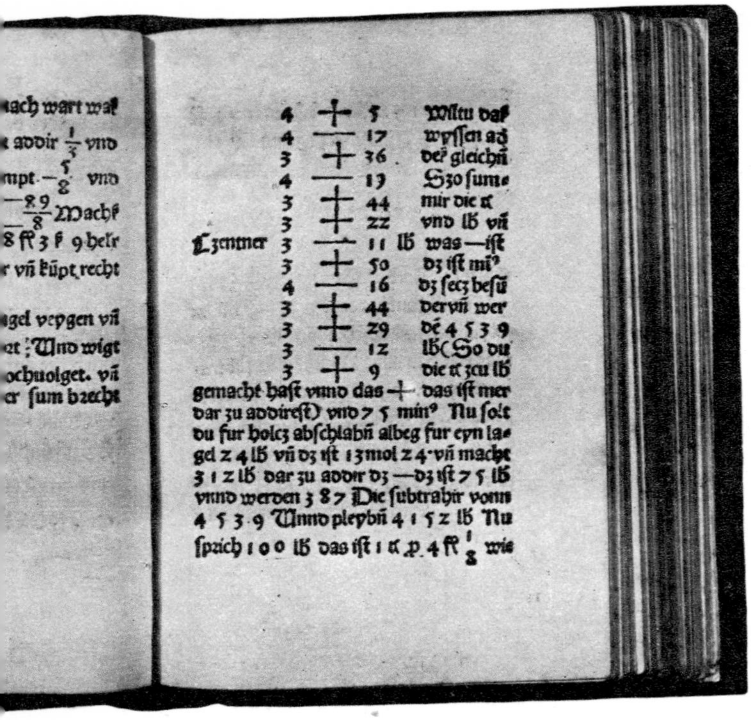
gemacht hast vnnd das + das ist mer
dar zu addirest vnd 7 5 min⁹ Nu solt
du fur holcz abschlahn albeg fur eyn lagel 24 lb vnd dz ist 13 mol 24 vnd macht
3 1 2 lb dar zu addir dz — dz ist 7 5 lb
vnnd werden 3 8 7 Die subtrahir vonn
4 5 3 9 Unnd pleybn 4 1 5 2 lb Nu
sprich 1 0 0 lb das ist 1 ℔ p 4 fl $\frac{1}{8}$ wie

Johann Widman's Behede und hubsche Rechnung auf allen Kauffmanschafft (*1489*), *the first printed book to employ the operational signs + and −.*

Table showing the development of modern algebraic symbolism in the sixteenth and seventeenth centuries.

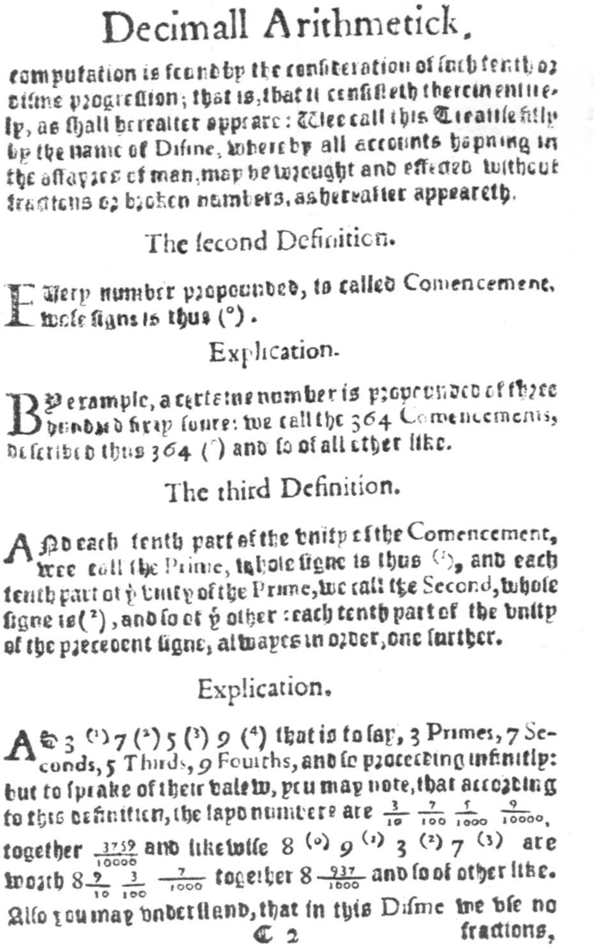
Decimall Arithmetick.

computation is found by the consideration of such tenth or Disme progression; that is, that it consisteth therein entirely, as shall hereafter appeare: Wee call this Treatise fitly by the name of Disme, whereby all accounts hapning in the affayres of man, may be wrought and effected without fractions or broken numbers, as hereafter appeareth.

The second Definition.

Every number propounded, is called Comencement, whose signe is thus (0).

Explication.

By example, a certaine number is propounded of three hundred sixty foure: we call the 364 Comencements, described thus 364 (0) and so of all other like.

The third Definition.

And each tenth part of the unity of the Comencement, wee call the Prime, whose signe is thus (1), and each tenth part of the unity of the Prime, wee call the Second, whose signe is (2), and so of the other: each tenth part of the unity of the precedent signe, alwayes in order, one further.

Explication.

As 3 (1) 7 (2) 5 (3) 9 (4) that is to say, 3 Primes, 7 Seconds, 5 Thirds, 9 Fourths, and so proceeding infinitly: but to speake of their valew, you may note, that according to this definition, the sayd numbers are $\frac{3}{10}$ $\frac{7}{100}$ $\frac{5}{1000}$ $\frac{9}{10000}$, together $\frac{3759}{10000}$ and likewise 8 (0) 9 (1) 3 (2) 7 (3) are worth $8\frac{9}{10}$ $\frac{3}{100}$ $\frac{7}{1000}$ together $8\frac{937}{1000}$ and so of other like. Also you may understand, that in this Disme wee use no fractions,

C 2

Page from the 1608 English edition of Simon Stevin's The Art of Tenths, or Decimall Arithmetick.

date	mathematician	symbolism used	modern form
1494	Pacioli	"Trouame .1.n°. che giõto al suo q̄drat° facia .12."	$x + x^2 = 12$
1514	Vander Hoecke	4 Se.−51 Pri.−30 N. dit is ghelijc 45⅗.	$4x^2 - 51x - 30 = 45\frac{3}{5}$
1521	Ghaligai	1 □e 32 c°−320 numeri	$x^2 + 32x = 320$
1545	Cardan	cub⁹ p: 6 reb⁹ aeq̄lis 20	$x^3 + 6x = 20$
1556	Tartaglia	"Trouame uno numero che azontoli la sua radice cuba uenghi ste, cioe .6."	$x + \sqrt[3]{x} = 6$
1559	Buteo	1 ◊ P6ϱP9[1 ◊ P3ϱP24	$x^2+6x+9=x^2+3x+24$
1577	Gosselin	12LM1QP48 aequalia 144M24LP2Q	$12x - x^2 + 48 = 144 - 24x + 2x^2$
1585	Stevin	3② + 4 egales à 2① + 4	$3x^2+4=2x+4$
1586	Ramus & Schoner	1q+8l aequatus sit 65	$x^2+8x=65$
1629	Girard	1 (4)+35 (2)+24=10 (3)+50 (1) or with the several exponents inclosed in circles	$x^4 + 35x^2 + 24 = 10x^3 + 50x$
1631	Oughtred	$\frac{1}{2}$ Z ± $\sqrt{}_q{:}\frac{1}{4}$ Z_q − AE = A	$\frac{1}{2}Z \pm \sqrt{\frac{1}{4}Z^2 - AE} = A$
1631	Harriot	aaa−3·bba ═ +2·ccc	$x^3 - 3b^2x = 2c^3$
1637	Descartes	yy ∞ cy − $\frac{cx}{b}$ y + ay − ac	$y^2 = cy - \frac{cx}{b}y + ay - ac$
1693	Wallis	$x^4 + bx^3 + cxx + dx + e = 0$	$x^4 + bx^3 + cx^2 + dx + e = 0$

changing world. A curious survival of the period when there was as yet no anticipation of the adoption of an internationally uniform symbolism illustrates the importance of one such task. The need for better nautical almanacs in the context of westerly navigation signifies also the need for tables of trigonometrical ratios with increasingly greater refinement of the interval. Thus the invention of logarithms by Napier for facilitating the computation of sines derives its *motif* from an eminently practical challenge.

Independently, a Swiss mathematician, Burgi, invented a system which in retrospect embodies the same principle; but the exposition of what is now to us so simple a device by both authors is almost unintelligible in the notation they employ. As we now conceive it, the notion of a logarithm is inherent in the rules of indices expounded by Oresme in the fourteenth century and developed more fully by Stevinus (Simon Stevin, 1548–1620) who first among European writers systematically developed the use of decimal fractions already understood in the East. Neither of the authors last named used the index notation of today. What follows could not become transparently clear till the influence of Wallis and Newton universalised the use of negative and fractional exponents as in $0{\cdot}01 = 10^{-2}$ and $10^{\frac{1}{3}} = \sqrt[3]{10}$. The long delay before the widespread adoption of a common index notation side by side with the persistence of the so-called surd signs ($\sqrt{\ }$, $\sqrt[3]{\ }$, $\sqrt[4]{\ }$) in place of fractional exponents is one explanation of the incredibly tortuous approach to the construction of tables of logarithms by Napier and Briggs (1614–24), of the obscurity of early expositions of their use and of the intrusion into the vocabulary of mathematics of the word itself when *index*, *exponent* and *power* were currently available synonyms.

To expound the rules of exponents set forth against the background of the abacus model in Chapter 2, arithmeticians such as Appianus (*circa* 1527) and Stifel (*circa* 1544) used the device of placing in parallel rows two series such as the first two in the table which follows.

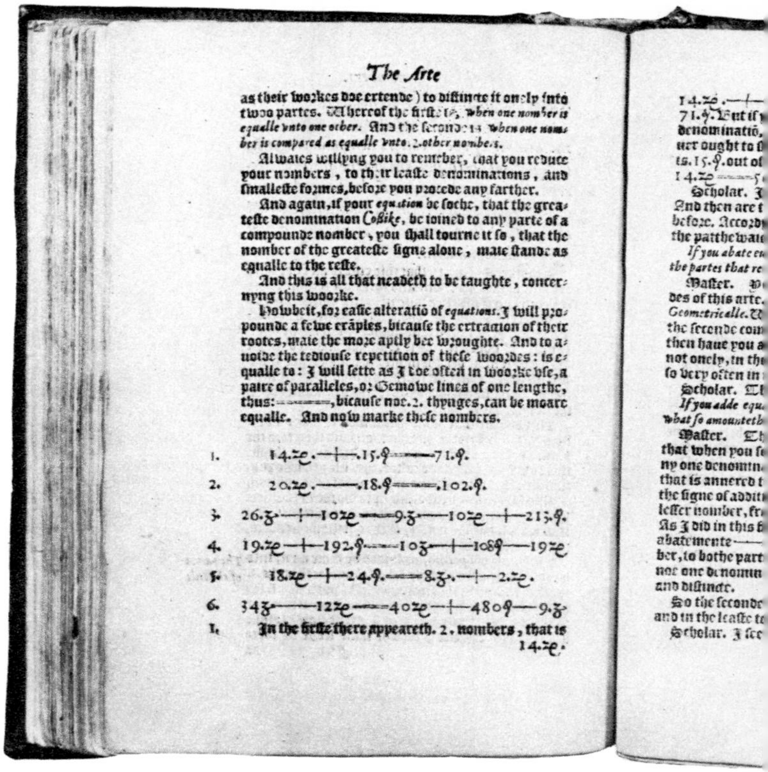

The sign = made its first appearance in print in Robert Recorde's book The Whetstone of Witte, *1557.*

Stifel's Arithmetica Integra *displayed the rules of exponents by placing two series in parallel rows.*

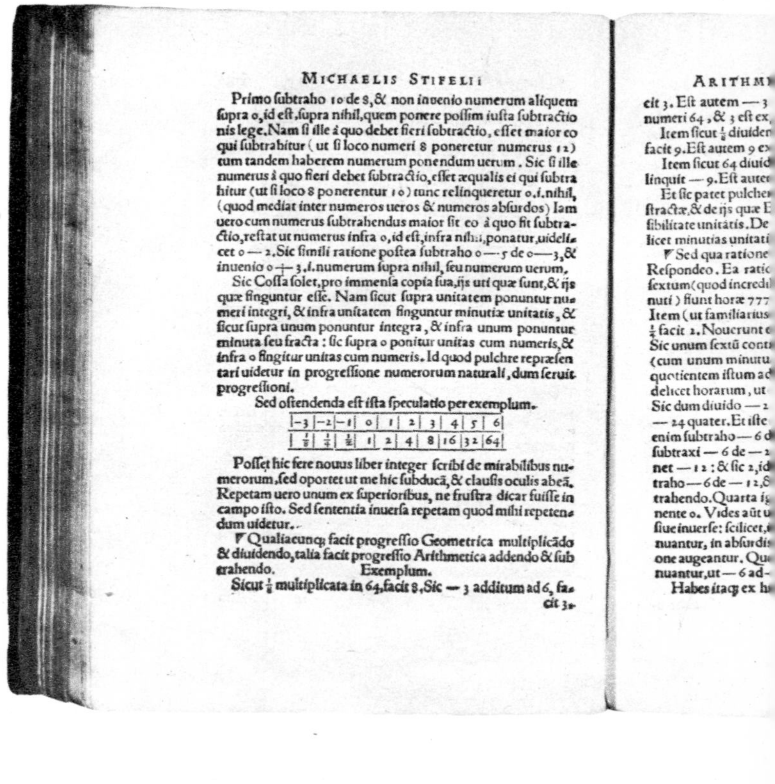

−4	−3	−2	−1	0	+1	+2	+3	+4	...
0·0625	0·125	0·25	0·5	1·0	2·0	4·0	8·0	16·0	...
$=2^{-4}$	$=2^{-3}$	$=2^{-2}$	$=2^{-1}$	$=2^{0}$	$=2^{1}$	$=2^{2}$	$=2^{3}$	$=2^{4}$	...

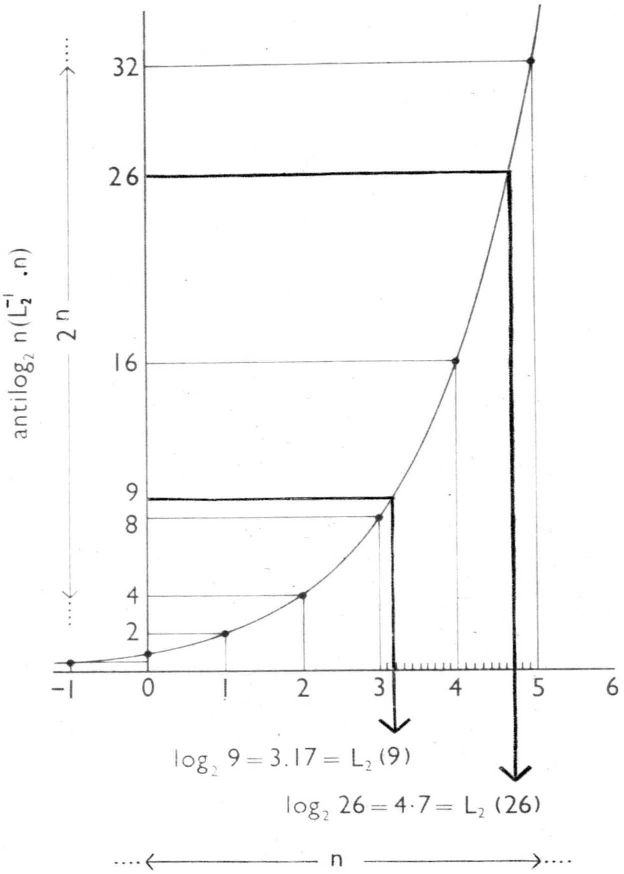

Simple log-antilog graph.

We now speak variously of the *indices* (superscripts) in the bottom line as *exponents*, as *powers* of the base 2 or as *logarithms* to the base 2. In the idiom of logarithms we also speak of any number in line 2 as the *antilogarithm* of the number above it. This saddles the vocabulary of mathematics with two new ways of stating verbally the identity $2^5=32$:

(a) 5 is the logarithm of 32 to the base 2, *meaning*: 5 is the exponent (x) which satisfies the equation $2^x=32$;

(b) 32 is the antilogarithm of 5 to the base 2, *meaning*: 32 is the number (y) which satisfies the equation $2^5=y$.

Symbolically it is still customary to write (a) as $log_2\,32=5$ and (b) as $antilog_2\,5=32$. Hereafter, we shall write more briefly:

$$L_2.32=5\equiv L_2^{-1}.5=32$$

We now see at once that

$$L_2.(L_2^{-1}.5)=5 \text{ and } L_2^{-1}.(L_2.32)=32$$

More generally in this notation, if $b^x=n$:

$$L_b.(n)=x \text{ and } L_b^{-1}.x=n$$
$$L_b^{-1}.(L_b.n)=n \text{ and } L_b.(L_b^{-1}.x)=x$$

Before using this notation to show the relevance of the rule of exponents to the use of tables of logarithms, let us look again at the series 2^b set out (as did Stifel) in the form above.

First, we notice that we can multiply two numbers in the second row by adding the corresponding numbers in the top one and picking out the number in the second row below their sum.

Slide rule in action.

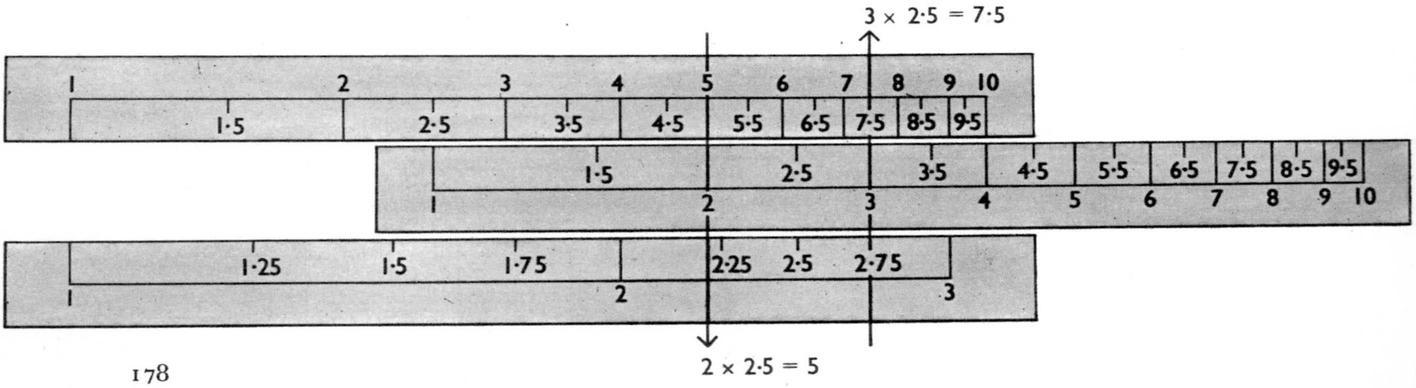

This makes use of the Archimedean rule $2^n.2^m = 2^{n+m}$. If $x=b^n$ and $y=b^m$, so that $xy=b^{m+n}$, we may write alternatively:

$$L_b x = n;\ L_b y = m;\ L_b(xy) = m+n$$
$$\therefore\ L_b(xy) = L_b x + L_b y \ \text{ and } \ xy = L_b^{-1}(L_b x + L_b y)$$

Similarly:

$$x \div y = L_b^{-1}(L_b x - L_b y)$$

Next, notice that we can raise to the power n any number in the second row if we multiply the number above it by n and pick out the number below where the product occurs in the top series. This makes use of the rule $(b^m)^n = b^{mn}$. If $x=b^m$ so that $m=L_b x$ and $x^n=b^{nm}$ we may write $nm = L_b . x^n$, whence

$$n.L_b x = L_b x^n \ \text{ and } \ x^n = L_b^{-1}(n.L_b x)$$

Set out as above, the two series based on 2^n constitute a very crude table of the logarithms of numbers in the second row and their corresponding antilogarithms of numbers in the first. The conspicuous advantage of what might otherwise seem to be a redundant notation is that it makes more explicit how to read off the appropriate items in the table as a means of shortening the labour of multiplication, division or (since n in $L_b x^n$ may be fractional) extracting roots. Needless to say, the table as it stands is of no use unless we can make the gaps vastly smaller. To accomplish this by the most elementary procedure, we note that:

(a) the middle term of any consecutive triplet a_1, a_2, a_3, in the top (arithmetic) series is the arithmetic mean of the other two, *i.e.* $a_2 = \frac{1}{2}(a_1+a_3)$;

(b) the middle term of any consecutive triplet g_1, g_2, g_3, in the second (geometric) series is the geometric mean of the other two, *i.e.* $g_2 = (g_1 \cdot g_3)^{\frac{1}{2}}$.

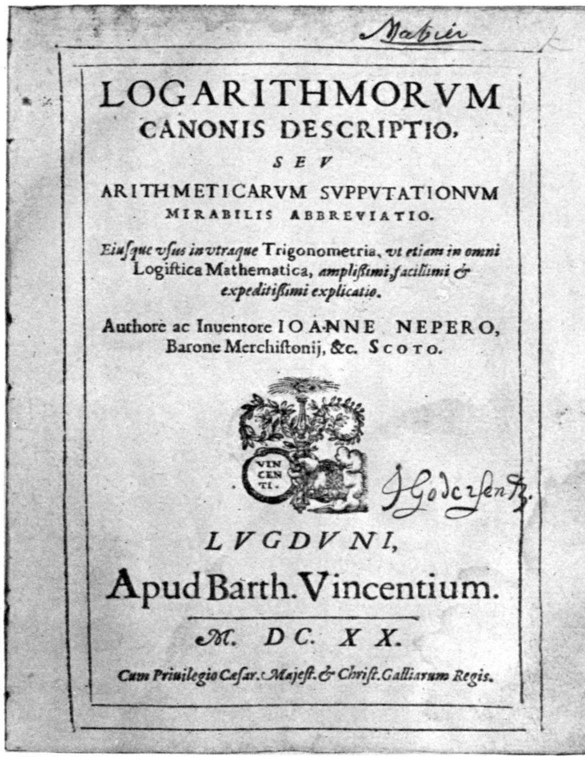

LOGARITHMORVM
CANONIS DESCRIPTIO,
SEV
ARITHMETICARVM SVPPVTATIONVM
MIRABILIS ABBREVIATIO.

Eiusque vsus in vtraque Trigonometria, *vt etiam in omni* Logistica Mathematica, *amplissimi, facillimi & expeditissimi explicatio.*

Authore ac Inuentore IOANNE NEPERO, Barone Merchistonij, &c. SCOTO.

VIN CEN TI.

LVGDVNI,
Apud Barth. Vincentium.
M. DC. XX.

Cum Priuilegio Cæsar. Majest. & Christ. Galliarum Regis.

Title-page (above) and a table (below) from John Napier's work on logarithms (1620).

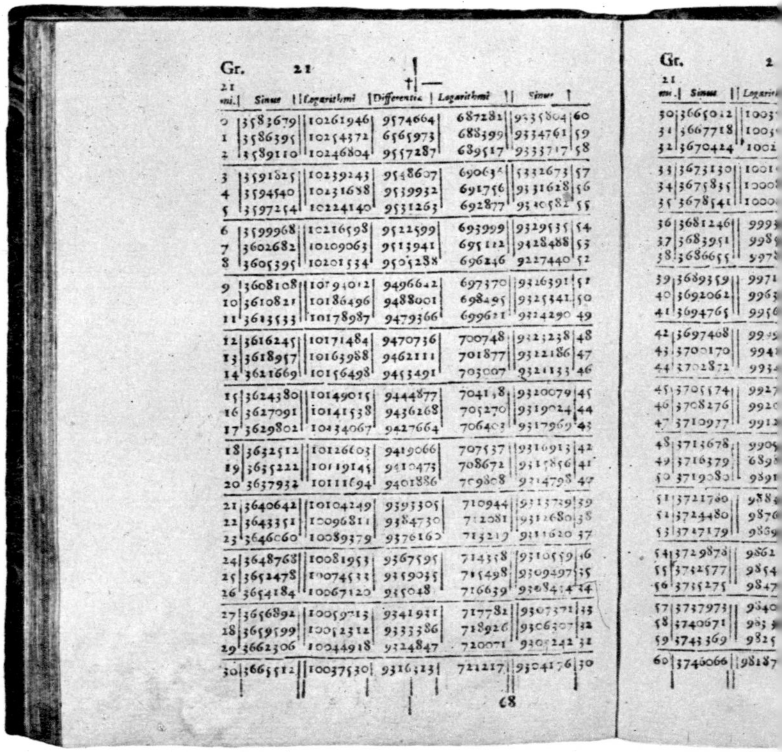

Gr. 21

21 mi.	Sinus	Logarithmi	Differentia	Logarithmi	Sinus	
0	3583679	10261946	9574664	687281	9335804	60
1	3586395	10254372	6565973	688399	9334761	59
2	3589110	10246804	9557287	689517	9333717	58
3	3591825	10239243	9548607	690636	9332673	57
4	3594540	10231688	9539932	691756	9331628	56
5	3597254	10224140	9531263	692877	9330582	55
6	3599968	10216598	9522599	693999	9329535	54
7	3602682	10209063	9513941	695112	9328488	53
8	3605395	10201534	9505288	696246	9327440	52
9	3608108	10194012	9496642	697370	9326391	51
10	3610821	10186496	9488001	698495	9325341	50
11	3613533	10178987	9479366	699621	9324290	49
12	3616245	10171484	9470736	700748	9323238	48
13	3618957	10163988	9462111	701877	9322186	47
14	3621669	10156498	9453491	703007	9321133	46
15	3624380	10149015	9444877	704138	9320079	45
16	3627091	10141538	9436268	705270	9319024	44
17	3629802	10134067	9427664	706403	9317969	43
18	3632512	10126603	9419066	707537	9316913	42
19	3635222	10119145	9410473	708672	9315856	41
20	3637932	10111694	9401886	709808	9314798	40
21	3640642	10104249	9393305	710944	9313739	39
22	3643351	10096811	9384730	712081	9312680	38
23	3646060	10089379	9376162	713219	9311620	37
24	3648768	10081953	9367595	714358	9310559	36
25	3652478	10074533	9359035	715498	9309497	35
26	3654184	10067120	935048	716639	9308434	34
27	3656892	10059715	9341911	717782	9307371	33
28	3659599	10052312	9333386	718926	9306307	32
29	3662306	10044918	9324847	720071	9305242	31
30	3665512	10037530	9316323	721217	9304176	30

68

If we wish to obtain the *antilog* of $1\cdot5$, *i.e.* interpolate in the second line a number corresponding to half the gap between 1 and 2 in the top one, the required number is $[(2)(4)]^{\frac{1}{2}}=\sqrt{8}$. This merely expresses the identity $(2^1.2^2)^{\frac{1}{2}}=2^{\frac{3}{2}}=2^{1\cdot5}$. Having found $\sqrt{8}$, we can now halve the gap on either side in the same way. Thus $(1.\sqrt{8})^{\frac{1}{2}}=L_2^{-1}(0\cdot75)$ and $(2.\sqrt{8})^{\frac{1}{2}}=L_2^{-1}(1\cdot25)$. Similarly, $(4.\sqrt{8})^{\frac{1}{2}}=L_2^{-1}(1\cdot75)$. By continuing the process, we can make the interval between the entries on the top line as small as we wish. We then speak of the outcome as a *table of antilogarithms*, *i.e.* one which exhibits the antilogarithms of numbers set out in equal intervals. For convenience of computation, it is desirable to use a separate *table of logarithms*, *i.e.* one in which the items of the second row are equally spaced. To do this, we may proceed by a process of averaging; but we shall later see that there are less laborious ways of doing so.

If we choose a suitable base, there are also less laborious ways of making a table of antilogarithms than the continuous repetition of square roots. Since the most convenient base (10) for use is not suitable in this sense, it is important to have a rule for converting logarithms (or antilogs) of one base to those of another. We derive it as follows. Let $x=b^y$ and $b=10^a$, so that $x=10^{ay}$, whence

$$L_bx=y; \quad L_{10}b=a; \quad L_{10}x=ay$$
$$\therefore L_{10}x=L_{10}b\,.\,L_bx$$

In this expression $L_{10}b$ is a *constant* factor by which we multiply the logarithm of any number to base b in order to get the logarithm of the same number to base 10. The great practical advantage of the base 10 is that a table exhibiting logarithms in the range $L_{10}(1)=0$ to $L_{10}(10)=1$ suffices for all purposes. One example will make this clear:

$$L_{10}(5\cdot5)=0\cdot7033$$
$$L_{10}(550)=L_{10}(5\cdot5)+L_{10}(100)=0\cdot7033+2$$
$$=+2\cdot7033$$
$$L_{10}(0\cdot55)=L_{10}(5\cdot5)-L_{10}(10)=0\cdot7033-1$$
$$=-0\cdot2967$$

We have already seen that the alphabetic numeral system of the Alexandrian mathematicians made the task of devising a convenient algebraic shorthand exploiting literal symbolism for common use well-nigh superhuman. Since the Chinese used an ideographic script, anything comparable to our notation was unattainable in the Far East. With an Indic battery of alphabetic signs at their disposal, the Hindus did independently make some advance towards an algebraic symbolism such as we use; but this could have little influence on Moslems who exclusively used the Arabic script of the Koran. Thus Western Christendom in the two centuries which followed the rise of the printing industry and witnessed the emergence of dynamical notions was a peculiarly novel setting for the elaboration of tools of communication to elevate mathematical exploration beyond the level of pictorial representation or verbal disputation. In the first intoxication of novelty, those who exploited them left unanswered many paradoxes for the great mopping-up offensive of the past century to dispose of.

Perhaps in no small measure because the mathematicians of the seventeenth century now had at their disposal widely-used symbols for the operations denoted by the negative and positive signs, they indulged freely in a licence to which their Moslem predecessors, unlike their Hindu teachers, were never prone. They assumed that it was meaningful to speak of -5 as a number on all fours with 5 and that it is immaterial whether we do or do not put $+$ in front of the latter. On the other hand, they made no attempt to justify the identity $(-5)\times(-5)=+25$ in terms of the *modus operandi* of the abacus.

The Hindu algorithms make it possible to explore quickly and to exhibit such identities as:

$$(10-3)\ (7+4)=10(7)+10(4)-3(7)-3(4)$$
$$(8-5)\ (11-2)=8(11)-8(2)-5(11)+5(2)$$

Here, to be sure, $(-c)\ (-d)$ as a member of a sequence is replaceable by $+cd$; but the abacus rationale of computation confers no justification on the change of sign, still less on the operation performed on two such symbols *in isolation*. If those who first formulated the rules of signs and their successors who exploited them before the middle of the eighteenth century A.D. sought a rationale for them, they relied on a static Euclidean model illustrating the build-up and break-down of rectangles labelled as below:

$$(a-b)\ (c+d)=ac+ad-bc-bd$$
$$(a-b)\ (c-d)=ac-ad-bc+bd$$

Perhaps to the Hindus, and certainly to their European successors before our own century, it seemed obvious that this justifies the assertion: $(-x)\ (-x)=+x^2$, whence that every number has two square roots, *e.g.* $\sqrt{16}=+4$ or -4. Actually, the Euclidean model gives no justification for this. It can represent $(a-b)$ only if $a>b$ and $(c-d)$ only if $c>d$, in which event the multipliers are always positive. Nor does the metaphor of debt for a negative number, as suggested by the Hindu algebraists and adopted by Leonardo Fibonacci in the *Liber Abaci*, get us any further. One may add one overdraft to another; but there is no intelligible sense in which we can multiply one overdraft by another. Still less can we envisage the outcome of any such manipulation as a credit balance.

Unless, therefore, we do not shrink from what to a pure mathematician is the unforgivable sin against the third person of the Trinity, we must redefine the meaning of the minus sign in terms *consistent with the naïve notion of Euclidean or abacus subtraction but able to accommodate an entirely novel meaning.* Fortunately, it is no longer difficult for us to side-step this dilemma because we are nowadays familiar with instruments whose scales have divisions on either side of a zero mark. It is therefore a suggestive fact that the practice of dividing what we wrongly call the Cartesian framework into four quadrants like two thermometer scales crossing at right angles did not become common till a generation after the introduction of the first instrument ever calibrated in negative and positive units. The parental model was the Fahrenheit scale (*circa* A.D. 1710), followed shortly after by the Reaumur and Centigrade scales.

A little thought about this innovation intensifies

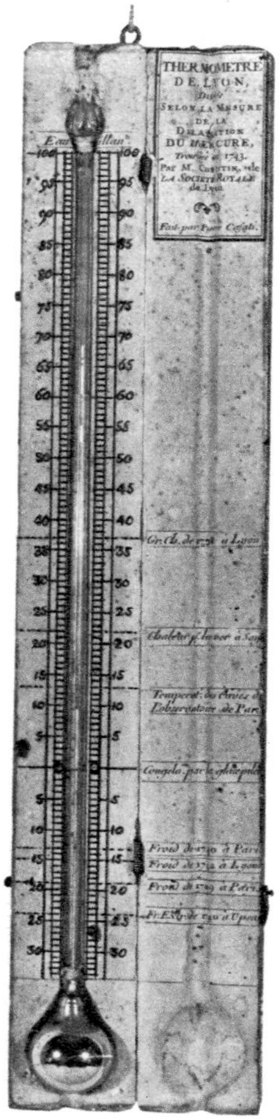

On this eighteenth-century thermometer, numbers above and below the zero respectively denote temperatures above and below the melting-point of ice. Before the introduction of such thermometers, -5 and -10 had no meaningful claim to rank as numbers if we regard numbers as labels for counting or measuring.

the dilemma foreshadowed above and discloses the need to re-examine our terminology. If we write $n(-t)=-nt$, we imply the operation of multiplying the numerical value (t) of the temperature (in degrees) by a number n on the understanding that the $-$ sign before t locates it in the lower end of the scale, just as B.C. and *Before the Hegira* label dates in the Christian and Moslem calendars respectively. Here we do not depart from the primitive approach to multiplication as n-fold addition. What is entirely different is writing: $-n(+t)=-nt$ or *a fortiori*: $-n(-t)=+nt$.

The conventions B.C. and B.H. make no explicit provision for a *zero* origin of reference. Before there was a scale of mensuration with marks left and right of zero, -5 had indeed no intelligible claim to rank as a number, if we mean by number a label for counting or measuring things. Hitherto, indeed, the only meaningful classification of numbers in either domain had been to separate them as integers (even and odd, prime or otherwise), rational (terminating or periodic) fractions and irrational (nonperiodic and nonterminating basal) fractions. We have, however, anticipated how the Hindu Arabic notation made easy the transition to another use of numbers as *labels of operations*, *e.g.* we write $10^{-2}=0{\cdot}01$; and in that domain -5 has a meaning different from but entirely consistent with the primitive use of the minus sign for subtraction. Given a scale graduated both left and right, we are ready to extend the domain of real numbers to embrace meaningfully *directed* numbers, both rational and irrational, in two directions. This is not inconsistent with the way in which we have already come to interpret a directed number such as 6 in 10^{-6} as an instruction which combines in itself an *operation* and a label of counting or measurement.

An initial obstacle to clarifying the use of signs is a defect of our notation, *i.e.*, that we drop $+$ at the beginning of an expression (or part thereof enclosed in brackets) regardless of whether we are referring to a naked (undirected) multiplier like n in $n(-t)\equiv -nt$ or a positive directed number like $+t$ in $-n(+t)\equiv -nt$. Having recognised this source of remediable misunderstanding, all that remains to ask is: what operations do we now associate with the attachment of either sign to the naked number? We cannot visualise a meaningful answer to this unless we cease to look at the problem within the static domain of Euclid's geometry; but we can do so if we take a kinematical viewpoint emergent in the setting of Galileo's treatment of the trajectory of the cannon ball. A naïve answer to our problem suffices for the purposes of interpreting addition and subtraction if we envisage a thermometer placed horizontally with its bulb to the left:

(a) $+t$ means count off t divisions towards the right from the starting point (*s.p.*);
(b) $-t$ means count off t divisions towards the left from the *s.p.*

However, this throws no light on the tie-up with multiplication as *repetitive addition* in the domain of the naked numbers of our abacus model. To probe deeper, we may conceptualise our thermometer model as a line (so-called *real* axis) extending indefinitely right and left of a point corresponding to the zero mark of the scale. We shall then: (a) interpret the signless t to mean the length of a line segment (linear magnitude in Euclidean jargon) between two points p_1 and p_2 ($>p_1$) to the right of the zero mark; (b) consider what happens when we rotate the segment through 180° or 360° about the point p_1. If we rotate the segment of length t about p_1 through 2π ($=360°$), 4π or any even multiple of π radians, we get back to where we were before – in fact, the result is as if we did not rotate it. If we rotate it through π ($=180°$), 3π or any odd number of radians, we may distinguish between two possibilities: (a) the point (p_3) at which we arrive will lie on the same (*positive*) half of the axis if the distance between p_1 and p_2 is less than the distance (p_1) between p_1 and the zero mark; (b) the point p_4 at which we arrive will be in the alternative (*negative*) half of the axis if the distance t is greater than the distance p_1. Accordingly, we shall interpret: (a) an operation we may call multiplying by $(+1)$ as a rotation through 0, 2π, *etc.* radians leaving our direction unchanged; (b) another operation we may call multiplying by (-1) as a rotation through π, 3π, *etc.* radians reversing our direction without affecting the magnitude of the segment involved. To make clear that these are operators, in contradistinction to numbers, we shall preface them with a dot, thus: $\cdot(+1)$ and $\cdot(-1)$. To signify that we have operated with $\cdot(-1)$ twice in a sequence of operations, we shall write $\cdot(-1)^2$. Since this brings us back to where we were at the start, $\cdot(-1)^2 \equiv \cdot(+1)$.

On this understanding, we can now give a consistent meaning to the naked number, the directed (real) numbers of each sort and to the operations which we may perform with one or the other or both:

(i) $+a$ or "a" $= a\cdot(+1)$

meaning: lay out the magnitude a on the real axis ($a°$ on the positive side of thermometer zero) to the right of the starting point (*s.p.*) and rotate the segment of length a about the *s.p.* through 0°, 360°, 720° . . . *etc.*, *i.e.* do not rotate at all.

(ii) $-b = b\cdot(-1)$

meaning: lay out the magnitude b on the real axis ($b°$ on the positive side of thermometer zero) to the right of the *s.p.* and rotate the segment of length b about the *s.p.* through 180°.

(iii) $n\cdot(\pm x) = n\cdot[x\cdot(\pm 1)]$

meaning: repeat the outcome of the sequence of operations here denoted as $(\pm x)$ another $(n-1)$ times additively, *i.e.* starting each repetition where the predecessor ends without change of direction.

(iv) $-n\cdot(\pm x) = \{n\cdot[x\cdot(\pm 1)]\}\cdot(-1)$

meaning: having performed the *undirected* repetitive operation here denoted by $n\cdot$, perform the rotary operation denoted by $(+1)\cdot$ or $(-1)\cdot$ as indicated.

We are now ready to interpret the sign rule $(-)\times(-)=+$. Having invoked the first time the operation $\cdot(-1)$ in a sequence, we have effected a rotation through π radians ($=180°$). If we do so a second time, we have rotated our resulting line segment through 2π radians ($=360°$), and have got back on the scale to the side where we were before. Thus the operation $(-1)\,(-1) \equiv (-1)^{1+1} \equiv (-1)^2 \equiv (+1)$. So we may set out part of (iv) in greater detail thus:

$$\text{(v)}\quad -n(-x) = \{n\cdot[x\cdot(-1)]\}\cdot(-1)$$
$$= [nx(-1)]\cdot(-1)$$
$$= nx(-1)^2 = nx(+1)$$

The use of 2 in $\cdot(-1)^2$ is here consistent with the use of numbers as labels of operation. Thus $10^{-2} = (10^{-1})\,(10^{-1})$ as a multiplier means perform twice the operation (10^{-1}) of dividing by 10. The sequence thus outlined makes it possible to exhibit explicitly the relation between multiplication and repetition in the domain of directed real numbers, provided that we draw a clear distinction between two operations commonly written in the confusing form $-n(+x) = -nx = n(-x)$, *i.e.*:

$$n\cdot(-x) \equiv n\cdot[x\cdot(-1)]$$

and

$$-n(x) \equiv [n\cdot(x)]\cdot(-1)$$

Having satisfied themselves that quadratic equations may have two positive (*e.g.* $x^2-4x+3=0$), two negative (*e.g.* $x^2+4x+3=0$) or one positive and one negative (*e.g.* $x^2+2x-3=0$) solutions without clarifying the meaning conferred on a negative in contradistinction to a positive or an undirected number, Hindu algebraists did not take what now seems to be a reckless, though none the less rewarding, step into the dark. Their European successors of the seventeenth century took it for granted that every quadratic must have a solution. On this assumption, the Moslem rule confers on any equation of the form $x^2+bx+c=0$ two solutions involving the square root of a negative number if $b^2<4c$, *e.g.*:

$$x^2-2x+4=0 \text{ when } x=1\pm\sqrt{-3}=1\pm\sqrt{3}\,.\,\sqrt{-1}$$

Here a new inconsistency emerges. If the square

of either a positive number or a negative number is itself a positive number, in what conceivable sense can the square root of a negative number exist? Eighteenth-century writers used the symbol $i=\sqrt{-1}$ to sidestep a firm answer to the question: is $\sqrt{-1}$ a number? Thus the "solution" would be $x=1\pm i\sqrt{3}$. We shall not here anticipate an answer to the question last stated, except to say that the use of $(-1)\cdot$ and $(+1)\cdot$ as labels for rotational operations is the clue to what we can meaningfully do with such so-called complex numbers as $1\pm i\sqrt{3}$ now more properly referred to as number couples. What is more relevant in this context is the great practical importance of being clear that no equation involving an unknown x admits of a solution unless we first state the class of numbers to which x belongs.

Though Diophantos himself and Chinese or Hindu mathematicians who later developed so-called indeterminate analysis to a higher level of sophistication contributed little to natural science, their preoccupation with equations which are at least relevant to counting herds or flocks, and as such can have by initial agreement only integers

$$3(-4) = 3[4.(-1)] = -12$$

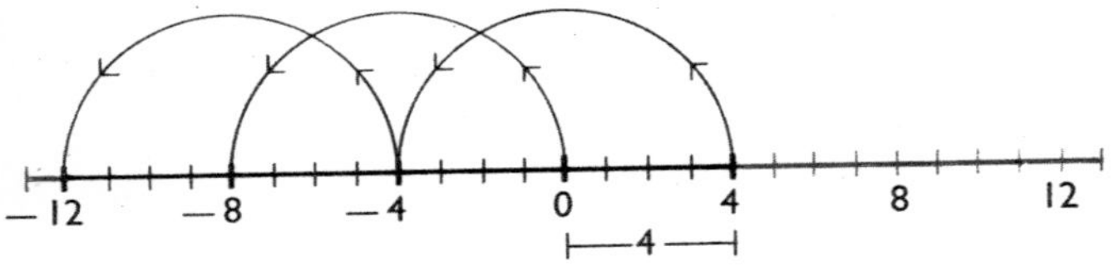

We here visualise the signless 4 as the length of a line segment and the multiplier $\cdot(-1)$ as an instruction to rotate it through 180° anticlockwise about its starting point at zero on the scale. We interpret the signless multiplier 3 as an instruction to repeat the operation 3 times additively by starting each repetition where its predecessor ends without change of direction. The multiplier $\cdot(+1)$ below is an instruction to rotate the line segment through 360° back to its original position.

$$3(+4) = 3[4.(+1)] = +12$$

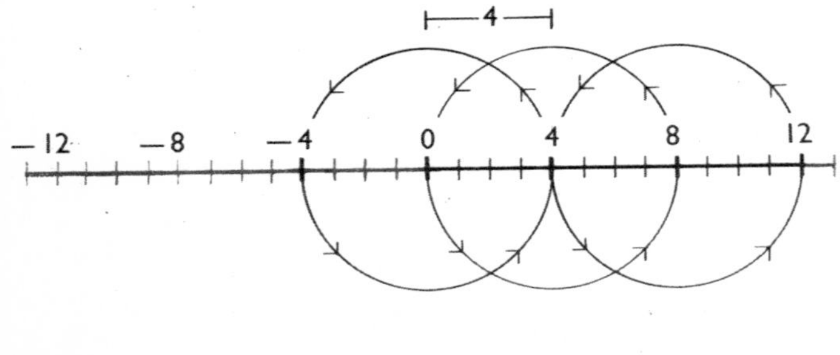

$$-4.(+3) = \{4.[3.(+1)]\}.(-1)$$
$$= -12$$

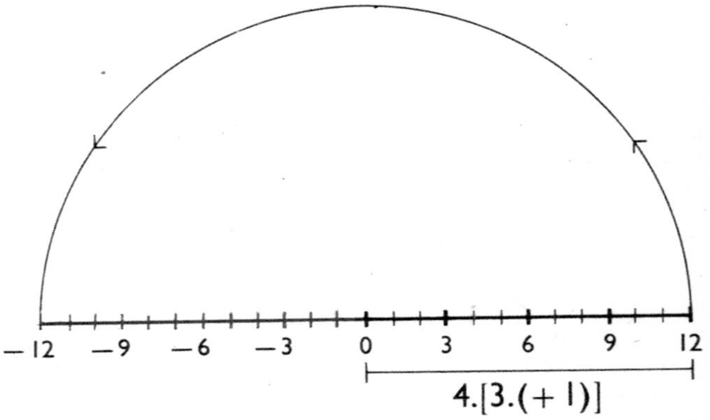

In any expression which takes the form $-n\cdot(\pm x) = \{n\cdot[x\cdot(\pm 1)]\}\cdot(-1)$, we may interpret $n\cdot$ as an instruction to perform an undirected operation on a line segment x, and the multipliers $\cdot(-1)$ or $\cdot(+1)$ as instructions to rotate it through 180° or 360°.

$$-4.(-3) = \{4.[3.(-1)]\}.(-1)$$
$$= +12$$

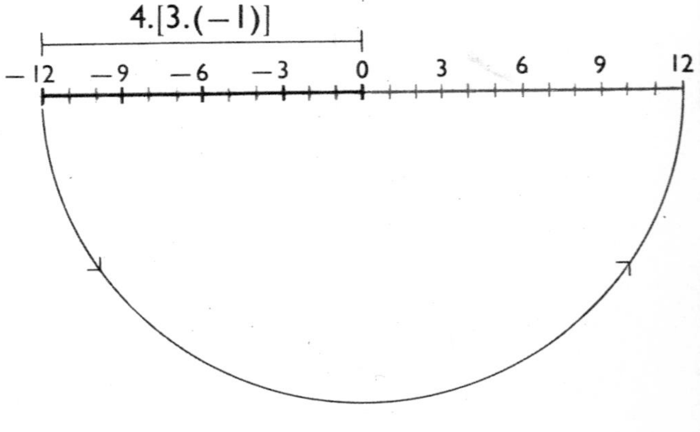

as solutions, did in fact register a clear recognition that what solution or solutions an equation admits depends upon the concessional limitations imposed on the *number domain.* This is very patent in the world's work, as the following examples pinpoint:

1. A farmer sells one third of his flock. Was it correctly reported that the product of the initial and the residual numbers of sheep was 300? Here only an undirected integer exactly divisible by 3 is a factually admissible solution. The reported assertion implies $2n^2 \div 3 = 300$, so that $n = 3\sqrt{50}$. This is exactly divisible by 3, but the residue $\sqrt{50}$ is not an integer. Hence the report was false.
2. Think of a number. Multiply it by 3 and subtract 4. If the answer is 10, what was the number? If we here mean by number an integer the equation ($3n - 4 = 10$) has no solution; but if we admit any real (directed or naked) number, the answer is $n = 4\cdot\dot{6}$.
3. The height of a rectangle is 3 times its base (b). What is the ratio of the diagonal (d) to the height (h)? Here $d^2 = 10b^2$ and $d{:}h = \pm\sqrt{10}{:}3$. An irrational magnitude is here admissible; but the negative root has no meaning in the context of the problem.
4. A workman earns two dollars a day more than his mate, and the numerical product of their daily earnings (dollars) is 15. What is the wage of the mate? Here we are dealing with undirected numbers, and a negative solution is inadmissible. A positive solution is admissible only if the fractional part (if any) is expressible in terms of some integer n in the form $n.10^2$ (to look after the cents). The usual drill gives $w(w+2) = 15$, so that $(w+5)(w-3) = 0$. Whence the only solution which satisfies either condition is $w = 3$ (dollars per day).
5. At a second reading 5 minutes later, the temperature of a liquid was two degrees higher. If the product of the two readings was $+15$, what was the initial reading? In cookery-book terms, the formal statement is the same for this as for the last example, *i.e.* $t_0(t_0+2) = 15$; but the problem admits of a real solution positive or negative, rational or irrational, and the two solutions $t_0 = -5°$ or $+3°$ are equally valid in the real world.
6. Mr. Jones (in Welsh) asked his bank manager about the state of his account. The manager replied (in English) as follows: If you were to add £5 to the entry in our books and multiply the sum by the correct amount, you would be £50 to the good. Was Mr. Jones any the wiser? Here we may write the formal statement as $x(x+5) = 50$, so that $x^2 + 5x - 50 = 0$. We may also properly interpret a negative solution as balance shown on the debit side of the account, and the only restriction we need to place on an admissible solution in the field of all real numbers arises from the circumstance that the bank would not credit sums less than one penny (240 to the £). Hence only integers and fractions expressible in the form $(n \div 240)$ are admissible. Both formal solutions satisfy this

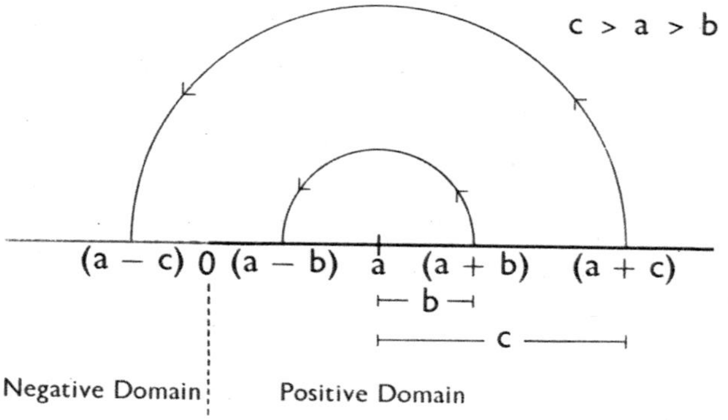

The interpretation of $-x$ as the result of operating once with $\cdot(-1)$ on x implies rotating a line of length x through 180° anticlockwise. In the positive domain of the abacus this is consistent with the elementary approach to addition and subtraction.

condition and both are real. Consequently, Mr. Jones could not decide whether he was £5 to the good or £10 in the red.

Before the writings of three Frenchmen, Descartes (already mentioned), Fermat (1601–1665) and Pascal (1623–1662), initiated an outburst of new activity in the first half of the seventeenth century, the invention of logarithms was by far the most outstanding mathematical innovation made by Western Christendom; but a French mathematician, Vieta (1540–1603), and two Italians, Tartaglia (1500–1557) and Cardano (1501–1576), broke new ground by their treatment of the general cubic equation ($x^3+Ax^2+Bx+C=0$). This episode is specially notable because it eventually provided the first valid reason for taking complex numbers seriously. Tartaglia was first in the field. Before examining his recipe, which is redundant if the expression equated to zero is recognisably factorisable, it is relevant to remind ourselves that a cubic may have 1, 2 or 3 *distinct* real roots and no complex ones, as illustrated by the following:

(1) $x^3-3x^2+3x-1=0$ if $(x-1)^3=0$
and $x=+1$

(2) $x^3-x^2-x+1=0$ if $(x-1)^2\,(x+1)=0$
and $x=+1$ or -1

(3) $x^3+2x^2-x-2=0$ if $(x+1)(x-1)(x+2)=0$
and $x=+1$ or -1 or -2

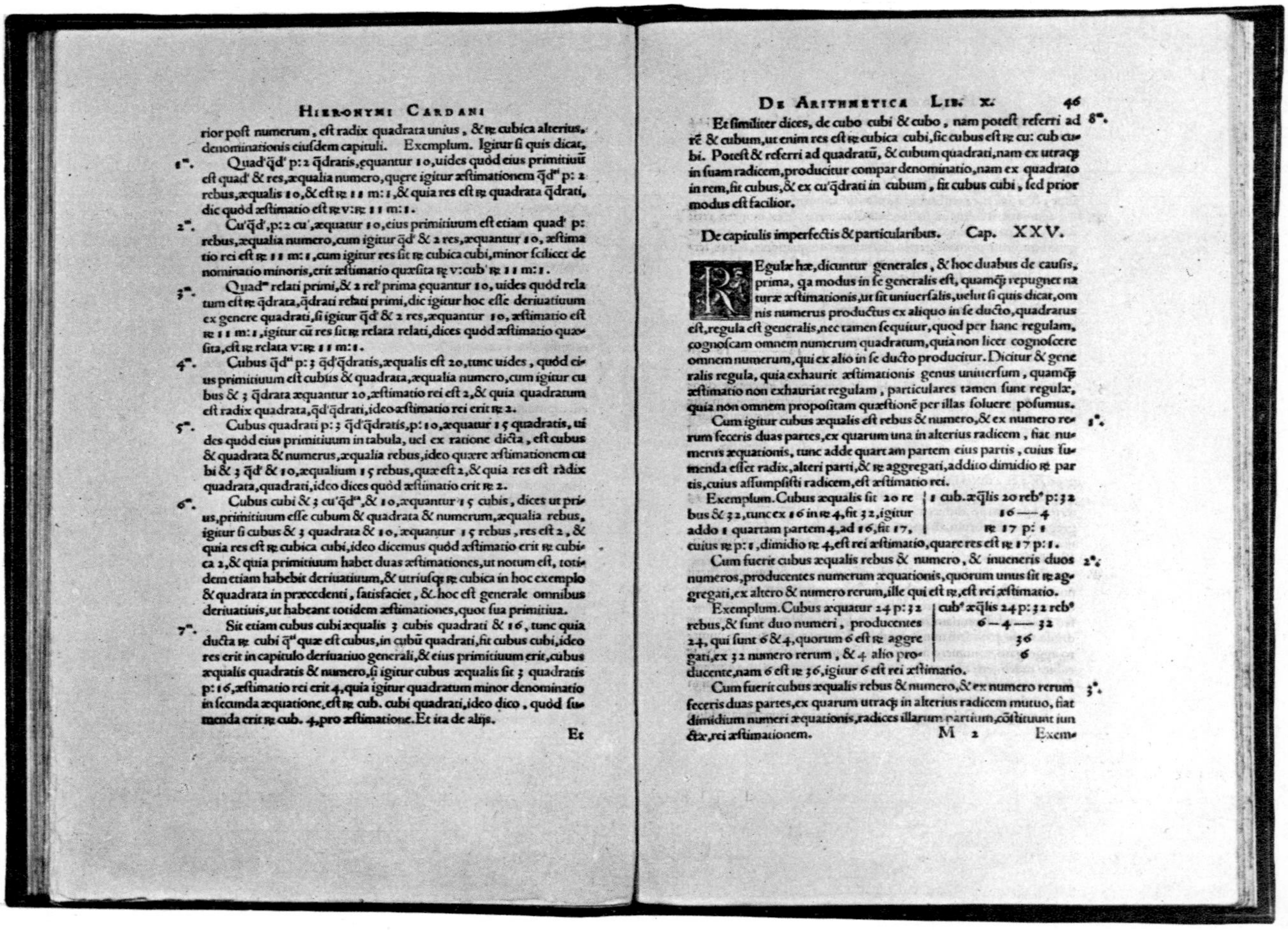

HIERONYMI CARDANI

rior post numerum, est radix quadrata unius, & ℞ cubica alterius, denominationis eiusdem capituli. Exemplum. Igitur si quis dicat,

1ᵐ. Quad' q̄d' p: 2 q̄dratis, equantur 10, uides quòd eius primitiuū est quad' & res, æqualia numero, quere igitur æstimationem q̄dᵗⁱ p: 2 rebus, æqualis 10, & est ℞ 11 m: 1, & quia res est ℞ quadrata q̄drati, dic quòd æstimatio est ℞ v: ℞ 11 m: 1.

2ᵐ. Cu' q̄d', p: 2 cu', æquatur 10, eius primitiuum est etiam quad' p: rebus, æqualia numero, cum igitur q̄d' & 2 res, æquantur 10, æstimatio rei est ℞ 11 m: 1, cum igitur res sit ℞ cubica cubi, minor scilicet denominatio minoris, erit æstimatio quæsita ℞ v: cub' ℞ 11 m: 1.

3ᵐ. Quadᵐ relati primi, & 2 rel' prima equantur 10, uides quòd relatum est ℞ q̄drata, q̄drati relati primi, dic igitur hoc esse deriuatiuum ex genere quadrati, si igitur q̄d' & 2 res, æquantur 10, æstimatio est ℞ 11 m: 1, igitur cū res sit ℞ relata relati, dices quòd æstimatio quæsita, est ℞ relata v: ℞ 11 m: 1.

4ᵐ. Cubus q̄dᵗⁱ p: 3 q̄d' q̄dratis, æqualis est 20, tunc uides, quòd eius primitiuum est cubus & quadrata, æqualia numero, cum igitur cubus & 3 q̄drata æquantur 20, æstimatio rei est 2, & quia quadratum est radix quadrata, q̄d' q̄drati, ideo æstimatio rei erit ℞ 2.

5ᵐ. Cubus quadrati p: 3 q̄d' q̄dratis, p: 10, æquatur 15 quadratis, uides quòd eius primitiuum in tabula, uel ex ratione dicta, est cubus & quadrata & numerus, æqualia rebus, ideo quære æstimationem cubi & 3 q̄d' & 10, æqualium 15 rebus, quæ est 2, & quia res est radix quadrata, quadrati, ideo dices quòd æstimatio erit ℞ 2.

6ᵐ. Cubus cubi & 3 cu' q̄dᵗⁱ, & 10, æquantur 15 cubis, dices ut prius, primitiuum esse cubum & quadrata & numerum, æqualia rebus, igitur si cubus & 3 quadrata & 10, æquantur 15 rebus, res est 2, & quia res est ℞ cubica cubi, ideo dicemus quòd æstimatio erit ℞ cubica 2, & quia primitiuum habet duas æstimationes, ut notum est, totidem etiam habebit deriuatiuum, & utriusq; ℞ cubica in hoc exemplo & quadrata in præcedenti, satisfaciet, & hoc est generale omnibus deriuatiuis, ut habeant totidem æstimationes, quot sua primitiua.

7ᵐ. Sit etiam cubus cubi æqualis 3 cubis quadrati & 16, tunc quia ducta ℞ cubi q̄ᵗⁱ quæ est cubus, in cubū quadrati, fit cubus cubi, ideo res erit in capitulo deriuatiuo generali, & eius primitiuum erit, cubus æqualis quadratis & numero, si igitur cubus æqualis sit 3 quadratis p: 16, æstimatio rei erit 4, quia igitur quadratum minor denominatio in secunda æquatione, est ℞ cub. cubi quadrati, ideo dico, quòd sumenda erit ℞ cub. 4, pro æstimatione. Et ita de alijs.

Et

DE ARITHMETICA LIB. X. 46

Et similiter dices, de cubo cubi & cubo, nam potest referri ad rē & cubum, ut enim res est ℞ cubica cubi, sic cubus est ℞ cu: cub cubi. Potest & referri ad quadratū, & cubum quadrati, nam ex utraq; in suam radicem, producitur compar denominatio, nam ex quadrato in rem, fit cubus, & ex cu' q̄drati in cubum, fit cubus cubi, sed prior modus est facilior. 8ᵐ.

De capitulis imperfectis & particularibus. Cap. XXV.

Regulæ hæ, dicuntur generales, & hoc duabus de causis, prima, qa modus in se generalis est, quamq; repugnet naturæ æstimationis, ut sit uniuersalis, uelut si quis dicat, omnis numerus productus ex aliquo in se ducto, quadratus est, regula est generalis, nec tamen sequitur, quod per hanc regulam, cognoscam omnem numerum quadratum, quia non licet cognoscere omnem numerum, qui ex alio in se ducto producitur. Dicitur & generalis regula, quia exhaurit æstimationis genus uniuersum, quamq; æstimatio non exhauriat regulam, particulares tamen sunt regulæ, quia non omnem propositam quæstionē per illas soluere possumus.

Cum igitur cubus æqualis est rebus & numero, & ex numero rerum feceris duas partes, ex quarum una in alterius radicem, fiat numerus æquationis, tunc adde quartam partem eius partis, cuius sumenda esset radix, alteri parti, & ℞ aggregati, addito dimidio ℞ partis, cuius assumpsisti radicem, est æstimatio rei. 1ᵐ.

Exemplum. Cubus æqualis sit 20 rebus & 32, tunc ex 16 in ℞ 4, fit 32, igitur addo 1 quartam partem 4, ad 16, fit 17, cuius ℞ p: 1, dimidio ℞ 4, est rei æstimatio, quare res est ℞ 17 p: 1.

| 1 cub. æq̄lis 20 rebᵘ p: 32
| 16 — 4
| ℞ 17 p: 1

Cum fuerit cubus æqualis rebus & numero, & inueneris duos numeros, producentes numerum æquationis, quorum unus sit ℞ aggregati, ex altero & numero rerum, ille qui est ℞, est rei æstimatio. 2ᵐ.

Exemplum. Cubus æquatur 24 p: 32 rebus, & sunt duo numeri, producentes 24, qui sunt 6 & 4, quorum 6 est ℞ aggregati, ex 32 numero rerum, & 4 alio producente, nam 6 est ℞ 36, igitur 6 est rei æstimatio.

| cubᵘ æq̄lis 24 p: 32 rebᵘ
| 6 — 4 — 32
| 36
| 6

Cum fuerit cubus æqualis rebus & numero, & ex numero rerum feceris duas partes, ex quarum utraq; in alterius radicem mutuo, fiat dimidium numeri æquationis, radices illarum partium, cōstituunt iunctæ, rei æstimationem. 3ᵐ.

M 2 Exem-

Page dealing with cubic equations, from a work by Cardano. During the sixteenth century, the work of Tartaglia, Vieta and Cardano on the general cubic equation ($x^3+Ax^2+Bx+C=0$) first provided valid reasons for taking complex numbers seriously.

However, it is also easy to see that a cubic might admit of only one real and two complex "solutions". For instance, $x^3-5x^2+17x-13=0$ if $(x-1)(x^2-4x+13)=0$, *i.e.* if either $x-1=0$ so that $x=+1$ or $x^2-4x+13=0$ for which – meaningfully or otherwise – our quadratic formula gives $x=2+3\sqrt{-1}$ or $2-3\sqrt{-1}$. If the cubic is not amenable to solution by factorisation, the formal solution by the method of Tartaglia and Cardano proceeds as follows. Given an equation such as $x^3+6x^2-6x-63=0$, we first reduce it as prescribed in Chapter 3 to the form $X^3+pX+q=0$. In this case, our original equation reduces to $X^3-18X-35=0$ if we put $X=(x+2)$. Now we may write $(u+v)^3=0$ in the form $u^3+v^3+3uv(u+v)=0$, whence if $X=(u+v)$,

$$X^3=u^3+v^3+3uvX$$
$$X^3-3uvX-(u^3+v^3)=0$$
$$3uv=18 \text{ and } u^3+v^3=35$$

We have now 2 equations involving only 2 variables and may solve them thus:

$$(u^3+v^3)^2=35^2 \text{ so that } u^6+2u^3v^3+v^6=1225$$
$$u^3v^3=(18\div 3)^3=216 \text{ so that } 4u^3v^3=864$$
$$u^6-2u^3v^3+v^6=1225-864=361=19^2$$
$$u^3-v^3=19 \text{ and } u^3+v^3=35$$

From this we obtain $2u^3=54$ so that $u^3=27=3^3$ and $2v^3=16$ so that $v^3=8=2^3$ and $u+v=2+3=X$. Thus $X=5=x+2$ and $x=+3$, whence $(x-3)=0$ is a factor of $(x^3+6x^2-6x-63)=0$ $=(x-3)(x^2+9x+21)$, and $(x^2+9x+21)=0$ furnishes a solution. If we can give a meaning to a so-called complex number such as $a+b\sqrt{-1}$ (henceforth written as $a+bi$), the quadratic formula equips the last equation with two solutions, $x=\frac{1}{2}(-9+3i)$ and $x=\frac{1}{2}(-9-3i)$.

Of itself, the circumstance last named inspires no confidence in the utility of complex numbers; but what eventually, *i.e.* by the beginning of the eighteenth century, showed that they had a future emerged from the discussion of what had previously been recognised as the *irreducible* case, *i.e.* when the equation has 3 roots all real. As an example we may consider the following:

$$x^3-6x^2+11x-6=0=(x-1)(x-2)(x-3)$$

The roots (solutions) of the above, as shown by equating to zero each factor on the right, are $x=1$, $x=2$, $x=3$. By the substitution $x=X+2$, we can replace it by $X^3-X=0=X(X+1)(X-1)$, so that $x=X+2$ has the values given. If we could not recognise the factor of the reduced equation and therefore proceeded in accordance with Tartaglia's recipe, we should set $X=(u+v)$, so that $u^3+v^3=0$ and $u^3v^3=\frac{1}{27}$.

This yields:

$$u^3-v^3=\left(\frac{-4}{27}\right)^{\frac{1}{2}}=\frac{2i}{3\sqrt{3}}$$

The reader who has a nodding acquaintance with complex numbers can easily check that the cube of $\frac{1}{2}(i\pm\sqrt{3})$ is i, whence

$$u=\frac{i\pm\sqrt{3}}{2\sqrt{3}} \text{ and } v=\frac{-i\mp\sqrt{3}}{2\sqrt{3}}$$

Thus both components of X are complex; but the enigmatic i drops out of the expression $X=u+v$, leaving three possibilities, $X=+1$ or 0 or -1. Since $x=X+2$, this gives the solutions $x=1$ or 2 or 3. Here then we obtain the correct real solutions by applying Tartaglia's recipe only if we also invoke complex numbers, and operate with them in conformity with the rules of elementary arithmetic. Having done so, we have at least seemingly shown that we can set them to do useful work. Henceforth it was inevitable that mathematicians would explore the possibility of other tasks which they could accomplish without involving the user in any inconsistency.

From the number of steps involved in the foregoing, the reader will infer that a discussion of the general solution of the cubic – let alone that of equations involving higher powers than 3 – is well-nigh inconceivable unless we can condense the argument by reliance on some sort of shorthand; and further exploration of algebraic equations of a higher order proceeded *pari passu* with the standardisation of algebraic symbolism. After Vieta solved the quartic, the outcome was disappointing. In the end, *i.e.* early in the nineteenth century, it was possible to show that no exact general method of solution is attainable if an equation involves a power of n greater than 4; but any algebraic equation admits of a solution as exact as we need in terms of the only language the machine understands, *i.e.* that of *iteration.* Any iterative method of solution signifies first making a good initial guess, *e.g.* with the aid of a graph which will disclose how many real solutions exist, then determining the error approximately and successively making the latter less by adjusting the guess accordingly.

The simplest of all such iterative procedures invokes the binomial theorem. Suppose that our first guess of the real solution in the equation $x^3+6x^2-6x-63=0$ had been $x\simeq2$. If the true error is e we may then write $x=2+e$, so that the equation becomes

$$(2+e)^3+6(2+e)^2-6(2+e)-63=0$$
$$(8+12e+6e^2+e^3)+6(4+4e+e^2)-6(2+e)$$
$$-63=0$$

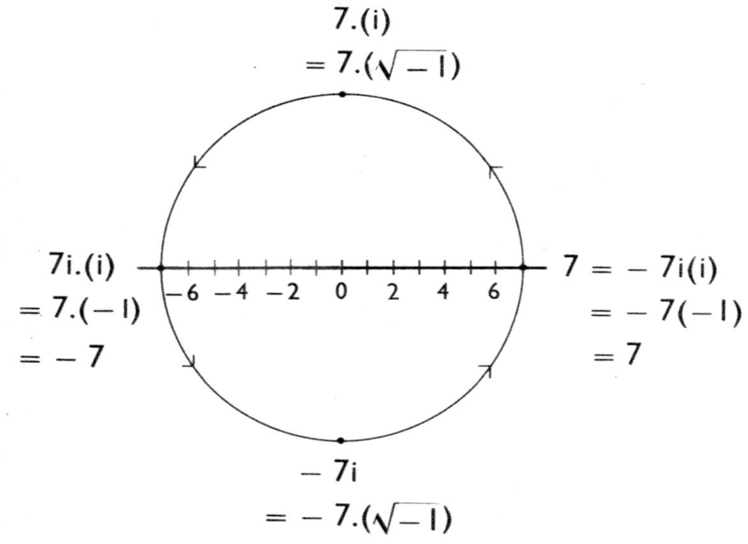

We may interpret .(i) as an instruction to rotate a line segment anticlockwise through 90°. The result of such an operation locates a point on a plane.

We now assume that e is small in the sense that e^2 and e^3 are negligible and write tentatively:

$$(8+12e)+6(4+4e)-6(2+e)-63=0$$
$$\text{and } e=1\cdot4\dot{3}$$

We conclude that our original guess was too small by an amount not exceeding $1\cdot4\dot{3}$, and may next try $x\simeq2+1\cdot4\dot{3}$. To simplify the work, it will serve our purpose to consider the initial error as $+1\cdot4$ and write as a second approximation $x=3\cdot4+e_2$. Rejecting e_2^3 and e_2^2 as before, we next write

$$3\cdot4^3+3(3\cdot4)e_2+6(3\cdot4^2+6\cdot8e_2)-6(3\cdot4+e_2)$$
$$-63=0$$

This yields $e_2\simeq-0\cdot36$, so that $3\cdot4$ is too large and our next approximation will be $x\simeq3\cdot4-0\cdot36=3\cdot04$. We therefore write $x=3\cdot04+e_3$ and proceed as before. At each repetition of the procedure we get closer to the correct answer $x=3$.

The hints already given with a view to restating the rules of signs in a way consistent with the naïve notions of subtraction and multiplication in the domain of the abacus or of the Euclidean figure, but also able to accommodate operations involving directed numbers, provide the clue to an interpretation of the use of $i^2\equiv-1$ in an algebra whose domain embraces the complex number consistently with the rules of an algebra which recognises only real numbers. Oddly enough, however, nobody took the rules of signs seriously until what we are really doing when we manipulate with the mysterious i had received a quasi-kinematical interpretation.

We have seen that the laws of signs make sense if we interpret $-n$ to mean a naked number n associated with an operation $\cdot(-1)$ which we loosely call multiplying by -1 but interpret as a rotation of a segment n of the real axis through 180°. If we do this twice, we get back to where we started, so that the notation $\cdot(-1)\cdot(-1)=\cdot(-1)^2=\cdot(+1)$ embodies the usual statement of the rule $(-)\times(-)=(+)$. What then is the operation we imply if its *twofold* performance is the same as a rotation through 180°? The answer is that it is the operation of rotating through 90°. Let us call it $\cdot(j)$ so that $\cdot(j)\cdot(j)=\cdot(j)^2=\cdot(-1)$. In the older notation $(\sqrt{-1})^2=i^2=\cdot(-1)$. In short $i=j$ (as used above) and $3i$ means measuring 3 units of distance at right angles to the starting point. If so, the only meaning we can attach to a complex number such as $4+3i$ is:

measure 4 linear units to the right (positive) on the real axis and then proceed vertically through 3 units.

In taking this step, we have conceptualised all directed numbers as instructions for mapping points in a plane. Real numbers such as $\pm a$ and $\pm b$ in $\pm a\pm bi$ then represent distances right and left of the zero mark on the horizontal axis, imaginary numbers such as $\pm bi$ as distances above or below the zero mark along an axis vertical to it, and complex numbers such as $a+bi$ itself as distances from the zero mark to a point situated at a distance $r=\sqrt{(a^2+b^2)}$ from it. We can then define the rules of addition, subtraction, multiplication and division in a way which is consistent with the rules of the abacus and of the Euclidean figure, if we treat i^2 as the equivalent of multiplying by -1 and manipulate i in accordance with the familiar pattern:

$$(a+bi)\pm(c+di)=(a\pm c)+(b\pm d)i$$
$$(a+bi)\ (c+di)=ac+adi+bci+bdi^2$$

This does not answer the question: in what factual sense can a quadratic equation have a solution expressible in the form $a\pm bi$ unless $b=0$? Here we shall not anticipate an answer. It suffices to say that we at least have a clue to what we are doing when, as in the solution of the so-called irreducible cubic, we invoke such *couples* to get a solution in terms of real numbers. We shall meet them again in Chapters 11 and 14.

Chapter 9 Framework and Function

Our last chapter began with what some might call the rebirth of mathematics in Europe, though in fact Italy alone of the countries in which mathematics developed rapidly under the impact of the Moslem synthesis had participated in the enlightenment of the Greek-speaking world before the final collapse of the Western Roman Empire. In tracing the sources of the outstanding innovation of the sixteenth and seventeenth centuries of the Christian era, namely the creation of an international sign language of algebra, we touched on several themes which did not receive clarification much before the middle of the nineteenth; and it is inevitable that we shall make no attempt to follow a consistently chronological pattern in the four chapters which follow. The themes of each of them cover progress over the greater part of the period 1640–1840; and progress in each of those dealt with in this and the two succeeding chapters is a story of cross-fertilisation.

None the less, one event is of antecedent interest to all three. We may call this the emergence of the *function concept* in the geometrical domain. In more familiar language this means especially, but not exclusively, the recognition of the possibility of exploring the properties of solid or plane figures with the aid of algebraic manipulations. Clearly, this could not proceed rapidly before the widespread adoption of a compact notation had advanced the art of algebraic manipulation to a far higher level than hitherto attained; but the possibility also signifies the adoption of a convention which deserves preliminary comment before we can do justice to the notion of a function.

When mathematicians of the Greek-speaking world dealt with the properties of curves, they proceeded from initial assumptions by recourse to dissection of figures in accordance with Euclidean principles and with little or no attempt to exploit the advantage of having what we may call *a fixed framework of reference*. In effect, this is what we do when we chart a ship's course on a flat map or on a globe. Had Alexandrian mathematics long survived Ptolemy's excursions into cartography, the recognition of this advantage might well have dawned earlier. Actually, the outline of a figure conceived as the *locus* (*i.e.* track) of a point in latitude and longitude makes its first appearance explicitly in the writings of Oresme (*circa* 1360) when Moslem cartography was beginning to kindle European thought; but the systematic development of the possibility of analysing the geometrical properties of figures from this viewpoint had to await the availability of better algebraic tools than those at his disposal.

Every reader of this book will be familiar with

Captain Cook's chart of the voyage of the Resolution *and the* Adventure *(1772–75), with an entry from the log. For two centuries algebraic symbolism had developed rapidly; and, with better means of finding longitude, the practice of interpreting a figure as the locus, or track, of a point against a framework of lines of latitude and longitude became increasingly congenial.*

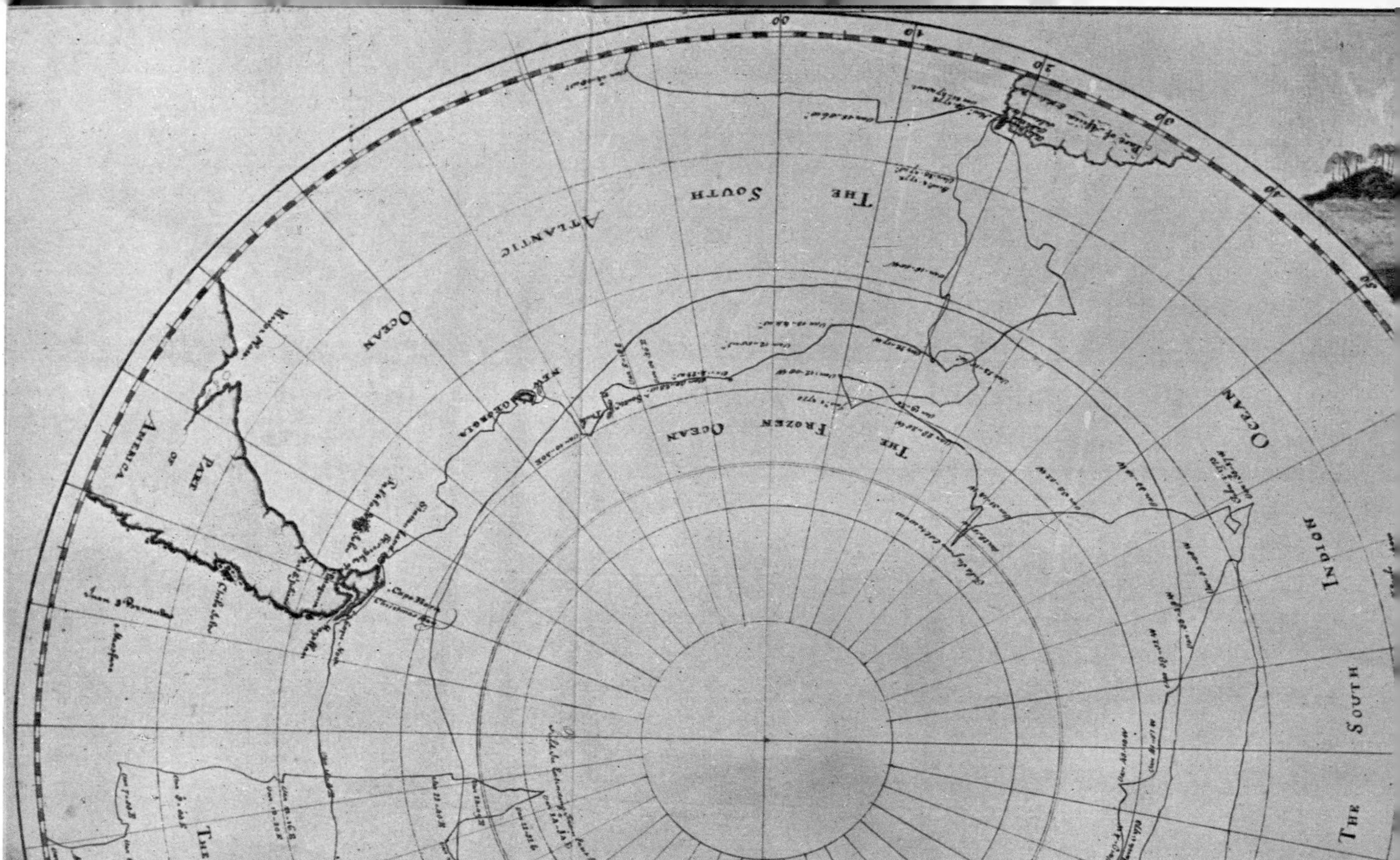

The South Atlantic Ocean
The Frozen Ocean
The South Indian Ocean
Part of America
The Great South Seas
The Antartic Circle

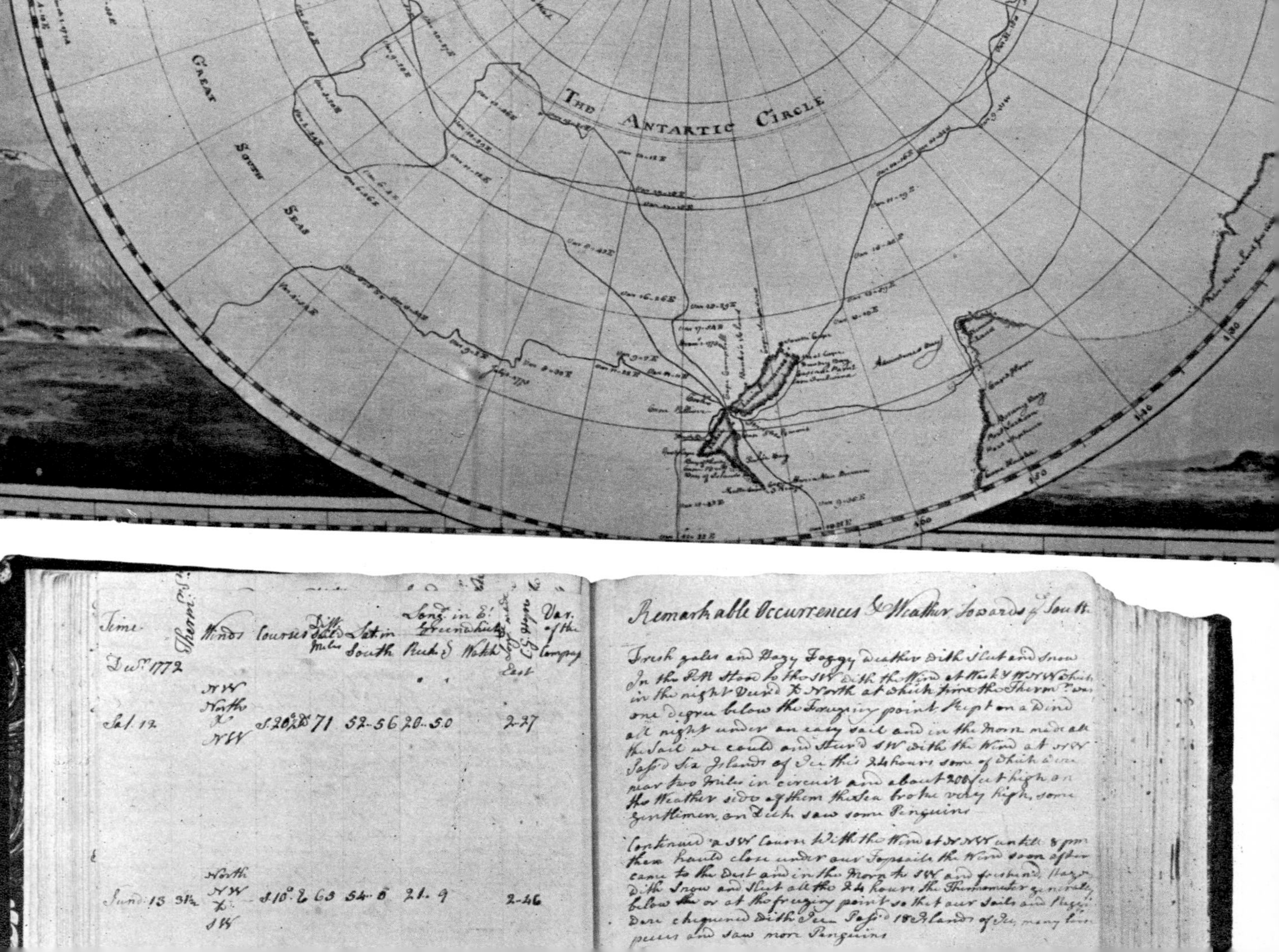

Time
Winds
Courses
Dist. Sailed Miles
Lat. in South
Long. in E. Greenwich
Var. of the Compass
Decr 1772
Remarkable Occurrences & Weather towards ye South
Fresh gales and Hazy Foggy weather with Sleet and Snow In the PM stood to the SW with the Wind at West & WNW which in the night Veerd to North at which time the Therm. was one degree below the Freezing point. Kept on a Wind all night under an easy sail and in the Morn made all the sail we could and Steerd SW with the Wind at NNW Pass'd Six Islands of Ice this 24 hours some of which were near two Miles in circuit and about 200 feet high on the Weather side of them the Sea broke very high, some Gentlemen on Deck saw some Penguins
Continued a SW Course with the Wind at N NW untill 8 pm then hauld close under our Topsails the Wind soon after came to the West and in the Morn to SW and freshend, Hazy with Snow and Sleet all the 24 hours the Thermometer generally below the or at the freezing point so that our Sails and Rigging were chequered with Ice Pass'd 18 Islands of Ice, many loose pieces and saw more Penguins

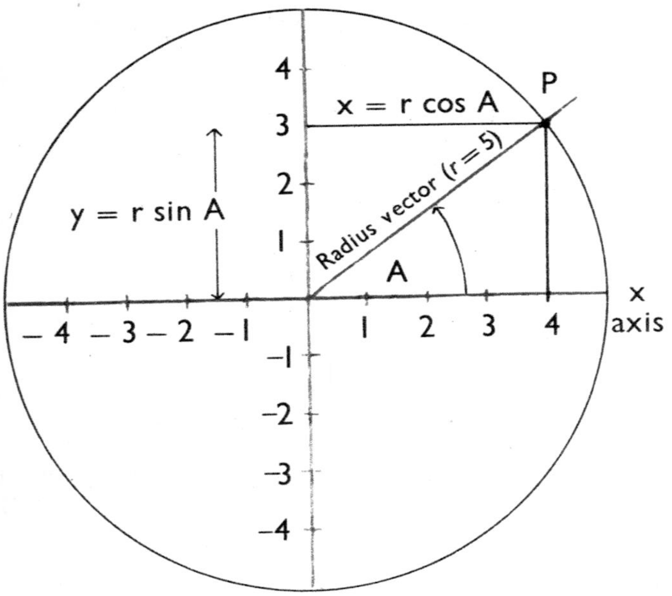

Co-ordinates of P

a. rectangular $x = 4,\ y = 3$

b. polar $r = 5,\ A = \tan^{-1}\frac{3}{4}$

RECTANGULAR AND POLAR CO-ORDINATES OF THE PLANE (*See p. 195.*)

Notice that for a circle whose centre is the origin, r is constant for all A and $x^2+y^2=r^2$ *so that* $y=\sqrt{r^2-x^2}$.

the convention of the fixed framework at the level propounded by Oresme. When we draw the graph of our schooldays, each point is specifiable in terms of two co-ordinates, *i.e.* a distance x from the prime meridian (Y-axis) and a distance y from the equator (X-axis) of a map projection in which the meridians and parallels are equally spaced. To be sure, there is no limit to the number of frameworks we can conceive; but only a few of them have conspicuous advantages. For description of plane figures, one method of mapping the locus of a point is comparable to the type of map which gives a polar view of a hemisphere showing meridians as lines radiating from one centre (the pole) and parallels of latitude as circles with a common (polar) centre. In contradistinction to equally-spaced rectangular (so-called *Cartesian*) co-ordinates (*i.e.* measurements from the base lines) of the framework, we speak appropriately of the co-ordinates of such a system as *polar*. The origin (*i.e.* the pole) is the *start* of the base line running towards the right along the prime meridian. The co-ordinates which fix any point are: (a) the angle (A) between the base line and the meridian which passes through it; (b) the length (r) between the point and the pole along this meridian, usually called the *radius vector*.

We have indeed already used the polar system to explore the properties of several curves which interested the Greek-speaking world of antiquity, *e.g.* quadratrix, spiral, conchoid. Since the two systems mentioned are useful for different purposes, it is important to be able to translate from one to the other. We must then define how one fixed framework fits into the other. The usual and simplest convention is that: (a) the origins of the

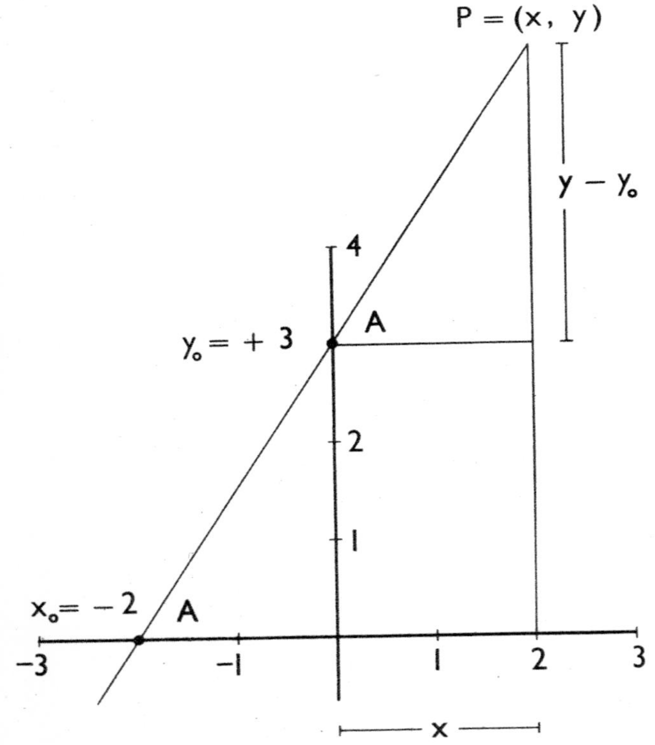

CARTESIAN EQUATION OF THE STRAIGHT LINE IN THE PLANE

If we consider any point $P=(x, y)$ *whose co-ordinates are x and y, tan* $A=(y-y_0)\div x$. *If we write tan* $A=m$, *the equation of the straight line in a plane becomes* $y=mx-y_0$ *in which* y_0 *is the intercept of the line by the Y axis, i.e. the value of y for* $x=0$. *We may express tan A in terms of* y_0 *and* $-x_0$ *which is the intercept of the X axis, viz.,* $\tan A=-y_0\div x_0$, *whence* $y_0x=-x_0(y-y_0)$, *so that* $y_0x+x_0y-x_0y_0=0$. *In this alternative formulation, the only constants are the intercepts, and the slope is* $\tan^{-1}(y_0\div x_0)$. *The general pattern is* $Ax+By+C=0$ *in which* $A=y_0$, $B=x_0$ *and* $C=-x_0y_0$.

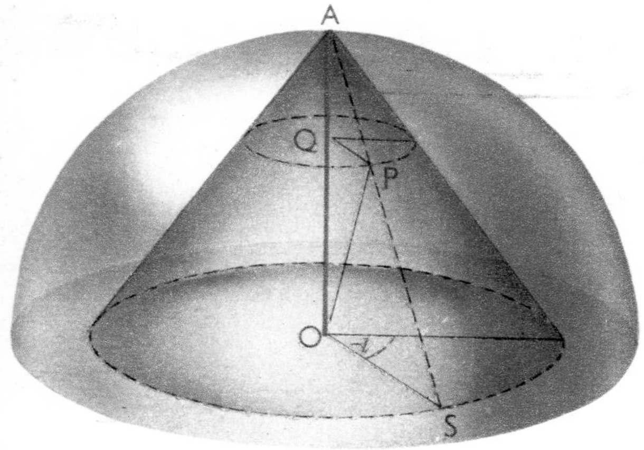

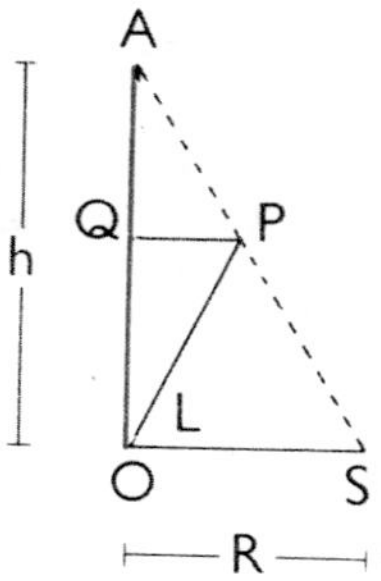

$$r\cos L = r\sin(90-L) = PQ$$

$$PQ = AQ\tan PAQ = AQ\,(R \div h)$$

$$AQ = h - OQ = h - r\cos(90-L) = h - r\sin L$$

$$\therefore\ r\cos L = \frac{R\,(h - r.\sin L)}{h}$$

$$\therefore\ r = \frac{R}{\cos L + \frac{R}{h}.\sin L}$$

THE CONE IN SPHERICAL CO-ORDINATES

Notice that r (=OP) is constant for all values of l if we fix L. When L=0=sin L and cos L=1, r=R. When L=90° so that sin L=1 and cos L=0, r=h.

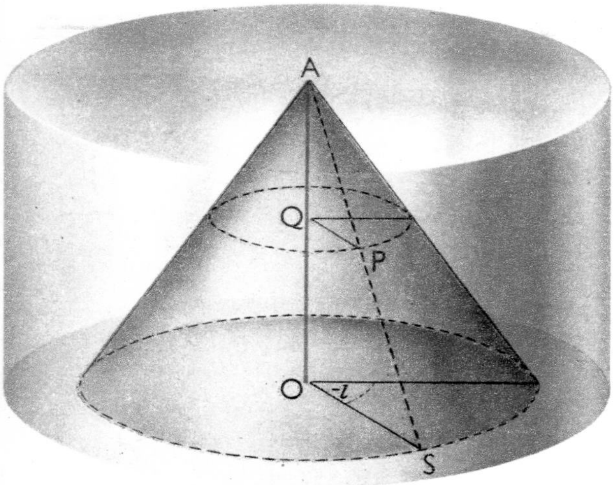

THE CONE IN CYLINDRICAL CO-ORDINATES

Again note that l is fixed for r (=OP) if we fix L. In the figure z=OQ. As before, r cos L=AQ(R÷h) but AQ=h−z.

$$\therefore\ r\cos L = (h-z)\frac{R}{h} = R - \frac{R}{h}z$$

$$\therefore\ r = \frac{R - \frac{R}{h}z}{\cos L}$$

When z=0=L so that cos L=1, r=R.
When L=90° so that r cos L=0, z=h.

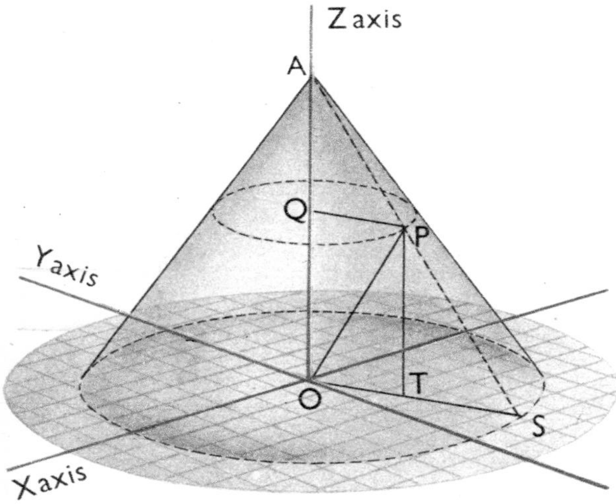

THE CONE IN RECTANGULAR CO-ORDINATES

Here QP:AQ=R:h and QP=OT. Since AQ=h−z,

we derive $OT = \dfrac{R(h-z)}{h}$ *also* $OT^2 = x^2 + y^2$

$$\therefore\ \frac{R^2(h-z)^2}{h^2} = x^2 + y^2$$

Whence if z=0, $R^2 = x^2 + y^2$*; if x=0=y, z=h*

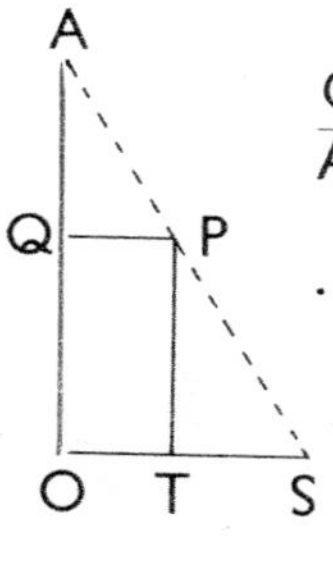

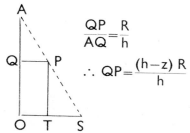

$$\frac{QP}{AQ} = \frac{R}{h}$$

$$\therefore\ QP = \frac{(h-z)\,R}{h}$$

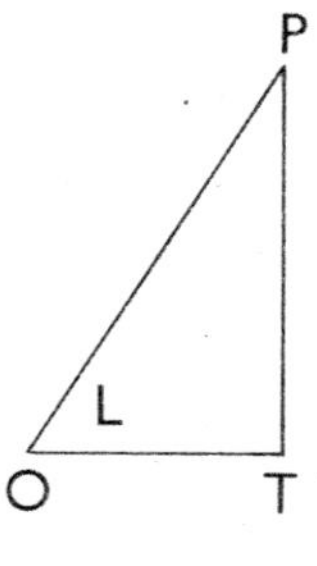

$$PT^2 = z^2$$
$$= r^2 - OT^2$$

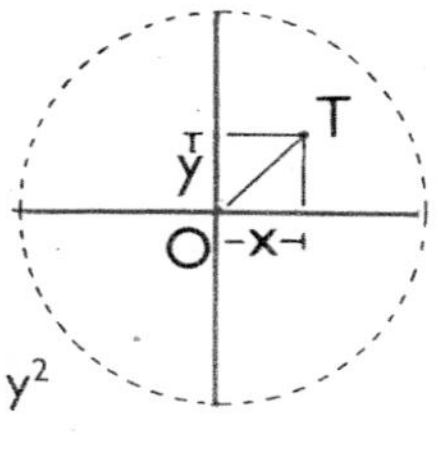

$$OT^2 = x^2 + y^2$$

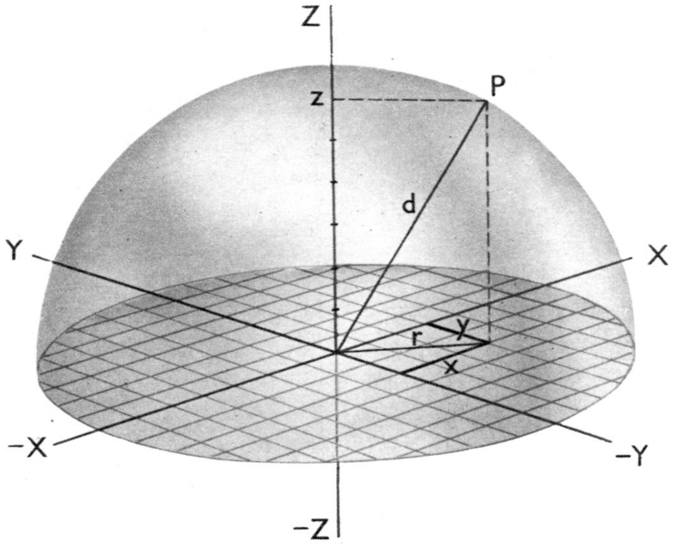

CARTESIAN EQUATION OF THE SPHERE WITH CENTRE AT THE ORIGIN

Here $r^2=x^2+y^2$ *and* $d^2=r^2+z^2$

$\therefore x^2+y\ +z^2=d^2$

The distance d from the centre of the circle to any point P on the surface of the sphere is everywhere the same, being the length of the radius R, so that

$x^2+y^2+z^2=R^2$

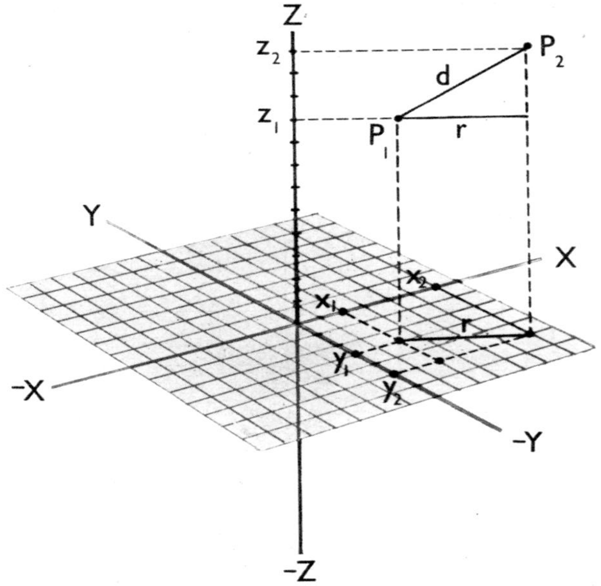

The Pythagorean relation in three dimensions embodies the distance between two points (P_2 *and* P_1) *in solid space. Here we see:*

$(x_2-x_1)^2+(y_2-y_1)^2=r^2$

$(z_2-z_1)^2+r^2=d^2$

$\therefore d^2=(x_2-x_1)^2+(y_2-y_1)^2+(z_2-z_1)^2$

$\therefore d=\sqrt{[(x_2-x_1)^2+(y_2-y_1)^2+(z_2-z_1)^2]}$

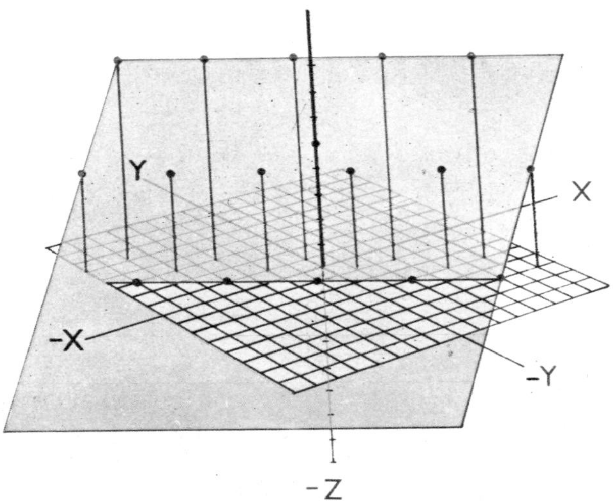

The Cartesian equation of the plane in Euclidean space is of the general form $Ax+By+Cz+D=0$. *When* $z=0$ *this reduces to the equation of a straight line in the XY plane, being the line of intersection therewith.*

two coincide; (b) the base line of the polar system coincides with the X-axis; (c) the radius vector rotates anti-clockwise from the position $y=0=A$ and $x=r$ on the positive (right-hand) side of the axis. Our figure (p. 192) shows that if we want to translate from polar into plane rectangular co-ordinates, we substitute:

$$x=r\cos A$$
$$y=r\cos(90°-A)=r\sin A$$

Conversely, we may write:

$$x^2+y^2=r^2(\cos A)^2+r^2(\sin A)^2$$
$$=r^2[(\sin A)^2+(\cos A)^2]=r^2.$$

Likewise

$$\frac{y}{x}=\frac{r\sin A}{r\cos A}=\tan A$$

$$r=\sqrt{(x^2+y^2)} \text{ and } A=\tan^{-1}\left(\frac{y}{x}\right)$$

Of the various more or less convenient frameworks for locating a point in three-dimensional space and hence for defining the surface of solid figures, only three deserve attention here:

(a) *Spherical Co-ordinates* (r, L, l) are the most primitive of all, because they correspond to the way in which we define our position on earth or locate a celestial body. Think of the origin as the earth's centre and the two planes of reference as: (a) the equatorial; (b) one we may call the prime meridional, passing through the meridians 0° and 180°. The radius vector is a line drawn from the origin to the point, this being equivalent to the earth's radius if the point is on the earth's surface. Its length (r) together with the angle (L) which it makes with the equatorial plane (*i.e.* latitude) and the angle (l) it makes with the prime meridional in a plane parallel to the equatorial (*i.e.* longitude) suffice to locate any point in three-dimensional space in relation to an arbitrarily-fixed origin, *e.g.* an observer's position.

(b) *Cylindrical Co-ordinates* (r, z, l) are easy to envisage if we think of a plane of reference corresponding to the equatorial and a line (polar axis) vertical to it at the origin (centre of the equatorial). The radius vector, its vertical height (z) parallel to the polar axis, and its inclination to the prime meridional in a plane parallel to the equatorial define the position of any point.

(c) *Rectangular Co-ordinates* (x, y, z) are more easily defined in terms of the school-book graph. We imagine a third (Z) axis passing through the origin at right angles to the XY plane. Our third co-ordinate is the vertical distance (z) from the latter. When translating to or from this system and either of the foregoing, we assume that the origin corresponds to the centre of the equatorial, and the base line for the angle l is the positive half of the X-axis or its projection on a plane parallel to the equatorial and vertically above it.

For translation from spherical to cylindrical co-ordinates or *vice versa* on the assumption that we

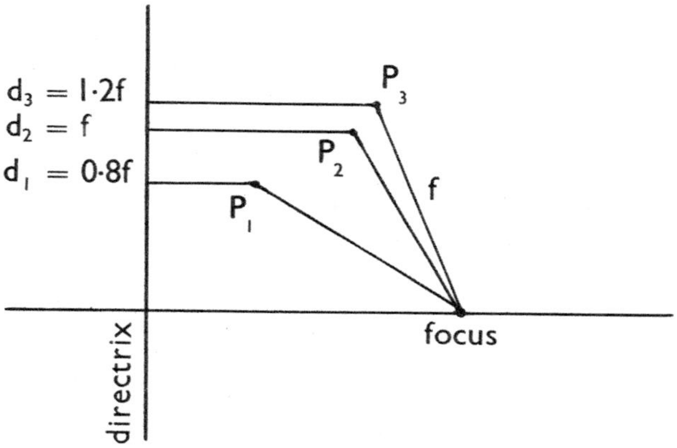

For any conic whose radius vector is f and the distance of the point at its extremity from the directrix is d, the ratio $f:d=e$ is constant. If $e<1$ so that $d>f$, the figure traced by the rotation of the radius vector is an ellipse. If $e=1$ so that $d=f$, it is a parabola. If $e>1$ so that $d<f$, it is an hyperbola. In the above figure P_1 is a point on the boundary of an hyperbola, P_2 on that of a parabola, P_3 on that of an ellipse. (*See p. 198.*)

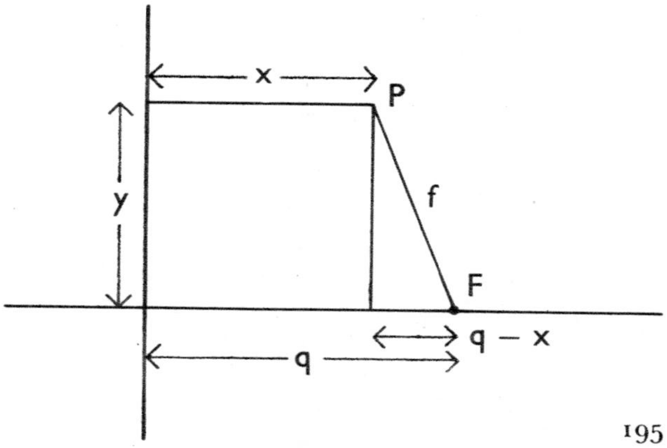

P is a point on the conic. The directrix corresponds to the y axis, being distant q from the focus along the x axis. By definition $f:x=e$ so that $f^2=e^2x^2$ and $f^2=y^2+(q-x)^2=e^2x^2$
$\therefore y^2=(e^2-1)x^2+2qx-q^2.$

measure r and l in the same units, we merely need the relations:

$$z = r \sin L \text{ or } L = \sin^{-1}\left(\frac{z}{r}\right)$$

For translation from cylindrical to rectangular co-ordinates in the same unit as z or *vice versa* on the assumption that our base line for l measured anti-clockwise is the X-axis, we need:

$$x = \sqrt{(r^2 - z^2)} \cos l \text{ and } y = \sqrt{(r^2 - z^2)} \sin l$$

$$\text{or } l = \tan^{-1}\left(\frac{y}{x}\right)$$

For translation from spherical to rectangular co-ordinates on the assumption that we measure r, x, y, z in the same units, we need:

$$z = r \sin L \text{ and } \sqrt{(r^2 - z^2)} = r \cos L$$
$$x = r \cos L . \cos l \text{ and } y = r \cos L . \sin l$$

Conversely:

$$l = \tan^{-1}\left(\frac{y}{x}\right);\ L = \sin^{-1}\left(\frac{z}{r}\right);\ r = \sqrt{(x^2 + y^2 + z^2)}$$

We may call the last relation on the right the Pythagorean Theorem in three dimensions. Since $r = R$ (the radius) is constant for all L and l if the figure is a sphere, the equation of the sphere in spherical co-ordinates is

$$R^2 = x^2 + y^2 + z^2$$

In what has gone before, the notion of a continuous function is implicit; but the function concept is of wider significance. To define it, we have to assume that the relationship it expresses contains symbolic ingredients of two types: (a) referable to quantities called variables (such as x, y, z above) which can assume different values; (b) referable to quantities called constants (such as R in the above) which have a single fixed value in the context. The equation last cited for R, x, y, z expresses such a functional relationship but does not make explicit any one so-called *dependent variable* as what we call a function of one or more *independent variables*. It does so for the dependent variable y if we recast it in a form which exhibits y as a function of x and z, as when we write: $y = \sqrt{(R^2 - x^2 - z^2)}$. Likewise, we may also exhibit x as a function of y and z when we write $x = \sqrt{(R^2 - y^2 - z^2)}$ or z as a function of x and y when we write $z = \sqrt{(R^2 - x^2 - y^2)}$.

To say this means that we can tabulate x for all admissible values of y and z, y for all admissible values of x and z, z for all admissible values of x and y. If we confine ourselves to functions of not more than two variables, a satisfactory definition of a function in general is indeed that we can exhibit as a table corresponding numerical values of each of the variables involved. What values are *admissible* depends on the number domain to which the function is relevant. If the functional relationship suffices to trace the outline of a figure, the domain embraces all real numbers, negative or positive, rational or irrational, within a specifiable range. If the dependent variable between the upper and lower limits it assumes when we permit the independent variable (or variables) to assume every possible value in that range can itself take at least once every possible value between such limits, we call it a *continuous* function. In contradistinction to such a function, we may speak of a *discrete* function if the independent variable or variables can assume only regularly-spaced values, as when the number domain is the integers. Below is a table of the discrete function $z = 2x + 3y + 1$ in the range of integers $x = \pm 5$ and $y = \pm 5$.

y \ x	−5	−4	−3	−2	−1	0	+1	+2	+3	+4	+5
−5	−24	−22	−20	−18	−16	−14	−12	−10	−8	−6	−4
−4	−21	−19	−17	−15	−13	−11	−9	−7	−5	−3	−1
−3	−18	−16	−14	−12	−10	−8	−6	−4	−2	0	+2
−2	−15	−13	−11	−9	−7	−5	−3	−1	+1	+3	+5
−1	−12	−10	−8	−6	−4	−2	0	+2	+4	+6	+8
0	−9	−7	−5	−3	−1	+1	+3	+5	+7	+9	+11
+1	−6	−4	−2	0	+2	+4	+6	+8	+10	+12	+14
+2	−3	−1	+1	+3	+5	+7	+9	+11	+13	+15	+17
+3	0	+2	+4	+6	+8	+10	+12	+14	+16	+18	+20
+4	+3	+5	+7	+9	+11	+13	+15	+17	+19	+21	+23
+5	+6	+8	+10	+12	+14	+16	+18	+20	+22	+24	+26

When y is an unspecified continuous function of x, or z of x and y, *etc.*, it is customary to express this symbolically in the form $y=f(x)$ or $z=f(x, y)$, *etc.*; but if one is dealing with several functions in the same context, it may be convenient to use F or its Greek equivalent Φ in the same sense. When one is dealing with a discrete function of an integer n it is convenient to represent u as a function of n in the form u_n. In that event our table of values of u_n corresponding to successive values of n gives the layout of a series of terms corresponding to a particular rank n; and it is convenient to specify as a constant the term of rank 0 (u_n for $n=0$), *i.e.* u_0. In terms of the constant u_0, undefined constants C or K and explicit numerical constants (such as 3 or 8) we may recall the following examples of discrete functions, hitherto designated series when set out in rank sequence:

(1) *Arithmetic Series* $u_n = u_0 + n.C$
which corresponds to the natural numbers when $u_0=0$ and $C=1$

(2) *Triangular Numbers* (see notation near end of Chapter 3)
$u_n=(n+1)_{(2)}+u_0$; $(u_0=0)$

(3) *Tetrahedral Numbers* $u_n=(n+2)_{(3)}+u_0$; $(u_0=0)$

(4) *Pyramidal Numbers*

$$u_n = \frac{n(n+1)\ (2n+1)}{3!} + u_0;\ (u_0=0)$$

(5) *Geometric Series* $u_n=u_0 . C^n$

We may add the *Harmonic Series*:

(6) $u_n = \dfrac{K}{C.n+u_0} = \dfrac{1}{n}$ when $K=1=C$ and $u_0=0$

If we admitted all real values of n in the foregoing formulae, they would correspond to functional relationships referable to rectangular co-ordinates of the plane respectively specifying (1) the straight line, (2) the parabola, (3) the cubic, (4) the cubic, (5) a so-called exponential, (6) the hyperbola. However, the general definition of a function does not imply that it is necessarily expressible by what we alternatively designate as an equation or formula. Though there is no known formula for it, the

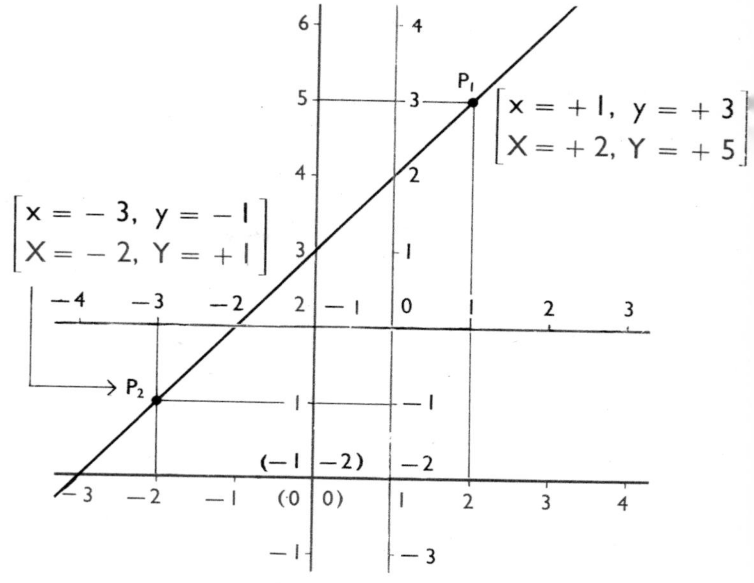

TRANSLATION
Here the primary framework (grey) fixes co-ordinates x, y. If we shift the origin from $x=0$, $y=0$ to $x=-1$ and $y=-2$, the relation between the new framework which fixes co-ordinates X, Y is such that $X=x+1$ and $Y=y+2$.
If we replace $(x+a)$ by X and $(y+b)$ by Y in an equation, we merely shift our x and y axes so that the new corresponding axes are parallel with them.
The origin of the secondary (XY) framework has the co-ordinates $x=-a$, $y=-b$ in the initial system.

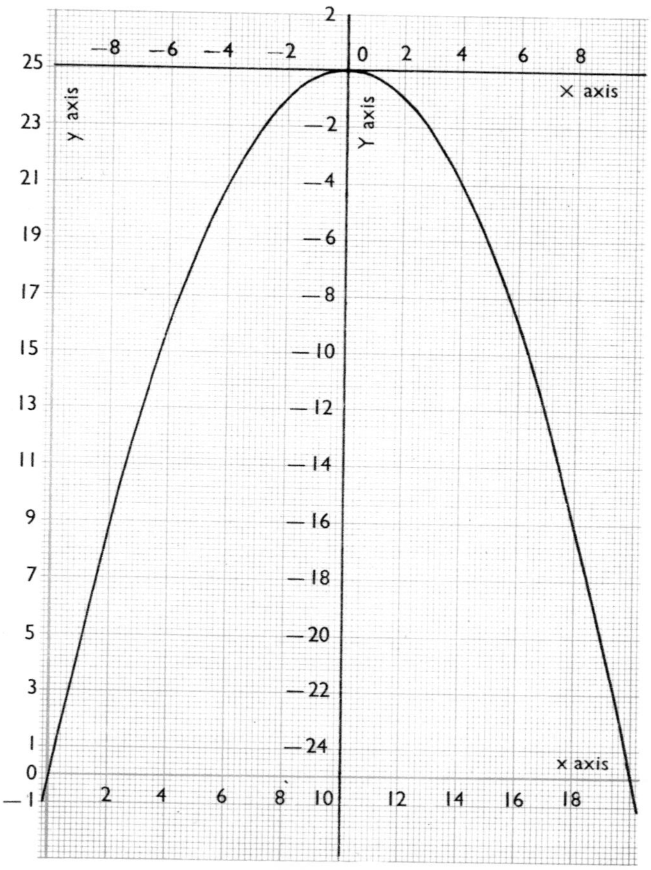

The same parabola in two translational frameworks,
$Y=-\frac{1}{4}X^2$ *and* $y=5x-\dfrac{x^2}{4}$.
The translation formulae are: $X=x-10$ *and* $Y=y-25$.

following is clearly a function of n being an integer:

$$u_0=0;\ u_1=1;\ u_5=1+\frac{1}{2^2}+\frac{1}{3^2}+\frac{1}{4^2}+\frac{1}{5^2}$$

It should now be clear that a functional equation referable to a particular fixed framework, *i.e.* one expressing the relation between the dependent and independent variable (or variables), can give us a complete description of a figure, plane or solid, if we put no restriction on what real values (within a prescribed range) the variables can assume. The particular form of the function will then necessarily depend on:

(a) what framework of reference we choose;
(b) the orientation of the figure within the framework;
(c) whether the scale, *i.e.* the unit of measurement, is or is not the same for co-ordinates of the same dimensions (*i.e.* angular if angles, linear if distances). For the time being, we shall assume that it is.

We have already disposed of (a). Before we examine (b) and (c), a digression may be worth while. Let us scrutinise the functional implications of the definition of a conic section as transmitted to posterity from the work of Apollonius, who demonstrated by Euclidean methods that any conic section is the locus of a point which moves so that the ratio of its distance (f) from a fixed point (the *focus*) to its distance (d) from a straight line (the *directrix*) is constant. (See diagrams on p. 195.) Whether this constant (e) is greater than, less than or equal to unity determines which of the three types of curve the function represents:

$e<1$ *ellipse*
$e=1$ *parabola*
$e>1$ *hyperbola*

To analyse the implications of the definition, we shall first suppose that the focus is at a distance q to right of the origin on the X-axis and that the Y-axis coincides with the directrix, so that

$$d=x \quad \text{and} \quad f^2=y^2+(q-x)^2$$

Since $f:d=e$,

$$f^2=d^2e^2=y^2+(q-x)^2 \quad \text{and} \quad e^2x^2=y^2+(q-x)^2$$
$$\therefore\ y^2=(e^2-1)x^2+2qx-q^2$$

We may replace the constants e and q by the substitutions:

$$A=(e^2-1);\ B=2q;\ C=-q^2$$

We now have the general equation of the conic in the form

$$y^2=Ax^2+Bx+C$$

This reduces to the equation of the circle with origin as centre and radius R if $A=-1$, $B=0$ and $C=R^2$. Since $A=(e^2-1)$ and $e=1$ when the curve is a parabola, the parabola is more simply expressible as

$$y^2=Bx+C$$

Since q is positive in our construction, B $(=2q)$ is necessarily positive, C $(=-q^2)$ necessarily negative, but A will be positive if $e>1$ (hyperbola) and negative if $e<1$ (ellipse).

The expressions cited exhibit the functional relations of the two rectangular co-ordinates of a point whose locus is a conic section on two assumptions with reference to its orientation within the fixed framework:

(a) the directrix coincides with the Y-axis;
(b) the focus lies on the X-axis to the right of the origin.

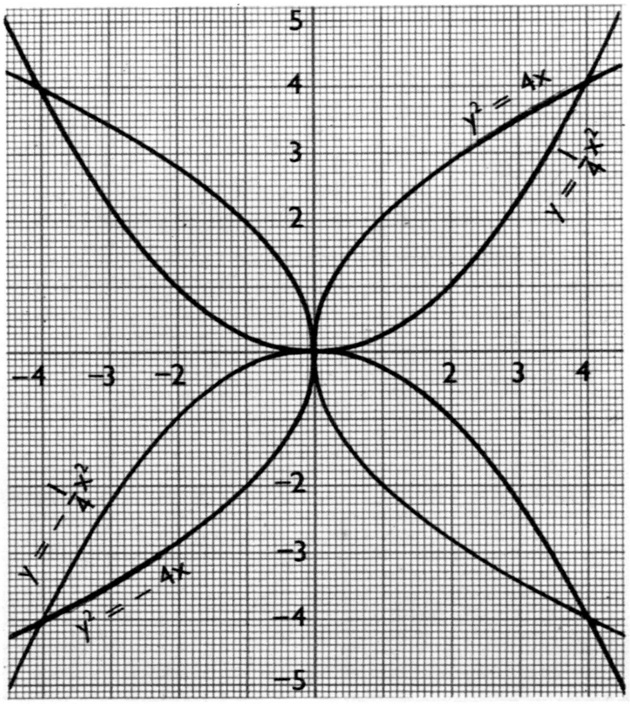

The same parabola rotated through 90°, 180° and 270° $\left(\frac{\pi}{2},\ \pi \text{ and } \frac{3\pi}{2} \text{ radians}\right)$.

More or less convenient ways of describing one and the same curve in the same framework will depend on the angular orientation of the directrix to the Y-axis, its distance from the origin along a line connecting the focus thereto and the co-ordinates of the focus ($x=q, y=0$ above) itself. To explore other functional relations from this view-point we may conveniently distinguish two major types of reorientation:

(a) *translation*, *i.e.* shifting the origin vertically, horizontally or both;

(b) *rotation*, *i.e.* rotating the framework through an angle A ($<360°$) about the origin.

If we perform either of the above, we shall require to define new co-ordinates which we shall call X and Y related to the initial framework (x, y) by the following relations in a new translation of the origin to the point whose co-ordinates are $x=p, y=q$ in the latter:

$$X=x-p \text{ and } Y=y-q$$

If therefore we put $X=x+\frac{C}{B}$ in the equation of the parabola ($y^2=Bx+C$) as given above, what we have done is to change the origin of the horizontal axis from $x=0$ to $x=-\frac{C}{B}$ without changing that of the Y-axis, and our equation becomes $y^2=BX$. To transform the general equation of the conic in the same way, we first cast it in a simpler form thus:

$$y^2=Ax^2+Bx+C=A\left(x^2+\frac{B}{A}x+\frac{B^2}{4A^2}\right)+C-\frac{B^2}{4A}$$

$$y^2=A\left(x+\frac{B}{2A}\right)^2-\frac{B^2-4AC}{4A}$$

For brevity, we may replace the term free of x on the right by a single constant K. If we now transfer the origin from $x=0, y=0$ to $x=\frac{-B}{2A}, y=0$, by the substitution $X=x+\frac{B}{2A}$, the equation reduces to

$$y^2=AX^2-K$$

Since $A=e^2-1$, both A and K are negative if $e<1$, in which event the curve is an ellipse, and we

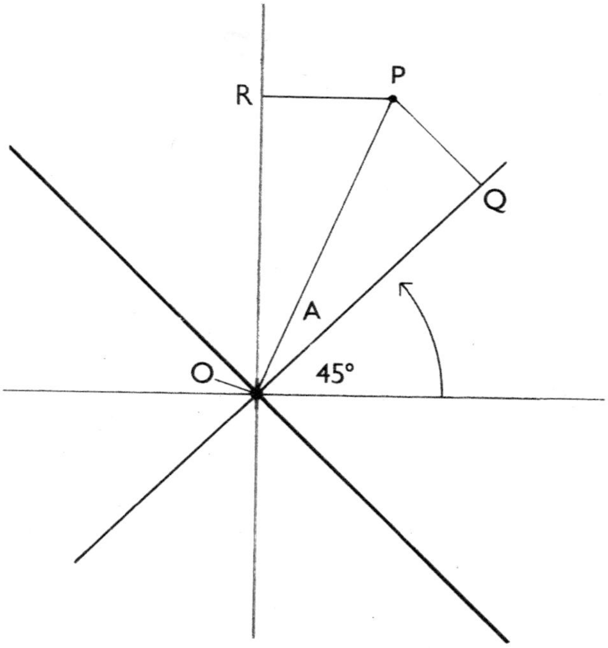

ROTATION OF AXES THROUGH 45° $\left(\frac{\pi}{4}\text{ RADIANS}\right)$

$OQ=X$; $PQ=Y$; $PR=x$; $OR=y$; $OP=r$.

$$\sin 45°=\frac{1}{\sqrt{2}}=\cos 45°. \quad X=r\cos A;\ Y=r\sin A.$$

$$x=r\cos(A+45°);\ y=r\sin(A+45°)$$

$$\therefore\ x=r\cos A\,.\cos 45°-r\sin A\,.\sin 45°$$

$$y=r\sin 45°\,.\cos A+r\sin A\,.\cos 45°$$

$$\therefore\ \sqrt{2}\,.x=X-Y \text{ and } \sqrt{2}\,.y=X+Y$$

$$\therefore\ X=\frac{y+x}{\sqrt{2}} \text{ and } Y=\frac{y-x}{\sqrt{2}}$$

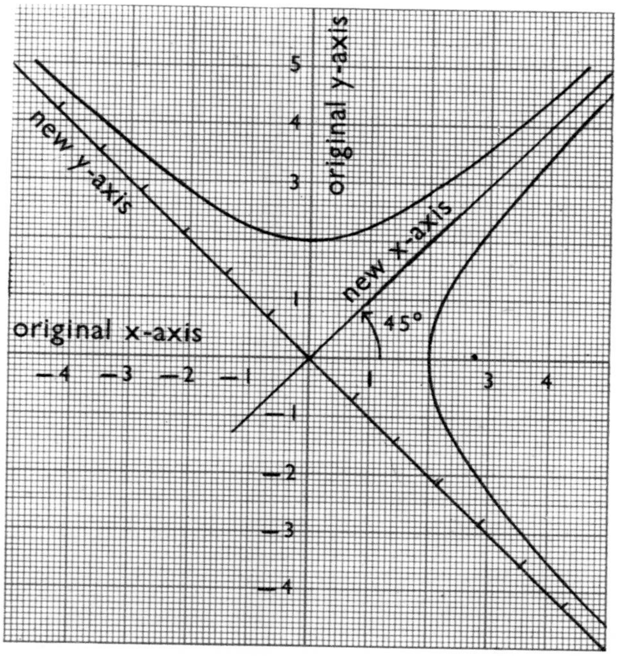

Rotation of the focus-directrix form of the hyperbola through 45° $\left(\frac{\pi}{4}\text{ radians}\right)$. The same equation of the hyperbola describes two curves, each of which is the mirror image of the other located in the opposite quadrant.

may write this in a form which contains only positive constants by putting $p=-A$ and $q=-K$, *i.e.*

$$pX^2+y^2=q$$

Otherwise, if $e>1$ both A and K are positive; and we may therefore write for the hyperbola:

$$Ax^2-y^2=K$$

How rotation of the axes about the origin affects the form of the functional relationship for one and the same curve in the same fixed framework depends on the angle (A); and two cases are of special interest, *viz.* $A=90^\circ$ and $A=45^\circ$. If $A=90^\circ$ (anti-clockwise):

$$Y=x \text{ and } -X=y$$

Thus the equation of the parabola expressed in the form $y^2=Bx$ becomes $Y=\frac{1}{B}X^2$. By clockwise rotation through 90°, it becomes $Y=-\frac{1}{B}X^2$. If we transfer the origin from $X=0$, $Y=0$ to $X=10$, $Y=25$, when $Y=-\frac{1}{4}X^2$ we may write it in the form:

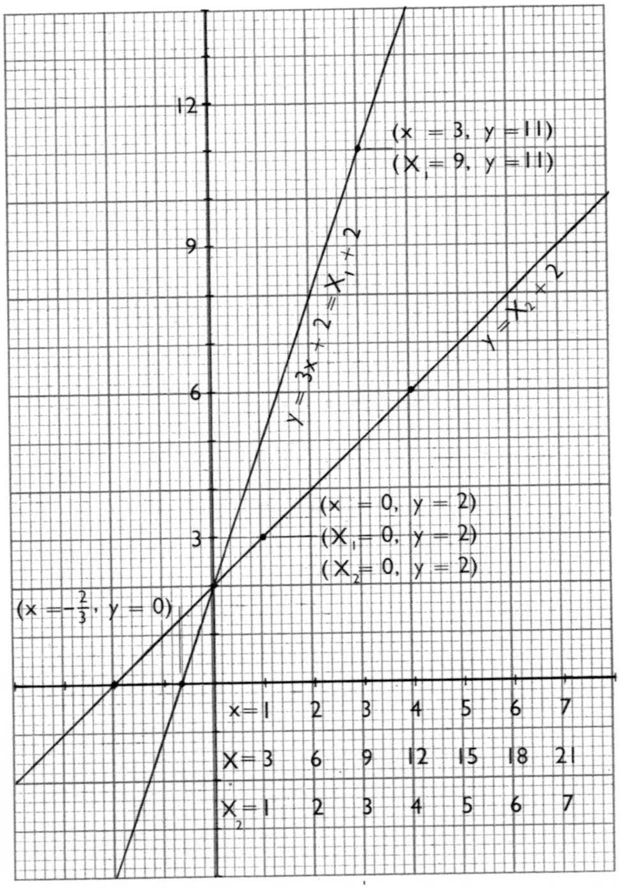

$$(y-25)=-\tfrac{1}{4}(x-10)^2 \text{ and } y=5x-\frac{x^2}{4}$$

The general form ($y=ax-bx^2$) of the equation on the right above gives the correct orientation implied when we speak of the trajectory of a cannon ball as a parabola. More precisely, we should say that the path of the missile tallies with that segment of such a parabola above the X-axis.

For rotation (anti-clockwise) of the *primary* axis through 45°, the appropriate formulae are:

$$X=\frac{y+x}{\sqrt{2}} \text{ and } Y=\frac{y-x}{\sqrt{2}}$$

When $A=1$ in our last equation ($Ax^2-y^2=K$) for the hyperbola, the limbs of the curve approach indefinitely near two lines (*asymptotes*) at right angles and inclined to the X-axis at 45°. The original equation being $x^2-y^2=K$ when this is so, a rotation of the axes through 45° signifies:

$$\frac{(y+x)^2}{2}-\frac{(y-x)^2}{2}=K \text{ so that } y=\frac{K}{2x}$$

The axes of the framework are now the asymptotes of the curve, and the functional relation shows that the corresponding discrete function is a harmonic series, defined as above.

We have spoken of the circle as a particular (so-called *degenerate*) form of the conic. So also is the straight line. By a suitable change of origin, we can express the parabolas $Y=KX^2$ whose two limbs reach vertically upwards or downwards in the form:

$$y=Kx^2+Px+Q$$

When $K=0$, this degenerates into $y=Px+q$. It is easy to see that the fundamental property of the line inclined at A° to the X-axis is expressed by:

$$\frac{y_2-y_1}{x_2-x_1}=\tan A=\frac{(Px_2+Q)-(Px_1+Q)}{x_2-x_1}$$

$$\text{if } P=\tan A$$

Thus $y=Px+Q$ defines a line inclined at $A=\tan^{-1}p$ to the X-axis; and since $y=Q$ when $x=0$, it crosses the X-axis at $y=Q$. The last statement is true of any function $y=f(x)$ in which an additive constant

If we make the substitution $X=3x$ in $y=3x+2$ without changing the scale, the slope of the line changes. The slope remains the same, however, if we make the unit of X one third the unit of x.

Right: A simple geometrical figure (top) is distorted first by re-plotting on oblique co-ordinates, second by scalar changes. The bottom distortion is produced by reducing the bottom horizontal line of the original grid to a single point and all other horizontal lines to concentric circles described from that point. The vertical lines of the original grid become equally-spaced radii of those circles.

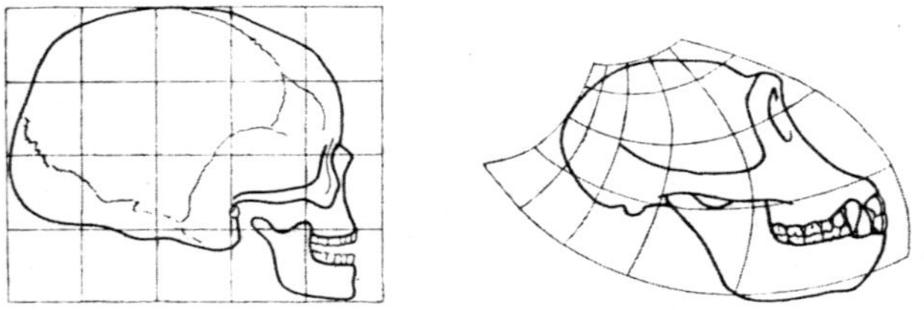

Man's skull drawn on Cartesian co-ordinates and skull of chimpanzee obtained by re-plotting point by point in a curvilinear framework. (From Darcy W. Thompson's Growth and Form, *C.U.P.)*

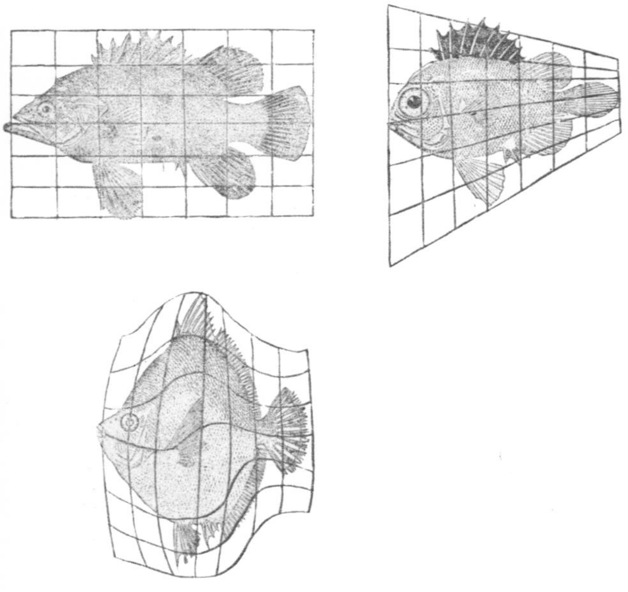

Here we see the outline of a fish, Polyprion, *superimposed on a Cartesian grid. By plotting corresponding points on a suitably chosen triangular or curvilinear framework we obtain the outlines of two other fishes,* Pseudopriacanthus (*above*) *and* Antigonia (*below*). (*Darcy W. Thompson*, op. cit.)

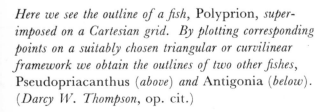

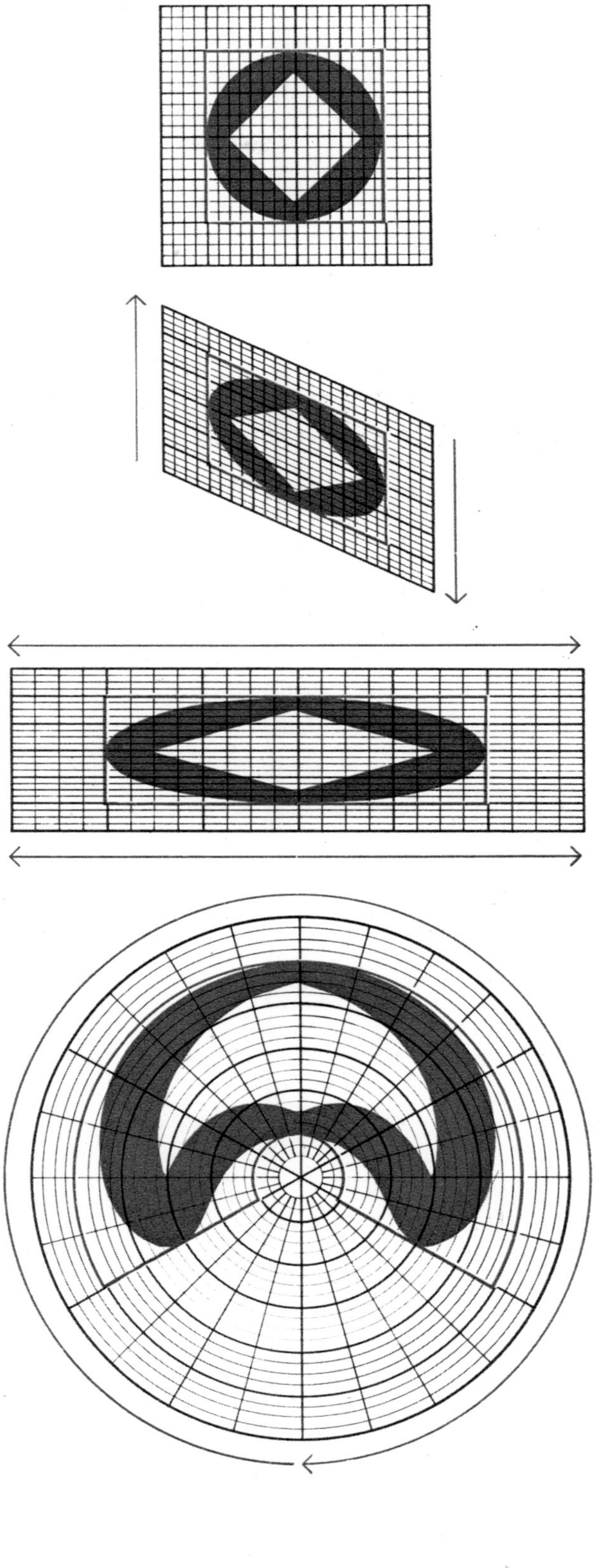

Q is unattached to a term involving x itself. The reader should now find no difficulty in seeing that the constant P in the linear equation and the constant K in the preceding quadratic are *invariant* under translation, *i.e.* respectively reappear as the coefficient (multiplier) of x and of x^2 in the outcome of any change of origin without rotation. This property of a curve leads one to ask: what do we imply if we make a substitution such as $X=Px$ in $Y=Px+Q$ or $X=x\sqrt{K}$ in $Y=Kx^2+Px+Q$?

To answer this question, we recall that we have hitherto assumed that distances from the origin, if represented by the same number, *e.g.* $x=5=-y$, are equivalent. However, we are free to label our grid so that points along the x-axis, if represented by the same numbers as points on the y-axis, are not equidistant from the origin. Let us suppose that $y=3x+2$ is the equation of a line in a grid whose x and y co-ordinates have the same scale of measurement. The slope to the x-axis is then $\tan^{-1}3 \simeq 71{\cdot}5°$. If we make the substitution $X=3x$, the equation $y=X+2$ leaves us where we were before, only if we also change our scale so that the mark $X=3$ corresponds to the mark $x=1$, $X=6$ to $x=2$, *etc.* However, this equation would represent a different line if the X-scale of distances were, as hitherto assumed, the same as the y-scale. It would still be a line which cuts the y-axis at $y=2$; but the slope would be $\tan^{-1}1$ $(=45°)$, *i.e.* it would incline equally to both axes.

The substitution $X=3x$ without scalar change here represents a three-fold stretch of the x-axis, pulling the line downwards to slope less steeply to the x-axis. Without a corresponding scalar change, the substitution of $Y=3y$ is equivalent to stretching the y-axis to the same extent; and the second equation now becomes $Y=3X+3Q$. Such a symmetrical substitution restores the original slope of a curve when the x and y scales are the same; but the constant which defines the point where the line crosses the Y-axis is not as before invariant under the transformation. An algebraic substitution of the type $X=K.x$ tilts the line upwards or downwards; but it remains a straight line, none the less. A similar substitution without scalar change in the form of the function which describes any other figure changes its shape in one way or another. The case of the ellipse sufficiently illustrates this. On the assumption that the two scales are identical the equation of the ellipse is $pX^2+y^2=q$. If we make the substitution $x=X\sqrt{p}$, it takes the form $x^2+y^2=q$. On the same assumption, this is the equation of a circle whose radius is $\sqrt{q}$.

When we cite an equation to exhibit a functional relationship definitive of a geometrical figure, we therefore commonly imply that we measure all distances in terms of the same unit and all angles in the same units. Otherwise, it has no geometrical meaning unless we explicitly state the relation between scalar units of the co-ordinates. It is important to emphasise this, because few of us make our first acquaintance with a graph as a means of analysing the properties of figures. More commonly, we first think of it as a means of solving one or more equations; and when this is the end in view, we are at liberty to make any scalar change which makes the relevant intersection more easy to read off.

The pattern of such a solution for one variable is as follows. If $Kx^2+Px+Q=0$, the solution is equivalent to the value or values of x which make $y=0$ in the function $Kx^2+Px+Q=y$, *i.e.* the value or values of x where the curve cuts the x-axis. When our concern is with simultaneous equations such as $3x-4y+2=0=4x+3y+5$, we can write the first and second respectively in the forms:

$$y=\tfrac{3}{4}x+\tfrac{1}{2} \text{ and } y=-\tfrac{4}{3}x-\tfrac{5}{3}$$

These represent two straight lines which cut at the point where the functions are equal, and the co-ordinates of this point are the only values of x and y consistent with this condition.

What else the new geometry of the seventeenth century can tell us about the natural history of equations will emerge when we consider its first-born offspring, the Newton-Leibniz calculus. Much of what the intrusion of the new algebraic symbolism can tell us about the geometrical properties of a figure we are now ready to recognise. Let us first recall that we have been able to derive a functional relation for the form of several figures – plane or solid – from a simple initial statement about the track of a point along their boundary without invoking any geometrical considerations other than the Pythagorean relation. How much dividend there is in this depends on what the function can tell us without further reliance on geometrical constructions. On the assumption that we measure x and y in the same units throughout, let us examine from this viewpoint a few results already cited in the domain of plane figures:

(1) *Line:* $y=3x+2$

We have seen that it slopes to the x-axis at an angle $A=\tan^{-1}3$ and cuts the y-axis at $y=+2$. It cuts the x-axis when $y=0$, whence $x=-\frac{2}{3}$.

(2) *Circle:* $(y-3)^2+(x-4)^2=25=5^2$

By transferring the origin to $X=0$, $Y=0$ by the substitution $X=x-4$, $Y=y-3$, we may put this in the form $X^2+Y^2=5^2$ or $Y=\pm\sqrt{(5^2-X^2)}$, whence the function has equal values of Y for each negative or positive numerical value of X, being therefore symmetrical about the Y-axis. Similarly, it is everywhere symmetrical about the X-axis because $X=\pm\sqrt{(5^2-Y^2)}$. When $X=0$, $Y\pm5$ and when $Y=0$, $X=\pm5$. The original equation thus describes a closed symmetrical curve with equal diameters (of length 10 units) corresponding to the principal axes and centre at $x=4, y=3$. By rotating the axes, we can show that the semi-diameters are equal in all directions.

(3) *Parabola:* $y=5x-3x^2+2$

By a simultaneous change of scale and origin we can recast this in the form $-Y=X^2$, whence it is symmetrical about the Y-axis. When $x=0$ in the original function, $y=2$, so that the curve cuts the y-axis at $y=+2$. As x becomes great, the limbs extend further apart below the x-axis. When $y=0$, so that $3x^2-5x-2=0$:

$$x=\frac{5\pm\sqrt{(25-24)}}{6} \text{ so that } x=1 \text{ or } \frac{2}{3}$$

Thus the parabola cuts the x-axis at $x=+1$ and $+\frac{2}{3}$. Since it is symmetrical about a line parallel

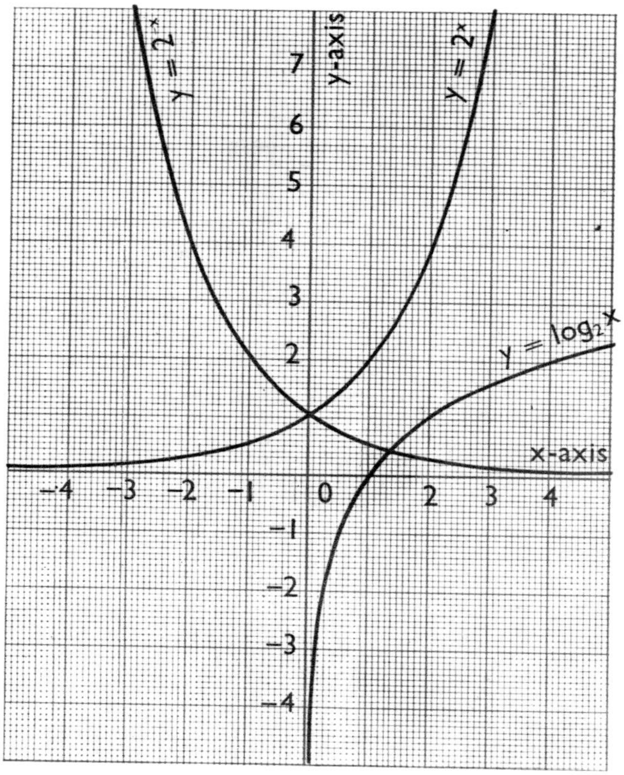

EXPONENTIAL AND LOGARITHMIC CURVES

Extract from La Géométrie *in which Descartes describes how he applies algebraic symbolism to the analysis of geometrical figures.*

LIVRE PREMIER. 299

gnes ſur le papier, & il ſuffiſt de les deſigner par quelques lettres, chaſcune par vne ſeule. Comme pour adiouſter la ligne B D a G H, ie nomme l'vne *a* & l'autre *b*, & eſcris *a* + *b*; Et *a* -- *b*, pour ſouſtraire *b* d'*a*; Et *a b*, pour les multiplier l'vne par l'autre; Et $\frac{a}{b}$, pour diuiſer *a* par *b*; Et *a a*, ou a^2, pour multiplier *a* par ſoy meſme; Et a^3, pour le multiplier encore vne fois par *a*, & ainſi a l'infini; Et $\sqrt{a^2 + b^2}$, pour tirer la racine quarrée d'$a^2 + b^2$; Et $\sqrt{C.\, a^3 -- b^3 + abb}$, pour tirer la racine cubique d'$a^3 -- b^3 + abb$, & ainſi des autres.

vſer de chiffres en Geometrie.

Où il eſt a remarquer que par a^2 ou b^3 ou ſemblables, ie ne conçoy ordinairement que des lignes toutes ſimples, encore que pour me ſeruir des noms vſités en l'Algebre, ie les nomme des quarrés ou des cubes, &c.

Il eſt auſſy a remarquer que toutes les parties d'vne meſme ligne, ſe doiuent ordinairement exprimer par autant de dimenſions l'vne que l'autre, lorſque l'vnité n'eſt point determinée en la queſtion, comme icy a^3 en contient autant qu'*abb* ou b^3 dont ſe compoſe la ligne que i'ay nommée $\sqrt{C.\, a^3 -- b^3 + abb}$: mais que ce n'eſt pas de meſme lorſque l'vnité eſt determinée, a cauſe qu'elle peut eſtre ſouſentendue par tout ou il y a trop ou trop peu de dimenſions: comme s'il faut tirer la racine cubique de *aabb* -- *b*, il faut penſer que la quantité *aabb* eſt diuiſée vne fois par l'vnité, & que l'autre quantité *b* eſt multipliée deux fois par la meſme.

Pp 2 Au

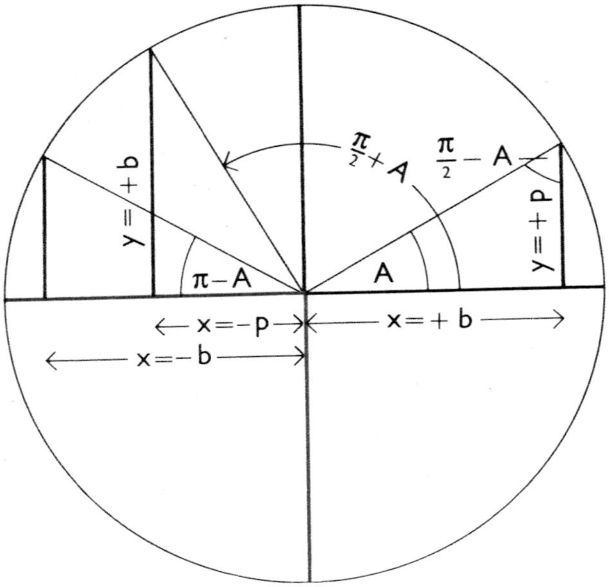

In a circle of radius $r=1$, $\sin A=y=\cos\left(\frac{\pi}{2}-A\right)$;

$\cos A=x=\sin\left(\frac{\pi}{2}-A\right)$.

$\sin\left(\frac{\pi}{2}+A\right)=+b=+\cos A$

$\cos\left(\frac{\pi}{2}+A\right)=-p=-\sin A$

$\sin(\pi-A)=+p=+\sin A$

$\cos(\pi-A)=-b=-\cos A$

$\sin\left(\frac{3\pi}{2}-A\right)=-b=-\cos A$

$\cos\left(\frac{3\pi}{2}-A\right)=-p=-\sin A$

$\sin(2\pi-A)=\sin(-A)=-p=-\sin A$

$\cos(2\pi-A)=\cos(-A)=+b=+\cos A$

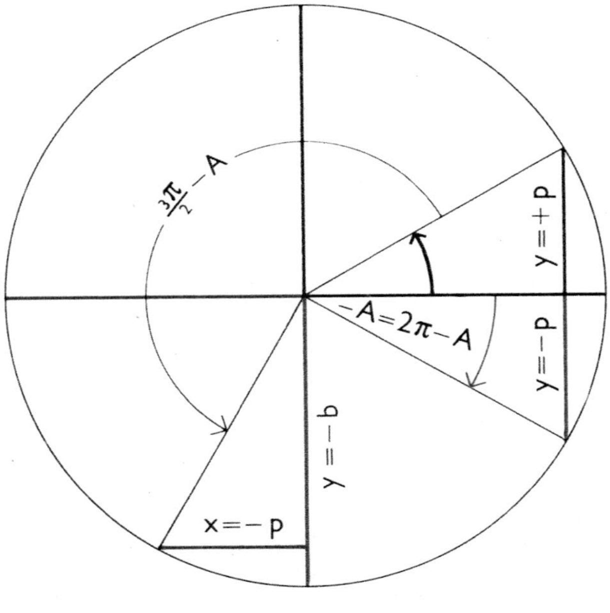

with the y-axis, this line must cut the x-axis mid-way between these two points, *i.e.* $x=\frac{5}{6}$, when

$$y=\frac{25}{6}-\frac{3(25)}{36}+2=\frac{49}{12}$$

Thus the curve has a turning point above the X-axis at the point $x=\frac{5}{6}$ and $y=\frac{49}{12}$.

(4) *Ellipse:* $4x^2+y^2=3$

We may write this alternatively in the forms:

$$x=\pm\tfrac{1}{4}\sqrt{(3-y^2)} \text{ or } y=\pm\sqrt{(3-4x^2)}$$

It is thus a closed curve symmetrical about the origin in relation to both rectangular axes. When $x=0$, $y=\pm\sqrt{3}$ and when $y=0$, $x=\pm\frac{1}{2}\sqrt{3}$. Since it is symmetrical about the axes, these represent maximum values and the lengths of its two diameters are respectively $2\sqrt{3}$ and $\sqrt{3}$, *i.e.* the vertical diameter is twice as long as the horizontal one.

(5) *Hyperbola:* $xy=4$, *i.e.* $y=\frac{4}{x}$ or $x=\frac{4}{y}$

Clearly, y approaches 0 as x becomes indefinitely large and conversely x approaches 0 as y becomes indefinitely large. For positive values of x and y, the curve thus has two limbs which become more and more nearly parallel to the two main axes in the right-hand top quadrant. We express this by saying that the two limbs are asymptotic to the main axes. The same equation describes a mirror image of the curve in the left-hand lower quadrant, since $(-x)(-y)=xy$. At $x=1, y=4$, at $x=4, y=1$, and at $x=2, y=2$. Here the curve tends from a nearly horizontal to a nearly vertical course; and it is symmetrical about a line going through $x=0$, $y=0$ and $x=2=y$.

So far we have spoken of the class of curves which are the topic of Fermat's pioneer work on co-ordinate geometry early in the seventeenth century. The function 2^x which we speak of as an exponential type when x may assume any real value provides us with another example of what we can infer about the shape of a curve from its equation. Since $2^0=1$, the curve cuts the vertical axis at $y=1$ when $x=0$. For positive values of x, corresponding values of y are positive, and increase

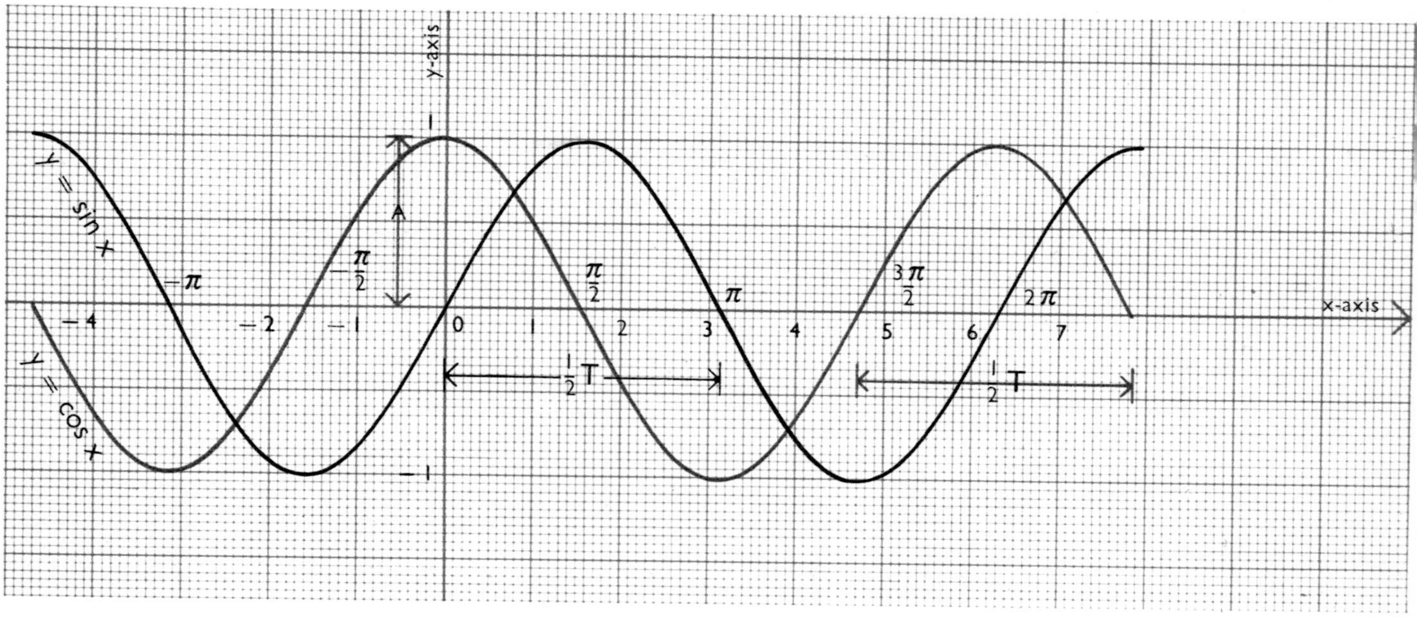

The curves of $y=\cos x$ and $y=\sin x$ have the same amplitude $A=2$ and the same period $T=2\pi$, if we measure x in radians.

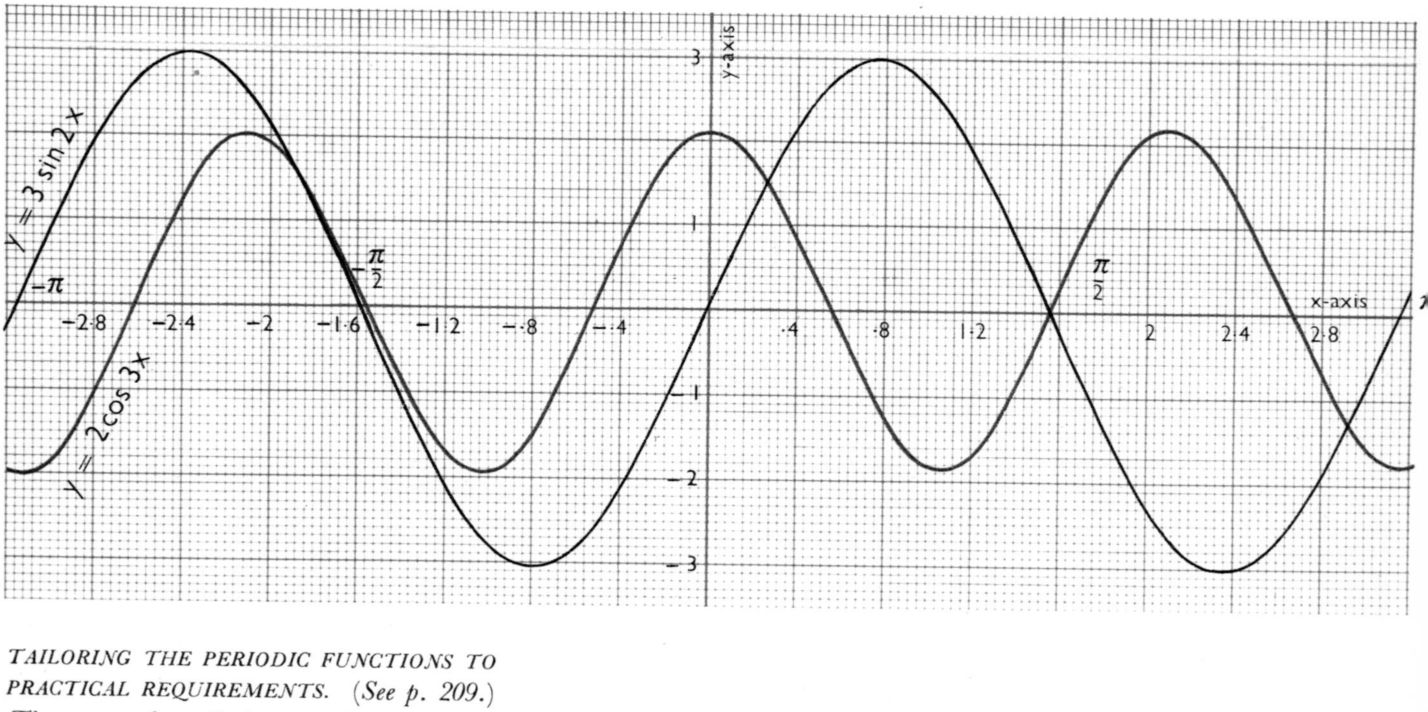

TAILORING THE PERIODIC FUNCTIONS TO PRACTICAL REQUIREMENTS. *(See p. 209.)* *The curves of $y=K \sin mx$ and $y=K \cos mx$ follow the same general course as $y=\sin x$ and $y=\cos x$ but their amplitudes and periods are different. When $mx=\frac{\pi}{2}$, $y=K$ if $y=K \sin mx$ and when $mx=\frac{3\pi}{2}$, $y=-K$. Thus the amplitude is $2K$. The function completes its cycle when x increases from mx to $mx+2\pi=m\left(x+\frac{2\pi}{m}\right)$, that is to say when x increases from x to $x+\frac{2\pi}{m}$. Thus the period is $\frac{2\pi}{m}$. This figure shows these functions: $y=3 \sin 2x$ with maxima at ± 3 and amplitude 6; $y=2 \cos 3x$ with maxima at ± 2 and amplitude 4.*

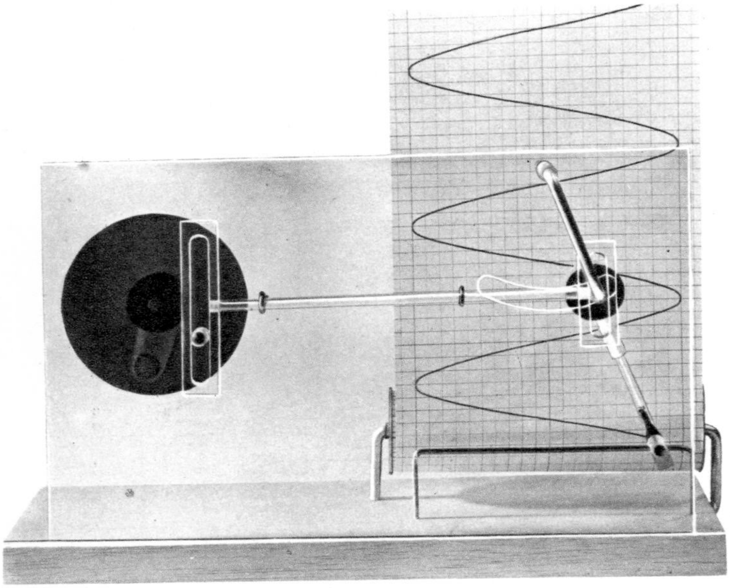

The horizontal and vertical displacements of a point moving at fixed speed in a circle of radius R vary periodically with amplitude 2R. The model here shown translates the circular motion into the corresponding wave form on a surface moving at fixed speed parallel with the vertical axis.

two-fold for unit increase of x. When x is negative, y is a positive fraction (*e.g.* $2^{-3}=0{\cdot}125$) approaching zero as x becomes indefinitely large. Thus the curve lies wholly above the x-axis, being asymptotic thereto in the left-hand quadrant. After cutting the y-axis (at $x=0, y=1$) it bends sharply up, becoming more steep. The reader can verify the following: (a) the curve of the function $y=\log_2 x$ lies wholly on the positive side of the x-axis, being asymptotic to the y-axis for negative values of y; and it cuts the x-axis at $x=1$ and bends upwards with diminishing steepness; (b) the curve of the function $y=2^{-x}$ lies wholly above the x-axis, so that y is always positive; and it descends steeply towards the origin for negative values of x, being fractional for $x>0$ and asymptotic to the x-axis on the right-hand side of the origin.

Our story so far covers somewhat more than a century of growth. Though it is customary to regard Descartes as the father of what we call *co-ordinate geometry* when our main concern is with the build-up of a figure or *analysis* when our main concern is with the properties of its equation, the illustrations of *La Géométrie*, published in 1637, have very little resemblance to a modern introduction to the same topics. In a letter to Roberval, written a year before the publication of the treatise of Descartes, Fermat had anticipated the representation of the straight line, circle, parabola, ellipse and hyperbola in a functional form referable to two axes at right angles; but his notation was, even for his own time, archaically verbose. For instance, the following (with modern

A DAMPED OSCILLATION ($y=3^{-\frac{x}{3}}\sin 6x$)

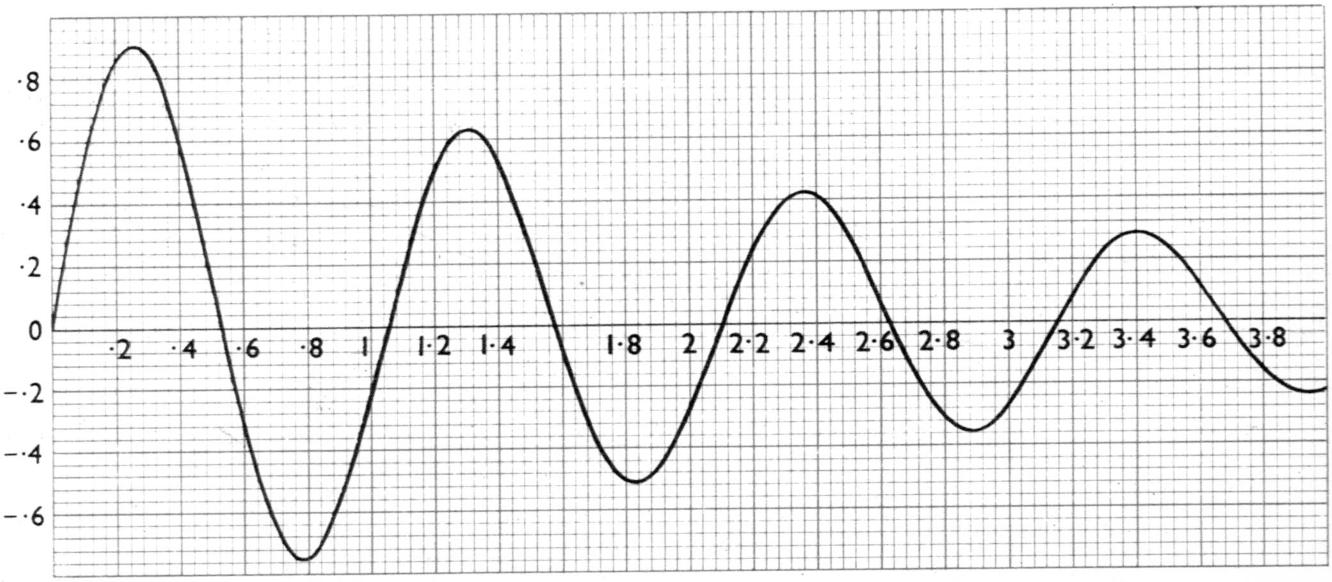

translation below) is the equation of the straight line:

ut B ad D, ita R — A ad E

$a : b = (c - x) : y$ so that $ay = b(c - x)$

What makes the treatise of Descartes conspicuously one step ahead of Fermat is that his symbolism has the New Look; but neither of the two operates simultaneously in the now familiar four quadrants of the x-y grid. Thus Descartes allows x to take negative or positive values on either side of a fixed point, but there is little indication of the possibility of permitting y to take negative values. He does not deal explicitly with solid figures, which were indeed first treated extensively from the same viewpoint by the Swiss mathematician Euler (1748). Neither Descartes nor his immediate successors made use of the polar form. This is comprehensible because the accepted definition of the trigonometrical ratios in terms of the sides of a right-angled triangle made no provision for angles greater than 90°. To make much use of polar co-ordinates, we need in short a redefinition of sin A, cos A, *etc.* to accommodate angles of any size in a way which is consistent both with the Hindu specification and with the addition or subtraction rules, which we may here recall:

$$\sin (A \pm B) = \sin A . \cos B \pm \sin B . \cos A$$
$$\cos (A \pm B) = \cos A . \cos B \mp \sin A . \sin B$$

Though it is difficult to say how far familiarity with a fixed framework of four quadrants encouraged the liberation of trigonometry from the strait-jacket of the Hindu triangle – or *vice versa* – such a redefinition was a simple step to take when mathematicians who used analytical methods began to formulate problems against the background of a grid in which x and y can each simultaneously have numerically equal negative and positive values. To reformulate the elements of trigonometry to meet the requirements stated above, we assume that the radius of a circle whose centre is the origin of the x-y grid rotates anti-clockwise from the position $y=0$, $x=r$ when the angle (A) it makes with the x-axis is zero. (See diagrams on p. 204.) We define for all values of A with due regard to the signs of x or y in the relevant quadrant of the complete rotation through 360°:

$$\sin A = \frac{y}{r}; \quad \cos A = \frac{x}{r}; \quad \tan A = \frac{y}{x}$$

If $180° > A > 90°$, y is positive but x is negative, so that sin A is positive but cos A and tan A are both negative. When $A = 180°$, $\sin A = 0 = \tan A$ and $\cos A = -1$. This reversal of sign is consistent with the addition rules, since $\cos 90° = 0$ and $\sin 90° = 1$ and:

$$\cos 180° = \cos (90° + 90°)$$
$$= \cos 90° . \cos 90° - \sin 90° . \sin 90°$$
$$= -1$$

If $270° > A > 180°$, both x and y are negative, so that tan A is positive, sin A and cos A being both negative. When $A = 270°$, $\cos A = 0$ and $\sin A = -1$. This again is consistent with the addition rules, since $\cos 180° = -1$ and $\sin 180° = 0$ and:

$$\sin 270° = \sin (180° + 90°)$$
$$= \sin 90° . \cos 180° + \cos 90° . \sin 180°$$
$$= -1$$

If $360° > A > 270°$, y is negative but x is positive, so that sin A and tan A are both negative and cos A positive. When $A = 360°$, $\sin A = 0 = \tan A$ and $\cos A = 1$. We have now got back to where we started, and all the values of sin A, *etc.* between 0° and 360° repeat themselves between 360° and 720°, 720° and 1080°.

We can appreciate this periodicity better if we express the angle in circular measure (2π radians

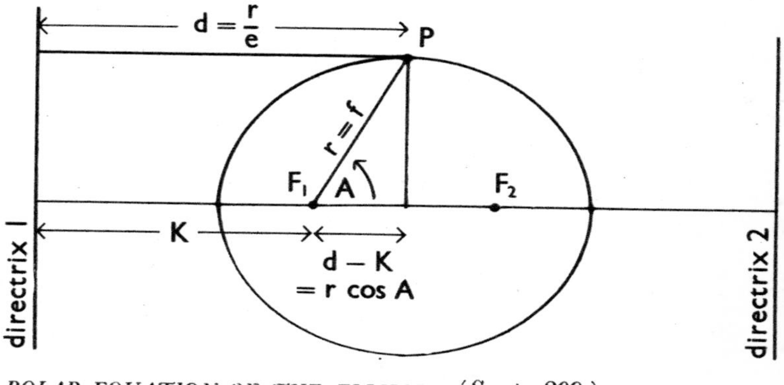

POLAR EQUATION OF THE ELLIPSE. (*See p. 209.*) *We may generate one and the same ellipse from the rotation of the radius vector (r) about either of two foci (F_1 and F_2) distant K from its own directrix. For any point P distant d from the directrix r : d = e. The figure shows that d — K = r cos A so that*

$$\frac{r}{e} - K = r \cos A \text{ and } r = \frac{eK}{1 - e . \cos A}$$

By the mid-sixteenth century, warfare in Europe meant cannon warfare, but there was still as much miss as hit in artillery practice.

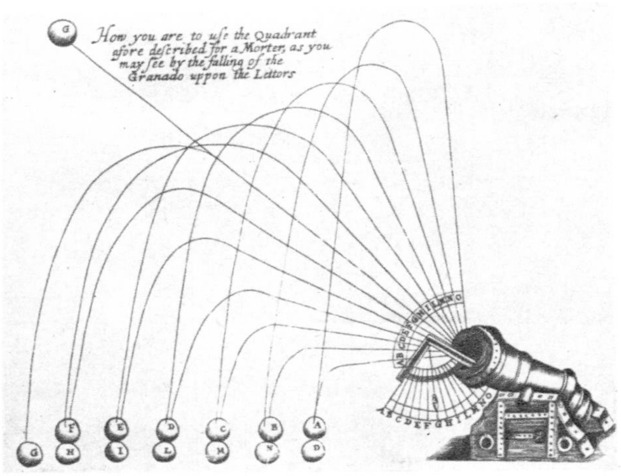

The Complete Gunner, *compiled by Thomas Venn in 1672, includes extracts from Galileo on ballistics, but nevertheless shows trajectories incorrectly.*

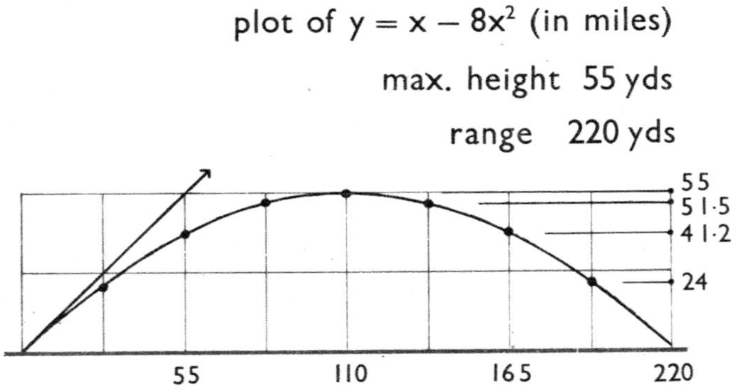

Path of the Galilean cannon ball. If the unit is one mile, the equation is $y=x-8x^2$.

$=360$ degrees). Thus $\sin A=\sin(A+2\pi)=\sin(A+4\pi)$ *etc.* One commonly expresses this by writing $\sin A=\sin(A+2n\pi)$ for $n=1, 2, 3$ *etc.* Similarly for $\cos A$ and $\tan A$. From tabular values of the ratios in the range 0 to $\frac{\pi}{2}$ ($0°$ to $90°$), we can then obtain all other values by recourse to either of the following series of formulae consistent both with the geometry of the grid and (as the reader may check) with the addition or subtraction rules.

$\sin\left(\frac{\pi}{2}+A\right)=+\cos A$	$\sin\left(\frac{\pi}{2}-A\right)=+\cos A$
$\sin(\pi+A)=-\sin A$	$\sin(\pi-A)=+\sin A$
$\sin\left(\frac{3\pi}{2}+A\right)=-\cos A$	$\sin\left(\frac{3\pi}{2}-A\right)=-\cos A$
$\sin(2\pi+A)=+\sin A$	$\sin(2\pi-A)=-\sin A$
$\cos\left(\frac{\pi}{2}+A\right)=-\sin A$	$\cos\left(\frac{\pi}{2}-A\right)=+\sin A$
$\cos(\pi+A)=-\cos A$	$\cos(\pi-A)=-\cos A$
$\cos\left(\frac{3\pi}{2}+A\right)=+\sin A$	$\cos\left(\frac{3\pi}{2}-A\right)=-\sin A$
$\cos(2\pi+A)=+\cos A$	$\cos(2\pi-A)=+\cos A$

The reformulation of the meaning we attach to a trigonometrical ratio in terms which set no limit to the size of the angle and endow the latter with a sign referable to the direction of the rotating radius vector (negative if clockwise, positive if anti-clockwise) had vastly important consequences, of which two especially are worthy of comment. One

outcome was that it equipped dynamics with continuous periodic functions by recourse to which we can visualise in metrical terms the kind of motion (*e.g.* horizontal displacement of the pendulum) customarily called *simple harmonic* when we represent time (t) as the independent variable. By either or both of two scalar changes, we can tailor this function to accord both with the amplitude and with the period of the wave form. (See diagrams on p. 205.) For instance, the amplitude is $2A$ if $y=A \sin t$, since the maximum value of $\sin t$ in the first period $t=0$ to $t=2\pi$ is $y=A$ when $t=\frac{1}{2}\pi$ and the minimum is $y=-A$ when $t=\frac{3}{2}\pi$. Without reliance on such functions, it is difficult to conceive that the theory of the alternating current could have made much headway during the past century. A simple device makes it possible to represent damped periodic motions, such as the vibration of a violin string or the oscillatory discharge of a condenser. The reader can test it by plotting a few points on the curve of $2^{-x}.\cos x$, the first factor of which gives it a continuously diminishing amplitude.

A second consequence of the new trigonometry is of special interest in connexion with a theme already mentioned. Seemingly, the first person who made systematic use of polar co-ordinates was an Italian mathematician, Gregorio Fontana (1735–1803). That no one had exploited their use at an earlier date is easy to understand. Though the polar method is often the simplest way of expressing a functional relationship which specifies the outline of a closed curve, its use for that purpose commonly demands that the angle A in the function $r=f(A)$ can take all values from 0 to 2π radians. Thus the polar equation of the ellipse (see p. 207) referred to the focus as origin and a base line at right angles to the directrix to the left of it can be descriptive of only a small part of the curve unless we can give a meaning to $\cos A$, when $A>90°$ in the expression:

$$r = \frac{eK}{1-e.\cos A}$$

Commonly, the analytical representation of a geometrical figure in algebraic symbolism involves no more variables than the dimension involved – plane or two-dimensional expressible as a functional relation such as $y=f(x)$ and solid or three-dimensional expressible as a functional relation such as $z=f(x, y)$. However, it is sometimes convenient, if only as a half-way house to such a formulation, to express the dependent and the independent variable in a parametric equation, *i.e.* each in terms of a third one. The analysis of the trajectory of the cannon ball well illustrates this procedure. We shall assume:

(a) that the muzzle velocity is v at an angle of elevation a;

(b) that it retains this velocity in the same direction while falling with a fixed acceleration (g) earthwards in conformity with the Galilean principle of gravitation and composition of motions;

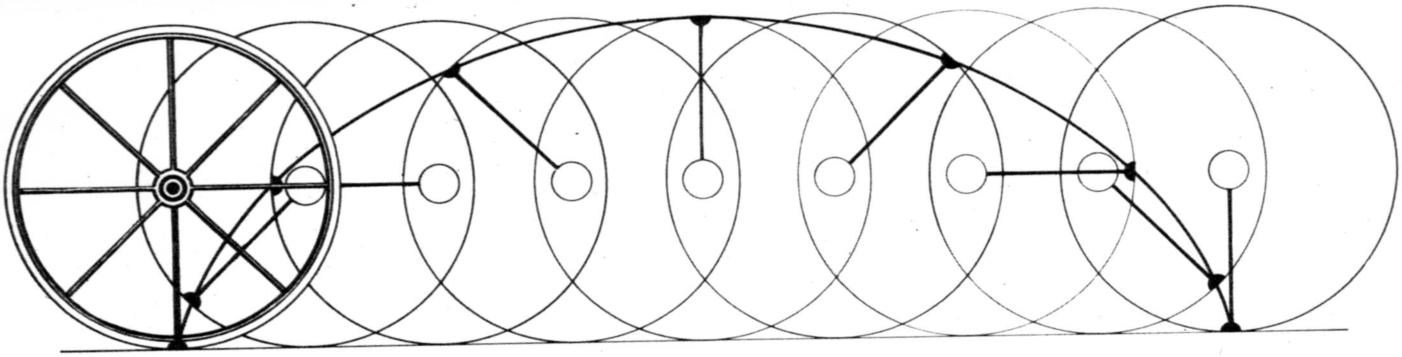

THE GENERATION OF THE CYCLOID

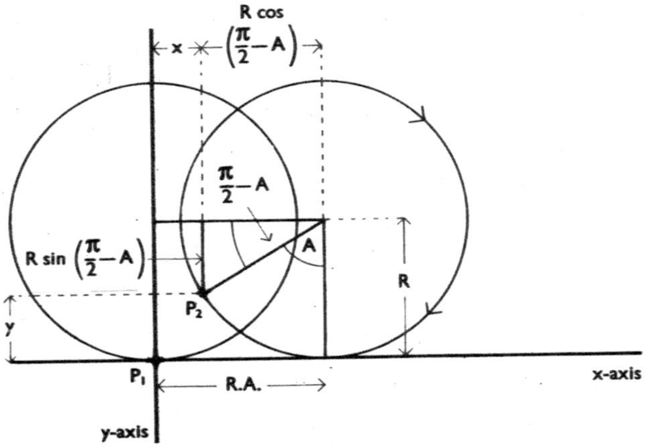

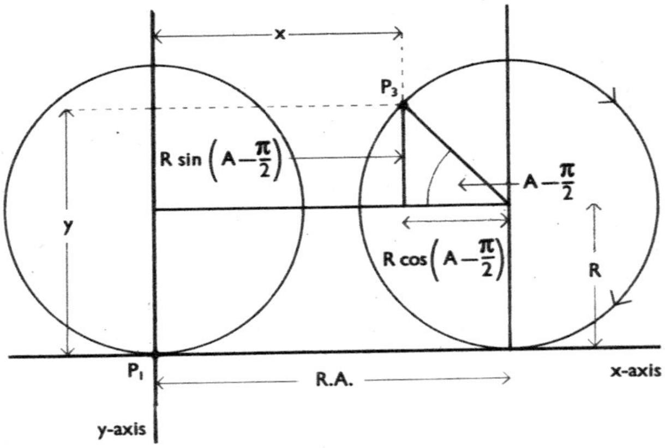

THE PARAMETRIC EQUATION OF THE CYCLOID

The generating circle of radius R rolls rightwards along the x axis. The initial position (P_1) of the point on the circumference is at the origin. When it has rotated through A radians, the horizontal displacement of the centre is R . A units of length.

$$A\left(=\frac{\pi}{3}\right) < \frac{\pi}{2}:—$$

$$x = R.A - R\cos\left(\frac{\pi}{2}-A\right) = R(A-\sin A)$$

$$y = R - R\sin\left(\frac{\pi}{2}-A\right) = R(1-\cos A)$$

$$A\left(=\frac{3\pi}{4}\right) > \frac{\pi}{2}:—$$

$$x = R.A - R\cos\left(A-\frac{\pi}{2}\right) = R(A-\sin A)$$

$$y = R + R\sin\left(A-\frac{\pi}{2}\right) = R(1-\cos A)$$

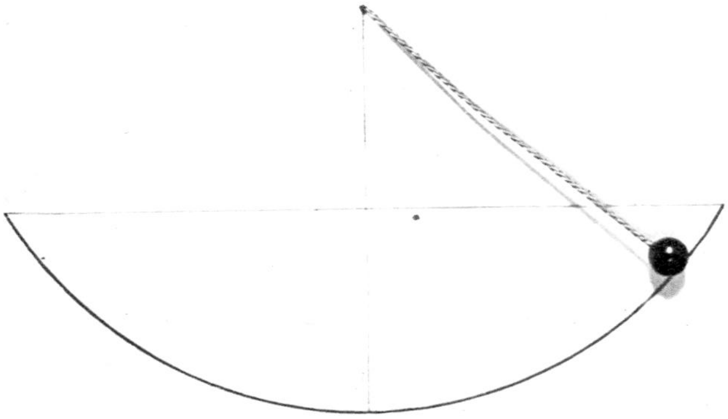

A free-swinging pendulum moves through a circular arc. If it swings through arcs varying considerably in amplitude, its time of swing will vary.

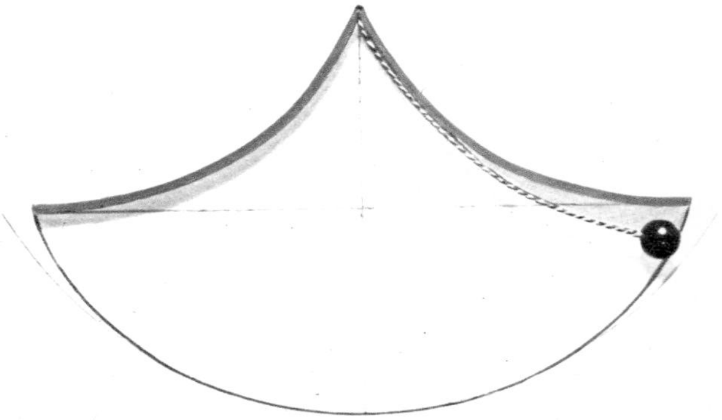

If curved jaws are so constructed as to make it swing through a cycloidal arc, the length of arc can vary more widely without affecting time of swing.

(c) that the X-axis lies on the horizontal plane and the Y-axis is the plumb-line.

If x is the horizontal and y the vertical displacement of the missile when t seconds have elapsed since the explosion, it then follows that:

$$x = vt . \cos a \text{ and } y = vt . \sin a - \tfrac{1}{2} gt^2$$

This is an example of a parametric functional relationship involving x and y. We can eliminate t in virtue of the fact that:

$$t = \frac{x}{v \cos a} \text{ so that } y = (\tan a) . x - \frac{gx^2}{2v^2 (\cos a)^2}$$

Since a, v and g are constants by definition, we may pack them up in the form $\tan a = A$ and $g \div 2v^2 (\cos a)^2 = B$, whence

$$y = Ax - Bx^2$$

We have seen that this is the equation of a parabola which has asymptotes on the negative side of the X-axis and its hump (where the cannon ball is at its highest) lies above it. Since $y = 0$ when $Bx^2 = Ax$, it cuts it at $x = 0$ or $\frac{A}{B}$; and since the curve is symmetrical about a line parallel with the Y-axis, its turning-point is at the half range, *i.e.* when $x = \frac{A}{2B}$ and

$$y = \frac{A^2}{4B}$$

Since $\tan 45° = 1$, $(\cos 45°)^2 = \frac{1}{2}$ and $g = 32$ foot-sec^2, the equation simplifies when $a = 45°$ to:

$$y = x - \frac{64x^2}{v^2}$$

Perhaps with a premonition that it describes the ideal arc for the swing of a pendulum with the appropriate suspension, Galileo himself made a preliminary attack in the social context of clock and cannon ball on the properties of another plane curve, which (as subsequently transpired) is easily expressible in parametric form, though not otherwise. Its name is the *cycloid*, and it is the locus of a point on the circumference of a rolling circle, *e.g.* the two-dimensional path of a pin-head at the rim of a car tyre in motion. If the radius of the generating circle (*i.e.* wheel) is R and the initial position of the pin-head at the origin, horizontal (x) and vertical (y) displacements of the latter with respect to the road surface uniquely depend on the angle (a) through which the wheel has rotated between two instants of time. For example, $y = 2R$ after a half turn, when $a = \pi$ (in radians), and $y = 0$ after a full turn, when $a = 2\pi$. Similarly, $y = R$ when $a = \frac{1}{2}\pi$ or $3(\frac{1}{2}\pi)$. The arc will repeat itself for corresponding intervals involving a complete revolution, during which the horizontal displacement is $2\pi r$, so that $y = 2R$ when $x = \pi r$.

While the circle rolls clockwise through a, the vertical displacement of the pin-head is the difference between the radius (R) and its downward displacement from the centre ($R \cos a$). Its horizontal displacement is the difference between that of the centre and its displacement (left) from the latter. The displacement (rightward) of the centre is the length of the arc through which the wheel has revolved. If we measure the angle (a) in radians, this is $R . a$. Whence we may write:

$$y = R(1 - \cos a) \text{ and } x = R(a - \sin a)$$

The shape of the curve is deducible from the parametric equations, and the foregoing redefinition of the meanings we attach to sin a or cos a accommodates every such cycle $a = 2n\pi$ to $a = 2(n+1)\pi$ as a repetition of its predecessor.

Chapter 10 The Newton - Leibniz Calculus

In so far as we may trace its beginnings to the work of Fermat, the start of our story in this chapter is coincident with that of its predecessor. One sometimes, and preferably, refers to it as the *infinitesimal* calculus, but not uncommonly as *the* calculus, a term equally appropriate to any technique applicable to the use of numbers. Since our concern will indeed be with three entities which we shall respectively refer to as the *derivative*, the *anti-derivative* and the *definite integral*, what is more misleading is to split it into two branches severally called the *differential* and the *integral* calculus.

To say this disposes sufficiently of a silly and undignified dispute about the respective priority claims of Newton and of Leibniz. Fermat is the *fons et origo* of what is essentially novel in the notion of a derivative as we here use the term. Newton first appreciated its dynamical implications and that of the inverse function here called the anti-derivative. Leibniz (1646–1716), whose writings on the composite theme appeared a little before those of Newton, may well have gained a few clues from personal contact with the latter while on a visit to Britain. What is quite certain is that he first clearly appreciated: (a) a tie-up between the anti-derivative and the definite integral; (b) a wider field of practical applications of the three classes of functions than the restricted domain of Newtonian dynamics.

THE DERIVATIVE. Our first task will be to trace to its source the class of functions now commonly known as *derivatives* but formerly referred to as *differential co-efficients*. As elsewhere in what follows, we shall here restrict ourselves to continuous functions such as: (a) involve only one independent variable; (b) are expressible in terms of a unique equation throughout the whole range under consideration. We speak of a continuous function which satisfies the last condition as *differentiable*. The implications of the term will be clear with the aid of the figure (p. 214, top) which exhibits a continuous function which is *not* differentiable. Here the vertical ordinates (y) of the triangle as a whole constitute a continuous function of x throughout the range $x=-3$ to $x=+12$. On the other hand, the form of the function changes within the range, *viz.*:

$$y=\frac{5x}{3}+5 \text{ from } x=-3 \text{ to } x=+1 \text{ inclusive}$$

$$y=7-\frac{7x}{12} \text{ from } x=+1 \text{ to } x=+12 \text{ inclusive}$$

The notion of a derivative could not emerge in its modern guise until the concept of function took shape within a fixed framework of reference. As such, it appears first in the work of Fermat, to whom Newton acknowledged his indebtedness. Fermat's own contribution was to recognise at least implicitly two principles: (a) maximum or minimum values of a differentiable function $y=f(x)$ are those at which the tangent to the curve is parallel to the x-axis; (b) the slope of the tangent to the x-axis is itself a continuous function of x

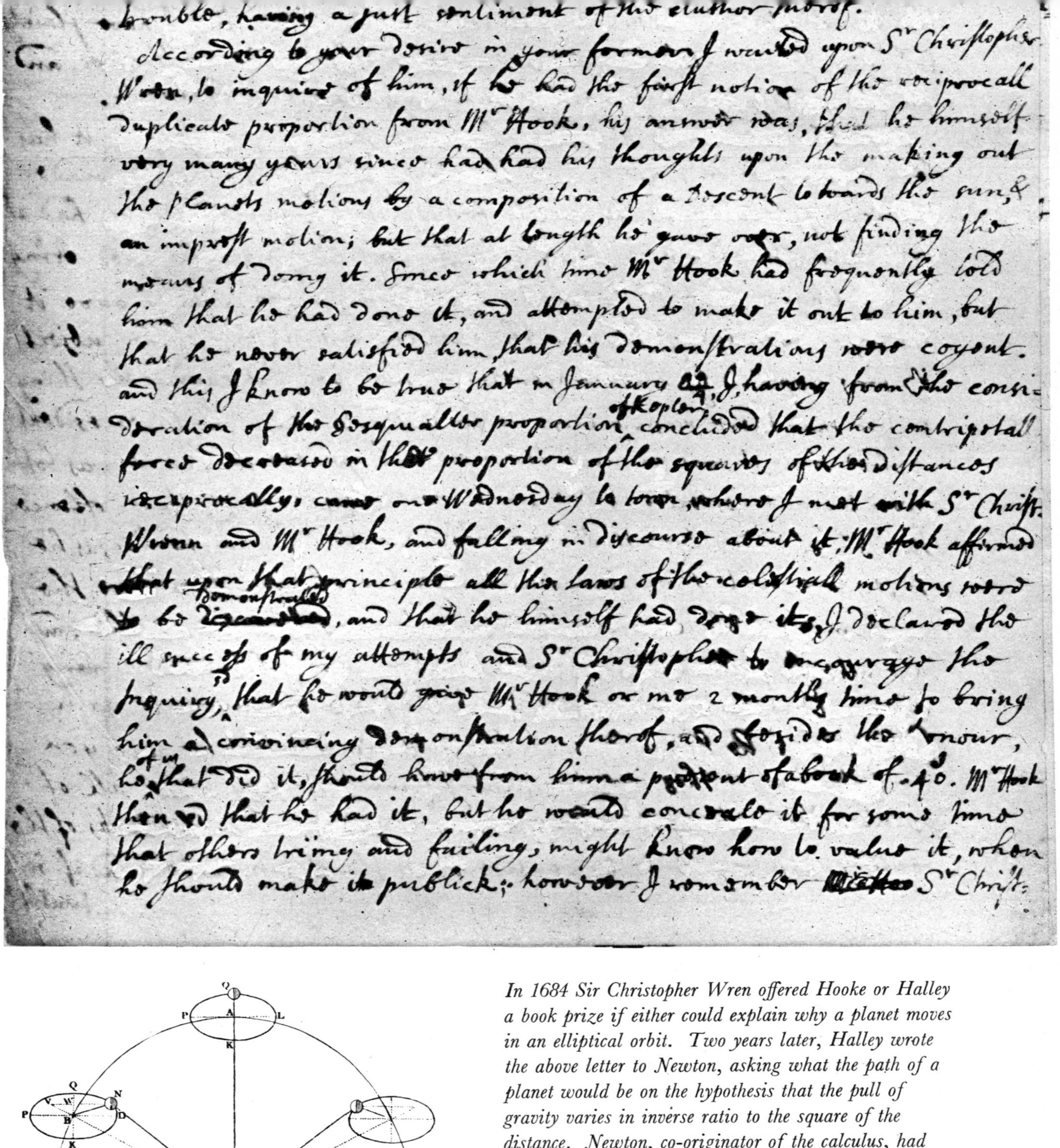

humble, having a just sentiment of the authors merit.

According to your desire in your former I waited upon Sr Christopher Wren, to inquire of him, if he had the first notion of the reciprocall duplicate proportion from Mr Hook, his answer was, that he himself very many years since had had his thoughts upon the making out the planets motions by a composition of a descent towards the sun, & an imprest motion; but that at length he gave over, not finding the means of doing it. Since which time Mr Hook had frequently told him that he had done it, and attempted to make it out to him, but that he never satisfied him, that his demonstrations were cogent. and this I know to be true that in January 83/4, I having from the consideration of the sesquialter proportion of Kepler concluded that the centripetall force decreased in the proportion of the squares of the distances reciprocally, came one Wednesday to town, where I met with Sr Christ. Wrenn and Mr Hook, and falling in discourse about it, Mr Hook affirmed that upon that principle all the laws of the celestiall motions were to be demonstrated, and that he himself had done it. I declared the ill success of my attempts and Sr Christopher to encourage the Inquiry sd, that he would give Mr Hook or me 2 months time to bring him a convincing demonstration thereof, and besides the honour, he of us that did it, should have from him a present of a book of 40s. Mr Hook then sd that he had it, but he would conceale it for some time that others triing and failing, might know how to value it, when he should make it publick; however I remember Sr Christ.

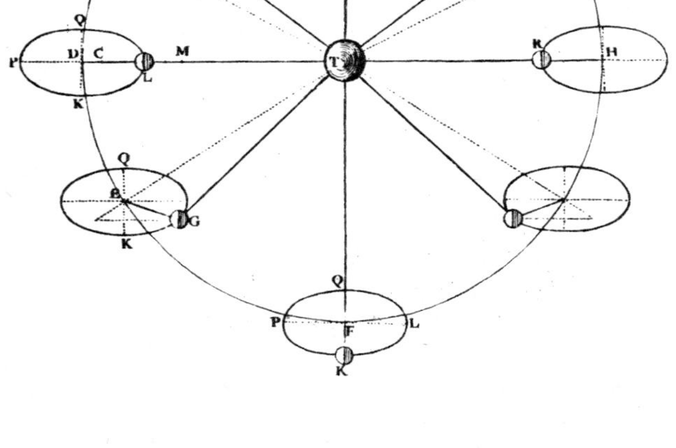

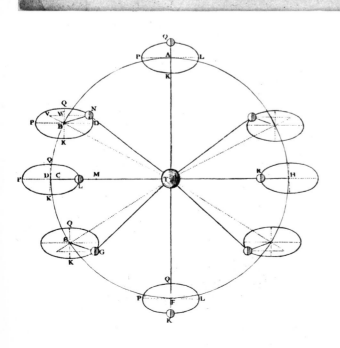

In 1684 Sir Christopher Wren offered Hooke or Halley a book prize if either could explain why a planet moves in an elliptical orbit. Two years later, Halley wrote the above letter to Newton, asking what the path of a planet would be on the hypothesis that the pull of gravity varies in inverse ratio to the square of the distance. Newton, co-originator of the calculus, had already done the necessary calculations and at once replied that it would be an ellipse. Left: A diagram from Newton's Principia.

which one writes in the notation of Leibniz as:

$$F(x) = \frac{dy}{dx}$$

The meaning of the second statement is easy to grasp if we sketch at equally-spaced intervals the slope of the tangent to a curve for any continuous function so far mentioned. Needless to say, we must then be clear about what we mean by its slope (or *gradient*) to the x-axis, *viz.* the *tangent* of the angle (A) it makes thereto. If we can find a formula for tan A in terms of x alone, it is thus the one labelled as $F(x)$ above. Henceforth, we shall speak of it as *the derivative of y with respect to x.* We shall dispense with the notation of Leibniz, and write it in a more modern form as $F(x) \equiv D_x.y$. This convention is more adaptable to dealing with functions of more than one independent variable, *e.g.* if $z = f(x, y)$ for a three-dimensional figure, $D_{x.y}(z)$ can stand for the tangent to the curve in a plane fixed by one value of y and $D_{y.x}(z)$ for the line tangent in a plane fixed by one value of x.

To understand more fully the problem which Fermat first clarified, it is necessary to define more precisely what we loosely call maxima and minima. In this context, they stand for *turning-points* on either side of which the slope of a curve changes from: (a) positive to negative, *i.e.* y increasing to y decreasing (*maximum*); (b) negative to positive,

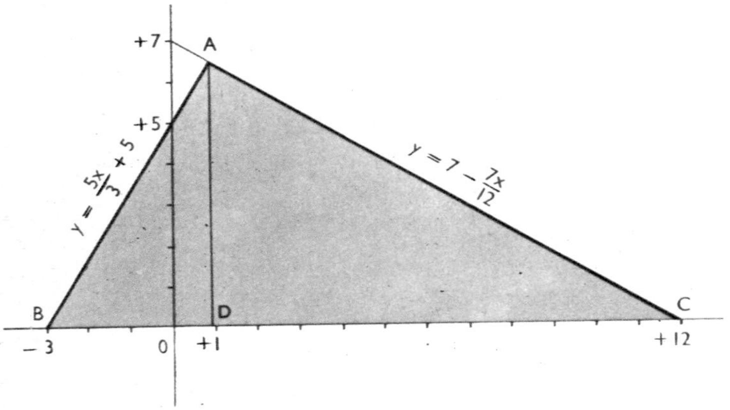

The vertical ordinates of AB constitute a continuous differentiable function of x in the range $x = -3$ to $x = +1$. Those of AC do so in the range $x = +1$ to $x = +12$. Those of the figure ABC as a whole constitute a continuous function of x from $x = -3$ to $+12$; but the function is not differentiable throughout the whole range.

Finding the greatest possible area of a rectangular figure which has a fixed perimeter. Here the perimeter is 28 linear units $= P$. $\therefore$ $P = 2(a + b)$ where a and b are sides. $\therefore$ $28 = 2(a + b)$ and $a = 14 - b$. $\therefore$ Area $A = ab = b(14 - b)$.

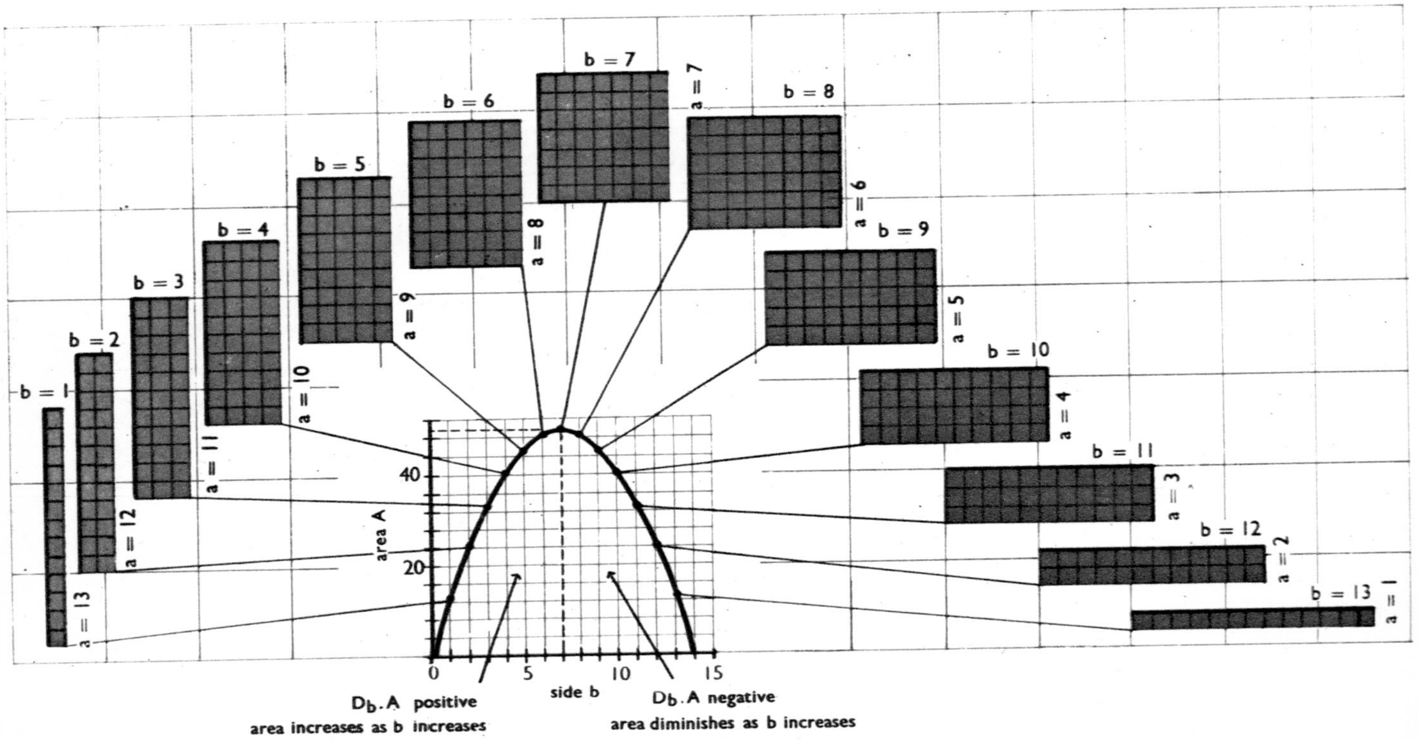

i.e. y decreasing to y increasing (*minimum*). A single figure suffices to make clear that A is zero at any such point. Since tan $0°=0$, so also is the slope of the tangent to the x-axis. If then we can find the derivative of y with respect to x, *i.e.* a function $F(x)=D_x.y$ which expresses tan A in terms of x and one or more constants, we may write $F(x)=0$. The solution of this equation gives us the x-co-ordinates of the points at which tan $A=0$, and the substitution of the corresponding numerical values of x in $y=f(x)$ gives us their y-co-ordinates.

For one reason or another, earlier mathematicians in the classical tradition of Apollonius had concerned themselves with the construction of tangents to curves by *ad hoc* Euclidean methods; but the notion of a descriptive equation of the sort last mentioned was alien to their outlook. Before the end of the sixteenth century, the issue had acquired a new interest. Eminently practical applications of the use of maxima or minima had emerged in the social context of cannon ball and clock; but the concept of function was still embryonic when Galileo did his best work. By approaching the issue within a fixed framework of reference, Fermat turned his back on the piecemeal solutions of a classical tradition which Galileo had not outgrown. Albeit without explicitly formulating it, his approach anticipates the notion that the function we are seeking is expressible as a *limiting ratio*. Newton's teacher, Isaac Barrow, developed the concept a little further without grasping, as did Newton, its dynamical implications.

It will be helpful if we now follow up the clues to a general solution of Fermat's problem both from the geometrical and from the kinematical viewpoint. The figure on page 216 exhibits the curve of a specified function $y=f(x)$ and the slope ($A°$) of the chord cutting the curve at (x_1, y_1) and (x_2, y_2) so that

$$\tan A = \frac{y_2-y_1}{x_2-x_1}$$

If t stands for time (*secs.*) and d for a *rectilinear* distance (*feet*) in this set-up, we see (top of p. 217) that tan A represents the fixed speed (*distance-time* ratio) at which a body would be moving over a distance (d_2-d_1) in time $\triangle t=(t_2-t_1)$. This is its *mean* speed throughout the interval. During one part of the interval $\triangle t$, the distance is increasing more steeply than at another. Thus the speed is not constant throughout the interval; and we may translate Fermat's problem into dynamical terms as follows: what is the speed at any point represented on the graph by P or Q? In more general terms, we may ask what is the rate of change of d per unit change of t at any point (t, d) on the curve?

It is here that the notion of a limiting ratio helps us. We assume (p. 217, centre) that we can make the chord QR in our second figure as small as we like. When Q and R coalesce at a point P, tan A will be the slope to the x-axis of the tangent to the curve at P, whose co-ordinates are x and y. We therefore imagine that a very small difference $\triangle x$ separates the x-co-ordinates of Q and R. We may then write them respectively as:

$$x-\tfrac{1}{2}\triangle x \text{ and } x+\tfrac{1}{2}\triangle x$$

For the corresponding y-co-ordinates of $y=f(x)$, we shall therefore write:

$$f(x-\tfrac{1}{2}\triangle x) \text{ and } f(x+\tfrac{1}{2}\triangle x)$$

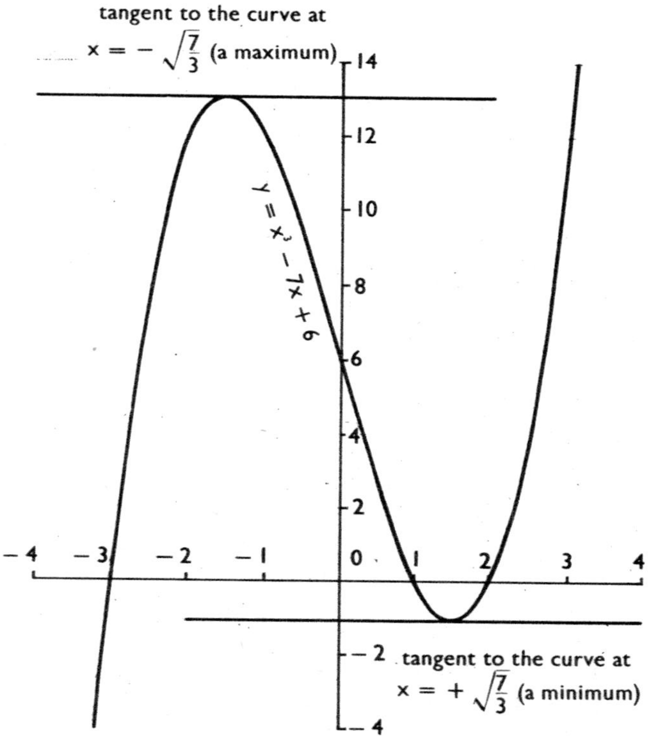

FERMAT'S PRINCIPLE

The continuous function $y=x^3-7x+6$ *has two turning-points: (i) at* $x=-\sqrt{\tfrac{7}{3}}$, *a maximum; (ii) at* $x=+\sqrt{\tfrac{7}{3}}$, *a minimum. At these two points (and nowhere else) the tangent to the curve is parallel to the x-axis and its slope to the latter is zero. (As here drawn, the unit of the y-axis is half that of the x-axis.)*

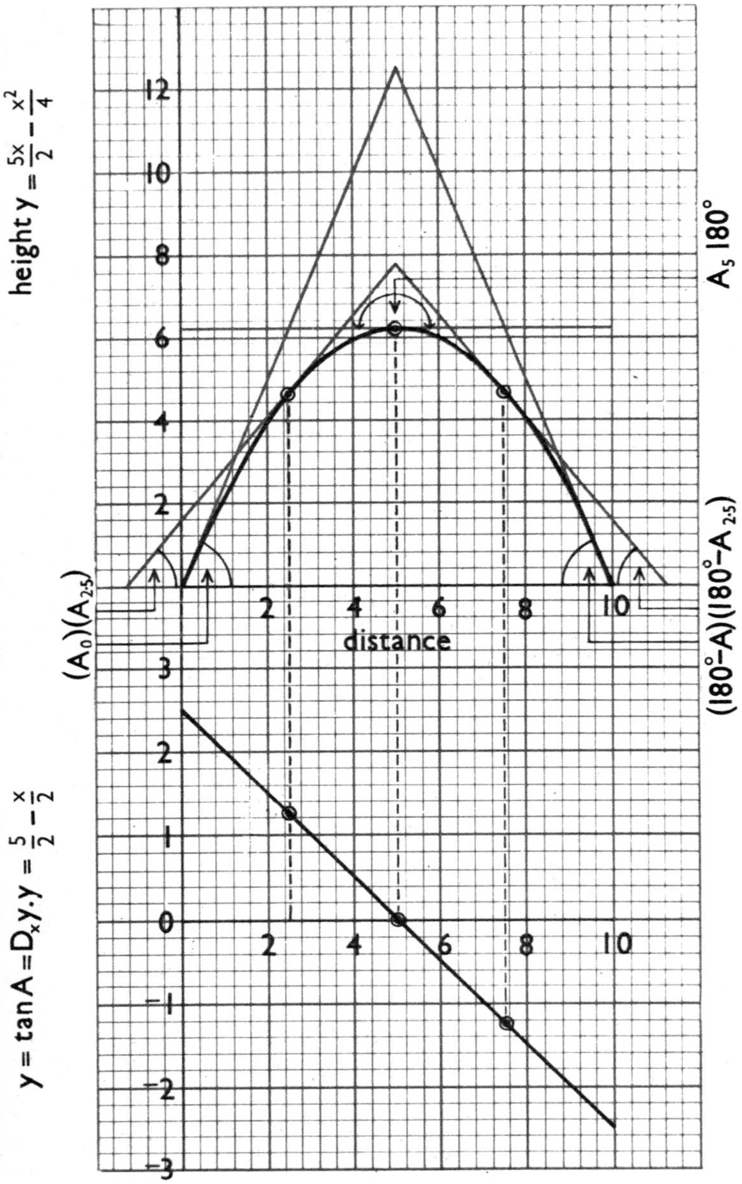

Turning-points of a big dipper track.

Implicit in Fermat's notion of finding a turning-point is the principle that the tangent of the slope to a point (x, y) on the curve of $y=f(x)$ which represents a differentiable function of x is itself a continuous function of x. For the trajectory of the cannon ball, the tangent is a linear *function of x, so that the vertical displacement is directly proportional to the horizontal throughout its path. One may here put in any plausible values for the units of x (distance) and y (height).*

The equation plotted left is $y=\frac{5x}{2}-\frac{x^2}{4}$.

The values exhibited by dots are:

x	y	$\tan A$
0	0	$\frac{5}{2}$
2.5	$4\frac{11}{16}$	$\frac{5}{4}$
5.0	$6\frac{1}{4}$	0
7.5	$4\frac{11}{16}$	$-\frac{5}{4}$
10.0	0	$-\frac{5}{2}$

Note the meaning of the minus sign of the slope in round-the-clock trigonometry:

$$\tan(180°-A)=\frac{\sin(180°-A)}{\cos(180°-A)}$$

$$=\frac{+\sin A}{-\cos A}=-\tan A$$

DISTANCE-TIME GRAPH

Here the independent variable is time (t). The dependent distance $d=f(t)$. An object moves from Q to P in time (t_2-t_1) seconds at increasing speed. If it moved the same distance at fixed speed, it would move through (d_2-d_1) feet in (t_2-t_1) seconds at a speed (s) such that

$$s=\frac{d_2-d_1}{t_2-t_1}=\tan A.$$

Thus the gradient of the chord PQ is its mean speed over the interval t_1 to t_2.

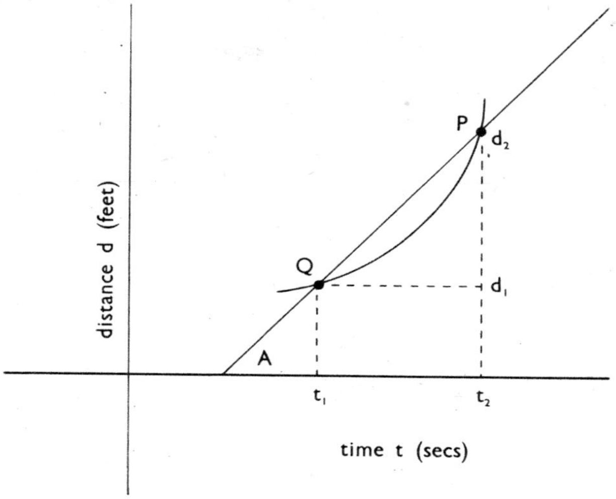

Here we imagine that two points Q and R lie close together on the curve of $y=f(x)$ in the neighbourhood of a point P so situated between Q and R that the x-co-ordinates of Q and R are equidistant from x. The tangent at P and the slope of the chord QR to the x-axis differ less and less as we make the interval $\triangle x$ smaller. In the neighbourhood of the point x, y we may write the mean gradient as

$$\frac{f(x+\frac{1}{2}\triangle x)-f(x-\frac{1}{2}\triangle x)}{\triangle x}=\tan A=\frac{\triangle y}{\triangle x}$$

This is approximately the slope of the tangent at P.

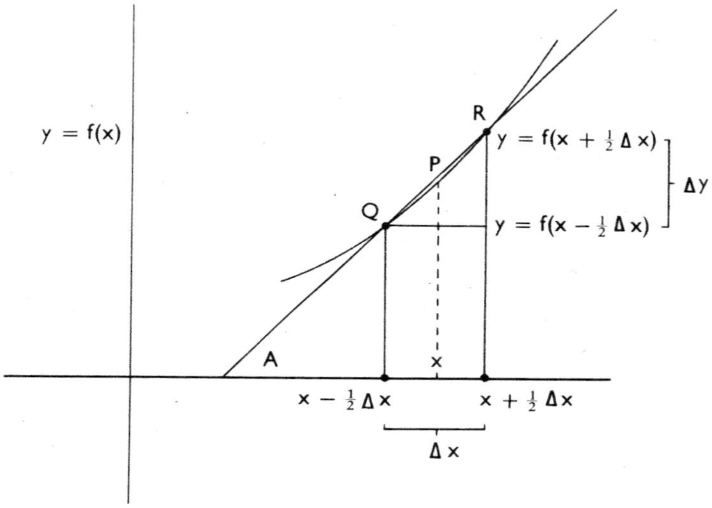

When the point R lies very near P, the gradient of the curve at P is approximately that of the chord PR. If the x-co-ordinate of R is $x+\triangle x$, we may write its gradient as

$$\frac{f(x+\triangle x)-f(x)}{\triangle x}=\tan A=\frac{\triangle y}{\triangle x}$$

In the limit, this is equivalent to

$$\frac{f(x+\frac{1}{2}\triangle x)-f(x-\frac{1}{2}\triangle x)}{\triangle x}$$

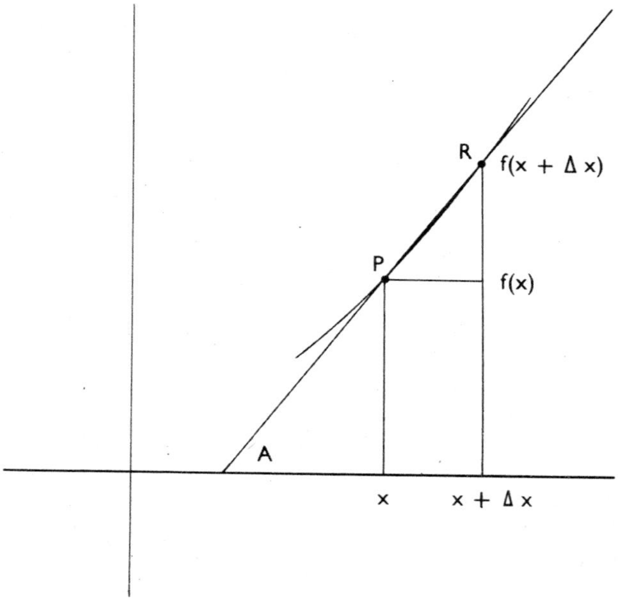

For the slope to the x-axis of the chord through Q and R we may therefore write:

$$\frac{f(x+\frac{1}{2}\Delta x)-f(x-\frac{1}{2}\Delta x)}{(x+\frac{1}{2}\Delta x)-(x-\frac{1}{2}\Delta x)} \equiv \frac{\Delta y}{\Delta x}$$

The slope to the x-axis of the tangent at P is the *limiting* value this ratio assumes as Δy and Δx approach zero. Thus we define symbolically the *derivative* in terms consistent with our verbal definition as:

$$D_x.y \equiv \underset{\Delta x \to 0}{Lt} \frac{\Delta y}{\Delta x}$$

We are now ready to exhibit the derivative of several simple functions which we have already met in Chapter 9 or elsewhere, and hence to take a more comprehensive view of the peculiarities of their descriptive curves without recourse to pictorial representation:

(i) $y=x^n$; (ii) $y=\sin x$; (iii) $y=\cos x$

In terms of our definition, for $y=x^n$:

$$\frac{\Delta y}{\Delta x}=\frac{(x+\frac{1}{2}\Delta x)^n-(x-\frac{1}{2}\Delta x)^n}{(x+\frac{1}{2}\Delta x)-(x-\frac{1}{2}\Delta x)}$$

If we write $a=(x+\frac{1}{2}\Delta x)$ and $b=(x-\frac{1}{2}\Delta x)$, the limit of this when we make Δx as near to zero as we care to is one already discussed in Chapter 4 on the assumption that n is a positive whole number, *viz.*:

$$\underset{b \to a}{Lt} \frac{a^n-b^n}{a-b} = na^{n-1}$$

On the foregoing assumption with reference to n, we may therefore write for the derivative of $y=x^n$:

$$\underset{\Delta x \to 0}{Lt} \frac{\Delta y}{\Delta x}=D_x(x^n)=nx^{n-1}$$

By painting in an undefined constant C throughout, it is easy enough to derive:

$$D_x(x^n+C)=nx^{n-1}$$

For instance, if $y=x^2+2$,

$$D_x(y)=2x$$

In terms of Fermat's principle, this has only one turning-point at $D_x(y)=0$, so that $2x=0$ and $x=0$. In that event $y=2$. The co-ordinates of the turning-point are therefore (0, 2).

We now turn to the periodic function $y=\sin x$, and write:

$$\frac{\Delta y}{\Delta x}=\frac{\sin(x+\frac{1}{2}\Delta x)-\sin(x-\frac{1}{2}\Delta x)}{(x+\frac{1}{2}\Delta x)-(x-\frac{1}{2}\Delta x)}$$

By recourse to the sum and difference formulae:

$$\sin(x+\tfrac{1}{2}\Delta x)=\sin x.\cos \tfrac{1}{2}\Delta x+\sin \tfrac{1}{2}\Delta x.\cos x$$

$$\sin(x-\tfrac{1}{2}\Delta x)=\sin x.\cos \tfrac{1}{2}\Delta x-\sin \tfrac{1}{2}\Delta x.\cos x$$

$$\frac{\Delta y}{\Delta x}=\frac{2 \sin \frac{1}{2}\Delta x.\cos x}{\Delta x}=\frac{\sin \frac{1}{2}\Delta x.\cos x}{\frac{1}{2}\Delta x}$$

Now we have seen (Chapter 5) that $\sin x \simeq x$ when $x \simeq 0$ (in radian measure), whence $\sin \frac{1}{2}\Delta x \div \frac{1}{2}\Delta x \simeq 1$ as $\Delta x \simeq 0$; and

$$\underset{\Delta x \to 0}{Lt} \frac{\Delta y}{\Delta x}=D_x(\sin x)=\cos x$$

We may leave it to the reader to solve $F(x)=D_x.\cos x$ with the clue:

$$\cos (x \pm \tfrac{1}{2}\Delta x)=\cos x.\cos \tfrac{1}{2}\Delta x \mp \sin x.\sin \tfrac{1}{2}\Delta x$$

Whence we obtain:

$$D_x(\cos x)=-\sin x$$

These results should not surprise us. The slope of the curve representing the periodic function $y=\sin x$ clearly varies in a regular way with the same period as $\sin x$, having turning-points at $x=0$ and $x=\frac{\pi}{2}$, or more generally $x=\frac{n\pi}{2}$ $(n=1, 2\ldots)$. Also, alternate turning-points correspond to positive and negative values of $y=\sin x$. If we apply Fermat's principle $\cos x=0$ when $x=\frac{\pi}{2}$ or $\frac{3\pi}{2}$, or more generally $x=\frac{(2n+1)}{2}\pi$ $(n=0, 1, 2\ldots)$. When $x=\frac{\pi}{2}$, $\sin x=+1$ and when $x=\frac{3\pi}{2}$, $\sin x=-1$. These represent the first two turning-points, a maximum when $\sin x=+1$ and a minimum when $\sin x=-1$.

Hereby hangs another tale. The solution of Fermat's equation ($D_x.y=0$) tells us the co-ordinates of the turning-point (or points) of a curve, but it leaves open the question: is a particular turning-point a maximum or a minimum in the sense defined above? To answer the question, let us first recall a rule of round-the-clock trigonometry, *viz.*: $\tan(180°-A)=-\tan A$. When $\tan A$ is positive, the tangent slopes to the x-axis upwards towards the right, and the value of y is increasing. When $\tan A$ is negative, it slopes upwards to the left and the value of y is diminishing. Thus $D_x.y$ changes from a positive to a negative value through zero in the neighbourhood of a maximum and from a negative to a positive value in the neighbourhood of a minimum.

This consideration leads us to explore the possibility of expressing in functional form the rate of change of $D_x.y$ itself. Since the slope of the tangent to the x-axis of the curve whose function is $y=f(x)$ is amenable to pictorial representation as the curve of a function $F(x)$, we may expect that it is possible to represent the slope to the x-axis of the latter by the curve of a third function $F^2(x)=D_x^2.y$. To preserve uniformity, we may here pause to write:

$D_x^0.y\equiv y=f(x)$ means we have not yet performed the operation of finding the derivative of $y=f(x)$;

$D_x^1.y\equiv D_x.y=F(x)$ means we have once performed the operation of finding the derivative of $y=f(x)$;

$D_x^2.y\equiv D_x.F(x)$ means we have twice performed the operation of finding the derivative of $y=f(x)$.

Similarly, we may define higher (third, fourth, *etc.*) derivatives $D_x^3.y$, $D_x^4.y$, *etc.* However, only one derivative higher than the first need concern us here, *viz.*:

$$D_x^2.y\equiv D_x.F(x) \text{ if } F(x)=D_x.y$$

At a maximum $F(x)$ has begun to decrease and $D_x^2.y$ will be negative; but at a minimum $F(x)$ is beginning to increase and $D_x^2.y$ will be positive. Consider the function $y=x^2+2$. Since $D_x.y=2x$ and $D_x.y=nx^{n-1}$ when $y=x^n$:

$$D_x^2.(x^2+2)=D_x(2x)=2$$

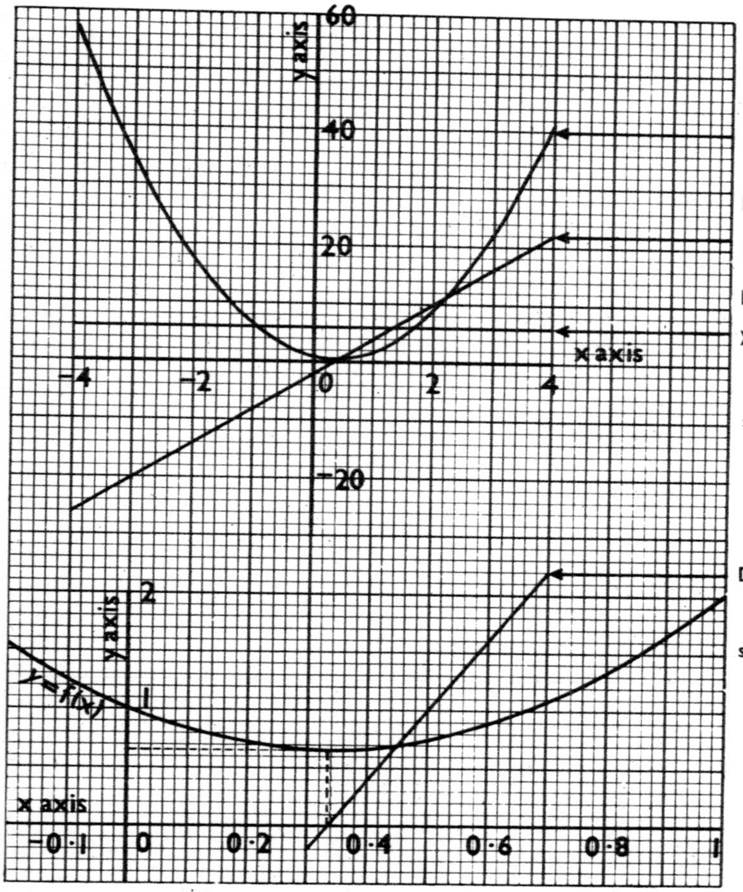

$D_x^1.y=0.\ x=0{\cdot}\dot{3}$ which is minimum for $y=f(x)$ when $0.67>y>0{\cdot}66$

We leave it to the reader to find out why the straight lines $y=6x-2$ and $y=6$ cut the curves where they do.

Since this is positive, the turning-point whose coordinates are (0, 2) is a minimum and the curve is symmetrical about the y-axis, bending upwards in either direction at $y=2$. If we look at the function $y=\sin x$ from the same viewpoint, we notice that:

$$D_x^2.\sin x = D_x.\cos x = -\sin x$$

We have seen that the first turning-point is at $\cos x=0,\ x=\frac{\pi}{2}$. Since the second derivative is negative, this is a maximum. It follows from the foregoing remarks that we may attach a unique geometrical meaning to the condition that the second derivative vanishes. The equation $D_x^2.y=0$ will be true at a *point of inflexion*, *i.e.* when the course of the curve is changing from concave upwards to concave downwards or *vice versa*.

In dynamical language, $D_x^2.y$ has a meaning of special interest, because $D_x.y$ signifies rectilinear speed when x stands for time and y for distance. With the same interpretation of x and y, the derivative of $D_x.y$ is the rate of change of $D_{x.y}$, *i.e.* its increase per unit increase of time (x). By definition, this is an acceleration. Thus the meaning of the three formulae discussed above is dynamically translatable as follows:

$D_x^0(x^2+2)=x^2+2$	rectilinear distance proportionate to the square of the time.
$D_x^1(x^2+2)=2x$	rectilinear speed proportionate to the time itself.
$D_x^2(x^2+2)=2$	acceleration constant.

The derivatives discussed so far depend on limits already disclosed. To proceed further, we shall need some general recipes in which C, K and m stand for unspecified real numerical values. The proofs of the first three follow the pattern of the foregoing derivations.

Rule 1 $\quad D_x.K.f(x) \equiv K.D_xf(x)$

Examples (i) $D_x(5x^3)=5.D_x(x^3)=15x^2$

(ii) $D_x(-\cos x)=(-1).D_x.\cos x = \sin x$

Rule 2 $\quad D_x[f(x)+C] \equiv D_x.f(x)$

Examples (i) $D_x(x^n+5)=D_x(x^n)=nx^{n-1}$

(ii) $D_x(7-\cos x)=D_x(-\cos x)=\sin x$

Rule 3 $\quad D_x[f_1(x)+f_2(x)+f_3(x)\ldots] \equiv D_xf_1(x)+D_xf_2(x)+D_xf_3(x)\ldots$

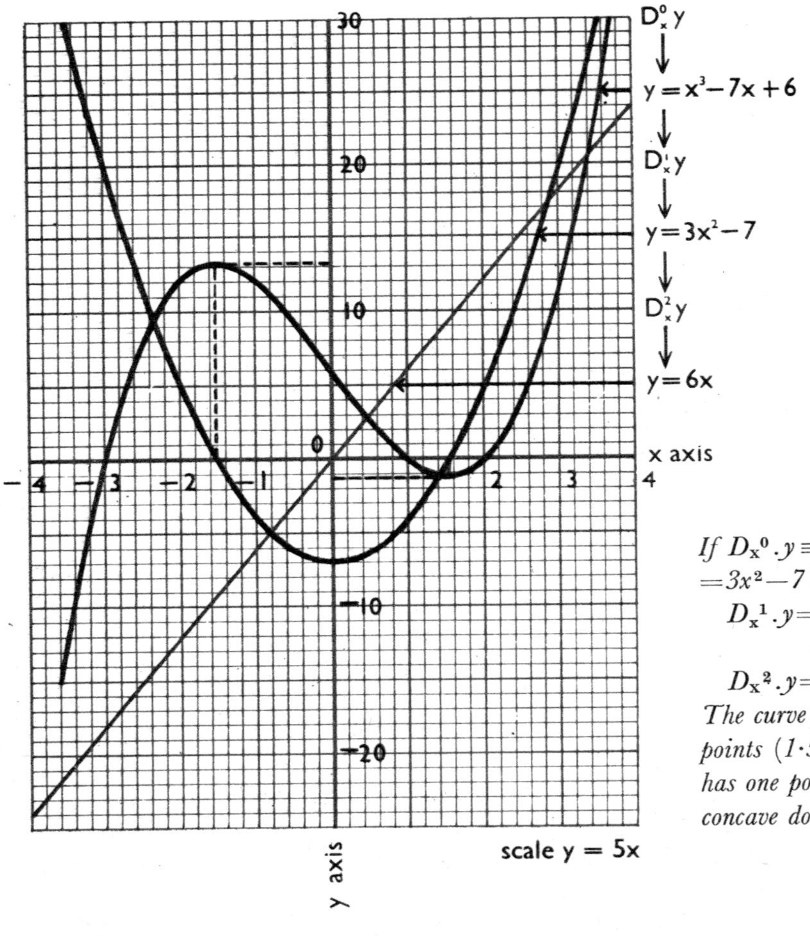

If $D_x^0.y \equiv y = x^3-7x+6$, so that $D_x^1.y \equiv D_x.y = 3x^2-7$ and $D_x^2.y=6x$:

$D_x^1.y=0$ when $x \simeq +1{\cdot}52$ (and $y \simeq -1{\cdot}2$) or $x \simeq 1{\cdot}52$ (and $y \simeq +13{\cdot}2$)

$D_x^2.y=0$ when $x=0$ (and $y=6$)

The curve $y=x^3-7x+6$ has therefore two turning-points ($1{\cdot}52$, $-1{\cdot}2$) and ($-1{\cdot}52$, $+13{\cdot}2$) and it has one point of inflexion (0, 6) when it changes from concave downwards to concave upwards.

Examples (i) $D_x(x^3+x^2)=3x^2+2x$

(ii) $D_x(\sin x+\cos x)=\cos x-\sin x$

Rule 4 If y and z are each functions of x, so that we may write: $z=f(y)$ and $y=f(x)$, and

$$\frac{\Delta z}{\Delta x}=\frac{\Delta z}{\Delta y}\cdot\frac{\Delta y}{\Delta x}$$

In the limit: $D_x.z\equiv(D_y.z)\ (D_x.y)$

We may illustrate this first by a result which we could obtain more simply from the formula $D_x(x^n)=nx^{n-1}$. If $y=x^2$ and $z=x^4=y^2$:

$$D_x(x^4)=D_y(y^2).D_x(y)$$
$$=2y.2x=4xy=4x^3$$

A simple application of Rule 4 is the following:

$$D.f(mx)=D_{mx}.f(mx).D_x.f(mx)$$
$$=m.D_{mx}.f(mx)$$

Example $D_x(\cos 3x)=D_{3x}.\cos 3x.D_x(3x)$

$$=(-\sin 3x)\ (3)$$
$$=-3\sin 3x$$

Rule 5 As the points A and B of our differential triangle approach more closely, it becomes less and less material whether we define the slope alternatively as:

$$\frac{f(x+\frac{1}{2}\Delta x)-f(x-\frac{1}{2}\Delta x)}{(x+\frac{1}{2}\Delta x)-(x-\frac{1}{2}\Delta x)}$$

or

$$\frac{f(x+\Delta x)-f(x)}{(x+\Delta x)-x}$$

If $y=f(x)$, we may thus write $f(x+\Delta x)=y+\Delta y$; and this is necessary to establish the rule for dealing with a product (yz) of two functions $y=f_1(x)$ and $z=f_2(x)$. We suppose that y becomes $y+\Delta y$ and z becomes $z+\Delta z$ while x increases to $x+\Delta x$, so that

$$\Delta(yz)=(y+\Delta y)\ (z+\Delta z)-yz$$
$$=y\Delta z+z\Delta y+\Delta y.\Delta z$$

$$\frac{\Delta(yz)}{\Delta x}=\frac{y\Delta z}{\Delta x}+\frac{z\Delta y}{\Delta x}+\Delta y.\frac{\Delta z}{\Delta x}$$

In the limit the last term vanishes, so that

$$D_x(yz)=y.D_x.z+z.D_x.y$$

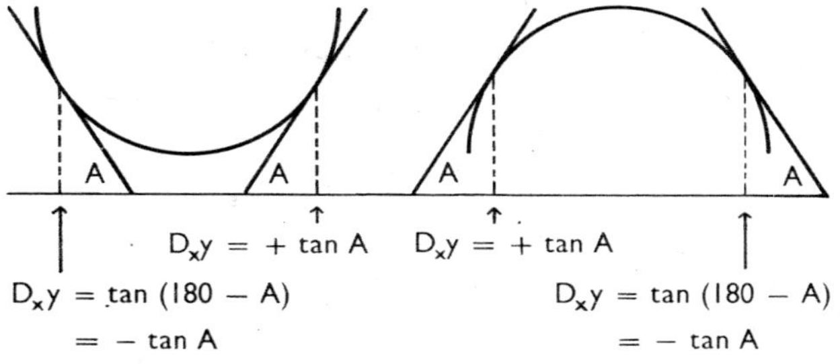

At a minimum
D_xy changes from negative to positive through zero ∴ $D_x{}^2y$ is positive.

At a maximum
D_xy changes from positive to negative through zero ∴ $D_x{}^2y$ is negative.

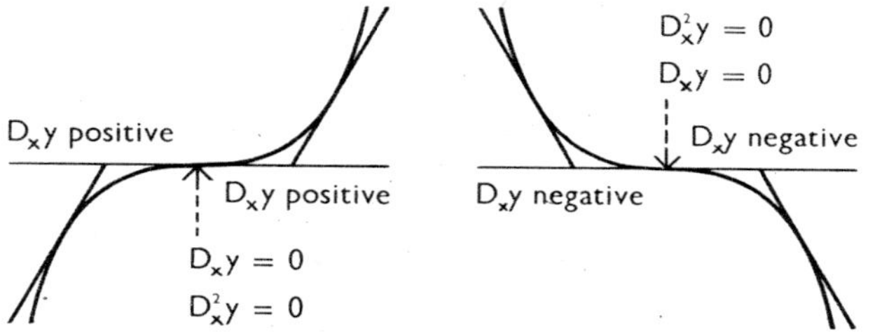

The first derivative may also vanish at a point of inflexion, *in which event D_xy changes from positive to positive or negative to negative through zero, and $D_x{}^2y=0$.*

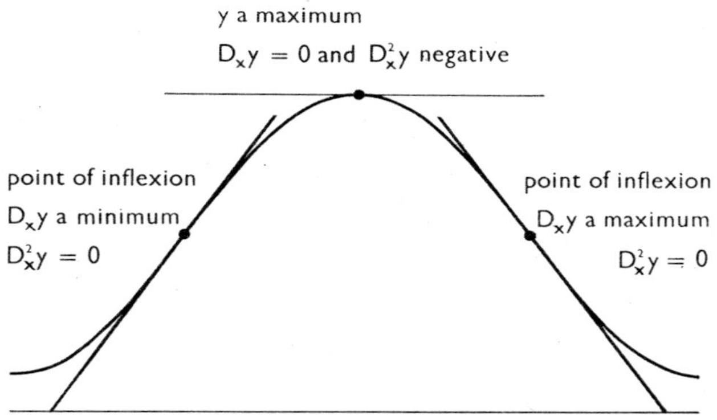

At a point of inflexion when the curve changes from concave upwards to concave downwards (or vice versa), the numerical value of the second derivative is zero.

As a simple example, we know that $D_x(x^5)=5x^4$, also $x^5=x^2.x^3$:

$$D_x(x^5)=x^3.D_x(x^2)+x^2.D_x(x^3)$$
$$=x^3(2x)+x^2(3x^2)=5x^4$$

In the same way, we can *differentiate*, *i.e.* find the x-derivative of, $(\cos x)^2$, *viz.*:

$$D_x(\cos x)^2=\cos x.D_x(\cos x)+\cos x.D_x(\cos x)$$
$$=-2\cos x.\sin x$$

We can however use Rule 4 to get the same result if we write $z=(\cos x)^2$ and $y=\cos x$, so that $z=y^2$ and

$$D_x(\cos x)^2=D_y(y^2).D_x.y=2y(-\sin x)$$
$$=-2\cos x.\sin x$$

Sometimes we can use Rules 4 and 5 with equal ease, *e.g.* if $y=\cos x.\sin x$, $y=\frac{1}{2}\sin 2x$ and by Rule 4

$$D_x.y=\tfrac{1}{2}D_{2x}(\sin 2x).D_x(2x)=\cos 2x$$

Similarly, by Rule 5

$$D_x.y=\cos x.D_x.\sin x+\sin x.D_x.\cos x$$
$$=(\cos x)^2-(\sin x)^2=\cos 2x$$

With the foregoing rules at our disposal, we may obtain numberless derivatives other than those mentioned. We can also remove a restriction on $D_x(x^n)$ hitherto interpreted in terms of n as a positive integer. We shall first consider its meaning when $n=-1$, so that

$$\Delta(x^{-1})=\frac{1}{x+\frac{1}{2}\Delta x}-\frac{1}{x-\frac{1}{2}\Delta x}=\frac{-\Delta x}{x^2-\frac{1}{4}\Delta x^2}$$

$$\frac{\Delta(x^{-1})}{\Delta x}=\frac{-1}{x^2-\frac{1}{4}\Delta x^2}$$

In the limit therefore:

$$D_x(x^{-1})=\frac{-1}{x^2}=-x^{-2}$$

This result is consistent with the rule for positive whole number values of n, *viz.* $D_x(x^n)=nx^{n-1}$. We now apply Rule 4 to the more general case of any negative integer, *i.e.* $y=x^{-n}$. If we write $z=x^n=y^{-1}$:

$$\frac{\Delta y}{\Delta x}=\frac{\Delta y}{\Delta z}\,\frac{\Delta z}{\Delta x}$$

In the limit, the second factor on the right becomes nx^{n-1}; and we have already shown that

$$D_z.y=-y^{-2}=-x^{-2n}$$

Thus we obtain

$$D_x(x^{-n})=-x^{-2n}.nx^{n-1}=-nx^{-n-1}$$

This conforms to precisely the same pattern as the rule for positive integers. In the same way, we can remove the restriction that n must be an integer. If n is a rational fraction, we may write $y=x^{\frac{p}{q}}$ and put $z=x^{\frac{1}{q}}$, so that $x=z^q$ and $y=z^p$, whence by Rule 4

$$D_x.y=D_z(y).D_x(z)=pz^{p-1}.D_x.z$$

The expression $D_x.z$ is the limit of

$$\frac{\Delta z}{\Delta x}=\frac{1}{\frac{\Delta x}{\Delta z}}\simeq\frac{1}{D_z(x)}=\frac{1}{qz^{q-1}}$$

Whence we have

$$D_x(x^{\frac{p}{q}})=\frac{pz^{p-1}}{qz^{q-1}}=\frac{p}{q}.\frac{x^{\frac{p-1}{q}}}{x^{\frac{q-1}{q}}}=\frac{p}{q}x^{\frac{p}{q}-1}$$

We shall defer to a later chapter the differentiation of some functions already mentioned, such as $y=\log_b x$; but shall here take stock of another useful application of Rules 4 and 5 above:

$$D_x(\tan x)=D_x(\sin x)(\cos x)^{-1}$$
$$=\sin x.D_x(\cos x)^{-1}$$
$$+(\cos x)^{-1}.D_x(\sin x)$$
$$=(\sin x)^2(\cos x)^{-2}+(\cos x)^{-1}(\cos x)$$
$$\therefore D_x(\tan x)=(\tan x^2)+1$$

If we write $y=\tan x$, this is equivalent to writing $D_x.y=1+y^2$. We have already seen that in the limit:

$$\frac{\Delta x}{\Delta y}=D_y.x$$

If $y=\tan^{-1}x$ so that $x=\tan y$:

$$D_y.x=(\tan y)^2+1=(x^2+1)\text{ and }D_x.y=\frac{1}{D_y.x}$$

$$\therefore D_x(\tan^{-1}x) = \frac{1}{1+x^2}$$

The reader may now be able to evaluate:

$$D_x(\sin^{-1}x) = \frac{1}{\sqrt{(1-x^2)}}$$

THE ANTI-DERIVATIVE. Let us now leave the derivative and examine the inverse operation. Accordingly, we shall ask the following question: *If given the equation for the slope of a curve, how much can we infer about the function represented by the latter?* When we consider the case $D_x . y=6x$, our question therefore signifies: what can we say about y in terms of x if $6x$ is its derivative? First, we know that $D_x . y=6x$ if $y=3x^2$; but Rule 2 signifies that $D_x(3x^2+C)=6x$. It is usual to write this in the form:

$$\int 6x . dx=3x^2+C$$

For a reason we shall see later, this notation, in which one calls the expression on the right the *indefinite* integral of $6x$, is misleading. The question to which the equation contains the answer is: what is the function of which $6x$ is the derivative, or by analogy with the *anti-logarithm*, what is the *anti-derivative* of $6x$? It is consistent with the use of the notation $\sin^{-1}x$, *etc.* to write this as $y=D_x^{-1}(6x)$, so that

$$D_x^{-1}(6x)=3x^2+C$$

If we interpret the sign for an operation as an instruction to do something, we may therefore say:

$D_x . y$ means find the function $F(x)$ which is the derivative of the function $y=f(x)$.

$D_x^{-1} . (y)$ means find the function $f(x)$ whose derivative is $y=F(x)$.

Before we examine the implications of the verb *find* in this context, let us ask what significance the anti-derivative has in the domain of Newtonian dynamics. With that end in view, we may return to the parabolic trajectory of a sixteenth-century cannon ball with a horizontal range of 220 yards and maximal vertical height of 55 yards. The appropriate equation in which the unit of both the x- and the y-axis is 1 yard is as follows:

$$y=x-\frac{x^2}{220}$$

$$\therefore D_x . y=1-\frac{x}{110}$$

$$D_x^{-1} . y=1-\frac{x}{110} \text{ if } y=x-\frac{x^2}{220}+C$$

The last expression on the right answers the question: what is the equation which connects the vertical with the horizontal displacement at every point in the range? As it stands, the answer is indefinite, because we have left on our hands an arbitrary (so-called *integration*) constant C. However, we can commonly dispose of this in a factual situation. For example, we may here agree to locate the mouth of the cannon for convenience at the origin of the curve. Then $y=0$ when $x=0$; and the implication is that $C=0$.

In the context of the Galilean cannon ball, it is instructive to consider the parametric equations connecting the two displacements with time in seconds, *viz.*:

$x=k.t$ so that $D_t(x)=k$ and $D_t^2(x)=0$
$y=C.t-B.t^2$ so that $D_t(y)=C-2B.t$
and $D_t^2(y)=-2B$

(a) the horizontal speed is constant, whence the horizontal acceleration is zero;
(b) the vertical speed is a linear function of the time, whence the vertical acceleration (downwards) is constant.

If we know that the vertical speed is $C-2B.t$ we can derive the parametric equation. Since $D_t(Bt^2)=2Bt$ and $D_t(Ct)=C$,

$$D_t^{-1}(y)=C-2Bt \text{ if } y=Ct-Bt^2+K$$

Here K is the so-called *integration constant*. If we place the origin so that $t=0$ at the ground-level explosion, $y=0=t$ and $K=0$. However, we are free to reckon time from 5 seconds before firing, in which event $y=0$ when $t=5$, so that

$$5C=25B-K=0 \text{ and } K=5(5B-C)$$

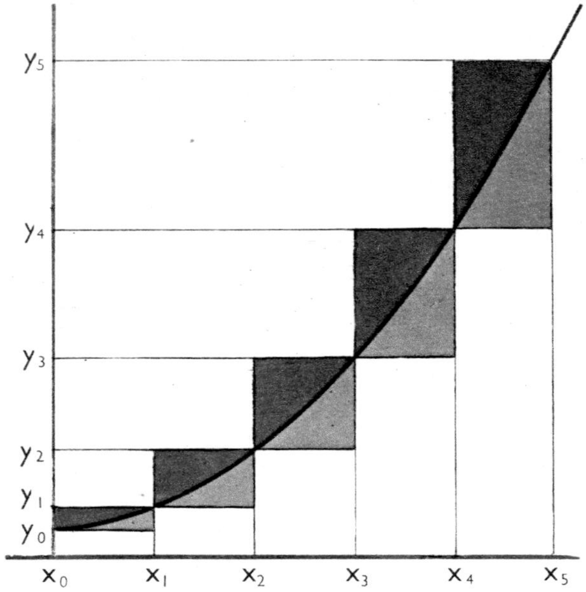

If A is the area bounded by the curve between the x-axis and the co-ordinates y_0 *and* y_5 *and* $\triangle x$ *is the width of each rectangular strip:*

$$A > y_0 \triangle x + y_1 \triangle x + y_2 \triangle x + y_3 \triangle x + y_4 \triangle x$$
$$A < y_1 \triangle x + y_2 \triangle x + y_3 \triangle x + y_4 \triangle x + y_5 \triangle x$$

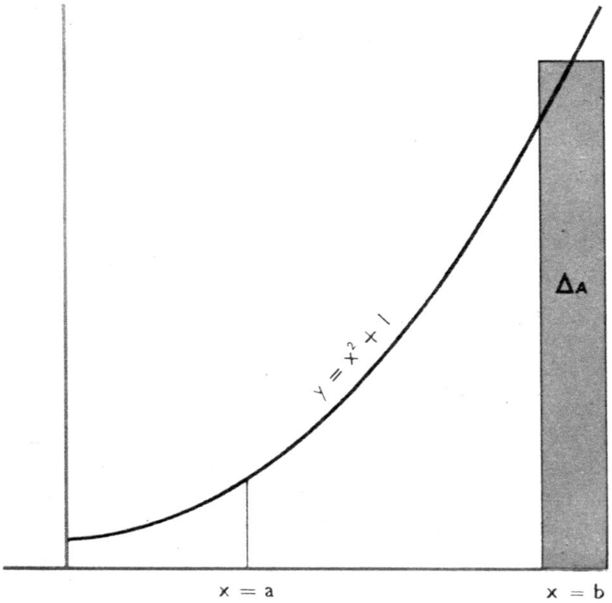

$$\triangle A = (x^2+1)\triangle x = x^2 \triangle x + \triangle x$$
$$A = D_x^{-1}(x^2+1)$$
$$= \frac{x^3}{3} + x + C$$

In a dynamical situation, the fact that the anti-derivative contains an undetermined constant need therefore cause no inconvenience. We can interpret it in terms of where we choose to locate the origin.

In connexion with the anti-derivative (so-called *indefinite integral*) what now especially remains for discussion is the meaning of the verb *find* when we say:

$D_x^{-1}(y) = F(x)$ means: find the function $f(x)$ whose derivative is $y = F(x)$.

In the last resort, we only know this by working backwards from a table of derivatives. For instance, we know that $D_x(x^n + C) = nx^{n-1}$, whence:

$$D_x . y = x^n \text{ if } y = \frac{1}{n+1}x^{n+1} + C$$

$$\therefore D_x^{-1}(x^n) = \frac{1}{n+1}x^{n+1} + C$$

By working backwards in this way, we may tabulate some results recorded to date as follows:

y	$D_x . y$	$D_x^{-1} . y$
$Ax^n + K$	Anx^{n-1}	$\frac{A}{n+1}x^{n+1} + Kx + C$
$A \sin (Kx)$	$A.K. \cos (Kx)$	$-\frac{A}{K} \cos (Kx) + C$
$A \cos (Kx)$	$-A.K. \sin (Kx)$	$\frac{A}{K} \sin (Kx) + C$
$\frac{1}{1+x^2}$	$\frac{-2x}{(1+x^2)^2}$	$\tan^{-1}x + C$

Having tabulated results of this sort, we have taken the first step towards solving what one calls a *differential equation*, *i.e.* an equation in which a derivative occurs as an unknown. Indeed, the task of *finding* the anti-derivative of a function is the task of solving an equation of this sort, and as such primarily the type of problem which one solves at the most elementary level by looking up the entries of a table. Such equations play an important role in dynamical, electromagnetic and acoustic theory. In accordance with the way we interpret x, the following may thus describe the horizontal speed of a pendulum, the speed of propagation of a region of compression in an organ pipe or the rate of voltage change in an alternating circuit:

$$D_t(x) = A \sin (Kt)$$

The solution is

$$x = D_x^{-1}(A \sin Kt) = -\frac{A}{K} \cos (Kt) + C$$

As explained, we can fix the constant C in the above by the definition of the origin with respect to t.

Though the task of evaluating the anti-derivative depends in the last resort on making a table with entries as above, it is often impossible to make use of such a table unless we rely on rules for breaking a function down into components recognisable in terms of table entries. The following are the most important. As before, k, K and C stand for numerical constants.

Rule 1 $\quad D_x^{-1}.K.f(x) = K.D_x^{-1}.f(x)$

Example $D_x^{-1}(3 \cos x) = 3.D_x^{-1}(\cos x) = 3 \sin x + C$

Rule 2 $\quad D_x^{-1}.f(x) + C = [D_x^{-1}.f(x)] + Cx + K$

Example $D_x^{-1}.\frac{1}{2}(1 + \cos x) = \frac{1}{2}x + \frac{1}{2} \sin x + K$

Rule 3 $\quad D_x^{-1}[f_1(x) + f_2(x) + f_3(x)\,..]$
$\quad = D_x^{-1}.f_1(x) + D_x^{-1}.f_2(x) + D_x^{-1}.f_3(x)\,..$

Example Since $(x^2+1)^{-1} = 1 - x^2 + x^4 - x^6 + x^8\,..$

$$D_x^{-1}(1+x^2)^{-1}$$
$$= D_x^{-1}(1 - x^2 + x^4 - x^6 + x^8\,..)$$
$$= x - \frac{x^3}{3} + \frac{x^5}{5} - \frac{x^7}{7} + \frac{x^9}{9}\,.. + K$$

Rule 4 $\quad D_x^{-1}.f(kx) = \frac{1}{k} D_{kx}.f(kx)$

Example $D_x^{-1}(\cos 5x) = \frac{1}{5} \sin 5x + C$

Rule 5 This depends on the rule that if z and u are both functions of x, such that

$$z = D_x^{-1}.f(x) \text{ whence } D_x.z = f(x)$$

By Rule 4 for the derivative:

$$D_u z = D_x z.D_u x = f(x).D_u x$$
$$z = D_u^{-1}.f(x).D_u x$$

To apply this successfully, we need to guess judiciously a function $u = F(x)$ which makes the product within the brackets of the expression in the middle a recognisable anti-derivative. We first note by definition $z = D_x^{-1}.f(x)$ is equivalent to $D_x.z = f(x)$.

Example (i) $D_x^{-1}.x(1+x^2)^{\frac{1}{2}} = z$

Here $f(x) = x(1+x^2)^{\frac{1}{2}} = D_x z$. We note that $2x = D_x(1+x^2)$ and accordingly try out $u = 1 + x^2$ so that $D_x u = 2x$ and $D_u x = \frac{1}{2x}$.

$$D_x z.D_u x = \frac{x(1+x^2)^{\frac{1}{2}}}{2x} = \frac{(1+x^2)^{\frac{1}{2}}}{2} = \frac{u^{\frac{1}{2}}}{2}$$

$$D_x^{-1}.x(1+x^2)^{\frac{1}{2}} = D_u^{-1}\frac{u^{\frac{1}{2}}}{2} = \frac{(1+x^2)^{\frac{3}{2}}}{3} + C$$

Example (ii) $D_x^{-1}(k^2 - x^2)^{\frac{1}{2}} = z = D_x^{-1}.k\left(1 - \frac{x^2}{k^2}\right)^{\frac{1}{2}}$

Since $1 - (\sin a)^2 = (\cos a)^2$, a possible substitution to explore is $x = k \sin a$, so that $a = \sin^{-1}\frac{x}{k}$. We shall then write:

$$1 - \frac{x^2}{k^2} = 1 - (\sin a)^2 = (\cos a)^2$$

$$D_x z = k \cos a \text{ and } D_a x = k \cos a$$
$$D_x z.D_a x = k^2(\cos a)^2$$

From the addition formulae, we have

$$(\cos a)^2 = \tfrac{1}{2}(1 + \cos 2a) \text{ and}$$
$$\sin 2a = 2 \sin a \sqrt{1 - (\sin a)^2}$$

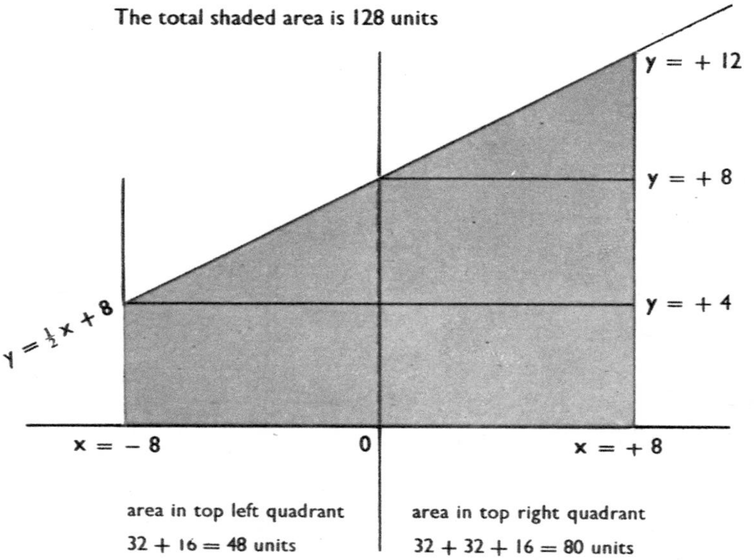

QUADRATURE IN MORE THAN ONE QUADRANT

Whence

$$z = D_a^{-1}(D_x z \, . \, D_a x) = \frac{k^2}{2} D_a^{-1}(1+\cos 2a)$$

$$= \frac{k^2 a}{2} + \frac{k^2}{2}\sin 2a + C$$

$$= \frac{k^2}{2}\sin^{-1}\left(\frac{x}{k}\right) + kx\sqrt{\left[1-\left(\frac{x}{k}\right)^2\right]} + C$$

Rule 6 If $v = f_1(x)$ and $w = f_2(x)$,

$$D_x^{-1} vw = vD_x^{-1}w - D_x^{-1}[D_x v \, . \, D_x^{-1}w]$$

Example $z = D_x^{-1} . x(\cos x)$

We write $v = x$ and $w = \cos x$, so that

$$D_x v = 1 \text{ and } D_x^{-1}w = \sin x$$

$$D_x^{-1}(x \, . \cos x) = x \sin x - D_x^{-1}(\sin x)$$

$$= x \sin x + \cos x + C$$

Rule 6 for manipulating the anti-derivative of a product function follows from the rule for manipulating that of the derivative. If u, v and w are each differentiable functions of x:

$$D_x(uv) = uD_x v + vD_x u$$

$$uv = D_x^{-1}(uD_x v) + D_x^{-1}(vD_x u)$$

If we put $D_x . u = w$ so that $u = D_x^{-1} . w$

$$v \, . \, D_x^{-1}w = D_x^{-1}(D_x^{-1} . w \, . \, D_x v) + D_x^{-1}(vw)$$

$$\therefore D_x^{-1}(vw) = v . D_x^{-1}w - D_x^{-1}(D_x^{-1}w \, . \, D_x v)$$

Only practice can confer *expertise* in manipulating these rules. Before we take leave of them, one application of Rule 3 above merits comment. We have seen that

$$\tan^{-1}x = D_x^{-1}(1+x^2)^{-1}$$

$$= D_x^{-1}(1 - x^2 + x^4 - x^6 \, . \, .)$$

$$= \left(x - \frac{x^3}{3} + \frac{x^5}{5} - \frac{x^7}{7} \cdots\right) + C$$

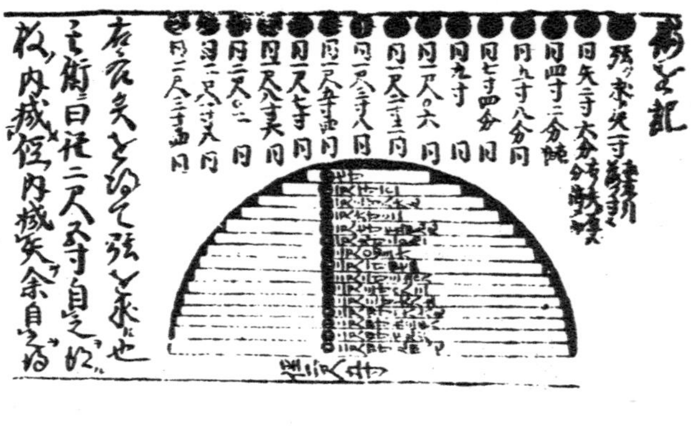

Late seventeenth-century Japanese diagram showing area measurement by integration.

Since $\tan x = 0$ if $x = 0$, we may put $\tan^{-1}0 = 0 + C$ so that $C = 0$. For reasons disclosed in our next chapter, where we take series for our theme, the last expression is convergent if $x \leqslant 1$. When $x = 1$, $\tan^{-1}x = \frac{\pi}{4}$, since $\tan 45° = 1$. Hence we may derive the following infinite series for π:

$$\frac{\pi}{4} = 1 - \tfrac{1}{3} + \tfrac{1}{5} - \tfrac{1}{7} + \tfrac{1}{9} \cdots$$

$$\pi = 4(1 - \tfrac{1}{3} + \tfrac{1}{5} - \tfrac{1}{7} + \tfrac{1}{9} \cdots)$$

This series converges very slowly, the sums of the first n terms being

n	Σ	n	Σ
2	$2.\dot{6}$	8	3.01708
4	2.89524..	....	....
6	2.97605..	18	3.08609

THE DEFINITE INTEGRAL. So far, our story discloses nothing different in principle from what Newton set forth in his theory of *fluxions*, equipped in his own writing with a notation which was unambiguous only because tailored exclusively for analysis of dynamical problems in which *time* always appears as the independent variable. Thus Newton would write as on the left of each of the following:

$$\dot{x}^2 = 2x = D_t(x^2) \text{ and } \ddot{x}^2 = 2 = D_t^2(x^2)$$

What the anti-derivative signifies in the same domain we have seen, but this has nothing to do with a third issue which Leibniz first clarified. From time immemorial geometers before him had used what one variously calls the method of *exhaustion* or *quadrature* to determine volumes and areas; but they had hitherto relied on *ad hoc* Euclidean constructions, as did Archimedes when he found a formula for the area of what we should now call the segment of a parabola enclosed between the curve itself, the x-axis and two lines y_a and y_b parallel to the y-axis.

Let us suppose, as he did also, that we can divide the x-axis into small equal segments $\triangle x$. Whence if $y_0 = x_0^2$, we may write: $y_1 = (x_0 + \triangle x)^2$, $y_2 = (x_0 + 2\triangle x)^2$ and more generally $y_n = (x_0 + n\triangle x)^2$. In any part of the curve, the area $(\triangle A)$ enclosed between it and the x-axis by y_0 and y_1 lies between $y_0 \triangle x$ and $y_1 \triangle x$. In the ascending limb $y_0 \triangle x < \triangle A < y_1 \triangle x$. Between y_0 and y_4 we may say that the total area (A) is such that $A_{0.3} < A < A_{1.4}$ if

$$A_{0.3} = y_0 \triangle x + y_1 \triangle x + y_2 \triangle x + y_3 \triangle x$$
$$A_{1.4} = y_1 \triangle x + y_2 \triangle x + y_3 \triangle x + y_4 \triangle x$$

VOLUME OF THE PYRAMID

Each rectangular strip represents the section of a half step of thickness $\triangle x$ and width $2y$, thus corresponding to a volume of $\triangle v = 2y^2 . \triangle x$. If A is the sectional half-angle of the vertex

$$y = (\tan A)x \text{ and } \tan A = \frac{\frac{1}{2}B}{h}.$$

In the limit, the total volume is:

$$2\int_o^h 2y^2dx = 4\int_o^h y^2dx = 4\int_o^h \left(\frac{B}{2h}\right)^2 x^2 . dx$$

$$\therefore V = \frac{B^2}{h^2}\left(\frac{x^3}{3}\right)_{x=o}^{x=h} = \frac{hB^2}{3}$$

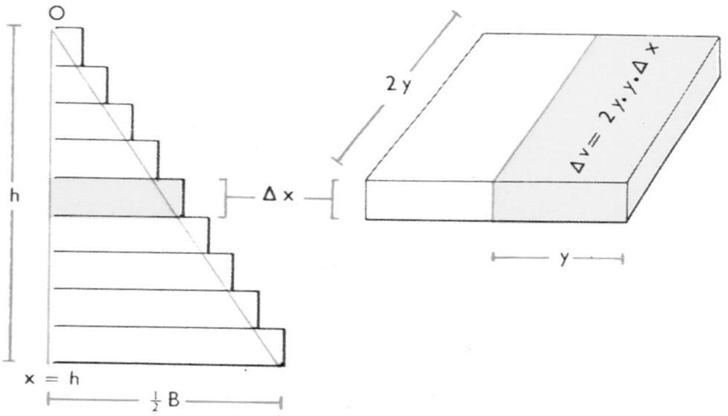

VOLUME OF THE CONE

Here $y = \frac{R}{h} . x$ *and the total volume is*

$$\int_o^h \pi y^2 . dx = \frac{\pi R^2}{h^2}\int_o^h x^2 . dx$$

$$= \frac{\pi R^2 h}{3}$$

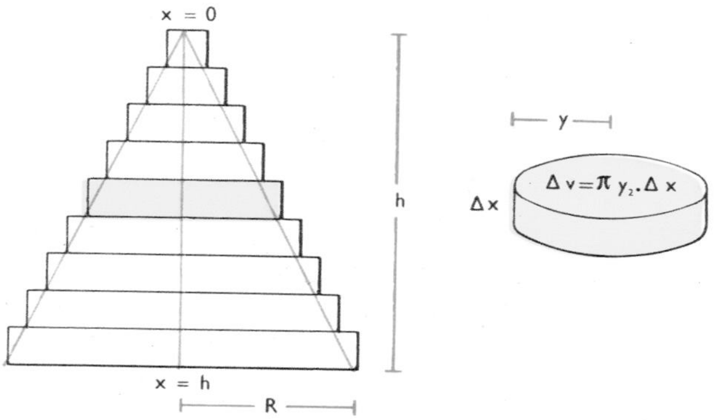

VOLUME OF SPHERE

As in the figure above, each rectangular strip corresponds to a cylinder of volume $\triangle v = \pi y^2 . \triangle x$. At every point on the circle $R^2 = x^2 + y^2$ so that $y^2 = R^2 - x^2$. The total volume is

$$2\int_o^R \pi(R^2 - x^2)dx = 2\pi\left(R^2x - \frac{x^3}{3}\right)_{x=o}^{x=R} = \frac{4\pi R^3}{3}$$

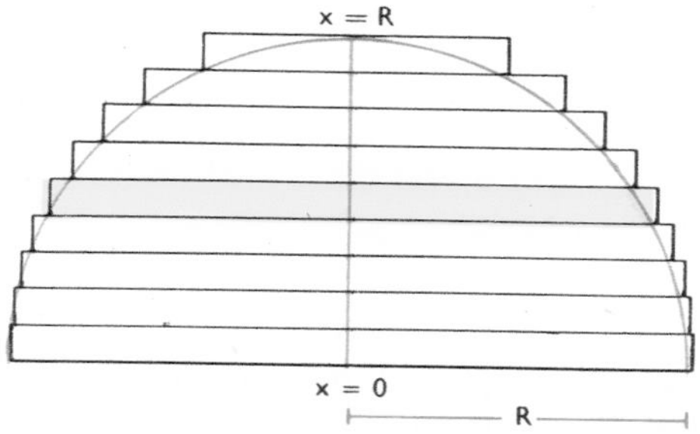

More generally, $A_{0(n-1)} < A < A_{1.n}$ if

$$A_{0.(n-1)} = \sum_{r=0}^{r=n-1} (x_0 + r\,\triangle x)^2 . \triangle x$$

$$A_{1.n} \quad = \sum_{r=1}^{r=n} (x_0 + r\,\triangle x)^2 . \triangle x$$

As our treatment of the volume of the pyramid has shown us in Chapter 3, the problem last stated is soluble without any premonition of the meaning of a derivative or anti-derivative. What Leibniz first clearly recognised is that it also admits of solution by a method of very wide application, if we invoke the latter. In essence, he argued as follows. For any rectangular strip $\triangle A \simeq y \triangle x$ in the foregoing, we may write:

$$y \simeq \frac{\triangle A}{\triangle x}$$

Since $\triangle A$ approaches $y \triangle x$ ever more closely as $\triangle x$ becomes smaller, we may write in the limit:

$$y = D_x . A \text{ so that } D_x^{-1} y = A$$

In the foregoing example, $y = x^2$, so that

$$A = \frac{x^3}{3} + C$$

This means that the whole area between the curve and the x-axis up to $x = b$ is:

$$A_b = \frac{b^3}{3} + C$$

Similarly the whole area up to $x = a$ $(< b)$ is:

$$A_a = \frac{a^3}{3} + C$$

The area enclosed by the ordinates $y = b^2$ and $y = a^2$ is therefore

$$A_{ab} = A_b - A_a = \frac{b^3}{3} - \frac{a^3}{3}$$

This interpretation of the method of exhaustion by summation of rectangular strips is what we now mean by *integration*; and we represent it as:

$$\int_a^b x^2 . dx = D_x^{-1}(x^2)_{x=b} - D_x^{-1}(x^2)_{x=a} = \left[\frac{x^3}{3}\right]_a^b$$

We call the expression on the left, written in the notation of Leibniz, the *definite integral* of x^2 between the limits b and a. The sign which precedes $x^2 . dx$ is an elongated S to suggest the notion of *summation* of rectangular strips and the alternative term *integration* for this process of exhaustion is fitting. What is not fitting is to use the term *indefinite integral* for the anti-derivative which we have seen to have an operational and dynamical meaning in its own right. The latter has no necessary connexion with summation.

We have here implicitly assumed that $x = a$ and $x = b$ lie in the *same quadrant* of the Cartesian grid. If not, the mechanical application of the procedure illustrated by the preceding example does not necessarily lead to a correct result. In the figure on page 225 we see that the area enclosed by the x-axis and the straight line $y = \frac{1}{2}x + 8$ between vertical lines at $x = \pm 8$ is 128 units, of which 80 lie in the top right and 48 in the left quadrant, in agreement with:

$$\int_{-8}^{8} (\tfrac{1}{2}x + 8)dx = \left[\frac{x^2}{4} + 8x\right]_{x=8} - \left[\frac{x^2}{4} + 8x\right]_{x=-8}$$
$$= 80 + 48$$

However, the area enclosed between the x-axis and by ordinates at $x = \pm 8$, when $y = \frac{1}{2}x$ is 32; and

$$\int_{-8}^{8} \tfrac{1}{2}x . dx = \left[\frac{x^2}{4}\right]_{-8}^{8} = 16 - 16 = 0$$

The balance sheet is here correct only if we: (a) integrate separately on either side of the origin *in*

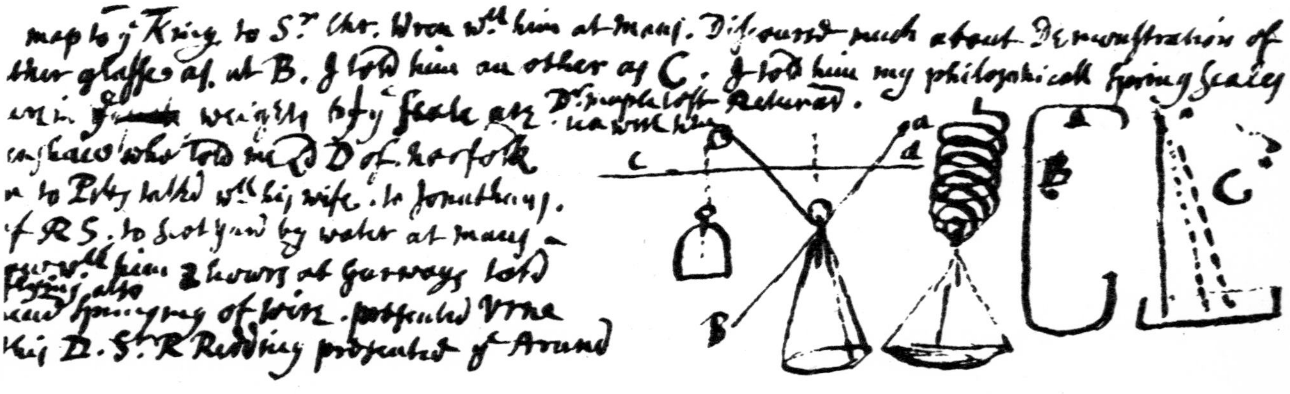

each quadrant; (b) treat all areas as *positive*. This is here equivalent to writing:

$$A=2\int_0^8 \tfrac{1}{2}x\,.\,dx=2(16)=32$$

Writers on the making of mathematics, when anxious to do justice to the role of Leibniz, sometimes state elliptically that we owe to him the recognition that integration is the operation inverse to differentiation (*i.e.* finding the derivative of a function). What Leibniz actually showed, if stated in the language of a later century, is that we can make use of the anti-derivative to accomplish integration in the sense of quadrature; but we can do so with the assurance of success only if we make a separate balance sheet area for each quadrant and treat each of the additive entries as positive quantities. The labelling of the grid by opposite signs on each side of the origin does not entitle us to treat an area as a directed number.

We have here followed the historic path by interpreting the summation process called *integration* by reference to a real measurement. The tie-up between the anti-derivative and the problem of quadrature, first recognised by Leibniz, who may or may not have taken over the notion of differentiation through personal contact with Newton, involves certain logical difficulties which did not emerge clearly till the mid-nineteenth century. We shall refer to these later. In this context it suffices to say that many geometrical and physical problems which do not involve quadrature as such are soluble by the use of the definite integral.

Among geometrical applications, one may mention especially the derivation of formulae for the surface areas and volumes of regular solids (*e.g.* the pyramid) and more especially of solids of revolution such as the cone and the sphere. These we may leave to our artists. As a simple illustration of the use of the definite integral in the dynamical domain, we may here consider the extension of a metal spring. By definition $W\ (=F.x)$ is the product of a constant force (F) applied through the distance (x) involved. When the force varies, we can treat it as approximately constant through a minute distance $\triangle x$ so that the work done is $\triangle W=F.\triangle x$. Now Hooke's law of the spring tells us that the force $(F=-kx)$ is proportional to the extension (x), so that $\triangle W=-kx.\triangle x$, and in the limit for a stretch from l to $L\ (>l)$ we may write

$$W=-k\int_l^L x\,.\,dx=\frac{-k}{2}(L^2-l^2)$$

The notation of the Newton-Leibniz calculus here makes explicit a familiar property of a spring when we apply a force to it, *e.g.* give a jerk to a weight vertically suspended from it. By Newton's definition, force is the product of mass (m) and acceleration (D_t^2x), so that

$$m.D_t^2x=-kx \quad \text{or} \quad D_t^2x=\frac{-k}{m}(x)$$

The reader will now recognise that this is true if $x=\sin.t\left(\frac{k}{m}\right)^{\frac{1}{2}}$, *i.e.* the equation is one which describes a simple periodic vibration.

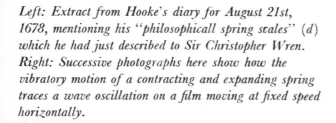

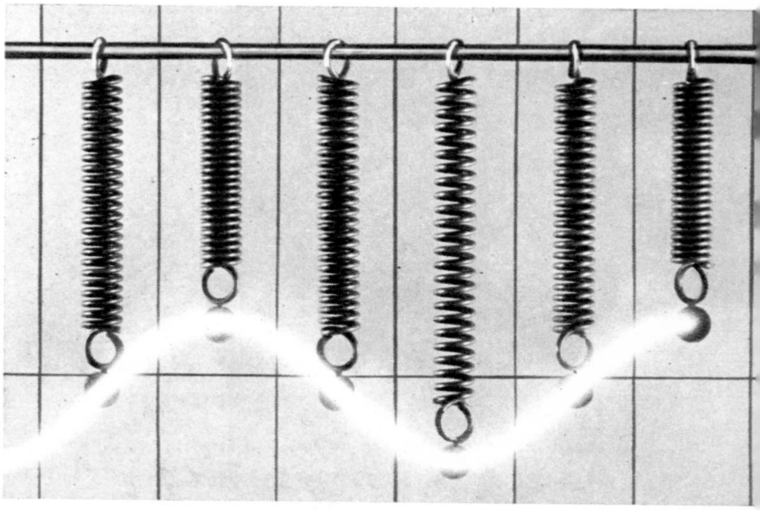

Left: Extract from Hooke's diary for August 21st, 1678, mentioning his "philosophicall spring scales" (d) which he had just described to Sir Christopher Wren. Right: Successive photographs here show how the vibratory motion of a contracting and expanding spring traces a wave oscillation on a film moving at fixed speed horizontally.

Chapter 11 A Century of Tabulation

The historical approach to any branch of worth-while study is profitable for two different, and in one sense contrary, reasons. Often what would otherwise seem on first acquaintance to be a remotely academic issue assumes in retrospect a lively topical relevance; and it is then encouraging to retrace the temporal sequence of discovery. As often, also, we can follow a logical sequence which is not intrinsically formidable only by side-stepping irrelevantly tortuous by-paths preserved by the inertia of tradition from a time when our forebears had not accustomed themselves to take full advantage of new tools at their disposal. To retrace the historic process is then a pastime of antiquarian interest with no pay-off in terms of lightening the load of learning. That is why we need not, and do not, here attempt to follow a strictly chronological plan in this chapter and in its two predecessors. All that history can usefully teach us about the three themes dealt with therein is how little we need to handicap ourselves by slavishly following the historic route of innovation.

Our theme in Chapter 9 was in one sense antecedent to first acquaintance with the Newton-Leibniz calculus; but the latter simplifies many problems of co-ordinate geometry otherwise amenable to manipulation only by *ad hoc* devices. Our theme in this chapter is the exploration of series during the period covered by the last two. In acquainting ourselves with new series first approached by piecemeal methods, we shall now be able to take full advantage of the derivative concept; but we can also enlarge our acquaintance with the infinitesimal calculus by using our knowledge of series to clarify the properties of functions commonly introduced by recourse to methods which are more laborious.

The emergence of the function concept which is the cornerstone of the Newton-Leibniz calculus came about when a compact algebraic symbolism made it possible to explore the properties of curves and solids against the background of a fixed framework of reference. From a geometrical point of view, this signifies a reorientation subsumed by the term *analytical*. When one uses it in a geometrical context, what one means primarily is the approach to a problem by first assuming a solution and then adapting it to the requirements of the initial statement; but one may use it in a wider sense to include the validification of an empirical discovery by the method of induction illustrated in Chapter 1. The representation of a function as the sum of a series invokes the analytical method in both senses. To get the meaning of the last sentence sharply into focus, let us consider the sum of the terms of the following nonterminating discrete series in which n is a

Among the spurs to mathematical discovery during the eighteenth century was the growing recognition that further technical progress demanded real precision engineering. James Watt's invention of the separate condenser (F)*, as applied to atmospheric pumping engines, achieved a great economy of fuel, but the further efficiency to be obtained from the use of direct steam pressure had to await the development of better cylinder-boring and piston-grinding.*

Mr WATT'S single ENGINE for pumping water,
for draining Mines 1788.

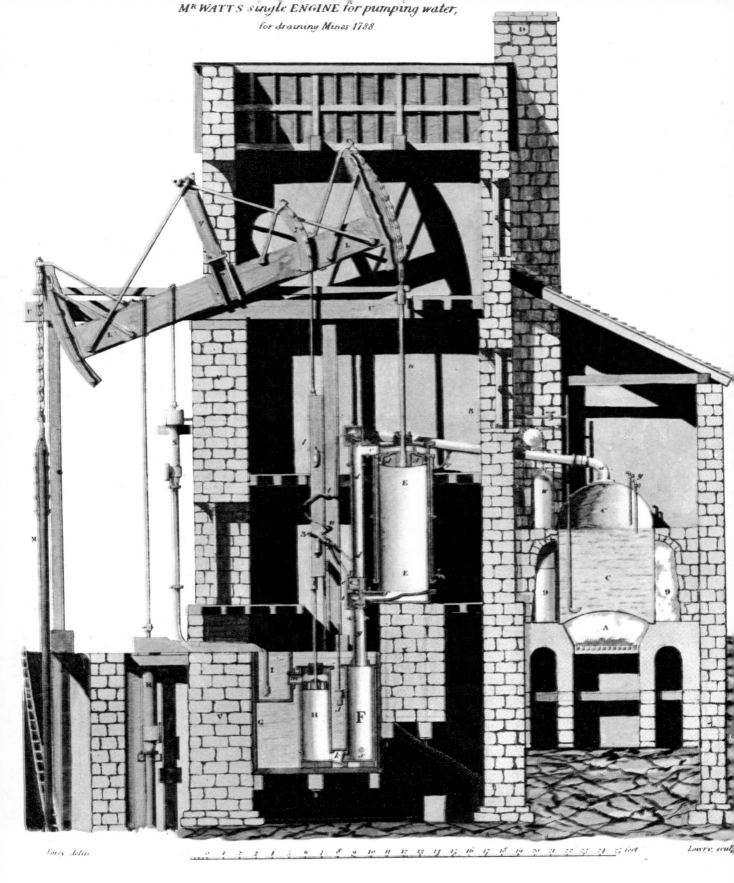

Farey delin. Lowry sculp.

positive integer:

$$s_n = 1 + \frac{1}{n} + \frac{1}{n^2} + \frac{1}{n^3} \cdots$$

We recognise this at once as the function:

$$s_n = \frac{n}{n-1}$$

The last is a more compact way of representing the function than the foregoing. It is also more explicit for purposes of numerical evaluation; but this would not be true of many continuous functions whose numerical evaluation is very laborious by methods so far mentioned, *e.g.* $y = L_{10}x$ or $y = L_{10}^{-1}x$, and $y = \sin x$ or $y = \tan^{-1}x$ *etc.* When we need to tabulate numerical values of any such function, the possibility of representing it as the sum of a series which converges like the foregoing to a finite limit ($s_n = 1.\dot{1}$ if $n = 10$) has a considerable merit which the decimal notation for fractions makes sufficiently explicit. By rejecting successively all terms after the first, second, *etc.*, it is apparent that:

$$1{\cdot}1 < s_n < 1{\cdot}2;\ 1{\cdot}11 < s_n < 1{\cdot}12;\ 1{\cdot}111 < s_n < 1{\cdot}112$$

We can thus attain any prescribed level of precision in numerical evaluation, if we can state that *the remainder after adding so many terms does not exceed a specifiable amount.* In an age when rapidly-advancing navigational science and astronomy demanded increasing refinement of such tabulated functions as logarithms and trigonometrical ratios, their representation by recourse to series had therefore an eminently practical *motif.* Meanwhile, and from the beginning of the quest, the undertaking progressively enriched the resources of the Newton-Leibniz calculus by the discovery of hitherto unknown functions and of new properties of already familiar ones. Conversely, the infinitesimal calculus provided a powerful new tool for the search.

The story begins in the latter half of the seventeenth century with the recognition of the possibility of exploiting the Binomial Theorem both for the study of discrete terminating (so-called *finite*) series and for the study of continuous non-terminating (so-called *infinite*) series. The names of an Englishman, Newton (1642–1727), and of two Scotsmen, Gregory (1638–1675) and Maclaurin (1698–1746), are of special interest in the formative stage; but early in the eighteenth century British mathematics went into decline. Thereafter, for more than a century, the development of the infinitesimal calculus and the exploration of series by exploiting its uses was mainly the work of Swiss mathematicians, notably the brothers Jacques (1654–1705) and Jean (1667–1748) Bernoulli who both taught at Basel, Daniel (1700–1782), the son of Jean, and, most outstanding of all, Leonhard Euler (1707–1783), a pupil of Jacques.

In retrospect, it is clear that many properties of series would have been easier to grasp if first approached by recourse to the infinitesimal calculus, and that many problems which arise in the domain of the latter yield more simply to solution if tackled by means of series. Gregory's formula for the general term of discrete finite series amenable to figurate representation discloses a new awareness of the uses of the Binomial Theorem and a preview of the possibility last mentioned. To explore the possibility of finding the general term of such a discrete series, we may examine the sequence whose terms are the sum of the cubes of the integers:

$$0;\ (0+1)=1;\ (0+1+8)=9;$$
$$(0+1+8+27)=36;$$
$$(0+1+8+27+64)=100 \textit{ etc.}$$

If familiar with figurate series, we shall recognise these as the squares of the triangular numbers starting with rank 0, *e.g.* $0=0^2$; $1=1^2$; $9=3^2$; $36=6^2$; $100=10^2$ *etc.* Since the general term of the triangular number series is $\frac{1}{2}n(n+1)$, we therefore suspect that the general term of the foregoing will be $\frac{1}{4}n^2(n+1)^2$. The proof by induction is very simple; but the initial step of finding a plausible formula to test by induction may be by no means so easy in a comparable situation. A procedure applicable to all such series has therefore much to commend it. To disclose its rationale, let us now lay out the foregoing series with the differences between successive terms below each line:

rank	**0**		**1**		**2**		**3**		**4**		**5**		**6**	...
series	0		1		9		36		100		225		441	...
first diff.		1		8		27		64		125		216		...
second ditto			7		19		37		61		91			...
third ditto				12		18		24		30				...
fourth ditto					6		6		6					...
fifth ditto						0		0						
sixth ditto							0							

We here see that the fifth and all subsequent differences vanish. That the differences do vanish at some level is a property of all number series amenable to figurate representation, being then expressible in the following form if N_0, N_1 *etc.* are rational and real (*i.e.* vulgar fractions or integers, negative or positive):

$$u_n = N_0 n^0 + N_1 n^1 + N_2 n^2 + N_3 n^3 \; \ldots$$

If we now denote the term of rank r in any such nonrecurrent series by u_r, we may define an operation (Δ) as that of *subtracting it from its successor, i.e.*

$$\Delta u_r = u_{r+1} - u_r$$

Accordingly, we interpret $\Delta^2 u_r = \Delta(\Delta u_r)$ as $\Delta u_{r+1} - \Delta u_r$ and $\Delta^3 u_r$ as $\Delta^2 u_{r+1} - \Delta^2 u_r$ *etc.* in accordance with the schema below:

$$u_0 \qquad\qquad u_1 \qquad\qquad u_2 \qquad\qquad u_3$$
$$(u_1 - u_0) = \Delta u_0 \qquad (u_2 - u_1) = \Delta u_1 \qquad (u_3 - u_2) = \Delta u_2$$
$$(\Delta u_1 - \Delta u_0) = \Delta^2 u_0 \qquad (\Delta u_2 - \Delta u_1) = \Delta^2 u_1$$
$$(\Delta^2 u_1 - \Delta^2 u_0) = \Delta^3 u_0$$

We may now define a second operation $(1+\Delta)$ on u_r as that of *adding to u_r the result of performing Δ on it*, so that:

$$(1+\Delta)u_r = u_r + \Delta u_r$$

Just as we define Δ^2 as an instruction to perform the operation Δ twice, we may denote by $(1+\Delta)^2$ an instruction to perform the operation $(1+\Delta)$ twice, so that:

$$\begin{aligned}(1+\Delta)^2 u_r &= (1+\Delta)(u_r + \Delta u_r)\\ &= (u_r + \Delta u_r) + (\Delta u_r + \Delta^2 u_r)\\ &= u_r + 2\,\Delta u_r + \Delta^2 u_r\end{aligned}$$

Similarly, we derive:

$$(1+\Delta)^3 u_r = u_r + 3\,\Delta u_r + 3\,\Delta^2 u_r + \Delta^3 u_r$$
$$(1+\Delta)^4 u_r = u_r + 4\,\Delta u_r + 6\,\Delta^2 u_r + 4\,\Delta^3 u_r + \Delta^4 u_r$$

Though Δ in this context is an operator, in contradistinction to the number x in $(1+x)^p$, it is not too difficult to see why the build-up of the coefficients on the right accords with the build-up of the coefficients of the expansion of $(1+\Delta)^p$ when $p=1$, 2, 3 *etc.* Hence in general:

$$(1+\Delta)^p u_r = u_r + p_{(1)}\,\Delta u_r + p_{(2)}\,\Delta^2 u_r + p_{(3)}\,\Delta^3 u_r \; etc.$$

Now the definition of $(1+\Delta)$ implies a step-up of rank by unity since:

$$u_r + \Delta u_r = u_r + u_{r+1} - u_r = u_{r+1}$$

If we raise the rank by unity each time we perform it

$$(1+\Delta)^p u_r = u_{r+p}$$

Given the initial term labelled as the term of rank zero ($r=0$), we can therefore write:

$$u_p = u_0 + p_{(1)}\,\Delta u_0 + p_{(2)}\,\Delta^2 u_0 + p_{(3)}\,\Delta^3 u_0 \; etc.$$

If we go back to the foregoing series whose terms are the sums of the cubes of the natural numbers, we then have:

$$u_0 = 0;\; \Delta u_0 = 1;\; \Delta^2 u_0 = 7;\; \Delta^3 u_0 = 12;\; \Delta^4 u_0 = 6$$
$$\Delta^n u_0 = 0 \quad if\; n > 4$$

Whence we derive

$$u_p = 0 + p(1) + \frac{p(p-1)}{1.2}(7) + \frac{p(p-1)(p-2)}{1.2.3}(12) + \frac{p(p-1)(p-2)(p-3)}{1.2.3.4}(6)$$
$$= \frac{p^2(p+1)^2}{4}$$

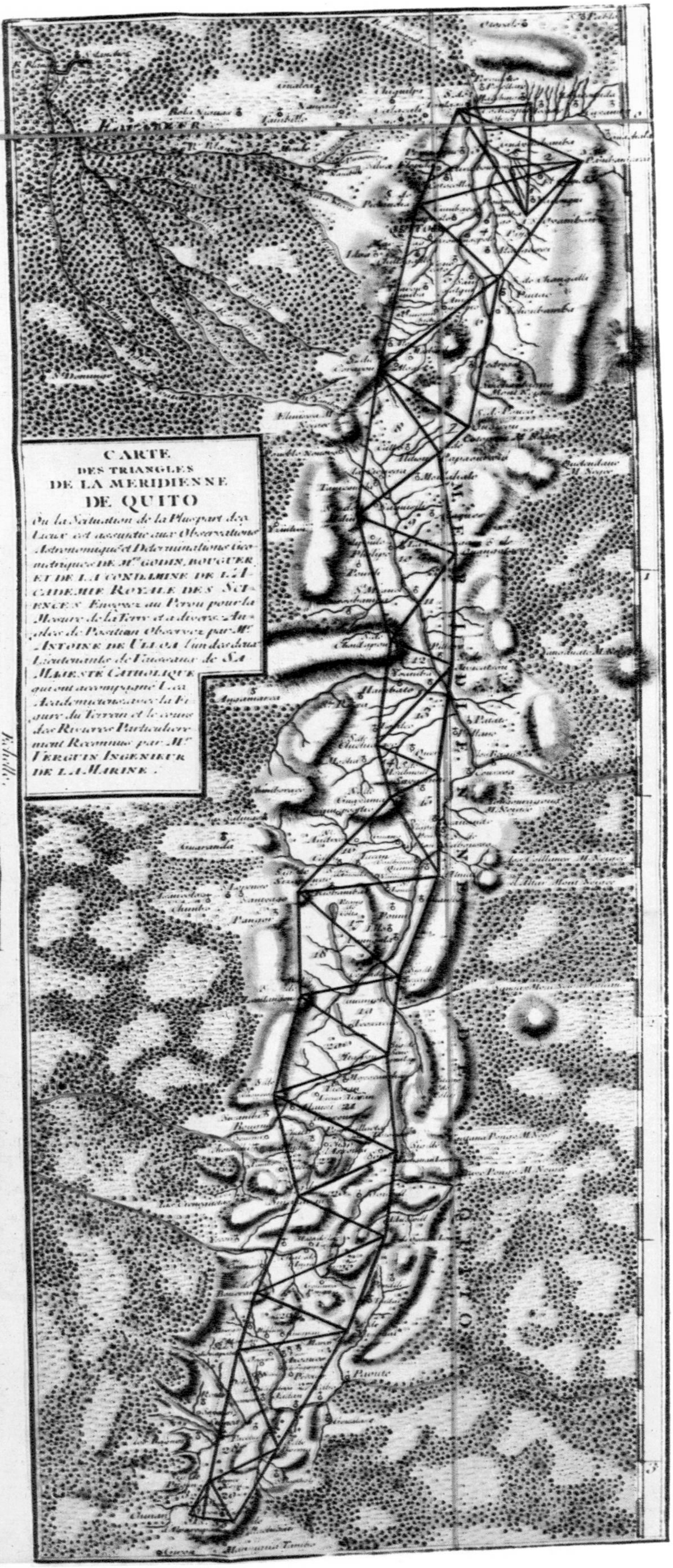

We may, of course, make use of the sequence:

$u_1=1$; $\triangle u_1=8$; $\triangle^2 u_1=19$; $\triangle^3 u_1=18$; $\triangle^4 u_1=6$ *etc.*

If we do this, we have to remember that we get u_p by stepping up u_1 with $(p-1)$ rank increments, so that

$$u=1+8(p-1)+\frac{19(p-1)(p-2)}{1.2}$$

$$+\frac{18(p-1)(p-2)(p-3)}{1.2.3}$$

$$+\frac{6(p-1)(p-2)(p-3)(p-4)}{1.2.3.4}$$

$$=\frac{p^2(p+1)^2}{4}$$

Not all series whose terms are integers yield a so-called vanishing triangle like the layout shown above. For instance, a so-called geometric series defined as $u_r=a^r$ (*e.g.* 1, 3, 9, 27, 81 *etc.*) does not do so. The differences conform to the pattern $\triangle^p u_r=(a-1)^p u_r$. In particular, if $a=2$, $\triangle^p u_r=u_r$ as is evident below:

	$2^0=1$		$2^1=2$		$2^2=4$		$2^3=8$		$2^4=16$		$2^5=32$
$\triangle u_r$		1		2		4		8		16	...
$\triangle^2 u_r$			1		2		4		8		...
$\triangle^3 u_r$				1		2		4			...

This layout of the series for 2^r invites us to look at the operator $\triangle$ from another viewpoint. The ranks of the terms of a series themselves constitute a series, *i.e.* that of the integers $u_r=r$, so that $\triangle r=(r+1)-r=1$. We may therefore write:

$$\frac{\triangle u_r}{\triangle r}=u_r \quad \textit{if } u_r=2^r$$

$$\frac{\triangle^p u_r}{\triangle r}=\triangle u_r \qquad (p=1,\ 2,\ 3\ \textit{etc.})$$

Expressed in words, this means that the geometric series $u_r=2^r$ increases in unit intervals by an amount equal to its value at the beginning of the interval. This raises the question: is there a continuous function whose rate of change (*derivative*) is equal to itself, *i.e.* one which satisfies the

Feats of surveying and geodesy hitherto unequalled called for better tables of trigonometrical ratios. This map was drawn in the 1730s, when a French expedition measured a geodetic arc of meridian in Peru.

equation $D_x . y = y$? Before exploring this possibility, we must examine the Binomial Theorem from a more general viewpoint than hitherto.

So far we have assumed that p is a positive integer in the Chinese formula for representing $(1+x)^p$ as a discrete series of $p+1$ terms. We shall later see that p may be a negative or positive whole number or rational fraction, in which event the series does not terminate and the expression is then meaningful only if the sum converges to a finite limit; but before we can profitably explore a topic which occupied so much intellectual energy during the latter half of the seventeenth century and throughout the century following, we must be clear about something else, *viz.* the representation of a function $y=f(x)$ as the sum of a nonterminating series is meaningful only within a range of values of x for which the series is convergent.

It will here suffice to mention two criteria of convergence. A necessary condition is that successive terms diminish without restriction, being therefore fractional beyond a specifiable rank. If all the terms after the one of specified rank diminish more rapidly than a sequence known to be convergent, *e.g.* $0.\dot{1}$, $0.0\dot{1}$, $0.00\dot{1}$ *etc.*, the sum of the series will then itself converge to a finite limit. We have applied such a *yardstick* test in Chapter 2 to the series:

$$e = 1 + 1 + \frac{1}{2!} + \frac{1}{3!} + \frac{1}{4!} + \frac{1}{5!} \ldots$$

However, a series of continuously diminishing terms is not necessarily convergent. For instance, the following harmonic series is not:

$$s_n = 1 + \tfrac{1}{2} + \tfrac{1}{3} + \tfrac{1}{4} + \tfrac{1}{5} + \tfrac{1}{6} \quad \ldots$$

It is easy to see that no term in this is equal to or less than a tenth of its successor. Whence it fails if our yardstick is a periodic decimal fraction. It is also possible to show that it cannot converge, if we signify the sum of the first n terms by s_n and lay out the summation in stages as follows:

$$1 + \tfrac{1}{2} + \tfrac{1}{3} + \tfrac{1}{4} > \tfrac{1}{4} + \tfrac{1}{4} + \tfrac{1}{4} + \tfrac{1}{4} \text{ so that } s_4 > 1 = \tfrac{2}{2}$$

$$\tfrac{1}{5} + \tfrac{1}{6} + \tfrac{1}{7} + \tfrac{1}{8} > \tfrac{1}{8} + \tfrac{1}{8} + \tfrac{1}{8} + \tfrac{1}{8} \text{ so that } s_8 > 1 + \tfrac{1}{2} = \tfrac{3}{2}$$

$$\tfrac{1}{9} + \tfrac{1}{10} + \tfrac{1}{11} + \tfrac{1}{12} + \tfrac{1}{13} + \tfrac{1}{14} + \tfrac{1}{15} + \tfrac{1}{16} > \tfrac{1}{2} \text{ so that } s_{16} > \tfrac{3}{2} + \tfrac{1}{2} = \tfrac{4}{2}$$

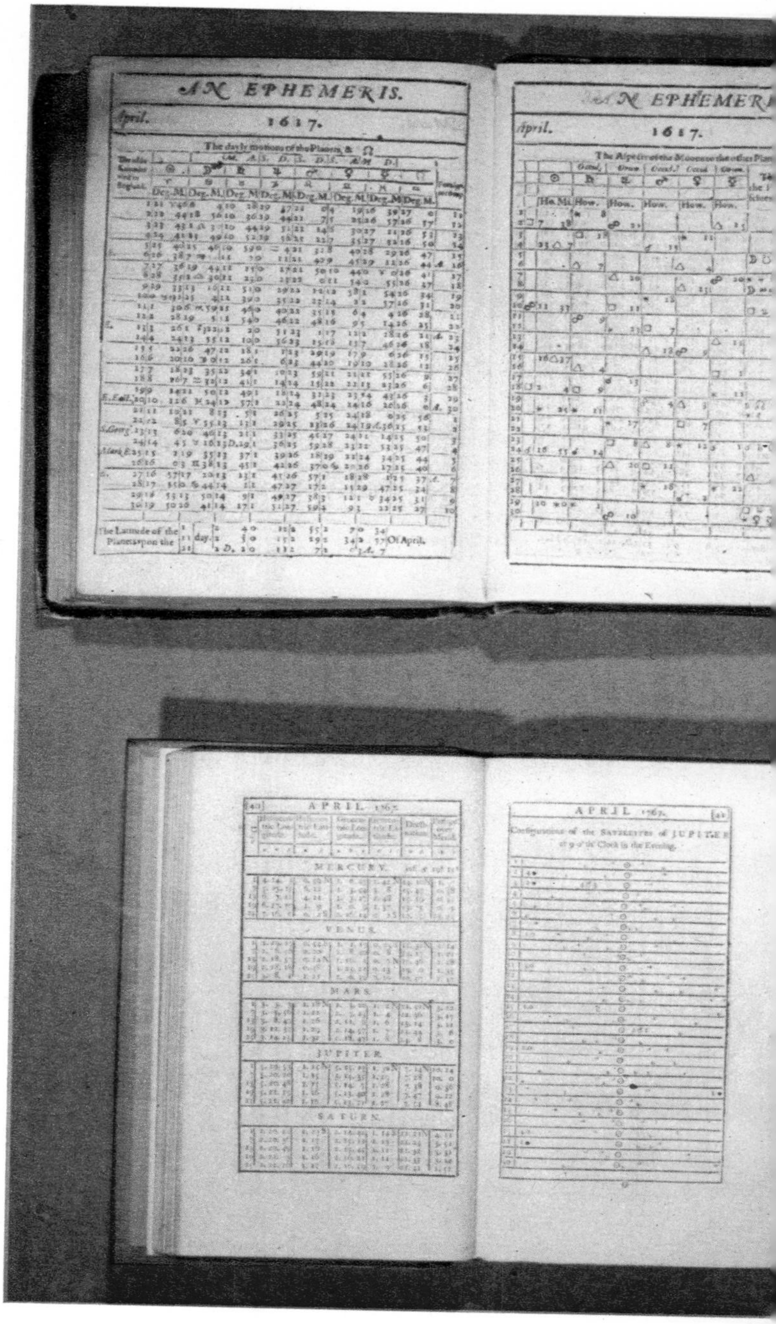

Table showing planetary motions for April, 1617 compared with part of a far more informative table for April, 1767. The book from which the latter is taken was published by the British Commissioners of Longitude. It gives information about the satellites of Jupiter and includes log tables.

By continuing in this way, we notice that for $n=2^p$, $s_n>\frac{1}{2}p$. Whence we can increase s_n without limit by increasing p accordingly.

Though a continuously-diminishing sequence of terms is not a sufficient condition of convergence when the signs of all terms are the same, a sufficient condition in that event is that the signs alternate. Consider the case below:

$$\begin{aligned} s_n &= 1-\tfrac{1}{2}+\tfrac{1}{3}-\tfrac{1}{4}+\tfrac{1}{5}-\tfrac{1}{6}+\tfrac{1}{7}-\tfrac{1}{8}+\tfrac{1}{9}\ \ldots \\ &= 1-(\tfrac{1}{2}-\tfrac{1}{3})-(\tfrac{1}{4}-\tfrac{1}{5})-(\tfrac{1}{6}-\tfrac{1}{7})-(\tfrac{1}{8}-\tfrac{1}{9})\ \ldots \\ &= 1-\tfrac{1}{6}-\tfrac{1}{20}-\tfrac{1}{42}-\tfrac{1}{72}\ \ldots \end{aligned}$$

also

$$\begin{aligned} s_n &= (1-\tfrac{1}{2})+(\tfrac{1}{3}-\tfrac{1}{4})+(\tfrac{1}{5}-\tfrac{1}{6})+(\tfrac{1}{7}-\tfrac{1}{8})\ \ldots \\ &= \tfrac{1}{2}+\tfrac{1}{12}+\tfrac{1}{30}+\tfrac{1}{56}\ \ldots \end{aligned}$$

Whence $1>s_n>\frac{1}{2}$. The reader should experience no difficulty with the derivation of the general rule that $t_0>s_n>t_0-t_1$, if

$$s_n=t_0-t_1+t_2-t_3+t_4\ \ldots\ etc.$$

The relevance of the foregoing digression will now be clear, if we take our last glimpse at what figurate series can teach us. So far, we have established the truth of the Binomial Theorem only for positive whole number values of n in the expression $(1+x)^n$; and we have seen that Chinese mathematicians have been familiar with it for a thousand years or more. When school-books loosely speak of Newton as its discoverer, they signify that Newton, Wallis and their circle first recognised its reliability for all negative or positive rational numbers. Actually, a proof wholly satisfactory to a modern mathematician could not emerge till Newton's successors clarified the notion of convergence. That it is valid when n is a negative integer might well have been recognised much earlier, if mathematicians before Wallis had used the explicit notation now current, *e.g.* $2^{-3}=0{\cdot}125$.

If we lay out as below horizontally in rank order from 0 onwards for dimensions 0, 1, 2, 3 *etc.* the figurate series of which the one-dimensional terms are the natural numbers, the diagonal terms from left to right upwards yield terminating series which tally with the rows of the Chinese (so-called Pascal) triangle. Though these series have no pictorial significance if extended backwards as here shown, it is instructive to examine what happens when we do so. The diagonal terms running from right to left downwards in the negative domain of the table are alternately negative and positive. Their numerical values reproduce those of successive terms of the horizontal rows in the positive half of the table. Like the latter, they do *not* terminate, and they tally with the series of coefficients obtained by substituting $-n$ for $+n$ in $n_{(r)}$, *e.g.*

$$(-3)_{-3}=\frac{-3(-3-1)(-3-2)}{3!}$$

$$=\frac{-3.4.5}{3!}=-10$$

$$(-3)_{-4}=\frac{-3(-3-1)(-3-2)(-3-3)}{4!}$$

$$=\frac{+3.4.5.6}{4!}=+15$$

r	**−6**	**−5**	**−4**	**−3**	**−2**	**−1**	**0**	**1**	**2**	**3**	**4**	**5**	**6**	...	**d**
...	1	1	1	1	1	1	1	1	1	1	1	1	1	...	0
...	−6	−5	−4	−3	−2	−1	0	1	2	3	4	5	6	...	1
...	15	10	6	3	1	0	0	1	3	6	10	15	21	...	2
...	−20	−10	−4	−1	0	0	0	1	4	10	20	35	56	...	3
...	15	5	1	0	0	0	0	1	5	15	35	70	126	...	4
...	−6	−1	0	0	0	0	0	1	6	21	56	126	252	...	5
...	1	0	0	0	0	0	0	1	7	28	84	210	462	...	6
...	...	...	...	...	...	...	...	...	...	...	...	...	...	...	...

Towards the close of the eighteenth century, a French team under Delambre and Méchain made an accurate measurement of the length of the great-circle arc between Dunkirk and Barcelona. Upon it was based the length of the standard metre, later to become an international unit of linear measurement. Here is part of the chaine des triangles *the team established.*

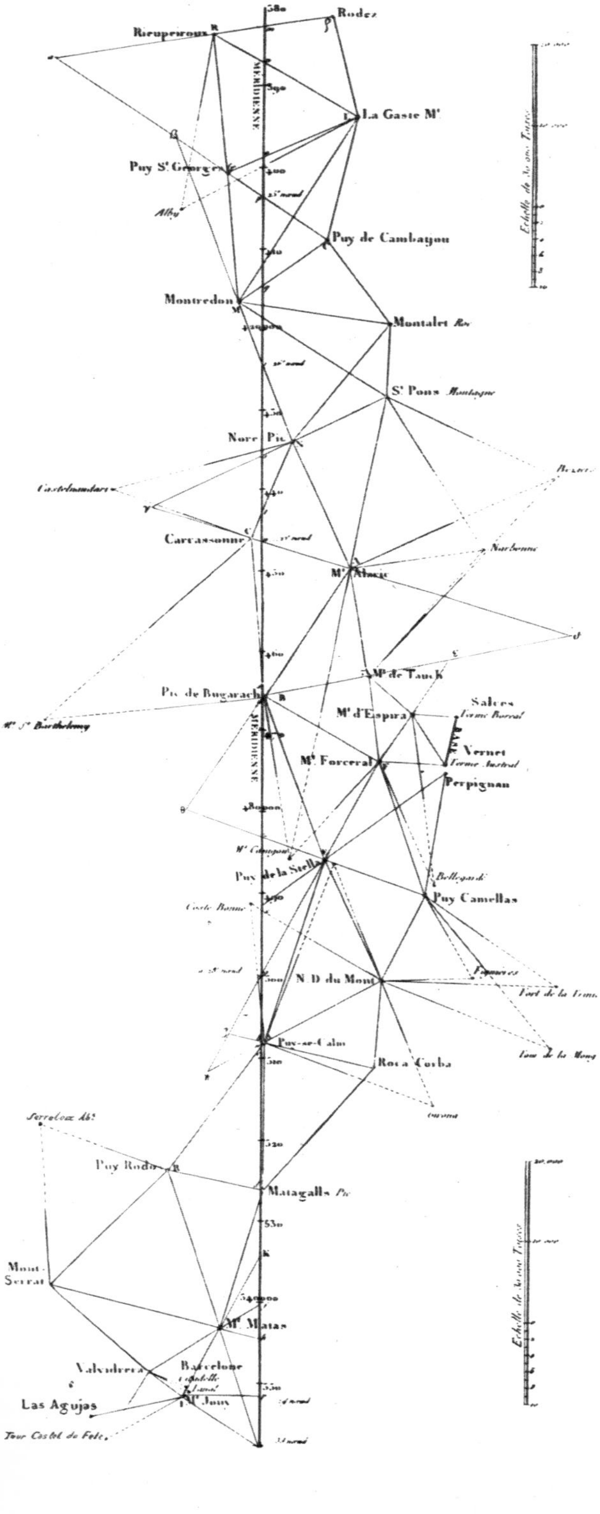

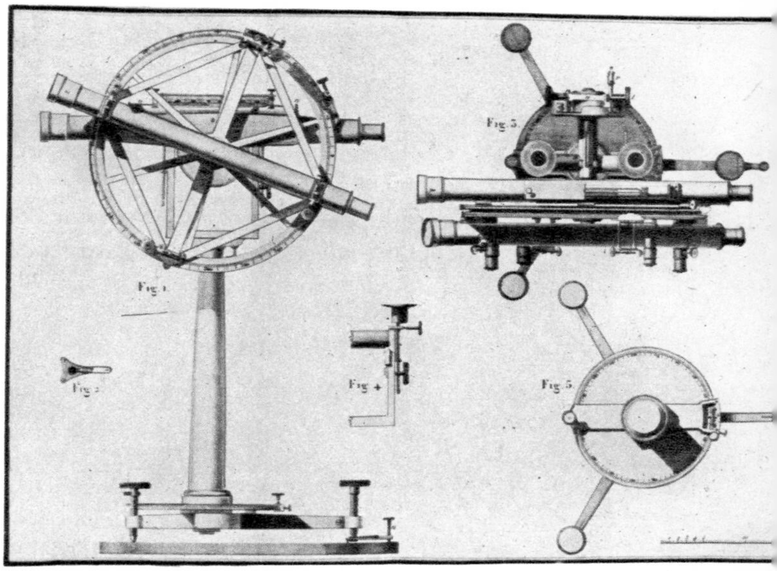

In the survey, Delambre and Méchain used the specially-constructed theodolite shown above. Terminal points of the base lines were marked with a pyramid-topped obelisk surrounded by a circle of stones.

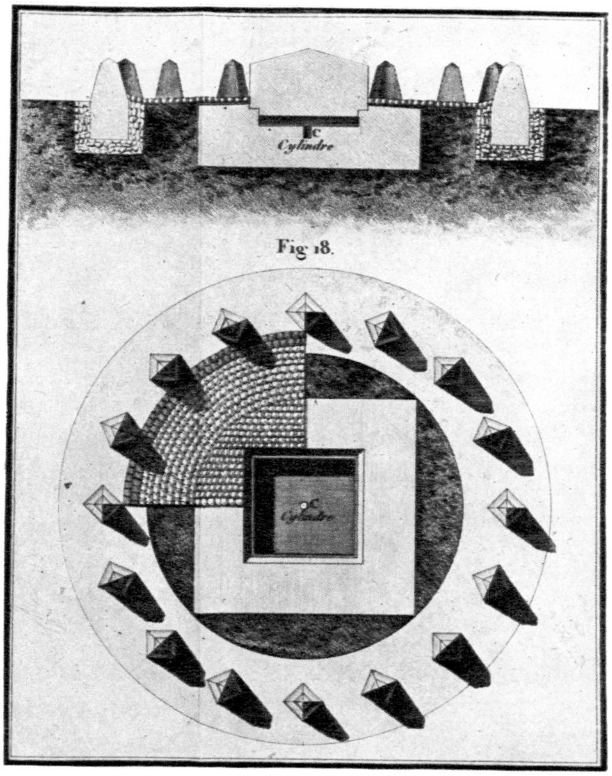

By direct division, we easily obtain the expansion of $(1+x)^{-1}$, whence of $(1+x)^{-2}$ *etc.* by multiplying as many terms as we like in the expansion $(1+x)^{-2}=(1+x)^{-1}(1+x)^{-1}$ *etc.* We then recognise a pattern now very familiar to the reader, *e.g.*

$$(1+x)^{-1}=1-x+x^2-x^3+x^4 \quad \ldots$$
$$(1+x)^{-2}=1-2x+3x^2-4x^3+5x^4 \quad \ldots$$
$$(1+x)^{-3}=1-3x+6x^2-10x^3+15x^4 \quad \ldots$$
$$(1+x)^{-4}=1-4x+10x^2-20x^3+35x^4 \quad \ldots$$

Being nonterminating series, the sum of any one of the above is finite only for values of x which satisfy one or other test for convergence. This is so, if $x<1$; but the mere fact that the Binomial Theorem is valid for a negative integer n in the expansion of $(1+x)^n$ for values of $x<1$ is not very noteworthy from a practical viewpoint except in so far as it: (a) makes a bridge between finite and nonterminating series; (b) invites us to ask a question which Newton first examined thoughtfully, *viz.* is it valid when n is a positive or negative rational fraction?

It is indeed a useful thing to know that the theorem is valid for fractional values of n when $x \leqslant 1$, and especially so in the social context of the emergent use of logarithms, whence also the need for making tables of fractional powers of a base as effortlessly as possible. Thus we can invoke its aid to extract roots of any order to the base 2 ($x=1$) or less by expanding:

$$(1+1)^{\frac{1}{2}}\ ; \quad (1+1)^{\frac{1}{3}}\ ; \quad (1+1)^{\frac{1}{4}} \ \textit{etc.}$$

For the extraction of square roots, we may tabulate once for all the coefficients in the expansion of $(1+x)^{\frac{1}{2}}$. If $n=\frac{1}{2}$, the coefficient (C_r) of the general term of rank r is

$$\frac{1}{r!}(\tfrac{1}{2})^r = \frac{1}{r!}\tfrac{1}{2}(\tfrac{1}{2}-1)(\tfrac{1}{2}-2) \ \ldots \ (\tfrac{1}{2}-r+1)$$

The numerical values of the first 11 coefficients (rank 0 to rank 10) are below, with omission of the initial term $t_0=1^{\frac{1}{2}}=1$:

r	$\mathbf{C_r}$	**r**	$\mathbf{C_r}$
1	−0·5	6	+0·02051
2	+0·125	7	−0·01611
3	−0·0625	8	+0·01309
4	+0·0306	9	−0·01091
5	−0·02734	10	+0·01030

If $x \leqslant 1$ and $r=\frac{1}{2}$, the series is clearly convergent since: (a) successive terms get smaller; (b) alternate terms after the initial one are of opposite sign. If $x=1$, so that $(1+x)^{\frac{1}{2}}=\sqrt{2}$, the summation of the above including the initial term (unity) yields 1·409, the square of which is 1·985281, a proportionate discrepancy of 0·75%.

When $x=1$, the calculation to a high level of precision is admittedly laborious; but such series converge rapidly when we make x a small fraction. With a little ingenuity, we can always cast $(1+x)^n$ in a form which satisfies this condition, *e.g.*:

$$1000+80=5(216)=5(6^3)$$
$$\therefore (1000+80)^{\frac{1}{3}}=5^{\frac{1}{3}}.6=1000^{\frac{1}{3}}(1+0{\cdot}08)^{\frac{1}{3}}$$
$$=10(1+0{\cdot}08)^{\frac{1}{3}}$$
$$\therefore 5^{\frac{1}{3}}=\tfrac{5}{3}(1+0{\cdot}08)^{\frac{1}{3}}$$

Here $x^2=0{\cdot}0064$, $x^3=0{\cdot}000512$, $x^4=0{\cdot}00004096$ and $x^5=0{\cdot}0000032768$. Thus the retention of only seven terms more than suffices to guarantee precision of six significant figures. By applying the Binomial Theorem in this way, all the results we may have the energy to calculate accord with the outcome of squaring, cubing, *etc.* to any order of precision we may demand.

This is a hopeful discovery; but it is not a proof. Indeed Newton, whose contributions to mathematical discovery are formidable, offered nothing to show conclusively why the rule will never let us down. His successors have done so in more than one way. What is perhaps the most economical one depends on a question prompted by the pattern of discrete series for u_n already discussed and discovered by a Scottish mathematician whom Newton outlived. The first person to answer the question was one of his fellow countrymen who outlived Newton.

THE LOGARITHMIC FUNCTION IN A MATTER OF LIFE AND DEATH

The following is a fictitious record of the fall of the infant death rate, roughly following the course of events in England and Wales from 1890 to 1960. The graph of the rates (per thousand born), ranging from 144 in 1890 to 25 in 1960, suggests that the tempo of improvement has fallen off, as it has in the sense that the absolute value of the decrement has declined decade by decade. Proportionately, it has not. In a finite interval of time $(t_2 - t_1) = \triangle t$, *the absolute decrement is* $(i_2 - i_1) = \triangle i$, *but the proportionate decrement is* $(i_2 - i_1) \div i = \triangle i \div i$. *The rate of decrease in a short interval is* $\triangle i \div \triangle t \simeq D_t . i$. *That of the proportionate decrease is* $\triangle i \div i$. *In the limit we may write the rates of change as:*

Absolute $D_t . i$ *Proportionate:* $\frac{1}{i} D_t . i$

If $y = \log_e i$ *so that* $i = e^y$ *and* $D_y . i = i$ *we can write* $D_i . y = i^{-1}$; *but by Rule 4, p. 221,* $D_i . y \times D_t . i = D_t . y$, *whence*

$$D_t(\log_e i) = \frac{1}{i} D_t(i).$$

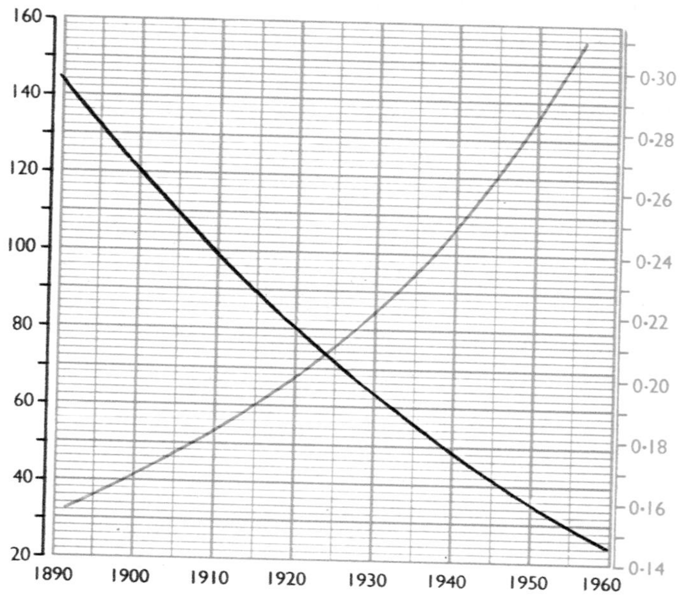

Thus the slope of the function $\log_e i$ *gives the correct picture of the proportional decrement. By plotting the infant death rate on semi-logarithmic paper* ($Y = \log_e y$), *we disclose what is happening to the* proportionate *decrement without recourse to calculation.*
The numerical data are here:

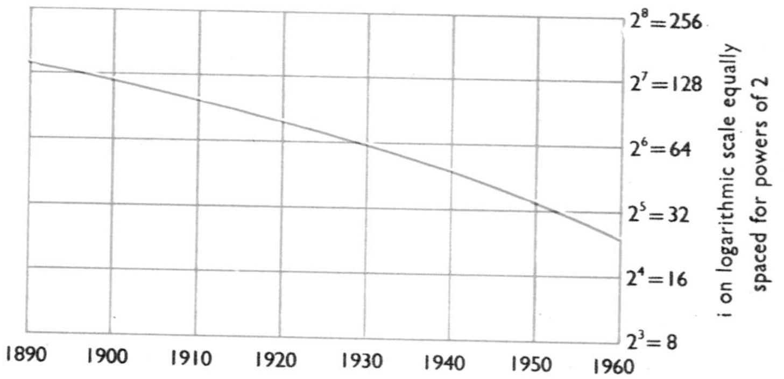

Date	*Deaths per 1000* (i)	*Decrement* ($\triangle i$)	*Proportionate Decrement* ($\triangle i \div i$)
1890	*144*	—	—
1900	*121*	*23*	$23 \div 144 \simeq 0{\cdot}16$
1910	*100*	*21*	$21 \div 121 \simeq 0{\cdot}17$
1920	*81*	*19*	$19 \div 100 = 0{\cdot}19$
1930	*64*	*17*	$17 \div 81 \simeq 0{\cdot}21$
1940	*49*	*15*	$15 \div 64 \simeq 0{\cdot}24$
1950	*36*	*13*	$13 \div 49 \simeq 0{\cdot}26$
1960	*25*	*11*	$11 \div 36 \simeq 0{\cdot}31$

If we use numerical constants A_0, A_1, A_2 *etc.* we may cast Gregory's series for the discrete function u_n of n in the more general form as follows:

$$u_n = A_0 \Delta^0 u_0 + A_1 \Delta^1 u_0 + A_2 \Delta^2 u_0 + A_3 \Delta^3 u_0 \textit{ etc.}$$

In this expression, we may write:

$$\frac{\Delta^m u_n}{\Delta n} = \Delta^m u_n \text{ since } \Delta n = 1$$

The limit of this ratio when we allow Δn to take any real value, however small, is the *m*th derivative $D_n^m u_n$ of the corresponding continuous function u; and $\Delta^m u_0$ in this context means $\Delta^m u_n$ when $n=0$. We may similarly write $D_x(y) = D_x(0)$ when $x=0$ for the continuous function $y=f(x)$. For instance, if $y = \sin x$, $D_x^1 y = \cos x$ and $\cos x = 1$ when $x=0$, so that $D_x^1(0) = 1$. Also $D_x^2 y = -\sin x = 0$ when $x=0$, or again for $y = \sin x + 3$ we may write $D_x^1(0) = 4$ and $D_x^2(0) = 3$. If $f(0) = y$ when $x=0$, it is consistent to write $f(0) = D_x^0(0)$.

We now have the clue to an exciting discovery of Maclaurin in the opening years of the eighteenth century. We ask ourselves: can we express the continuous function $y=f(x)$ as the sum of a nonterminating convergent series in a form comparable to the above:

$$y = A_0 D_x^0(0) + A_1 D_x^1(0) + A_2 D_x^2(0) + A_3 D_x^3(0) \textit{ etc.?}$$

To tailor the constants A_0, A_1 *etc.* to the requirements of a particular function, we first recall that $p_{(r)}$ in Gregory's formula for u_p is shorthand for $p^{(r)} \div r!$; and any $p^{(r)}$ is expressible in ordinary powers of p, *e.g.* $p^{(2)} = p(p-1) = p^2 - p$ and $p^{(3)} = p(p-1)(p-2) = p^3 - 3p^2 + 2p$. Since each $\Delta^m u_0$ is a numerical constant, we can express a finite series amenable to figurate representation in the general form:

$$u_n = B_0 n^0 + B_1 n^1 + B_2 n^2 + B_3 n^3 \textit{ etc.}$$

Let us therefore explore the possibility of expressing the continuous function $y=f(x)$ as a nonterminating series of powers of x, *i.e.*

$$y = B_0 x^0 + B_1 x^1 + B_2 x^2 + B_3 x^3 \textit{ etc.}$$

In the foregoing expression, we are free to equate to zero any value or values of B_n. Consequently, we can endow the curve of the function represented by the sum of such a series – if convergent – with as many or as few turning-points as we please. If we put $x=0$ in the above, so that $y = f(0) = B_0$, we may then successively differentiate as follows:

$$D_x^1 y = B_1 + 2B_2 x + 3B_3 x^2 + 4B_4 x^3 \ \ldots \textit{ etc.}$$
$$\text{and } D_x^1(0) = B_1$$
$$D_x^2 y = 2B_2 + 3.2.B_3 + 4.3.B_4 x^2 \ \ldots \textit{ etc.}$$
$$\text{and } D_x^2(0) = 2!B_2$$
$$D_x^3 y = 3.2.B_3 + 4.3.B_4 x^2 \ \ldots \textit{ etc.}$$
$$\text{and } D_x^3(0) = 3!B_3$$
$$D_x^4 y = 4.3.2.B_4 \ \ldots \textit{ etc.}$$
$$\text{and } D_x^4(0) = 4!B_4$$

In short: $D_x^n(0) = n!B_n$, so that

$$B_n = \frac{D_x^n(0)}{n!}$$

We may now rewrite our power series as:

$$y = f(0) + \frac{D_x(0)x}{1!} + \frac{D_x^2(0)x^2}{2!} + \frac{D_x^3(0)x^3}{3!} + \frac{D_x^4(0)x^4}{4!} \textit{ etc.}$$

From the foregoing formula, one may derive a host of differentiable functions. Consider the following (in radian measure):

$$\begin{array}{lll} y = \sin x & f(0) & = 0 \\ D_x y = \cos x & D_x(0) & = 1 \\ D_x^2 y = -\sin x & D_x^2(0) & = 0 \\ D_x^3 y = -\cos x & D_x^3(0) & = -1 \\ D_x^4 y = \sin x & D_x^4(0) & = 0 \end{array}$$

$$\begin{array}{lll} y = \cos x & f(0) & = 1 \\ D_x y = -\sin x & D_x(0) & = 0 \\ D_x^2 y = -\cos x & D_x^2(0) & = -1 \\ D_x^3 y = \sin x & D_x^3(0) & = 0 \\ D_x^4 y = \cos x & D_x^4(0) & = 1 \end{array}$$

etc.

By substitution in Maclaurin's formula, we therefore obtain:

$$\sin x = x - \frac{x^3}{3!} + \frac{x^5}{5!} - \frac{x^7}{7!} + \frac{x^9}{9!} \cdots$$

$$\cos x = 1 - \frac{x^2}{2!} + \frac{x^4}{4!} - \frac{x^6}{6!} + \frac{x^8}{8!} \cdots$$

Both the above are convergent throughout the whole range $x=0$ to $x=2\pi$ radians and they converge rapidly for values of $x<1$, *i.e.* less than 57°. From the viewpoint of the compiler of high precision tables, the practical advantage of expressing sin x and cos x in series form is therefore considerable. We are now ready to justify the validity of the Binomial Theorem for fractional or negative powers. We have earlier seen that $D_x x^n = nx^{n-1}$ for all rational values of n, positive or negative, fractional or whole, whence also $D_x{}^2 x^n = n(n-1)x^{n-2}$, $D_x{}^3 x^n = n(n-1)(n-2)x^{n-3}$ and more generally $D_x{}^m x^n = n^{(m)}.x^{n-m}$. If we now write $u = 1+x$, we derive:

$$D_x(1+x)^n = D_u u^n . D_x u = n(1+x)^{n-1}$$

More generally, $D_x{}^m(1+x)^n = n^{(m)}(1+x)^{n-m}$, whence $D_x{}^m(0) = n^{(m)}$. By substitution in Maclaurin's formula we therefore derive

$$(1+x)^n = 1 + nx + \frac{n^{(2)}}{2!}x^2 + \frac{n^{(3)}}{3!}x^3 + \frac{n^{(4)}}{4!}x^4 \text{ etc.}$$

$$= 1 + n_{(1)}x + n_{(2)}x^2 + n_{(3)}x^3 + n_{(4)}x^4 \text{ etc.}$$

This function terminates if n is a positive integer. Otherwise it does not. Either way, it exhibits the binomial expansion in a form which is applicable for any value of n within a range of x values which satisfy a convergence test.

Maclaurin's formula also equips us with the means of finding an answer to the question: does there exist a continuous function (E_x) of x whose derivative is itself? If so, $D_x{}^m . E_x = E_x$ for all values of m and $D_x{}^m(0) = E_0$ is constant so that the function which satisfies the condition in the range of x values for which the series converges is expressible in the form:

$$E_x = E_0\left(1 + x + \frac{x^2}{2!} + \frac{x^3}{3!} + \frac{x^4}{4!} \cdots\right)$$

We are free to give E_0 any rational value in this context and write for simplicity $E_0 = 1$ so that

$$E_x = 1 + x + \frac{x^2}{2!} + \frac{x^3}{3!} + \frac{x^4}{4!} \cdots$$

This must be convergent if $x \leqslant 1$ and when $x=1$ we can write

$$E_1 = 1 + 1 + \frac{1}{2!} + \frac{1}{3!} + \frac{1}{4!} \cdots$$

When applying the yardstick criterion for a diminishing series of terms with like signs, it is sometimes helpful to examine the ratio of the $(n+1)$th to the nth term, as for the foregoing series:

$$1 + \frac{x}{1!} + \frac{x^2}{2!} + \frac{x^3}{3!} \cdots \frac{x^n}{n!} + \frac{x^{n+1}}{(n+1)!} \cdots$$

Here we see that the ratio of t_{n+1} to t_n is

$$\frac{t_{n+1}}{t_n} = \frac{n!x^{n+1}}{(n+1)!x^n} = \frac{x}{n+1}$$

For any finite value of x we can here find a finite value of n such that the foregoing ratio is less than say 0·1, the ratio of successive terms of a series representing a recurring decimal. For instance, if $x=10$ and $n=100$ in the above:

$$\frac{t_{n+1}}{t_n} = \frac{10}{101} < \frac{1}{10} \text{ and } \frac{t_{n+2}}{t_{n+1}} = \frac{10}{102} < \frac{1}{10} \text{ etc.}$$

Thus the above is convergent for all finite values of x. It is now usual to denote E_1 as defined above by e (Euler's number), which we have already met. Correct to 25 decimal places,

$$e = 2{\cdot}7182818284590452353602874\ldots$$

The virtue of such a series as an aid to computation is apparent from the summation of the first ten terms. This suffices to guarantee a result correct to 7 significant figures as shown below:

$$1 + \frac{1}{1!} = 2$$

$$1 + \frac{1}{1!} + \frac{1}{2!} = 2{\cdot}5$$

x	0	2	4	6	8	10	20	30	40	50	60	70	80	90	100
$y^a = L_{10}x$	0	0·3010	0·6021	0·7782	0·9031	1·0000	1·3010	1·4771	1·6021	1·6990	1·7782	1·8451	1·9031	1·9542	2·0000
$y^b = L_2x$	0	1·0000	2·0000	2·585	3·0000	3·322	4·332	4·907	5·322	5·644	5·907	6·019	6·322	6·491	6·644
$y^c = L_ex$	0	0·7138	1·387	1·792	2·079	2·3026	2·995	3·401	3·688	3·899	4·095	4·248	4·382	4·499	4·605

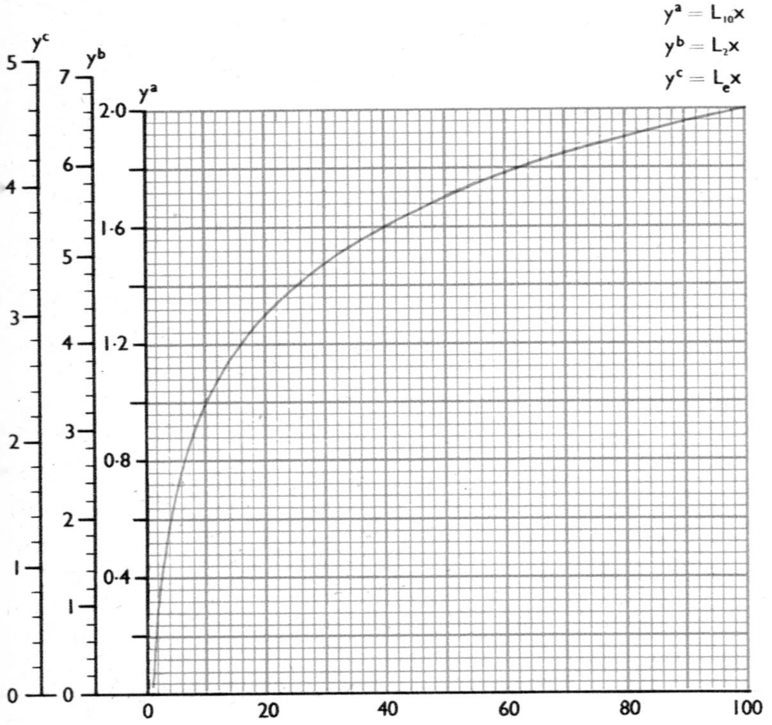

If $2^z = 10$ and $10^y = x$, $z = L_2 10$ and $y = L_{10}x$.
Also $10^y = 2^{yz}$ so that $yz = L_2x$.

$$\therefore L_2x = L_2 10 \,.\, L_{10}x \text{ or } L_{10}x = \frac{L_2x}{L_2 10}$$

If $y_a = L_{10}x$ and $y_b = L_2x$, $(L_2 10)y_a = y_b$. Thus the change of base involves merely a scalar change of the y-axis. Similarly:

$$L_{10}x = \frac{L_ex}{L_e 10}.$$

The value of $L_e 10$ is 2·3026

By definition, pressure (p) is force (F) per unit area (A) and work done (W) is the product of force and distance (l). During a small excursion ($\triangle l$) we may regard F as constant so that the increment of work is $\triangle W \simeq F.\triangle l$ or $\triangle W \simeq p.A.\triangle l$, and $A.\triangle l$ is the corresponding change of volume $\triangle v$. Thus $\triangle W \simeq p.dv$. The gas law states that $pv = RT$. At constant temperature T we may therefore write

$$\triangle W \simeq RT.\frac{1}{v}.dv.$$

In a change from v_2 to v_1

$$W = RT\int_{v_1}^{v_2} \frac{1}{v}.dv = RT\,(\log v_2 - \log v_1)$$

$$= RT \log \frac{v_2}{v_1}$$

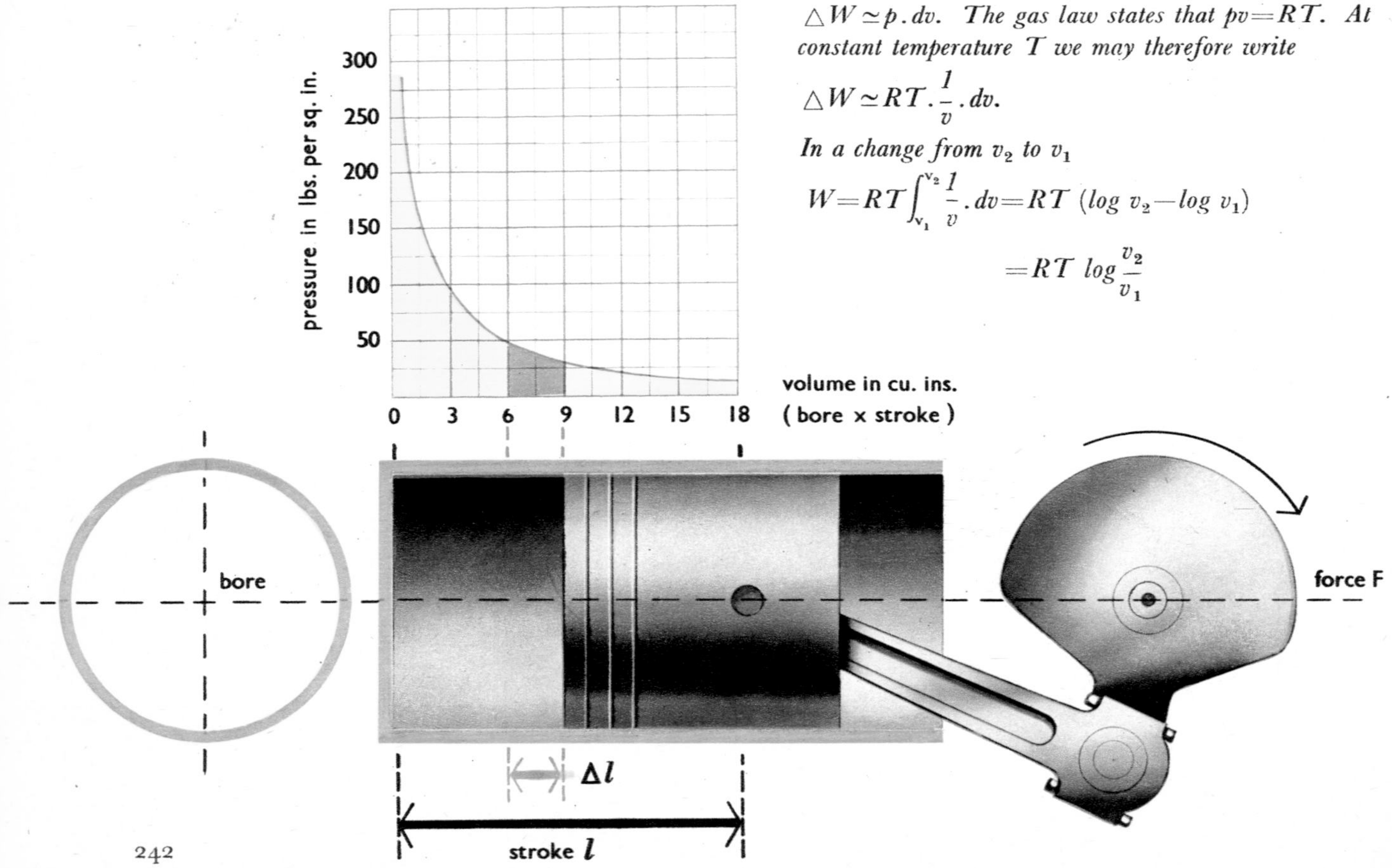

$$1+\frac{1}{1!}+\frac{1}{2!}+\frac{1}{3!}=2{\cdot}\dot{6}$$

$$1+\frac{1}{1!}\;\cdots\;\frac{1}{4!}=2{\cdot}708\dot{3}$$

$$1+\frac{1}{1!}\;\cdots\;\frac{1}{5!}=2{\cdot}71\dot{6}$$

$$1+\frac{1}{1!}\;\cdots\;\frac{1}{6!}=2{\cdot}7180\dot{5}$$

$$1+\frac{1}{1!}\;\cdots\;\frac{1}{7!}=2{\cdot}7182539683$$

$$1+\frac{1}{1!}\;\cdots\;\frac{1}{8!}=2{\cdot}7182787698$$

$$1+\frac{1}{1!}\;\cdots\;\frac{1}{9!}=2{\cdot}7182815256$$

So far, we have shown that there does exist a function of x with the entertaining property that any derivative of itself is also itself; and we have exhibited it as a convergent power series of x. We have not as yet a clue to its most useful property until we can express it alternatively in a more compact form comparable to $y=(1+x)^n$. To explore what makes it useful, let us first be clear about what meaning we may attach to E_{nx}, *i.e.* the series obtained if we replace x by nx in E_x, so that

$$E_{nx}=1+nx+\frac{n^2x^2}{2!}+\frac{n^3x^3}{3!}+\frac{n^4x^4}{4!}\;etc.$$

If we hold n constant and allow x to vary, we regard E_{nx} so defined as a continuous function of x alone. We can also tabulate it for different values of n and a fixed value of x, in which event we may regard it as a continuous function of n alone. On this understanding, let us explore the properties of another function of x and/or n, *viz.* $F_x=E_x{}^n$, so that:

$$D_x.F(x)=n.E_x{}^{n-1}.E_x=n.E_x{}^n$$
$$D_x{}^2.F(x)=n.D_x.E_x{}^n=n^2.E_x{}^n$$

Whence $D_x{}^3.F(x)=n^3.E_x{}^n$, and more generally

$$D_x{}^m.F(x)=n^m.E_x{}^n$$

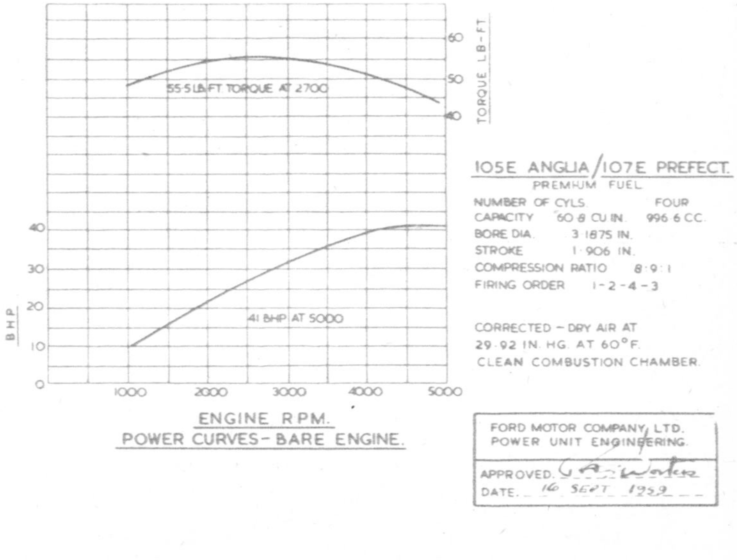

Top: Cut-through of a Ford engine. Bottom: Indication of mean effective pressure for the same engine.

Now $E_0=1=E_0^n$, so that

$$D_x^m . F(0)=n^m$$

If we now substitute $D_x.F(0)$, $D_x^2.F(0)$ *etc.* in the Maclaurin formula, we obtain

$$F(x)=1+nx+\frac{n^2x^2}{2!}+\frac{n^3x^3}{3!}+\frac{n^4x^4}{4!} \quad etc.$$

$$\therefore \ E_x^n=E_{nx} \quad \text{and} \quad E_1^n=E_n$$

$$\therefore \ e^n=1+n+\frac{n^2}{2!}+\frac{n^3}{3!}+\frac{n^4}{4!} \quad \ldots$$

The above is a pure function of n and it is immaterial what consonant we substitute for it, *i.e.* we may write

$$e^x=1+x+\frac{x^2}{2!}+\frac{x^3}{3!}+\frac{x^4}{4!} \quad \ldots$$

From the viewpoint of the compiler of a table of logarithms, this property of e makes it much more convenient than any other number for use as a base, and the more so because the series converges rapidly for $x \leqslant 1$. It also equips us with the means of differentiating a considerable assemblage of functions not dealt with in Chapter 10. Thus we may write a^x in the form e^{cx}, if we substitute $a=e^c$ so that $c=\log_e a$, whence

$$D_x . a^x=ce^{cx}=\log_e a . a^x$$

$$D_x^{-1} . a^x=\frac{1}{\log_e a} a^x+C$$

If $y=\log_e x$, we may write $x=e^y$, so that

$$D_y . x=x \quad \text{and} \quad D_x . y=\frac{1}{D_y . x}=\frac{1}{x}$$

$$\therefore \ D_x . \log_e x=\frac{1}{x} \quad \text{and} \quad D_x^{-1}\left(\frac{1}{x}\right)=\log_e x+C$$

The reader can check that

$$D_x . \log_e(1+x)=\frac{1}{1+x} \quad \text{and} \quad D_x^{-1}\left(\frac{1}{1+x}\right)$$

$$=\log_e(1+x)+C$$

In the idiom of Chapter 8, the determination of e^x for equally-spaced values of x yields a table whose cell entries are anti-logarithms to base e. For values of $n \leqslant 2$, we can obtain an expression for $\log n$, whence tables whose cell entries are logarithms to base e by using Maclaurin's recipe to find a series for $f(x)=\log_e(1+x)$. Since $\log_b 1=0$ for all b we here have $f(0)=0$. Also $D_x . f(x) =+(1+x)^{-1}$; $D_x^2 . f(x)=-(1+x)^{-2}$; $D_x^3 . f(x) =+2(1+x)^{-3}$; $D_x^4 . f(x)=-2.3(1+x)^{-4}$ *etc.* We thus obtain

$$f(0)=0;\ D_x f(0)=+1;\ D_x^2 f(0)=-1;$$
$$D_x^3 f(0)=+2!;\ D_x^4 f(0)=-3! \ etc.$$

Our series, which will be meaningful only if convergent, is thus:

$$\log_e(1+x)=x-\frac{x^2}{2}+\frac{x^3}{3}-\frac{x^4}{4}+\frac{x^5}{5} \quad \ldots$$

We might have reached the same result by another route; by direct division $(1+x)^{-1}=1-x+x^2-x^3$ *etc.*, so that

$$\log_e(1+x)+C_1=D_x^{-1}\left(\frac{1}{1+x}\right)$$

$$=D_x^{-1}(1-x+x^2-x^3+x^4 \ldots)$$

$$\log_e(1+x)+C_1$$
$$=\left(x-\frac{x^2}{2}+\frac{x^3}{3}-\frac{x^4}{4}+\frac{x^5}{5} \quad \ldots \right)+C_2$$

Since $\log_e(1+x)=\log_e(1)=0$ when $x=0$, the expression in brackets on the right vanishes when $x=0$. So the two arbitrary constants (C_1 and C_2) in the above are equal and cancel out. The foregoing series for $\log_e(1+x)$ consists of *diminishing* terms with alternating signs only if $x \leqslant 1$. As it stands, it therefore takes us up only to $\log_e 2$; but if we want $\log_e 3$ *etc.* we apply the rule:

$$L_e 3=L_e 2(1.5)=L_e 2+L_e(1.5)$$

The anti-derivative of $\log_e x$ is obtainable by recourse to Rule 6 (Chapter 10) if we write it as:

$$D_x^{-1}(1 . \log_e x)=\log_e x . D_x^{-1}(1)-D_x^{-1} . \frac{1}{x} . x$$

$$D_x^{-1}(\log_e x)=x . \log_e x-x+C$$

We may now also obtain the anti-derivative of tan x. With this end in view, it will be convenient to add three new names to our dictionary of trigonometrical ratios, *viz.* *cosec(ant)*, *sec(ant)* and *cot(angent)*, defined as: $\text{cosec } A = (\sin A)^{-1}$; $\sec A = (\cos A)^{-1}$; $\cot A = (\tan A)^{-1}$. Since $L_b . x^n = n . L_b . x$, we may always write $-L_b . x = L_b . x^{-1}$, whence $-\log \sin A = \log \text{cosec } A$; $-\log \cos A = \log \sec A$; $-\log \tan A = \log \cot A$. We obtain the now familiar relation $(\sin A)^2 + (\cos A)^2 = 1$, if we divide the terms of the Pythagorean identity $p^2 + b^2 = h^2$ by h^2. If we divide throughout by b^2, we get

$$1 + (\tan A)^2 = (\sec A)^2 \text{ so that } D_x . \tan A = (\sec A)^2$$

We obtain the anti-derivative of tan x by the use of Rule 5 (Chapter 10). If $u = \cos x$:

$$D_x^{-1} . \tan x = D_u^{-1}(\tan x . D_u x) = D_u^{-1}\left(\frac{\sin x}{u} . D_u . x\right)$$

In this expression

$$D_u . x = \frac{1}{D_x u} = -\frac{1}{\sin x}$$

$$\therefore D_x^{-1} . \tan x = D_u^{-1}\left(-\frac{1}{u}\right) = -\log_e u + C$$

$$D_x^{-1} . \tan x = -\log_e \cos x + C = \log_e . \sec x + C$$

As our diagram (p. 243) shows, the anti-derivative of the logarithmic function is the most important mathematical function for the study of gas and solution pressure; and our mathematical tool-bag now contains the wherewithal to describe a very familiar mechanical phenomenon of every-day life, *i.e.* the curve of a cord or chain suspended at both ends. For y (vertical) and x (horizontal), we now write the descriptive function in the form $y = A . \cosh(bx)$ or by appropriate change of scale $Y = \cosh(X)$, in which

$$\cosh X = \frac{e^X + e^{-X}}{2}$$

This is one of a class of functions of e^x with very important properties, *inter alia* for elucidating anti-derivatives by Rule 5 of Chapter 10. To get more insight into the family, we shall now adventurously leave that of real algebra for the imaginary domain, following a track first explored about A.D. 1700. In doing so, we shall assume, as mathematicians of that time did, what posterity justified later, *viz.* that the fundamental rules of real algebra (addition, subtraction, multiplication, division) still work if we write $i^2 = -1$. This need not commit us to the unwarrantable implication that there is a so-called number $i = \sqrt{-1}$. On the contrary, we adhere to the rule that i, like (-1) as a multiplier, is an operation. We may then (exploratively) write:

$$e^{ix} = 1 + ix + \frac{i^2x^2}{2!} + \frac{i^3x^3}{3!} + \frac{i^4x^4}{4!} + \frac{i^5x^5}{5!} + \frac{i^6x^6}{6!} \textit{ etc.}$$

$$= \left(1 + \frac{i^2x^2}{2!} + \frac{i^4x^4}{4!} + \frac{i^6x^6}{6!} \cdot\cdot\right)$$

$$+ i\left(x + \frac{i^2x^3}{3!} + \frac{i^4x^5}{5!} \cdot\cdot\right)$$

$$= \left(1 - \frac{x^2}{2!} + \frac{x^4}{4!} - \frac{x^6}{6!} \cdot\cdot\right) + i\left(x - \frac{x^3}{3!} + \frac{x^5}{5!} \cdot\cdot\right)$$

$$= \cos x + i \sin x$$

Similarly $e^{-ix} = \cos x - i \sin x$, so that

$$\frac{e^{ix} + e^{-ix}}{2} = \cos x\,; \quad \frac{e^{ix} - e^{-ix}}{2i} = \sin x$$

What, if anything, this means we need not here pause to discuss. Our only concern at this stage is to interpret cosh x as one of a battery of functions with many analogies to those of classical (circular) trigonometry. It is customary to call them the hyperbolic trigonometrical ratios: *sinh x* (pronounced *sinsh*), *cosh x* and *tanh x* (pronounced *tansh*). We define them as below:

$$\sinh x = \frac{e^x - e^{-x}}{2} \quad \cosh x = \frac{e^x + e^{-x}}{2} \quad \tanh x = \frac{\sinh x}{\cosh x}$$

It will suffice to exhibit step by step two of many similarities between a hyperbolic and a circular trigonometrical ratio:

(i) $D_x(\cosh . x) = \frac{1}{2}D_x . e^x + \frac{1}{2}D_x . e^{-x} = \frac{e^x - e^{-x}}{2} = \sinh . x$

Three stages in the construction of Brunel's famous suspension bridge at Clifton, Bristol, England. Work on the bridge was halted in the 1830s and not resumed until 1861, two years after Brunel's death. The top photograph shows the first chain in position. The coloured print (drawn from imagination before the work was completed, and therefore showing the towers as Brunel planned them and not as they were actually built) gives a direct view of the catenary curve which carries the road across the gorge.

(ii) $(\sinh . x)(\cosh . y)+(\cosh . x)(\sinh . y)$

$$=\frac{e^{x+y}+e^{x-y}-e^{-x+y}-e^{-x-y}}{4}$$

$$+\frac{e^{x+y}-e^{x-y}+e^{-x+y}-e^{-x-y}}{4}$$

$$=\frac{e^{x+y}-e^{-x-y}}{2}=\sinh (x+y)$$

The reader should now be able to check up the following similarities as a brush-up-your-maths exercise:

Hyperbolic	**Circular**
$\sinh 0=0$; $\cosh 0=1$	$\sin 0=0$; $\cos 0=1$
$(\cosh x)^2-(\sinh x)^2=1$	$(\cos x)^2+(\sin x)^2=1$
$\sinh(x+y)=\sinh x . \cosh y + \cosh x . \sinh y$	$\sin(x+y)=\sin x . \cos y + \cos x . \sin y$
$\cosh(x+y)=\cosh x . \cosh y + \sinh x . \sinh y$	$\cos(x+y)=\cos x . \cos y - \sin x . \sin y$
$\cosh 2x=2(\cosh x)^2-1$	$\cos 2x=2(\cos x)^2-1$
$\sinh 2x=2 \sinh x . \cosh x$	$\sin 2x=2 \sin x . \cos x$
$D_x . \sinh x=\cosh x$	$D_x . \sin x=\cos x$
$D_x . \cosh x=\sinh x$	$D_x . \cos x=-\sin x$
$D_x . \sinh^{-1} x=\frac{1}{\sqrt{(1+x^2)}}$	$D_x . \sin^{-1} x=\frac{1}{\sqrt{(1-x^2)}}$
$D_x . \tanh^{-1} x=\frac{1}{1-x^2}$	$D_x . \tan^{-1} x=\frac{1}{1+x^2}$
$D_x^2 . \sinh x=\sinh x$	$D_x^2 . \sin x=-\sin x$
$D_x^2 . \cosh x=\cosh x$	$D_x^2 . \cos x=-\cos x$

Let us now return to the imaginary domain where we picked up the so-called complex number $e^{ix}=\cos x+i \sin x$. If we substitute $na=x$, this becomes

$$\cos na+i \sin na=e^{ina}=(e^{ia})^n=(\cos a+i \sin a)^n$$

Similarly, since $e^{-ix}=\cos x-i \sin x$:

$$(\cos A+i \sin A)^{-n}=\cos nA-i \sin nA$$

This identity is the keystone of the algebra of the complex domain. Its discovery, though not by the same route as above, was due to de Moivre (1667–1754), a Huguenot who settled in London after the revocation of the Edict of Nantes. Conceivably, he hit upon it by direct multiplication from $(\cos a+i \sin a)^2$ and $(\cos a+i \sin a)^3$ with substitution of the familiar formulae for $\cos 2a$ and $\sin 2a$ or $\cos 3a$ and $\sin 3a$ in the products; but Cotes (circa 1710), a British mathematician, had already formulated the rule $ix=\log_e(\cos x+i \sin x)$. This is equivalent to $e^{ix}=\cos x+i \sin x$, and leads (as above) to the

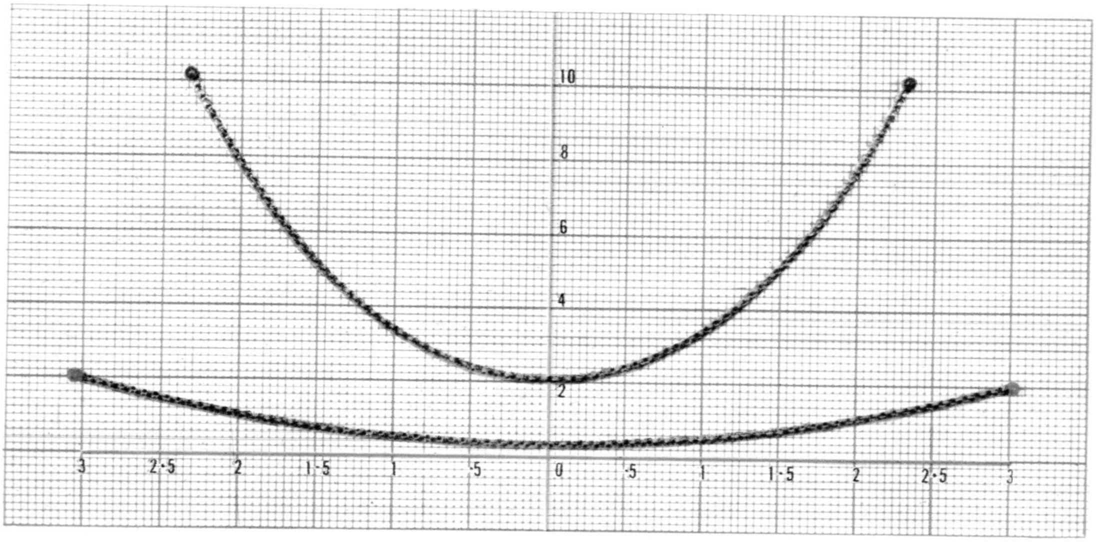

The catenary curve takes its name from the Latin catena, *a chain. The two chains above describe the curves $y=2 \cosh x$ and $y=\frac{1}{5} \cosh x$ respectively.*

same result. For positive integral values of n, the proof of de Moivre's theorem by induction is straightforward, *viz.*:

$$(\cos a+i\sin a)^{n+1}$$
$$=(\cos a+i\sin a)(\cos na+i\sin na)$$
$$=(\cos a.\cos na-\sin a.\sin na)+$$
$$i(\sin a.\cos na+\cos a.\sin na)$$
$$=\cos(n+1)a+i\sin(n+1)a$$

On the foregoing, though not as yet justified, assumption that we can mechanically apply the rules of real arithmetic to complex numbers by the substitution of -1 for i^2 wherever the latter occurs, the validity of the theorem for values of n which are rational fractions follows simply. If p and q are integers:

$$\left(\cos\frac{pA}{q}+i\sin\frac{pA}{q}\right)^q=\cos pA+i\sin pA$$
$$=(\cos A+i\sin A)^p$$
$$\therefore \cos\frac{pA}{q}+i\sin\frac{pA}{q}=(\cos A+i\sin A)^{\frac{p}{q}}$$

De Moivre's theorem has several useful applications from the viewpoint of the compiler of trigonometrical tables, as illustrated by the following evaluation which depends on the fundamental property of complex numbers expressed in the rule: equate real to real, imaginary to imaginary, *i.e.* $a+bi=c+di$ if (and only if) $a=c$, $b=d$:

$$(\cos 5a+i\sin 5a)=(\cos a+i\sin a)^5$$
$$=(\cos a)^5+5i(\cos a)^4.\sin a+10i^2(\cos a)^3.(\sin a)^2 \textit{ etc.}$$
$$=(\cos a)^5-10(\cos a)^3.(\sin a)^2+5(\cos a)(\sin a)^4$$
$$+i[5(\cos a)^4.\sin a-10(\cos a)^2.(\sin a)^3+(\sin a)^5]$$
$$\therefore \cos 5a=(\cos a)^5-10(\cos a)^3.(\sin a)^2+5(\cos a)(\sin a)^4$$
$$\text{and } \sin 5a=5(\cos a)^4.\sin a-10(\cos a)^2.(\sin a)^3+(\sin a)^5$$

The converse possibility of expressing $(\cos A)^n$, $(\sin A)^n$ in terms of a series of terms of the form $(\cos nA)$ has a less obvious utility. It is, in fact, a device by which we can obtain anti-derivatives, whence devices for integration, of functions which play an important role in physical problems involving wave motions and oscillatory phenomena.

A sufficient reason for regarding de Moivre's theorem as the cornerstone of complex algebra is that any complex number is expressible in trigonometrical form. In geometrical terms we visualise a complex number $a+bi$ as a point in a plane specified by a displacement of magnitude a along the real axis and a displacement b vertical thereto. The distance (d) of this point from the origin is the hypotenuse of a right-angled triangle, being the radius vector of the point inclined to the real axis at an angle $A=\tan^{-1}\left(\frac{b}{a}\right)$. We may thus write:

$$d.\sin A=b;\ d.\cos A=a;\ d=\sqrt{(a^2+b^2)}$$
$$\therefore a+bi=\sqrt{(a^2+b^2)}.(\cos A+i\sin A)$$

In conformity with the contemporary situation when the periodic concept of the trigonometrical ratios was at best nascent, we have displayed the theorem of de Moivre, as if $y=\sin A$ or $y=\cos A$ has a meaning uniquely determined by A. Actually, we now interpret $\sin A$ as $\sin(A+2k\pi)$ for $k=0, 1, 2$ *etc.* So also for $\cos A$; and it is usual to speak of $A+2k\pi=A$ as the principal value of the function, *i.e.* its value when $k=0$. This emendation has a particular interest in connexion with an issue which, though of no practical use, played a part in the eighteenth-century theory of algebraic equations. Since it is not without entertainment value, it is worthy of brief comment.

As we all know, $+1$ has two real square roots, *viz.* $+1$ and -1; but it has only one cube, fifth *etc.* real root, as also has -1. Let us consider how we can express either $+1$ or -1 in complex terms. We take advantage of the fact that $\cos 2\pi=1$, $\cos \pi=-1$ and $\sin 2\pi=0=\sin \pi$. More generally if $k=1, 2, 3$ *etc.*

$$\cos 2k\pi=+1;\ \cos(2k-1)\pi=-1$$
$$\sin 2k\pi=0=\sin(2k-1)\pi$$

Whence we shall write:

$$\cos 2k\pi+i\sin 2k\pi=+1$$
$$\cos(2k-1)\pi+i\sin(2k-1)\pi=-1$$

By de Moivre's theorem, therefore:

$$\sqrt[3]{+1}=(\cos 2k\pi+i\sin 2k\pi)^{\frac{1}{3}}$$

$$=\cos\frac{2k\pi}{3}+i\sin\frac{2k\pi}{3}$$

For $k=1, 2, 3$, after which we ring the changes on the same three values, this leads to

	$\frac{2\pi}{3}$	$\frac{4\pi}{3}$	$\frac{6\pi}{3}$
cos	$-\frac{1}{2}$	$-\frac{1}{2}$	1
sin	$+\frac{\sqrt{3}}{2}$	$-\frac{\sqrt{3}}{2}$	0

Thus we obtain as the cube roots of $+1$:

$$\frac{-1+i\sqrt{3}}{2}\;;\quad\frac{-1-i\sqrt{3}}{2}\;;\quad+1$$

Similarly, we derive for the cube roots of -1:

$$\frac{1+i\sqrt{3}}{2}\;;\quad\frac{1-i\sqrt{3}}{2}\;;\quad-1$$

The reader who wants to play with complex numbers may check the following four roots of $(4-3i)^{\frac{1}{4}}$:

$$5^{\frac{1}{4}}(0{\cdot}986\pm0{\cdot}168\,i)\qquad5^{\frac{1}{4}}(-0{\cdot}986\pm0{\cdot}168\,i)$$

(*Clue:* $\tan^{-1}\frac{3}{4}=38{\cdot}7°$)

In treating complex numbers as if we could manipulate them in accordance with the algorithms of our schooldays, we followed a course consistent with the entirely pragmatic temper of discovery in the century of de Moivre and of Leibniz; but we have not as yet given any justification for the intrusion of the signs $+$ or $-$ with a novel significance in $a\pm bi$. Nor does the fact that couplets linked by one or other of them turn up in the so-called complex solution of an otherwise insoluble quadratic equation dispel the aura of mystery which the word *imaginary* for the second component of the couplet so appropriately conveys. The eighteenth-century mathematicians pressed forward in the faith that they were on to a good thing; but they had indeed little more than faith to sustain their adventures. We shall return to this theme at a later stage.

One outcome of our survey of the study of series during the period A.D. 1670–1770 is that we can now add to our dictionary of derivatives and anti-derivatives some important additional entries as shown below:

y	$D_x.y$	$D_x^{-1}.y$
$K.e^{cx}$	$Kc.e^{cx}$	$\frac{K}{c}e^{cx}+C$
a^x	$a^x.\log_e a$	$\frac{a^x}{\log_e a}+C$
$\frac{1}{x}$	$\frac{-1}{x^2}$	$\log_e x+C$
$\log_e x$	$\frac{1}{x}$	$x.\log_e x-x+C$
$\sinh x$	$\cosh x$	$\cosh x+C$
$\cosh x$	$\sinh x$	$\sinh x+C$
$\frac{1}{(x^2+a^2)^{\frac{1}{2}}}$	$\frac{-x}{(x^2+a^2)^{\frac{3}{2}}}$	$\sinh^{-1}\frac{x}{a}+C$
$\frac{1}{(x^2-a^2)^{\frac{1}{2}}}$	$\frac{-x}{(x^2-a^2)^{\frac{3}{2}}}$	$\cosh^{-1}\frac{x}{a}+C$

Chapter 12 The Division of the Stakes

Contemporaneously with the themes of the last three chapters, but primarily relevant to problems of enumeration in contradistinction to those of measurement, a mathematical theory of probability took shape with little anticipation of what ambitious claims later generations would advance on behalf of its usefulness in the world's work. Strictly speaking, we must regard it as a branch of applied mathematics which invokes very few devices peculiar to its own domain; and its chief interest from the viewpoint of this book is how far its applications are justifiable.

The episode starts with a correspondence between Pascal and Fermat (1654) about the fortunes and misfortunes of a French nobleman in the social context of irresponsible gambling among the fashionable set of his age. The Chevalier de Méré was a great gambler and by that token *très bon esprit*; but alas, as Pascal wrote, *il n'est pas géomètre*. He made his pile by betting favourable odds on a score of at least one six in four tosses of a cubical die. He went bankrupt by betting small odds on getting at least one double six in twenty-four double tosses. There is thus no ambiguity about the practical issue from which what we may call the *classical* view emerged. It is the theme song of the later treatise of James Bernoulli and the main concern of all who discussed probability from a mathematical viewpoint during the period between 1680 and 1780. In brief: how should we adjust the stakes to be sure of winning, if we go on playing long enough?

Implicitly, we then invoke several factual assumptions which have nothing to do with mathematical sagacity. One is that we make *in advance* a rule to which we must adhere consistently in an endless succession of games. A second is that the toy, *i.e.* a pack of cards, roulette wheel, state lottery urn or die, remains the same throughout the unending game. What is no less important is that we assume a so-called *randomising* process, *e.g.* thoroughly shuffling the cards or tossing the die according to an agreed prescription. The implications of the last statement are difficult to state precisely. Provisionally, we may say that any such recipe imposes two conditions: (a) the sensory discrimination of the agent is powerless to influence the outcome of the programme; (b) no external agencies intervene to effect any orderly rhythm on the outcome of the agent's intervention. Whether the situation can or does conform to these requirements is a factual issue irrelevant to the domain of pure mathematics.

On this understanding, and against the background of a type specimen of real situations to which the Founding Fathers conceived a calculus of probability to be relevant, let us now examine Pascal's recipe for safeguarding the fortunes of a dissolute nobility in the casinos of his own time. We may choose a somewhat more sophisticated dilemma than that of the intrepid Chevalier by asking how to state a rule of division of the stakes to ensure *eventual* success to a gambler who bets on taking five cards simultaneously from a well-

A Darcis print of gaming in eighteenth-century France. It was against a background of gambling on games of chance that a mathematical theory of probability took shape in the seventeenth and eighteenth centuries. The prime concern of all the Founding Fathers of the theory was how to adjust the stakes so as to be sure of winning if we go on playing long enough. Today a more important question is how far the theory is at once valid and useful in other domains.

shuffled pack. Accordingly, we shall assume he wagers that: (a) three of the cards will be pictures; (b) two of the cards will be aces.

First, we must establish a master rule of the combinatory calculus with which we have at least gained a nodding acquaintance, if not previously, in Chapter 7. The rule, by no means obviously relevant to the practical issue, answers the following question: if we choose r objects among n different ones numerically classifiable as a, b, c, *etc.* by some criterion A, B, C, *etc.* (*e.g.* pictures, aces and others), what number of different ordered selections (linear permutations) laid out in a row will satisfy the condition that the number u will be of class A, v of class B and w of class C? In terms of a card pack of $n=52$ objects, a $(=12)$ may stand for picture cards, b $(=4)$ for aces and c $(=36)$ for others; and if our question is how many linear permutations of $r=5$ selections we can specify on the assumption that three cards will be picture cards and two will be aces: $u=3$, $v=2$ and $w=0$ in the foregoing statement of the problem. To make it wholly explicit, it is also necessary to define our criterion of choice. If we choose five cards from a pack, we may do so: (a) by removing them simultaneously or one at a time *without replacement*, in which event the size of the pack diminishes by unity after each withdrawal; (b) one at a time after random-wise insertion of its predecessor in the pack, in which event the size of the pack remains the same.

A few terms are useful in this context. We may speak of the n different objects as the *universe of choice* or *universal set*, of any selection (*subset*) of r objects therefrom as an r-fold *sample* or of a subset of r objects each specified in advance as the *event*. The event itself thus corresponds to a particular combination of objects, embracing a particular number of ordered sequences. By ordered sequences, we signify *linear* permutations as of cards laid face upwards in a row. We may note that the sampling process involved in tossing a die or in spinning a lottery wheel is intrinsically *repetitive*, *i.e.* like sampling *with replacement* from a card pack. Contrariwise, drawing numbered tickets from a so-called lottery urn is customarily a sampling process of the other type, if the regulations exclude replacement.

We may lay out a schema for the r-fold sample with respect to three classes as below:

Classes	A	B	C	*Total*
Universe	a	b	c	n
Sample	u	v	w	r

The rule is not hard to visualise. If we denote by $P_{r.uvw..}$ the number of linear permutations consistent with repetitive choice and by $P_{(r.uvw..)}$ the number consistent with non-replacement sampling:

$$P_{r.uvw..}=\frac{r!}{u!v!w!..}a^{u}.b^{v}.c^{w}...$$

$$P_{(r.uvw..)}=\frac{r!}{u!v!w!..}a^{(u)}.b^{(v)}.c^{(w)}...$$

Here the dots signify that we can extend the pattern for three (or two or one) classes to any number without ambiguity if we recall that $0!=1$. On this understanding, and in conformity with the usage of the Founding Fathers, we now define the *mathematical probability* of the r-fold event as the ratio of all linear permutations *consistent with its specification* to the *total* number of possible r-fold linear permutations of the n objects in the universal set. As we have seen in Chapter 7, this total is n^{r} if sampling is repetitive, otherwise $n^{(r)}$. For the event specified in the foregoing schema, we may thus write our ratio alternatively:

With replacement

$$p_{r}=\frac{r!}{u!v!w!}\cdot\frac{a^{u}b^{v}c^{w}}{n^{r}}=\frac{r!}{u!v!w!}\left(\frac{a}{n}\right)^{u}\left(\frac{b}{n}\right)^{v}\left(\frac{c}{n}\right)^{w}$$

Without replacement

$$p_{(r)}=\frac{r!}{u!v!w!}\cdot\frac{a^{(u)}b^{(v)}c^{(w)}}{n^{(r)}}$$

The reader not as yet sufficiently familiar with the notation used above will have opportunities for translation as a quiz challenge by turning back to the end of Chapter 3 and the latter part of Chapter 7. Two examples will suffice to remove any misunderstanding about the meaning of these formulae:

(1) What is the probability of getting 2 *fours* and 3 *others* in a five-fold toss of a cubical die? Here the universe consists of six different objects (*faces*) classifiable as A fours and B others, so that $n=6$, $r=5$, $a=1$, $b=5$, $u=2$ and $v=3$:

$$p_5 = \frac{5!}{2!3!}\left(\frac{1}{6}\right)^2\left(\frac{5}{6}\right)^3 = \frac{10(125)}{36(216)} = \frac{625}{3888}$$

(2) If a pack of ten cards has face scores with 1, 2, 3..9, 10 pips, what is the probability of drawing *simultaneously* a six-fold sample consisting of (a) *one* ace; (b) *three* cards with 3, 4 or 5 pips; (c) *two* others? Here $n=10$, $r=6$, $a=1$, $b=3$, $c=6$, $u=1$, $v=3$, and $w=2$, so that:

$$p_{(6)} = \frac{6!}{1!3!2!}\,\frac{1^{(1)}.3^{(3)}.6^{(2)}}{10^{(6)}} = \frac{(60)\ (1)\ (3.2.1)\ (5.6)}{10.9.8.7.6.5} = \frac{1}{14}$$

One's sympathies will be with any reader unable to discern what exiguous (if any) relevance a definition of probability in such terms as the above may have to the domestic economy of the gallant Chevalier; but we shall here pause to note one of its implications. If x is the number of linear permutations consistent with the specification of the r-fold event and y is the number of all linear permutations of r objects taken from the same universal set, $y-x$ is the number referable to the subsets which are *not* consistent with the

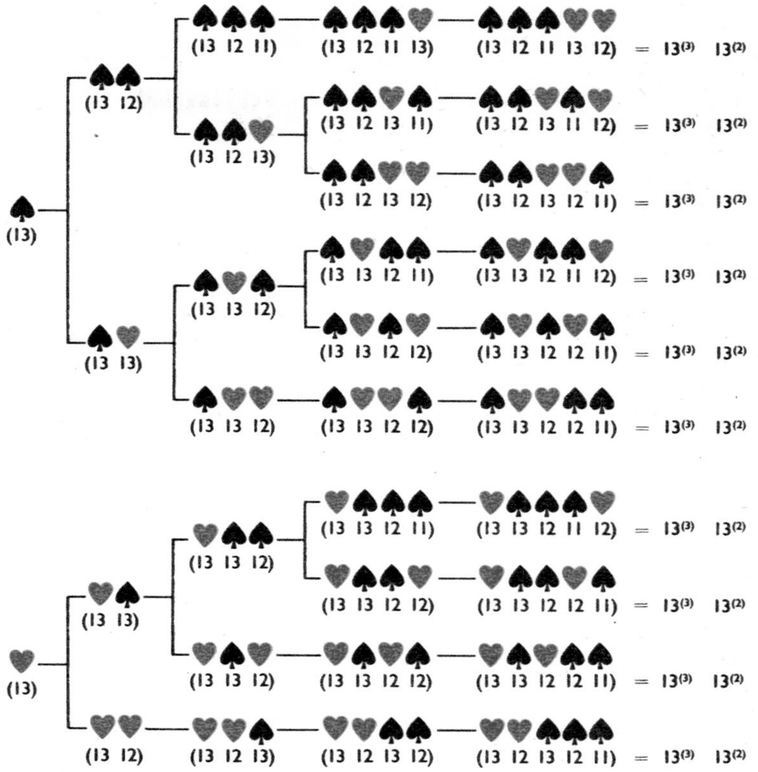

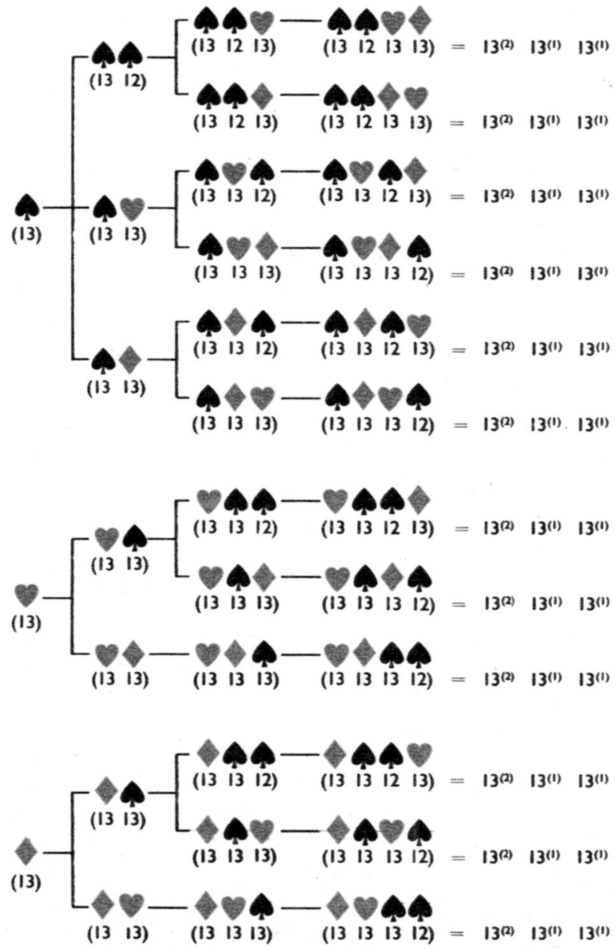

Top diagram shows how many different ways there are of taking three spades and two hearts from a full pack of fifty-two cards, without replacement. First, second, third, fourth and fifth choice of card are built up from left to right. Figures under suit-symbols denote chances remaining at each stage of choosing a card of that suit. The possibilities number $\frac{5!}{3!\ 2!}13^{(3)}.13^{(2)}$ $=2676960$. *Bottom diagram shows different ways of taking two spades, one heart and one diamond from a full pack, without replacement. The number of possibilities of doing so is* $\frac{4!}{2!\ 1!\ 1!}13^{(2)}.13^{(1)}.13^{(1)}=316368$.

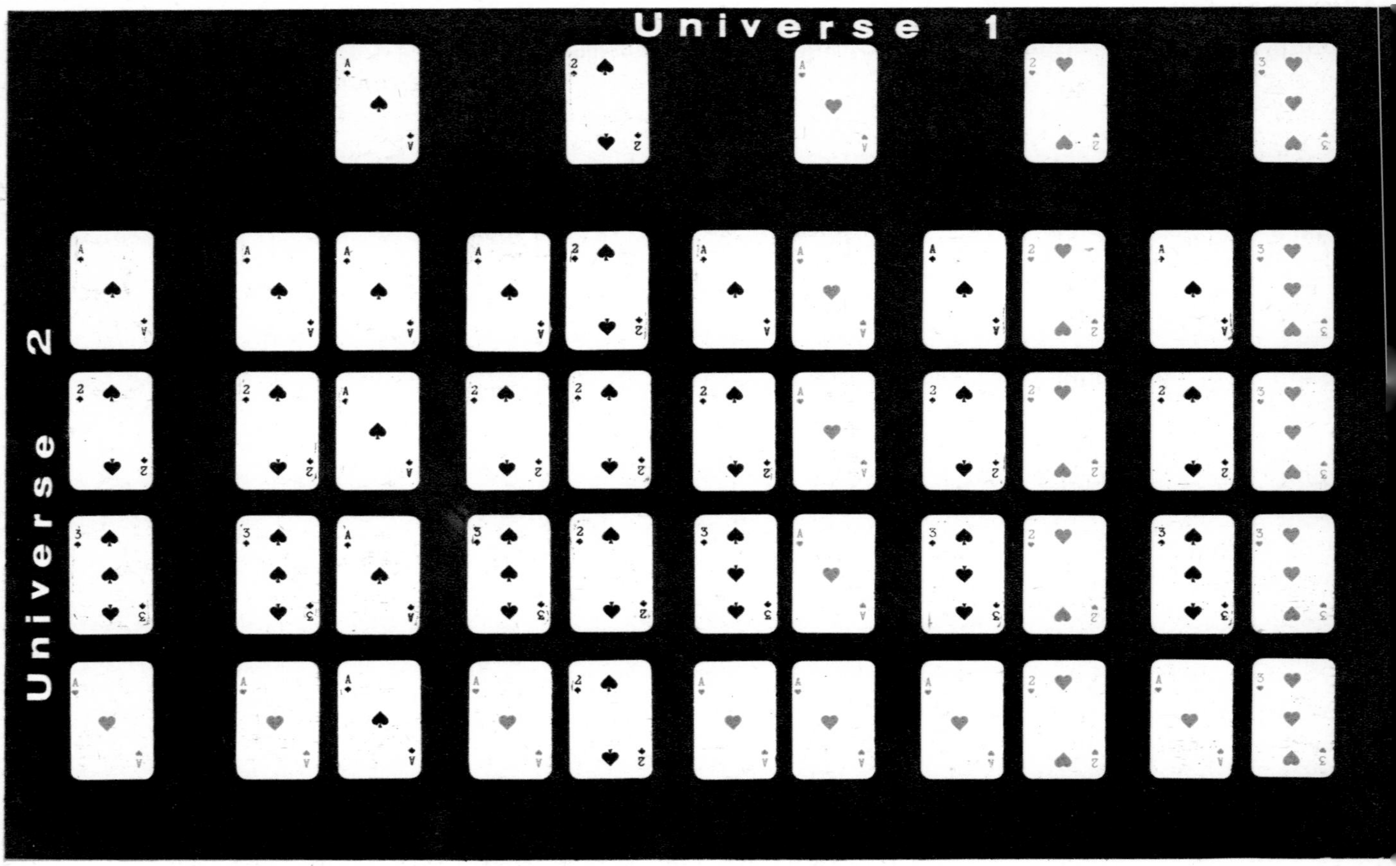

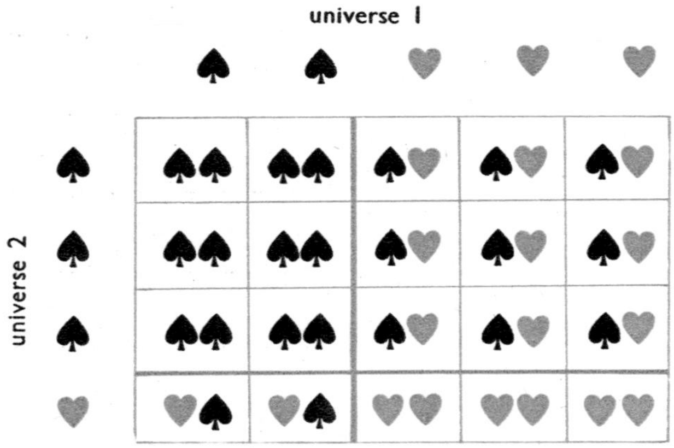

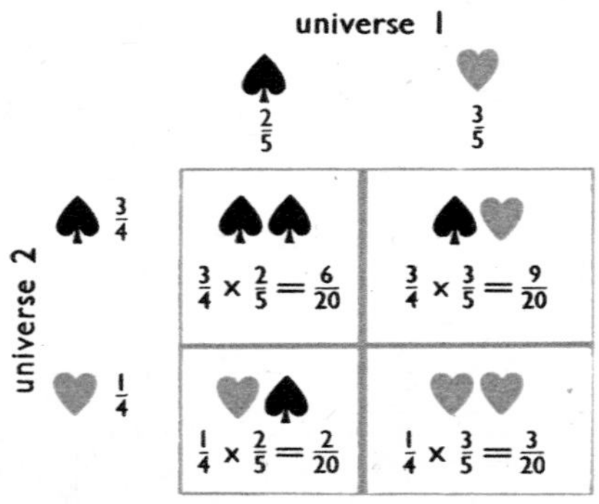

All possible 20 ways of choosing one card from a pack consisting of 2 spades and 3 hearts (universe 1) and another card from a second pack consisting of 3 spades and 1 heart (universe 2).

These two figures form a visual balance sheet of the top diagram when our only concern is with the suit. There are 6 possibilities of choosing 2 spades, 9 of choosing 1 spade then 1 heart, 2 of choosing 1 heart then 1 spade, and 3 of choosing 2 hearts.

pecification of the event, and the respective robabilities assigned to occurrence (*success*) and o non-occurrence (*failure*) are:

$$p_s = \frac{x}{y}; \; p_f = \frac{y-x}{y} = 1 - p_s$$

Thus one may say about the last example that:

$$p_s = \frac{1}{14}; \; p_f = \frac{13}{14}; \; \frac{p_s}{p_f} = \frac{1}{13}$$

n the jargon of the gaming table, one then says hat the odds are 13 to 1 against the occurrence of he event; and if our definition of mathematical robability has any relevance to the fortunes of he Chevalier, this means that he should expect o win *in the long run* if: (a) he agreed to forfeit dollar for every failure; (b) his opponent agreed o pay up a little more than 13 dollars in the event f a success. If the wager were 1 : 14, he would herefore eventually pay out 13 dollars per 14 lollars received. Needless to say, this is not a nathematical verdict, unless we also assume that very ordered *r*-fold sequence for *n*-fold universal et turns up in the long run as often as any other; nd experiment alone can settle whether this ssumption tallies as closely as one might expect vith the result of large-scale experiments. To Pascal, as to James Bernoulli and to d'Alembert, t seemed sufficiently obvious that this is so; but f we concede that theory and experiment do tally closely for well-constructed dice, roulette wheels, card packs, state lotteries, *etc.*, if with appro- priate prescriptions about the method of tossing, huffling and drawing, it does not necessarily ollow that theory has any relevance outside the lomain of gambling. Clearly then, mathematical probability so defined has no necessary connexion vith assessment of probability as a human judg- nent about the occurrence of events in general. The recognition of this has led to many re-defini- ions which lead to precisely the same algebraical esults.

Before commenting on them, it will be useful f we distinguish between two ways of numerically scoring the event in a so-called game of chance: (a) *classificatory*, as when we score the result of a five-fold selection of cards from an ordinary pack of 52 as 0, 1, 2, 3, 4 or 5 *hearts* or a twelve-fold selection as 0, 1, 2, 3 .. 10, 11, 12 picture cards; (b) *representative*, as when we score the result of a treble toss of a cubical die as either the total number of pips face upwards, *i.e.* 3, 4, 5, 6 .. 16, 17, 18, or the mean number, 1, 1·$\dot{3}$, 1·$\dot{6}$, 2 .. 5·$\dot{3}$, 5·$\dot{6}$, 6. Either way, we may visually exhibit three simple rules consistent with our definition of mathematical probability in the following terms:

Rule 1. If the probability of one event is p_1 and that of another *not influenced thereby* is p_2, the probability of simultaneous occurrence of both events *in that order* is the product $p_1 . p_2$. Similarly for three independent events, it is the product $p_1 . p_2 . p_3$.

For instance, if we toss a die twice, that of turning up a two or three at the first is $\frac{2}{6} = \frac{1}{3}$, the probability of turning up a six at the second

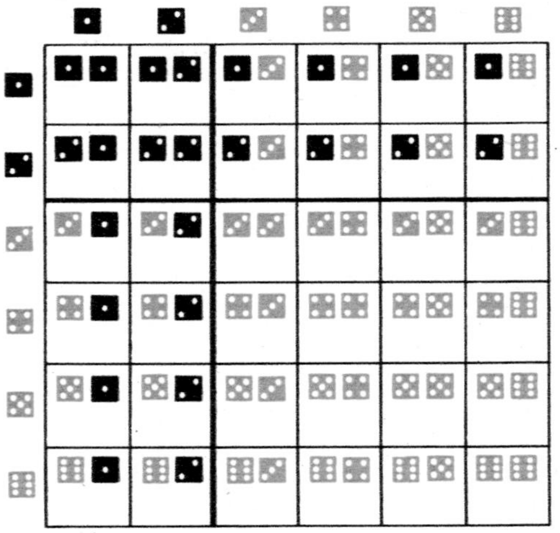

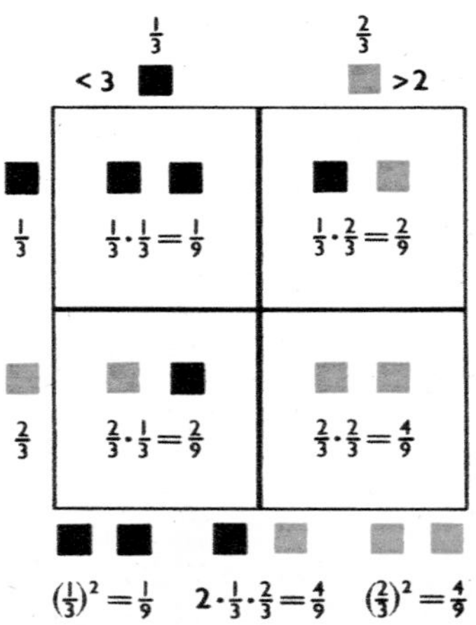

Table showing all 36 possible results of tossing two dice simultaneously or of tossing one die twice. The visual balance sheet below shows that $\frac{1}{9}$ *of these results occur when both dice show less than three pips,* $\frac{2}{9} + \frac{2}{9}$ *when only one or the other does so, and* $\frac{4}{9}$ *when neither does so.*

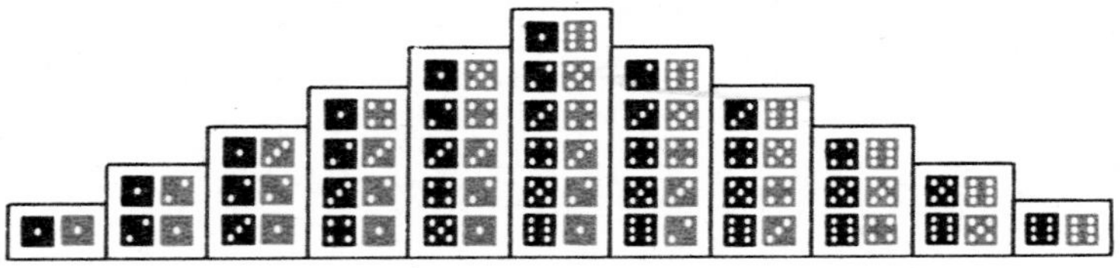

sum(x)	2	3	4	5	6	7	8	9	10	11	12
mean (x ÷ 2)	1	1·5	2	2·5	3	3·5	4	4·5	5	5·5	6
y_x	$\frac{1}{36}$	$\frac{2}{36}$	$\frac{3}{36}$	$\frac{4}{36}$	$\frac{5}{36}$	$\frac{6}{36}$	$\frac{5}{36}$	$\frac{4}{36}$	$\frac{3}{36}$	$\frac{2}{36}$	$\frac{1}{36}$

Above: Every possible way of making every possible score (from 2 to 12) in tossing two cubical dice. The probability of making each score is here given as a vulgar fraction with 36 as denominator.

The visual table below shows every possible result (216 in all) of tossing three cubical dice simultaneously or of tossing one such die three times.

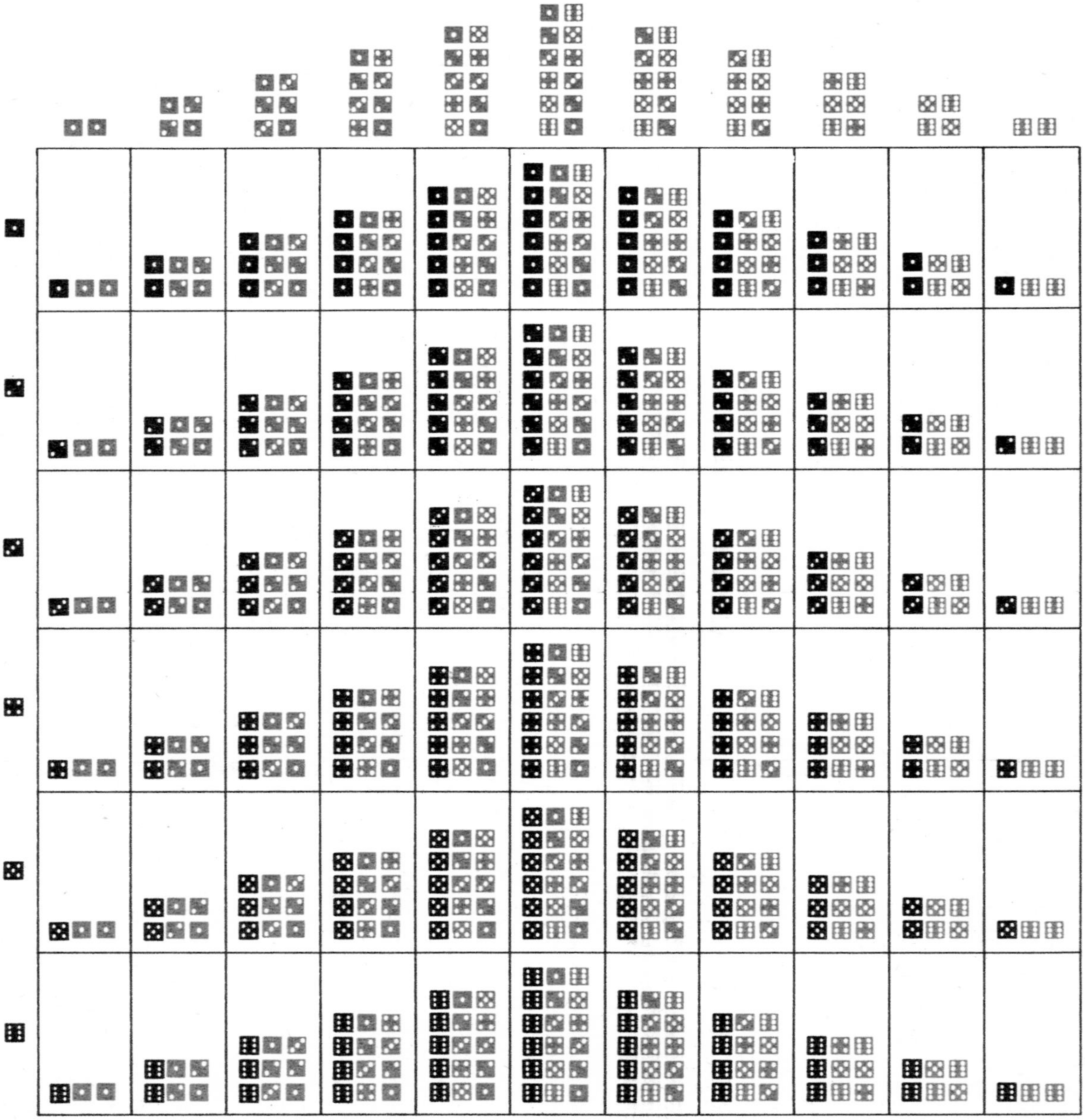

is $\frac{1}{6}$ and the probability that a score of two or three will precede a score of six is:

$$\tfrac{1}{3} \times \tfrac{1}{6} = \tfrac{1}{18}$$

Likewise, if we ask the probability of *simultaneously* extracting first a picture card, then an ace and last a card with 2 to 10 pips from a full (52) pack, the non-replacement restriction implies that the pack contains only 51 cards at the second trial and 50 at the third. Thus the answer is:

$$\tfrac{12}{52} \cdot \tfrac{4}{51} \cdot \tfrac{36}{50}$$

To use the rule intelligently, we must here pay attention to the words *in that order*. To get at one throw a score of a two or a three and at another throw a score of six does not mean the same thing as getting first a two or a three and second a six. The event as first defined implies the possibility of getting either of two ordered sequences. The alternative event is only one of the two.

Rule 2. If we denote the probability that any one of several events may happen by p_1, p_2, p_3, *etc.*, the probability that one or other will happen is $p_1 + p_2 + p_3$, *etc.*

For instance, the probability of scoring five or six in a toss of a die is $\frac{1}{3}$, that of scoring one, two or three is $\frac{1}{2}$, and the probability of scoring one, two, three, five or six is:

$$\tfrac{1}{3} + \tfrac{1}{2} = \tfrac{5}{6}$$

This is also the probability of *not* scoring four, which is the topic of the next rule. Rule 1 states that the probability of scoring a two or a three first and a six second is $p_{23} \cdot p_6$ or that that of scoring a six first and a two or a three second is $p_6 \cdot p_{23}$. Rule 2 implies that if we score in a two-fold toss both a two or a three and a six, the probability is:

$$p_{23.6} = p_{23} \cdot p_6 + p_6 \cdot p_{23} = 2p_{23} \cdot p_6$$

This is consistent with our general definition:

$$p_{23.6} = \frac{2!}{1!1!}\left(\frac{1}{3}\right)^1\left(\frac{1}{6}\right)^1$$

If we throw a die three times, the probability of getting a five *every time* in any such sequence is $\left(\tfrac{1}{6}\right)^3$; but no such sequence can differ from any other, *i.e.* the probability of the occurrence is also $\left(\tfrac{1}{6}\right)^3$. This again conforms to our definition, since

sum (x)	3	4	5	6	7	8	9	10	11	12	13	14	15	16	17	18
mean (x÷3)	1	1·$\dot{3}$	1·$\dot{6}$	2	2·$\dot{3}$	2·$\dot{6}$	3	3·$\dot{3}$	3·$\dot{6}$	4	4·$\dot{3}$	4·$\dot{6}$	5	5·$\dot{3}$	5·$\dot{6}$	6
y_x	$\frac{1}{216}$	$\frac{3}{216}$	$\frac{6}{216}$	$\frac{10}{216}$	$\frac{15}{216}$	$\frac{21}{216}$	$\frac{25}{216}$	$\frac{27}{216}$	$\frac{27}{216}$	$\frac{25}{216}$	$\frac{21}{216}$	$\frac{15}{216}$	$\frac{10}{216}$	$\frac{6}{216}$	$\frac{3}{216}$	$\frac{1}{216}$

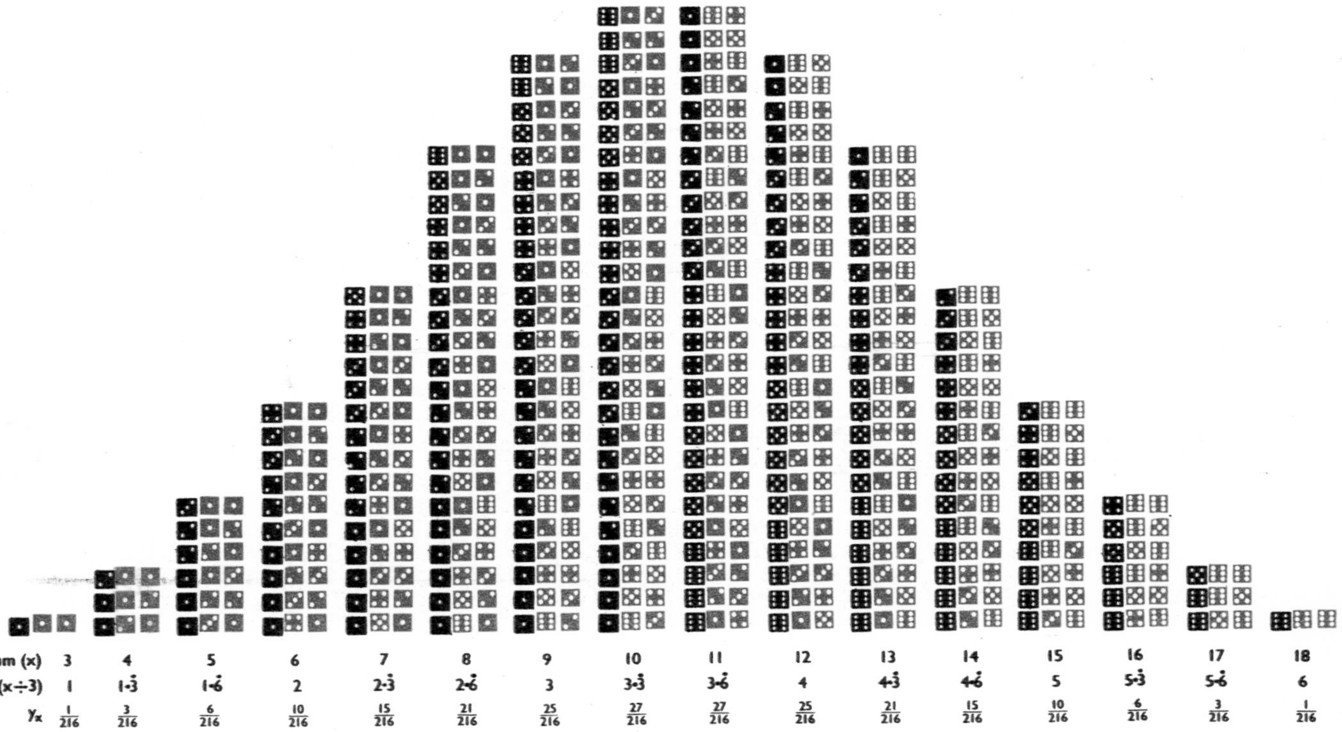

In an unending sequence of three-dice throws, the chances out of every 216 throws of making each score from 3 to 18 are (in rising score order):
1, 3, 6, 10, 15, 21, 25, 27, 27, 25, 21, 15, 10, 6, 3, 1.

$$\frac{3!}{3!0!0!}\left(\frac{1}{6}\right)^3 = \left(\frac{1}{6}\right)^3$$

Rule 3. If the probability of scoring at least one success in an r-fold trial is $p_{1.r}$ and that of scoring only failures is $p_{0.r}$

$$p_{1.r} = 1 - p_{0.r}$$

The probability of scoring at least one success in a 6-fold trial is that of not scoring six failures. If we denote the probability of scoring *at least one success* by $p_{1.5}$, that of scoring 5 failures is $1-p_{1.5}$. If p_f is the probability of scoring a failure in a single trial, that of scoring 6 in a 6-fold trial is p_f^6. Thus the probability of scoring *at least one success* in a 6-fold trial is $1-p_f^6$.

The last example brings us back to the harassed Chevalier. When he was in a position to provide for his wife, mistress and other dependants, he consistently and courageously adhered to the rule: bet small favourable odds on getting at least one six in four tosses of a cubical die. When, alas, unable to retrieve his fortunes and domestic obligations, his calamity was that he also bet

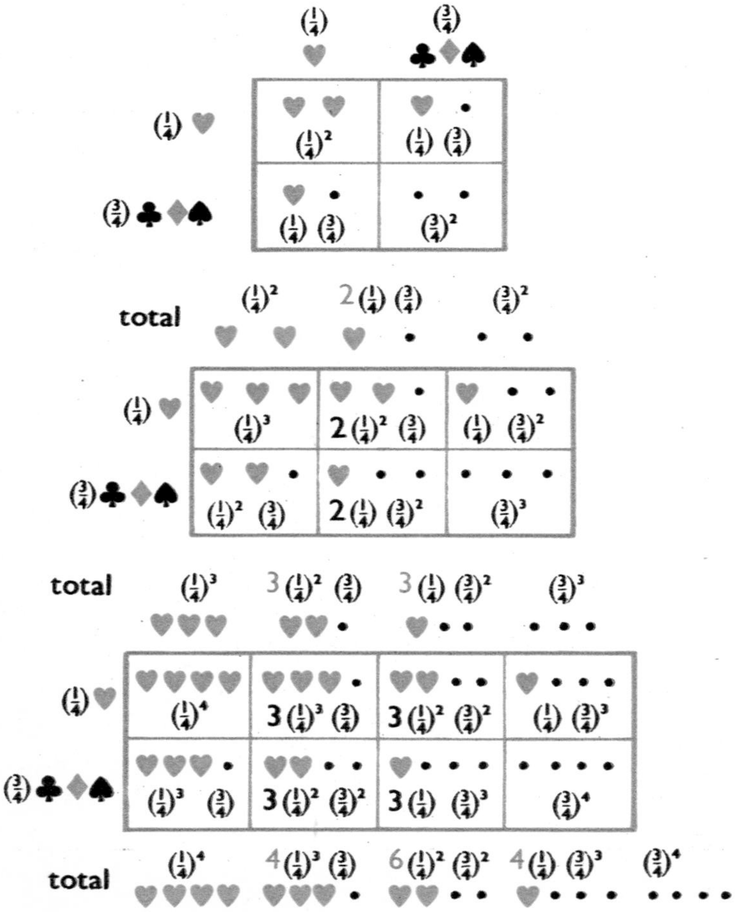

small favourable odds on getting a double six in 24 tosses. In the idiom of the Founding Fathers, the two problems are as follows:

Chance of getting at least one six in four tosses: $1-(\frac{5}{6})^4 = 0{\cdot}518$ (*small favourable odds*)
Chance of getting a double in 24 tosses: $1-(\frac{35}{36})^{24} = 0{\cdot}491$ (less likely than not)

Whether we score a result in the classificatory way as so many hearts (0, 1, 2 . . r) in an r-fold sample from a card pack or in the representative way as a total or mean score of an r-fold toss of a cubical die, we summarise conveniently all the information relevant to the classical theory of the division of the stakes by constructing a so-called *frequency distribution, i.e.* plotting against score x the probability (y) of the event. This is not a graph in the ordinary sense of the term, because x can advance only by discrete steps, unity for total scores, whether classificatory or representative, and fractional for mean scores. A convenient way of visualising this is to set up columns exhibiting the value of y equally spaced on either side of x bounded by a base from $x-\frac{1}{2}$ to $x+\frac{1}{2}$ if the interval is unity and appropriately otherwise. One calls this device for visualising a discrete distribution a *histogram*. Since y can have one value only for one value of x, the column as a whole signifies nothing in particular. None the less it can be a useful visual metaphor for what we do when we invoke the Newton-Leibniz calculus as a short cut to calculation.

When our concern is with only two possibilities, that an event can happen (*success*) or not (*failure*), we may denote the probability of success at a single trial by p and that of failure by $q=(1-p)$. Our preliminary definition then signifies for an r-fold trial from an n-fold universal set that the probability of getting x successes in the notation employed above is (*with replacement*):

$$y_x = \frac{r!}{x!(r-x)!} q^{r-x} p^x$$

When sampling *without replacement* from an n-fold universe in which the number of items whose

These three diagrams illustrate the binomial law of the distribution of heart scores for choice (with replacement) *of 2, 3 and 4 cards from a full pack.*

choice we classify as *success* is $b = (n - a)$:

$$y_{(x)} = \frac{r!}{x!(r-x)!} \quad \frac{a^{(x)} . b^{(r-x)}}{n^{(r)}}$$

We recognise the first of these two expressions as the general (xth) term of the expansion $(q+p)^r$. The reader may enjoy the satisfaction of discovering that the binomial theorem works for factorial indices as defined at the end of Chapter 3, *i.e.*

$$(b+a)^{(r)} = a^{(r)} + r_{(1)}a^{(r-1)}b^{(1)} + r_{(2)}a^{(r-2)}b^{(2)} \text{ etc.}$$

In either disguise, the binomial frequency distribution is easy to visualise without recourse to any considerations other than definition of the probability of the one-fold event either by a grid (*repetitive* sampling) or by a staircase (*without replacement*) diagram. As the size (r) of the sample increases (see figure at foot of p. 263), the contour of the frequency distribution histogram for repetitive sampling, as also for non-replacement sampling when n is very large compared with r, becomes more and more like a symmetrical curve with a peak and asymptotes to the x-axis in both directions. It is now customary to call this the *normal* curve, being indeed a specification of the limiting form a binomial distribution assumes as r becomes larger and larger. Its equation turns out to be: $y = Ae^{-kX^2}$, with the following interpretation of the constants in terms of M the mean score and x the crude score:

$$A = (\pi k)^{-\frac{1}{2}} \text{ and } k = \frac{1}{2\sigma^2}$$

σ^2 is the *mean* value of $X^2 = (x - M)^2$

When speaking of the average of the set of integers 1, 3, 3, 4, 4, 4, 7, 9, 9, 9, we may write the result alternatively as:

$$\frac{1+3+3+4+4+4+7+9+9+9}{10}$$

$$= \frac{1}{10}(1) + \frac{2}{10}(3) + \frac{3}{10}(4) + \frac{1}{10}(7) + \frac{3}{10}(9)$$

For the binomial frequency distribution of $x = 0, 1, 2 \ldots n$ successes whose probabilities (long-

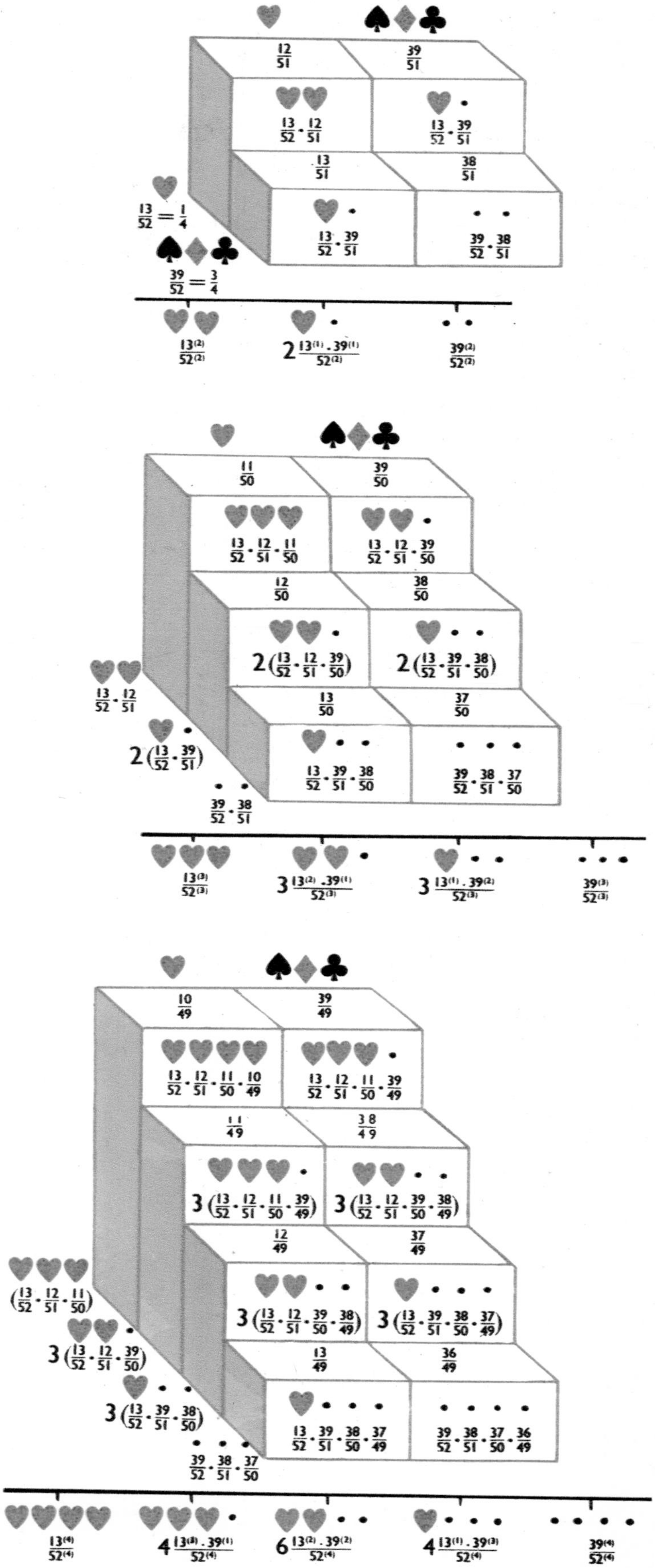

These diagrams illustrate the binomial law of the distribution of heart scores for choice (without replacement) *of 2, 3 and 4 cards from a full pack. In each case tread of stair shows result after a card is chosen, the highest tread representing the first choice. Risers show mathematical probability of next choice. Heavy dots indicate all suits other than hearts.*

run frequencies) are q^n, $n_{(1)}q^{n-1}p$, $n_{(2)}q^{n-2}p^2$, *etc.*, we may thus write the mean (M) as:

$$q^n(0)+nq^{n-1}p(1)+n_{(2)}q^{n-2}p^2(2)+n_{(3)}q^{n-3}p^3(3) \; ..etc.$$
$$=np(q^{n-1}+(n-1)q^{n-2}p+(n-1)_{(2)}q^{n-3}p^2...)$$
$$=np(q+p)^{n-1}=np.$$

By a similar procedure we find that the mean (σ^2) of $(x-M)^2$ is npq. So the equation of the normal curve becomes:

$$y=\frac{1}{\sqrt{(2\pi npq)}}e^{-\frac{(x-np)^2}{2npq}}$$

When $p=\frac{1}{2}=q$ (as for the tossing of an unbiassed coin) so that $M=\frac{1}{2}n$, this reduces to

$$y=\frac{1}{\sqrt{(\frac{1}{2}n\pi)}}exp\left[-\frac{(x-\frac{1}{2}n)^2}{\frac{1}{2}n}\right]$$

The reader should now be able to deduce that the curve: (a) has a maximum at $x=M$, $y=\sqrt{(M\pi)^{-1}}$; (b) is symmetrical about $x=M$ with $y=\sqrt{(M\pi)^{-1}e^{-M}}$ at $x=0$ or $x=n$.

Against the later background of the hypothetical infinite population of Laplace and of Quetelet's *urn of nature*, the so-called normal curve became the phallic symbol of a *mystique* which has no foothold in the classical theory. Within the framework of the latter, it is merely a calculating convenience for getting a good approximation to an otherwise laborious calculation of an essentially *discrete* distribution. With the frank admission that this commits us to an empirical enquiry for the outcome of which pure mathematics can furnish no conceivable justification, we shall ask the question: what is the probability that the score (x) will not exceed 8 in a non-replacement distribution of score values (*e.g.* x=number of cards which are *not* hearts) $x=0, 1, 2, 3..40$ specified by successive terms of $(q+p)^{40}$ when $q=\frac{1}{4}$ and $p=\frac{3}{4}$ so that $M=30$?

If y_0, y_1, y_2, *etc.* stand for the long-run frequencies of scoring $x=0, 1, 2$ .. *etc.* successes, the width of each column of the histogram is $\Delta x=1$ and its usefulness depends on the identity:

$$y_0+y_1+y_2..y_a=\sum_{x=0}^{x=a} y_x.\Delta x$$

In terms of the score deviation $X(=x-M)$ from its mean (M) and corresponding frequencies Y_X, we may write with equal propriety:

$$Y_0+Y_1+Y_2..Y_b=\sum_{X=0}^{X=b} Y_X.\Delta X$$

Thus we can represent the probability that a score (x) or score deviation ($X=x-M$) from its mean will lie in any specified range (*e.g.* 0 to a or b inclusive) by an area which we can evaluate with a high degree of precision by recourse to the integral of the normal curve, if we make due allowance, when stating the limits of integration, for the fact that the curve passes roughly through the mid-points of the upper extremities of the columns, *i.e.*

$$\sum_{X=a}^{X=b} Y_X.\Delta X\simeq A\int_{a-\frac{1}{2}}^{b+\frac{1}{2}} e^{-kX^2}.dX$$

Since we can construct, and therefore consult, tables of the values of this integral, we can sidestep the need for laborious calculations, when we have decided in what circumstances the error involved in the approximation is numerically trivial. We can do this empirically, but without any theoretical basis for a precise delimitation of the conditions in numerical terms. It turns out that the fit is very close for the distribution defined by $(\frac{1}{2}+\frac{1}{2})^n$ when $n=20$, and in general if $n > 10p$ (if $p<q$) or $10q$ (if $q<p$). This is a considerable convenience; but it is assuredly not *pure* mathematics in the sense of the term which purists would endorse.

In the foregoing paragraphs, we have made it sufficiently clear that there is no self-evident empirical justification for the tie-up of the frequency of an event in a game of chance with the formal definition of probability, as given by the Founding Fathers. This hiatus has been a headache to subsequent generations; but other definitions of mathematical probability concocted by posterity with the ostensible aim of closing the gap all lead their proponents to reach the conclusions embodied in Rules 1–3 above. We can state these rules in more than one way as tautologies of the calculus of choice, or (as is more modern to say) of *Set Theory*; but, if we do so, we have still no answer to the question: is there in any intelligible sense a theory of probability in the proper domain of pure mathematics? Let us look therefore at some definitions of probability other than the one already cited.

At the outset, we may dismiss briefly a current definition in terms of *sets*. With a notation of its own, *Set Theory* is a name which subsumes all that the calculus of choice embraced under such terms as *permutations* and *combinations*. In a universal set of n different discrete objects, what we used to call a combination is a *subset* whose constituents are unique. For instance (AB), (AC), (AD), (BC), (BD), (CD) are each two-fold subsets of the four-fold universal set $ABCD$. What we have hitherto called a linear permutation is a subset of any such subset, distinguished by an ordered relation, *e.g.* whether A comes first (AB) or second (BA) in the two-fold subset (AB) including A and B. The re-definition of mathematical probability in terms of set theory is therefore merely a process of putting old wine in new bottles. There remain three other attempts to justify the credentials of a mathematical theory of probability, two at different levels of sophistication quasi-empirical, the third in conformity with the predilection of the pure mathematician to build a superstructure of theory on a foundation of self-evident principles.

One claims prior attention *vis à vis* the question last stated. Its proponents rely on the so-called *principle of insufficient reason*. This amounts to saying that if we do not know whether A or B will happen, the probability that A (or B) will happen is $\frac{1}{2}$. To the reader, it may not be clear why this leads to precisely the same set of rules as the classical definition (*i.e.* that of the Founding Fathers). Suffice it to say that the Founding Fathers were, at least through a glass darkly, aware that a tie-up between theory and practice presumes certain elementary precautions (*e.g.* shuffling the pack) with no relevance whatsoever to the ignorance of the person who undertakes the task. If one concedes that a randomising process presupposes the intervention of an agent assigned a particular task such as shuffling, the naïveté of an attempt to relate mathematics to the fallibility of human judgments by appeal to the so-called principle of insufficient reason is almost ludicrous. What one does or does not know about a situation in which one does not participate, *e.g.* whether a penny one has not handled lies heads up or otherwise under a mat one did not place over it, has no bearing on what one may say about a situation in which one is an active participant, as when one actually tosses a penny in accordance with a prescribed ritual.

Though the view last stated had at least one powerful proponent in our own century, the main drift of opinion after the middle of the nineteenth century was in opposition to the attempt to erect a mathematical theory of probability on foundations which are not amenable to experimental verification. Thus one powerful school of thought led by Venn (1834–1923) defined the probability of an event in purely empirical terms somewhat as follows: if an event occurs r times in an n-fold sequence, the probability of its occurrence is the limit of the ratio $r \div n$ as n becomes indefinitely large. Without unobtrusively slipping in corollaries which have no necessary connexion with it, such a definition does not lead unambiguously to the classical calculus – or to any other. However R. von Mises (*b.* 1883) has made an at least more plausible attempt to found a theory on the factual concept of frequency by introducing a refinement into what Venn's school called the *series of the event*. According to von Mises, this must be an *irregular Kollektiv*, that is to say a completely *disordered* sequence of unit scores, such as for a universe of three score values:

1 3 1 1 2 3 3 1 2 1 2 3 3 3 3 1 1 . . .

There are mathematical difficulties about the definition of an *irregular Kollektiv* so conceived;

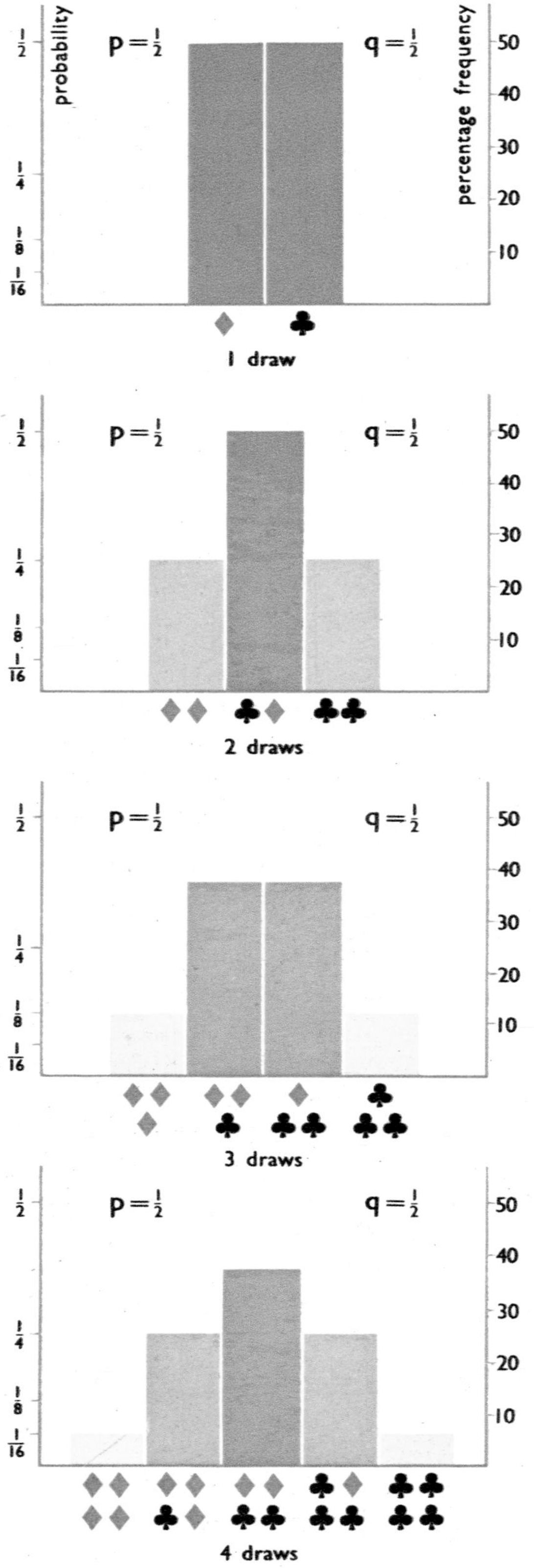

The probability of drawing at random (with replacement) either diamonds or clubs or any combination of both from a pack containing only clubs and diamonds in equal numbers after reshuffle before each draw.

but if one dismisses them by saying that von Mises at least recognised the inconsistency of the attempt to tie up the mathematical theory of probability with the notion of frequency without providing the frequency of occurrence with some sort of plausible background, the position still remains as it was in the days of the Founding Fathers. We still await a factual answer to the question: of what situations can we conceive the series of the event to be disordered in this sense? Let us therefore come down to earth awhile.

In one of Dickens' novels, the charity boy asks as he reaches the end of the alphabet: was it worth going through so much for so little? By the same token, the reader may well feel that the time is overdue for an honest attempt to answer the question: to what extent does experience endorse the factual relevance of the rules in the original domain of their application? If all we ask of the theory is that it works when the gambler adheres to the rule throughout a *sufficiently* long sequence of games, we must likewise concede that pure mathematics cannot lay down a criterion of *sufficiently* in this context. The best one can hope is that correspondence between theory and observation will be close in a sense which conforms to the sentiments of a physicist or a chemist. On that understanding, some recorded experiments have yielded encouraging results.

In the domain of non-replacement, Karl Pearson cites counts for 3400 hands (13 cards each) at whist. The theoretical expectation is calculable in terms of our definition and rules from the addition of appropriate terms of the expression $(13+39)^{(13)} \div 52^{(13)}$ when we group results as below:

TRUMPS PER HAND	HANDS OBSERVED	NUMBER EXPECTED
Under 3	1021	1016
3–4	1788	1785
Over 4	591	599

Assuredly the correspondence is here striking; and the same is true of one of several experiments recorded by Uspensky, *viz.* seven experiments each based on 1000 games in which the score is a success if a card of each suit occurs in a four-fold simultaneous withdrawal from a pack without picture cards. The mathematical probability of success is then

$$\frac{4!}{1!1!1!1!}\,\frac{10^{(1)}.10^{(1)}.10^{(1)}.10^{(1)}}{40^{(4)}}=0\cdot1094$$

The outcome of the seven (I–VII) 1000-fold trials was as below:

I	II	III	IV	V	VI
0·113	0·113	0·103	0·105	0·105	0·118

VII	Mean	Expected
0·108	0·1093	0·1094

In the domain of die and lottery models, the experimental data, as recorded by Uspensky (*Introduction to Mathematical Probability*) and Maynard Keynes (*Theory of Probability*), are copious, but Keynes himself points out that correspondence between theory and practice does not seem to become closer if the investigator perseveres beyond a certain level. Since indenting the pips on the surface of a cubical die entails a slight shift of the centre of gravity, this may be largely because the long-run frequency of falling on any one face is not exactly the same as that of falling on any other.

However, any such experiment involves the intervention of an agent with a prescribed task. From this viewpoint, the needle problem devised by the eighteenth-century naturalist Buffon is of special interest. The gamester has to drop a needle of length l on a flat surface ruled with parallel lines at a distance h apart. The score is a success if the needle falls across a line, a failure if it falls between two. By not very difficult

area under the curve

$$\int_{-2\cdot5}^{+2\cdot5} y \cdot dx$$

score x =	0	1	2	3	4	5	6	7	8
deviation X =	−4	−3	−2	−1	0	+1	+2	+3	+4
frequency y_x =	$(\tfrac{1}{2})^8$	$8(\tfrac{1}{2})^8$	$28(\tfrac{1}{2})^8$	$56(\tfrac{1}{2})^8$	$70(\tfrac{1}{2})^8$	$56(\tfrac{1}{2})^8$	$28(\tfrac{1}{2})^8$	$8(\tfrac{1}{2})^8$	$(\tfrac{1}{2})^8$
	·0039	·0313	·1094	·2188	·2734	·2188	·1094	·0313	·0039

$$|x| < 3$$

$$= \sum_{x=-2}^{x=+2} 8_{(x)}(\tfrac{1}{2})^8$$

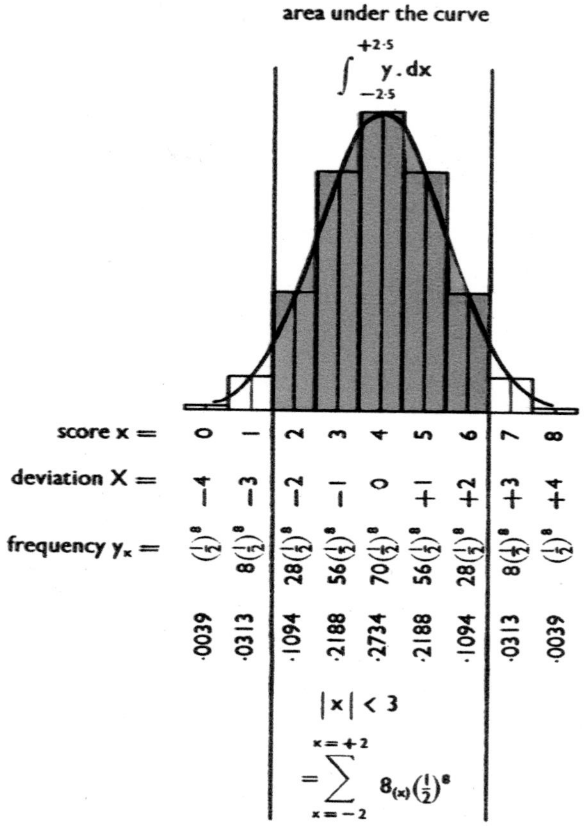

We may visualise the probability of getting a deviation from the mean number (4) of heads numerically less than 3 in an 8-fold toss of an unbiassed coin, i.e. the sum of the probabilities of getting deviations $X=0, \pm 1, \pm 2$, as the area of the histogram $Y=8_x(\tfrac{1}{2})^8$ in the range $X=\pm 2$. If $\triangle x=1$, we may write this as:

$$\sum_{x=-2}^{x=+2} 8_{(x)}(\tfrac{1}{2})^8 = \sum_{x=-2}^{x=+2} 8_{(x)}(\tfrac{1}{2})^8 \cdot \triangle X$$

The so-called normal curve ($y=Ae^{-kx^2}$) which passes through points approximately in the middle of the upper extremity of each column of the histogram and the true value of the area defined above is very close to that given by the definite integral of the normal function with a half-interval ($\tfrac{1}{2}\triangle X=0\cdot5$) correction, i.e. between the limits $X=\pm 2\cdot5$. The picture above shows why the half-interval correction is appropriate.

$(\tfrac{1}{2}+\tfrac{1}{2})^8$

$(\tfrac{1}{2}+\tfrac{1}{2})^{16}$

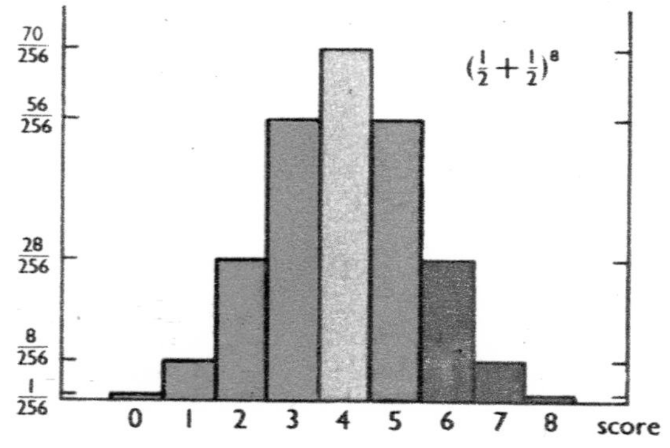

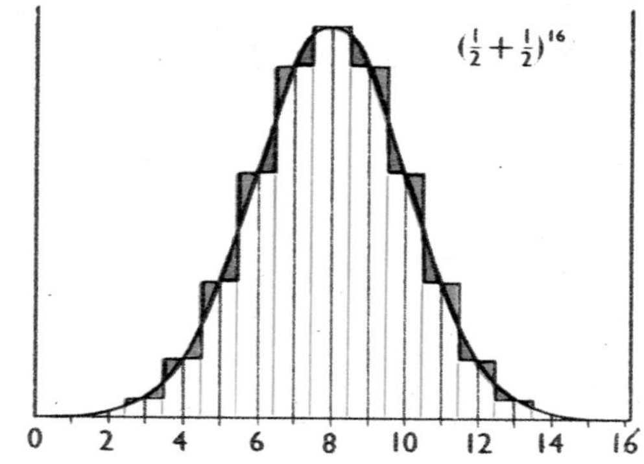

reasoning, which we need not here traverse, we arrive at the conclusion that the probability of success is $2l \div (h\pi)$. Since we can assign h and l, we can thus obtain a value for π as a criterion of correspondence between theory and observation. Two examples of such experiments will suffice:

	No. of throws	*Estimate of* π	*Error*
Wolf (1849) (*cited by* Uspensky)	5000	3·1596	0·019
Lazzerini (1901) (*ditto* Kastner and Newman)	3408	3·1415929	0·0000003

The sacred book of what we may call the classical period, *i.e.* before the concept of *inverse* probability raised its ugly head, was the *Artis Conjectandi* (1713) of James (Jacques or Jacob) Bernoulli. This formulates all there is of importance to say about the rationale of gambling. Here emerges a concept of fundamental importance to the theory of division of the stakes, *i.e. expectation* of gain. To clarify this, we shall suppose that A agrees to forfeit x dollars if an event whose probability is $p=1-q$ does not happen, and B agrees to forfeit y dollars if it does. If p and q respectively stand for the probabilities of success and failure in a sequence of unit trial, np is the mean number of successes in a limitless sequence of trials and $nq(=n(1-p))$ is the mean number of failures. In a limitless sequence of n games, A therefore forfeits nqx and B forfeits npy. Thus the net gain (so-called *expectation*) of A is $E=n(py-qx)$. Illustratively, we may suppose that A wagers 2 dollars that the score of a single toss of a die will be at least as great as 3, and B wagers 1·1 dollars that it will not. In this set-up

$$p=\tfrac{1}{6}(1+1+1+1)=\tfrac{2}{3} \text{ and } q=\tfrac{1}{3}$$

If A is prepared to waste 30 dollars, the worst that can happen is that he can continue to play only 15 games before he exhausts his outlay, and the best that can happen is that he will win 15(1·1) =\$16.50 per 15 games. Of an unlimited number of such gamblers, the mean earnings will be

$$E=15[\tfrac{2}{3}(1.1)-\tfrac{1}{3}(2)]=1 \text{ dollar.}$$

However the prospect of A is by no means as poor as this. That he can continue to play only 15 games is itself an occurrence of which the probability is less than one in ten million, being in fact

$$\left(\frac{1}{3}\right)^{15}=\frac{1}{14{,}348{,}907}$$

Hence it is extremely unlikely that the contest will finish at the end of the fifteenth game. The longer (n games) it goes on, the nearer will his winnings approach the mean figure:

$$E=n\left[\frac{2}{3}(1\cdot 1)-\frac{1}{3}\,(2)\right]=\frac{0\cdot 2n}{3}$$

Thus A's risk of bankruptcy after a specified number of games depends both on the stakes and on the probability of success assignable to the event; and the prospect of a net profit depends only on what stakes A can induce B to agree to. Let us call the forfeit of B

$$y=\frac{qx}{p}+d \text{ so that } E=n(qx+pd-qx)=np.d$$

The expectation of gain must therefore be positive if d is positive, otherwise negative (loss). If $n=15$, $p=\tfrac{2}{3}$ and $d=0\cdot 2$, the situation is that A may exhaust his capital at the end of the fifteenth game, with stupendous good fortune gain $15(1\cdot 2)=18$ dollars, or make the mean of a limitless number of gamblers playing 15 games for the same stakes, *i.e.* $15.\tfrac{2}{3}.0\cdot 2=2$ dollars. The possibility of losing all his capital (nx) depends on the number (n) of games which he plays at a forfeit of x per game, but the risk being q^n must diminish as n becomes larger, since $q<1$. If A has more capital, it thus becomes less and less likely (*i.e.* will happen less often) that he or others like him will have to stop at the end of the nth game.

A third gambler C with much more capital could therefore undertake to underwrite A's losses for a rake-off on his gains with a reasonable prospect of success. Let us assume that the honest broker C has a capital of 30,000 dollars, in which event he can pay 1000 times the 30 dollars forfeited by A for the loss of 15 games in succession. This means that A can continue to play *at least* a thousand sequences of 15 games with an expectation of 2 dollars per sequence if the stakes are 2 : 1·2. As before, we assume that A's capital is 30 dollars and that he agrees to pay out of every 2 dollars he wins 0·5 dollars to C for coverage of a fifteen-game sequence. A's net expectation per 15 games is therefore 1·5 dollars and for a thousand sequences of 15 games, 1500 dollars. For a thousand sequences of 15 games, the expectation of C is 500, and the risk of losing all his capital before A wins anything is 3^{-15000}. Thus the gambler with more capital has less risk of losing it if his expectation is positive. Having a better prospect of remaining in the contest, he can also afford to be content with a smaller offer from his opponent.

The theory of probability attracted widespread public attention through the publication in the *Encyclopédie* of an article on the subject by the mathematician d'Alembert, who was Diderot's co-editor. The issue was highly topical at the time, because the government of Royalist France reaped a handsome revenue from a state lottery. A citizen could purchase one or more tickets numbered 1 to 90 inclusive. The supervising officer drew randomwise five of a duplicate set; and the holder of a ticket corresponding to any one of those chosen could claim 15 times its cost. If N is the number of any such ticket, the probability that one out of five taken from 90 will be of the same denomination is $\frac{5}{90}=q=\frac{1}{18}$. On the assumption that the purchaser pays f francs for the ticket, the government gains f with probability $p=\frac{17}{18}$, and risks a net forfeit of $(15-1)f$. Thus the expectation of gain on single tickets would be:

$$E=\tfrac{17}{18}f-(\tfrac{1}{18}.14f)=\tfrac{1}{6}f$$

The holder of two winning tickets could claim 270 times the cost of each, in which event the forfeit of the government would be $(270)2f-2f=538f$, and the probability that the holder will have two tickets with different winning numbers is

$$\frac{5}{90}\cdot\frac{4}{89}=\frac{2}{801}$$

$$E=\frac{(799)2f}{801}-\frac{(538)2f}{801}=\frac{58}{89}f$$

The division of the stakes thus ensured a very high expectation on a large issue of tickets. This raises the question: what risk was there that the government would make no profit? To answer it in terms consistent with the theory, we shall suppose that the number of successes $(np-e)$ on n tickets scored by the government is smaller than the mean (np). The number of failures will then be $n-(np-e)=(nq+e)$. Thus the net gain

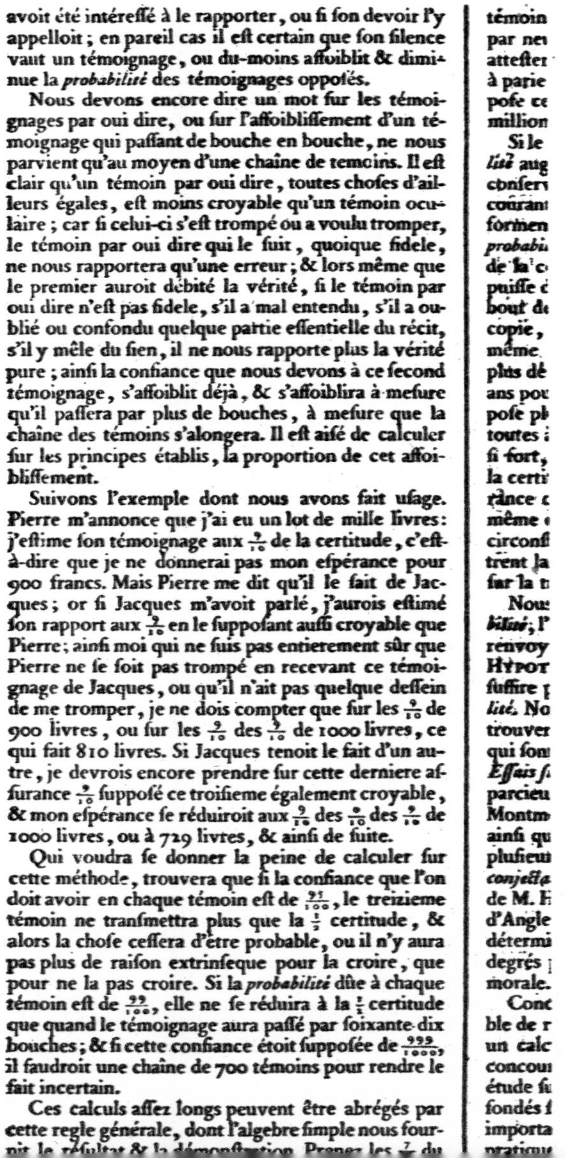
avoit été intéreſſé à le rapporter, ou ſi ſon devoir l'y appelloit; en pareil cas il eſt certain que ſon ſilence vaut un témoignage, ou du-moins affoiblit & diminue la *probabilité* des témoignages oppoſés.

Nous devons encore dire un mot ſur les témoignages par oui dire, ou ſur l'affoibliſſement d'un témoignage qui paſſant de bouche en bouche, ne nous parvient qu'au moyen d'une chaîne de temoins. Il eſt clair qu'un témoin par oui dire, toutes choſes d'ailleurs égales, eſt moins croyable qu'un témoin oculaire; car ſi celui-ci s'eſt trompé ou a voulu tromper, le témoin par oui dire qui le ſuit, quoique fidele, ne nous rapportera qu'une erreur; & lors même que le premier auroit débité la vérité, ſi le témoin par oui dire n'eſt pas fidele, s'il a mal entendu, s'il a oublié ou confondu quelque partie eſſentielle du récit, s'il y mêle du ſien, il ne nous rapporte plus la vérité pure; ainſi la confiance que nous devons à ce ſecond témoignage, s'affoiblit déjà, & s'affoiblira à meſure qu'il paſſera par plus de bouches, à meſure que la chaîne des témoins s'alongera. Il eſt aiſé de calculer ſur les principes établis, la proportion de cet affoibliſſement.

Suivons l'exemple dont nous avons fait uſage. Pierre m'annonce que j'ai eu un lot de mille livres: j'eſtime ſon témoignage aux $\frac{9}{10}$ de la certitude, c'eſt-à-dire que je ne donnerai pas mon eſpérance pour 900 francs. Mais Pierre me dit qu'il le fait de Jacques; or ſi Jacques m'avoit parlé, j'aurois eſtimé ſon rapport aux $\frac{9}{10}$ en le ſuppoſant auſſi croyable que Pierre; ainſi moi qui ne ſuis pas entierement ſûr que Pierre ne ſe ſoit pas trompé en recevant ce témoignage de Jacques, ou qu'il n'ait pas quelque deſſein de me tromper, je ne dois compter que ſur les $\frac{9}{10}$ de 900 livres, ou ſur les $\frac{9}{10}$ des $\frac{9}{10}$ de 1000 livres, ce qui fait 810 livres. Si Jacques tenoit le fait d'un autre, je devrois encore prendre ſur cette derniere aſſurance $\frac{9}{10}$ ſuppoſé ce troiſieme également croyable, & mon eſpérance ſe réduiroit aux $\frac{9}{10}$ des $\frac{9}{10}$ des $\frac{9}{10}$ de 1000 livres, ou à 729 livres, & ainſi de ſuite.

Qui voudra ſe donner la peine de calculer ſur cette méthode, trouvera que ſi la confiance que l'on doit avoir en chaque témoin eſt de $\frac{95}{100}$, le treizieme témoin ne tranſmettra plus que la $\frac{1}{2}$ certitude, & alors la choſe ceſſera d'être probable, ou il n'y aura pas plus de raiſon extrinſeque pour la croire, que pour ne la pas croire. Si la *probabilité* dûe à chaque témoin eſt de $\frac{99}{100}$, elle ne ſe réduira à la $\frac{1}{2}$ certitude que quand le témoignage aura paſſé par ſoixante-dix bouches; & ſi cette confiance étoit ſuppoſée de $\frac{999}{1000}$, il faudroit une chaîne de 700 témoins pour rendre le fait incertain.

Ces calculs aſſez longs peuvent être abrégés par cette regle générale, dont l'algebre ſimple nous fournit le réſultat & la démonſtration.

Extract from d'Alembert's article on probability in the Encylopédie. *Published when the government of Royalist France was reaping revenue from a state lottery, it attracted widespread public attention.*

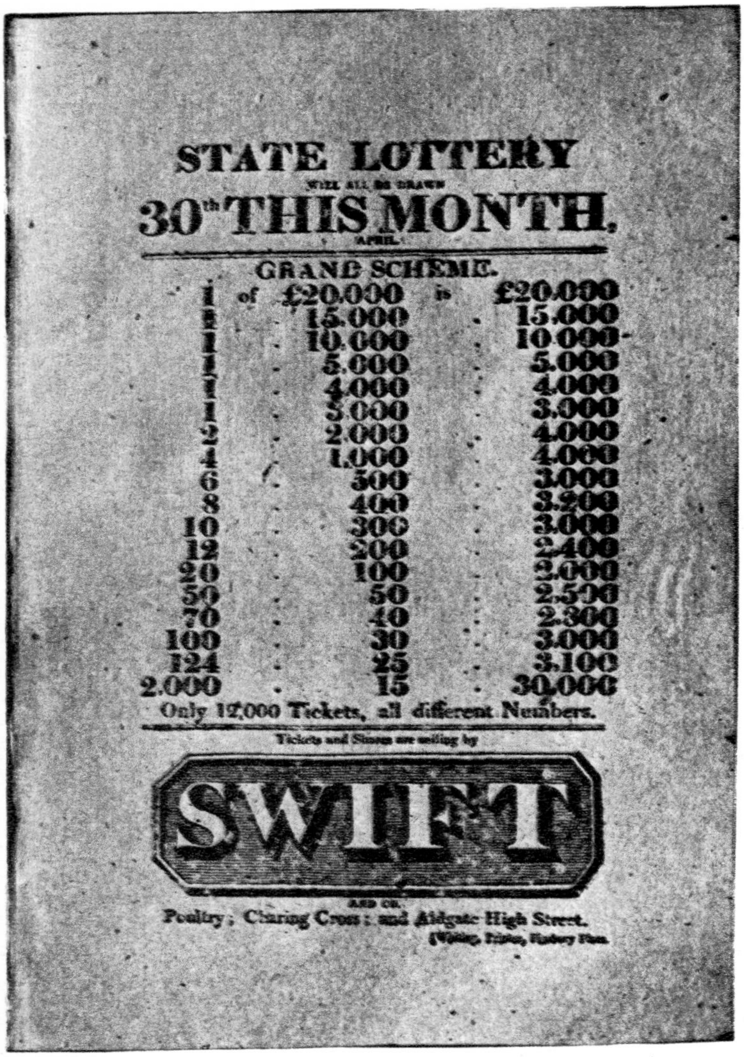

of the state will be $(np-e)f-(nq+e)14f$. This is zero if $np-e=14(nq+e)$, *i.e.* if

$$\frac{np-14nq}{15}=e$$

For single ticket holders, $np-14nq=\frac{1}{6}n$. Thus the probability of making no profit is the probability that the number of successes will not exceed:

$$np-\frac{n}{90}=\frac{17}{18}n-\frac{n}{90}=\frac{42}{45}n$$

It is here that the frequency distribution defined by successive terms of $(q+p)^n$ for 0, 1, 2..n successes becomes relevant to the gambler's risk. The probability that the number of successes will not exceed $42n\div45$ is in fact

$$\sum_{r=0}^{r=(42n\div45)} n_{(r)}\, q^{n-r}p^{r}$$

It was not only in Royalist France that state lotteries flourished. All these advertisements of well over a century ago remind us that the same was true in allegedly-puritanical Britain.

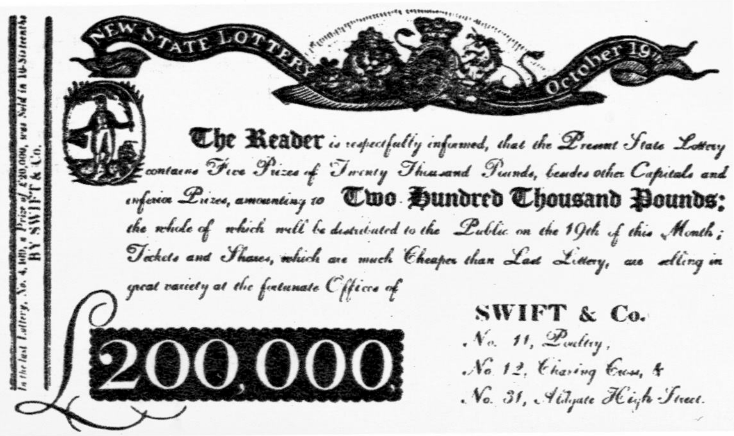

Strictly speaking, this holds only if n is exactly divisible by 45; but this is irrelevant if we use the normal integral as an approximation. If the government issued 100,000 tickets to the same number of individuals at the stake cited above, the risk of making no profit would be less than 10^{-10}. The moral we can draw from this, as from the foregoing parable of the three gamblers, is: to him that hath shall be given. According to the classical theory, the risk of bankruptcy is less when the amount of capital available is greater.

As an empirical judgment, this had been recorded by a sixteenth-century writer who commented on a practice from which the insurance business developed. At a time when dowries were a parental risk, merchants at medieval fairs would wager with willing clients on the sex of a forthcoming addition to the family. Data now available show that the frequency distribution of 0, 1, 2..r boys in a sibship of r offspring closely conforms to the terms of $(\frac{1}{2}+\frac{1}{2})^r$. There is therefore at least *some* plausibility in regarding gambling of this sort as a situation to which the classical theory may have some relevance. If so, it is by no means clear that it has any necessary connexion with life insurance, emergent in the same social context as the Royal Lottery. De Moivre, who earned a livelihood from actuarial work as a Protestant refugee in London, first tailored Bernoulli's treatment of expectation in a game of chance to the calculation of annuities for an insurance firm in his *Doctrine of Chances* (1718).

It is here important to insist on three implicit postulates of the classical theory: (a) we can assign in advance a definite value to the probability of an event; (b) the player must *adhere consistently to a rule stated in advance*, regardless of the changing fortunes of the game; (c) the event itself is the outcome of a *randomising process* which signifies an act of human intervention and particular changeless properties ascribed to the apparatus employed. Clearly, the practice of life insurance conforms to none of these conditions. It is therefore not surprising that the enlistment of the theory of probability in the service of the insurance broker demanded, rightly or wrongly, a revision and emendation of the classical theory. Thomas Bayes and Richard Price, both British nonconformist divines with the dissenter's enthusiasm for thrift combined with the dissenter's distaste for gambling and an understandable enthusiasm for an innovation which took over some of the financial obligations of the Anglican parish without transferring the burden to its competitors, took the initiative in this reorientation. To be sure, the rot had set in earlier. Thus a prize essay by D. Bernoulli (1734) solemnly discusses whether the narrow ecliptic zone in which the planets lie is (in Todhunter's words) "due to hazard". Clearly, the classical theory has no relevance to the issue. Since the planets lie where they do lie, in any conceivable sense endorsed by the classical theory, the probability that they do so is unity.

In 1763, the English Royal Society published a posthumous *Essay towards solving a problem in*

Modern Britain has quietened its conscience about offering big-money prizes against long odds by arranging things so that no stakes, as such, are placed. The investor in Premium Bonds patriotically waives his right to interest. A duly-grateful government rewards him with the chance of a big win.

the *Doctrine of Chance*, communicated and edited by Richard Price, who was the actuarial adviser for the *Equitable*, the first thereafter continuously successful Life Insurance Corporation. It seems that Price gave his services with little reward other than the satisfaction of promoting thrift, discouraging gambling and outdoing the parish. The *Equitable* flourished on his advice by basing their premiums on the *Northampton Life Table*, a compilation grossly defective because the source of its figures for births was baptismal registers in a city where a considerable segment of the citizenry disapproved of paedo-baptism. Meanwhile, the government of the day lost handsomely by basing its retirement annuities on the same Life Table.

However, the outcome was that probability seemed to be good business for a firm with large enough capital; and the *Equitable* changed its premiums to the advantage of the investor as it became more prosperous. Whatever one may say about its success, it assuredly therefore did not consistently adhere to a rule stated in advance about a situation in which one could numerically pre-assign the risk of failure. Though it is fair to add that the notion of a steadily improving standard of health was still totally alien to the generation to which Malthus belonged, we could not nowadays plausibly pretend that the card pack or die was the same throughout the game.

The statement of his problem by Bayes, annotated for publication by Price, is obscure; but the author's intention is clear about at least one thing. Its topic is a *two-stage* sampling process randomwise at each stage; and its implications will be easier to grasp if we examine a comparable model situation in which there is no ambiguity about the issues involved:

(1) an umpire equipped with a tetrahedral die with 1, 2, 3 or 4 pips on its faces and an ordinary cubical one with 1, 2 . . 6 pips on its faces draws *with replacement* one card at a time from a well-shuffled complete pack;

(2) he tosses the cubical die once if he draws a heart ($p_c = \frac{1}{4}$) and the tetrahedral die if he draws a card of another suit ($p_t = \frac{3}{4}$);

(3) he reports the score (s) as $s \leqslant 3$ or $s > 3$ without further details to two players A and B, who are otherwise ignorant of the performance, except in so far as they are familiar with the prescription of the game embodied in (1) and (2) above;

(4) neither A nor B bets on a result if $s \leqslant 3$ but A scores a success whenever $s > 3$ if the die chosen by the umpire was the tetrahedral one and B scores a success whenever $s > 3$ if the chosen die was the cubical one.

So stated, the issue is consistent with the classical approach. Since we then identify probability with long-run *proportionate* frequency, we may draw up a balance sheet of long-run proportionate frequencies in accordance with the multiplicative and additive rules as follows:

	$s \leqslant 3$	$s > 3$
Tetrahedral	$\frac{3}{4} \times \frac{3}{4} = \frac{9}{16}$	$\frac{3}{4} \times \frac{1}{4} = \frac{3}{16}$
Cubical	$\frac{1}{4} \times \frac{1}{2} = \frac{2}{16}$	$\frac{1}{4} \times \frac{1}{2} = \frac{2}{16}$

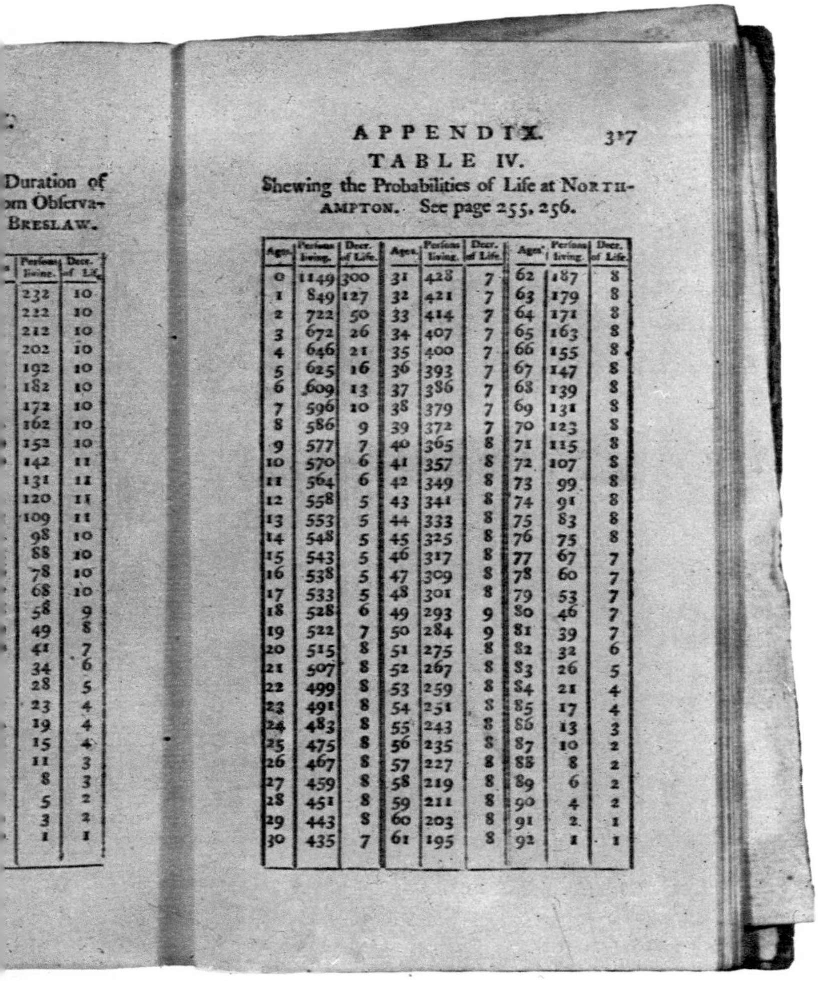

Duration of
Observa-
BRESLAW.

Persons living.	Decr. of Life.
232	10
222	10
212	10
202	10
192	10
182	10
172	10
162	10
152	10
142	11
131	11
120	11
109	11
98	10
88	10
78	10
68	10
58	9
49	8
41	7
34	6
28	5
23	4
19	4
15	4
11	3
8	3
5	2
3	2
1	1

APPENDIX. 317

TABLE IV.

Shewing the Probabilities of Life at NORTHAMPTON. See page 255, 256.

Ages.	Persons living.	Decr. of Life.	Ages.	Persons living.	Decr. of Life.	Ages.	Persons living.	Decr. of Life.
0	1149	300	31	428	7	62	187	8
1	849	127	32	421	7	63	179	8
2	722	50	33	414	7	64	171	8
3	672	26	34	407	7	65	163	8
4	646	21	35	400	7	66	155	8
5	625	16	36	393	7	67	147	8
6	609	13	37	386	7	68	139	8
7	596	10	38	379	7	69	131	8
8	586	9	39	372	7	70	123	8
9	577	7	40	365	8	71	115	8
10	570	6	41	357	8	72	107	8
11	564	6	42	349	8	73	99	8
12	558	5	43	341	8	74	91	8
13	553	5	44	333	8	75	83	8
14	548	5	45	325	8	76	75	8
15	543	5	46	317	8	77	67	7
16	538	5	47	309	8	78	60	7
17	533	5	48	301	8	79	53	7
18	528	6	49	293	9	80	46	7
19	522	7	50	284	9	81	39	7
20	515	8	51	275	8	82	32	6
21	507	8	52	267	8	83	26	5
22	499	8	53	259	8	84	21	4
23	491	8	54	251	8	85	17	4
24	483	8	55	243	8	86	13	3
25	475	8	56	235	8	87	10	2
26	467	8	57	227	8	88	8	2
27	459	8	58	219	8	89	6	2
28	451	8	59	211	8	90	4	2
29	443	8	60	203	8	91	2	1
30	435	7	61	195	8	92	1	1

The theory of probability, applied to life insurance, may or may not be good business. In the eighteenth century the Equitable *flourished by basing premiums on the table shown. By basing retirement annuities on it the British government lost handsomely.*

The long-run frequency of scoring $s > 3$ with the tetrahedral die is thus in the ratio 3:2 to that of scoring $s > 3$ with the cubical; and we may regard this ratio as that of the odds in favour of A's success. From the viewpoint of the classical theory, it is not exceptionable to state the same conclusion by citing the probability of A's success as:

$$P_a = \frac{3}{2+3} = \frac{3}{5}$$

Alternatively, we may say that the so-called *posterior* probability of the truth of the assertion that A consistently makes in the endless game is $P_a = \frac{3}{5}$. If we now look at the items on the balance sheet, we see that each consists of two factors:

(a) P_t $(=\frac{3}{4})$ and P_c $(=\frac{1}{4})$, designated the prior probabilities of the event, are probabilities assigned unconditionally to choice of one or other die at Stage 1 of the sampling process;

(b) L_t $(=\frac{1}{4})$ and L_c $(=\frac{1}{2})$, designated *likelihoods*, are the alternative conditional probabilities of the event $(s > 3)$ assignable if we *already know* which die the umpire chose.

Our balance sheet thus exhibits Bayes' theorem in terms consistent with the classical theory of the division of the stakes, *viz.*:

$$P_a = \frac{L_t . P_t}{L_t . P_t + L_c . P_c}$$

Bayes' own model (two balls rolling on a table) ascribes prior probabilities to Stage 1, each equal to $\frac{1}{2}$, as would be true of our model if the umpire followed a different plan, *i.e.* tossed one die if he turned up a red card and the other when he turned up a black one. The foregoing formula would then become:

$$P_a = \frac{\frac{1}{2}L_t}{\frac{1}{2}L_t + \frac{1}{2}L_c} = \frac{L_t}{L_t + L_c}$$

In that event, we should have $P_a = \frac{1}{3}$ or odds of 2 to 1 against A's decision.

The issues are straightforward till Bayes examines the question: what should A or B do if unfamiliar with the prescription for Stage 1 of the game, *i.e.* if they did not know the prior probabilities p_t and p_c? Here, perhaps tentatively, he introduces the so-called *principle of insufficient reason*, *i.e.* in the absence of knowledge to the contrary, we shall deem that all prior probabilities are equal, so that in this case $p_t = \frac{1}{2} = p_c$, an assumption which would be false unless the umpire adopted the second plan. There is no conceivable factual basis for embracing this axiom known as Bayes' postulate; and before he answered to his Maker, it seems that its author had not convinced himself that there is. However, Price made it the kingpin of his exposition. Whereafter, Laplace embraced it with boyish enthusiasm and built on so insecure a foundation a superstructure of doctrine usually referred to as that of *inverse* probability. The adjective signifies that the doctrine licenses one to draw conclusions about past occurrences. Among other exploits, K. Pearson, a modern disciple of Laplace, used it to prove that miracles cannot happen. None the less, one must concede the historical occurrence of at least one miracle in as much as many highly intelligent people have been willing to subscribe to the doctrine.

However much or little importance its author attached to the axiom, nothing Bayes propounded justified the verdict that he conceived Stage 1 of the two-stage process in other than factual terms. To exploit to the uttermost the licence the axiom confers, Laplace incorporated it in his own interpretation with an addendum peculiarly his own. Prior probabilities are not merely assignable to a factually specifiable randomwise sampling process involving the intervention of an agent and an apparatus. They are referable to a merely conceptual process to accommodate which we have to postulate a *hypothetical infinite population* from which we deem any selection of entities to be a random sample.

If we regard the theory of probability as relevant to life insurance, we must thus conceive – in effect, Laplace explicitly did so – that the actual population of Buffalo, N.Y., 1960 is a random sample from a limitless collection of conceptual populations of Buffalo, N.Y., 1960 stored in some inscrutable Platonic heaven of universals. There is here no need to examine in detail the credentials of a doctrine which makes such exacting demands on our credulity. It suffices to say that it encouraged the growth of a vast edifice of statistical theory throughout the nineteenth century; and, though few surviving mathematicians subscribe to it, few contemporary statisticians have fully faced up to what validity their claims can salvage if one abandons it.

What then is the indisputably legitimate usefulness of a mathematical theory of probability in the world's work? In its own domain, the classical theory itself does not provide an infallible recipe for any mortal gambler with insufficient prospect of longevity to continue playing an unlimited number of games. If they allow themselves enough safety margin to accommodate one limitation of the relevance of the theory to real life, it does however provide a rule which safeguards *most* gamblers who consistently adhere to a prescribed rule for the division of the stakes. Even this assertion calls for a *caveat*. We have seen that large-scale experiments with dice, roulette wheels, well-shuffled card-packs and urns containing balls or tickets disclose a very close correspondence with the implications of the concept of randomisation; but they do not encourage belief that experiments conducted on an ever larger scale would have a still more satisfactory outcome. In short, we must always allow for a safety margin in the probabilities we assign *a priori* to our score values, unless we adopt a system of scoring which (like the Wrighton system) eliminates possible sources of error on this account.

On this understanding, we may interpret certain physical phenomena in terms of a model endowed with the properties of a system to which the classical theory is applicable. For instance, we may take as our model for interpreting the long-run results of breeding from parents of specified genetic make-up randomwise sampling from a pair of urns with balls of different colours corresponding to ova (in one) and sperms (in the other) with different constituent genes. Whether such a model is appropriate, only experiment can decide; and experiment has abundantly confirmed the usefulness of statistical models both throughout the entire domain of genetics and throughout wide tracts of physical theory involving behaviour of hypothetical particles. To this extent, the mathematical theory of probability has now a very secure status in experimental science.

However, it made its first intrusion into physical science by another route, the so-called *Theory of Error* developed by Legendre (1752–1833), Laplace (1749–1827) and Gauss (1777–1855). In outline, this is as follows. In making repeated measurements of the same object or process, we do not

expect that the outcome of every observation will be the same; and in some types of measurement, especially astronomical, successive observations may range over many scale divisions. The simplest type of problem arising from variation due to so-called *instrumental error* which the investigator is powerless to eliminate is how to fix the value of one or more physical constants (*e.g.* k) for the formulation of a law expressed as an equation, *e.g.* the law ($pv=k$) connecting pressure (p) and volume (v) of a gas. In graphical form, our results then emerge ideally as a set of points which cluster closely round a hyperbola. We shall have several values of v for any one value of p; and it is not at all obvious how we can give due weight to more or less scatter in different regions of our ideal curve. Whence we face the problem: what are we to call the true value of k, *i.e.* the one which we should include in tables for the use of other investigators with least risk of introducing avoidable error in enquiries of another sort?

Here we have not space to outline the customary *least squares* technique. It will suffice to emphasise two considerations: (a) the underlying postulate, *i.e.* that uncontrollable instrumental errors are samples from a truly randomising process, is consistent with an important and already mentioned feature of a game of chance in as much as the situation sets the player (*i.e.* investigator) a task beyond his powers of sensory discrimination to fulfil; (b) a plot of the relative frequencies of errors of different magnitude accords with the form of a frequency distribution of the type $(q+p)^n$ sufficiently closely to encourage belief that the analogy with a game of chance is valid. Unfortunately, we have no certain *a priori* assignment of our *p*s and *q*s. Consequently, we have to rely on estimates which are themselves liable to error. In this context, it suffices to say that there are still many critics of any attempt hitherto devised to sidestep this dilemma.

The credentials of other aspects of statistical theory are still more debatable than the foregoing. We may call one the *calculus of exploration*, the other the *calculus of judgments*. Both of them have their source in the doctrine of inverse probability, though most contemporary exponents of their claims would be unwilling to subscribe to it. The *calculus of exploration* subsumes various fields of enquiry of which the ostensible aim is to disclose natural laws applicable to populations, human or otherwise, by exploiting the apparatus of the theory of error. Such a programme invites comment at two levels. At one, we observe that it relies on Quetelet's (1796–1874) teaching that nature is an urn. At the other, we detect a confusion between experimental error which obscures our view of a true law of nature and natural variation which is itself a manifestation of natural law. Quetelet's urn (*alias* the infinite hypothetical population of Laplace) does not suffice to endorse any randomising process unless we postulate in the background a Divine Dealer shuffling the pack and handing out the cards. There is indeed no justification for the analogy between natural variation and instrumental error, unless we invoke a supernatural agency.

What we have here called the *calculus of judgments* is an attempt to rationalise the validity of conclusions based on samples as illustrated by such a simple situation as the following: (a) out of 200 patients suffering from a particular ailment, 100 receive treatment A and 100 receive treatment B; (b) 45 patients under treatment A recover, as do 55 under treatment B. Can we conclude that on a larger scale treatment B would prove to be better than treatment A? There are at least three current views about the relevance of a mathematical-theory of probability to such questions. One school, represented by A. Wald and J. Neyman, have made an overdue attempt to formulate the issues in terms which do not invoke the doctrine of inverse probability; but this leads to certain mathematical difficulties as yet unsolved. While one must recognise the usefulness of a mathematical theory of probability in the domains of physics and genetics, one may legitimately question whether it can contribute much to the advancement of medicine, psychology and the social sciences.

The draw in a late eighteenth-century British state lottery. Randomising was achieved by mixing tickets in two huge circular drums. Boys from the famous Blue Coat School drew the winning numbers under the scrutiny of a panel of referees.

Chapter 13 Newer Geometries

In the last three chapters we have touched on the fringe of most of the major mathematical discoveries of the period which extends from 1650 to 1800. Two conspicuous omissions, especially in terms of their exploitation by nineteenth-century physics, are the prolific contributions of Euler to the study of differential equations and to the extension of the methods of the Newton-Leibniz calculus from the two-dimensional plane, in which we express one dependent $y=f(x)$ in terms of a single independent variable, to three-dimensional space, $z=f(x,y)$, whence by straightforward (though not visualisable) extension to manipulating a function of more than two independent variables. Here we must leave to our artist the task of conveying what we customarily speak of as *partial differentiation* and as *multiple integration* for the situation when $z=f(x,y)$ is a function of two independent variables (pages 273 and 274).

The dawn of the nineteenth century witnessed no deceleration of the tempo of discovery; but the dominant preoccupations disclose new paths radiating from a single focus in a new social setting. By A.D. 1800 a victorious European bourgeoisie which had nursed the science of the Newtonian age in its own adolescence had good reasons for hopeful gratitude, especially in the French setting where science was paying dividends to the armies led by Napoleon with not-as-yet dissipated revolutionary ardour and with a challenge to new exploits of military surveying. Five years after Waterloo, the *École Polytechnique* of Paris was the focus of a hitherto incomparable efflorescence of mathematics under the leadership of Laplace, Legendre and Lagrange. In the same setting, one may rightly mention Poncelet, for some years a prisoner after Napoleon's fiasco in the Russian campaign.

Several new social circumstances in the Napoleonic *milieu* – and throughout the next generation – set the scene for preoccupation from a new viewpoint with what we may call the mapping *motif*. A committee set up earlier by the National Assembly had adopted as the basis for what is now an international unit of length a newly-refined estimate of the earth's circumference in a great circle through Paris, after the French Navy had played its part in making measurements to confirm the Newton-Huygens prediction of a factually correct, though commonly exaggerated, flattening at the Poles. In Britain, leading a revolution of industrial mechanisation, canal construction had lately given to surveying an impetus which outlasted the construction of railroads, whereafter telegraphy disclosed new vistas of mapping the electromagnetic field. Meanwhile, the new steam technology had brought into being engineering draughtsmanship as a new, albeit at first elementary, application of projective geometry to subserve the needs of precision-instrument construction.

So far, the motive and incidentally the opportunity. What of the means in the threefold formula of historical detection? Between A.D. 1780 and

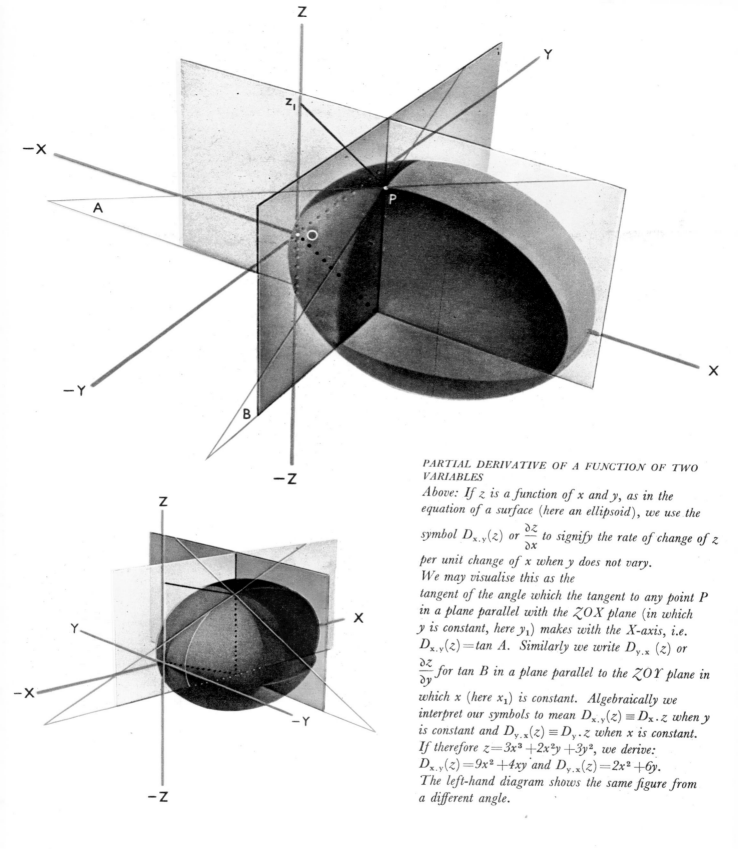

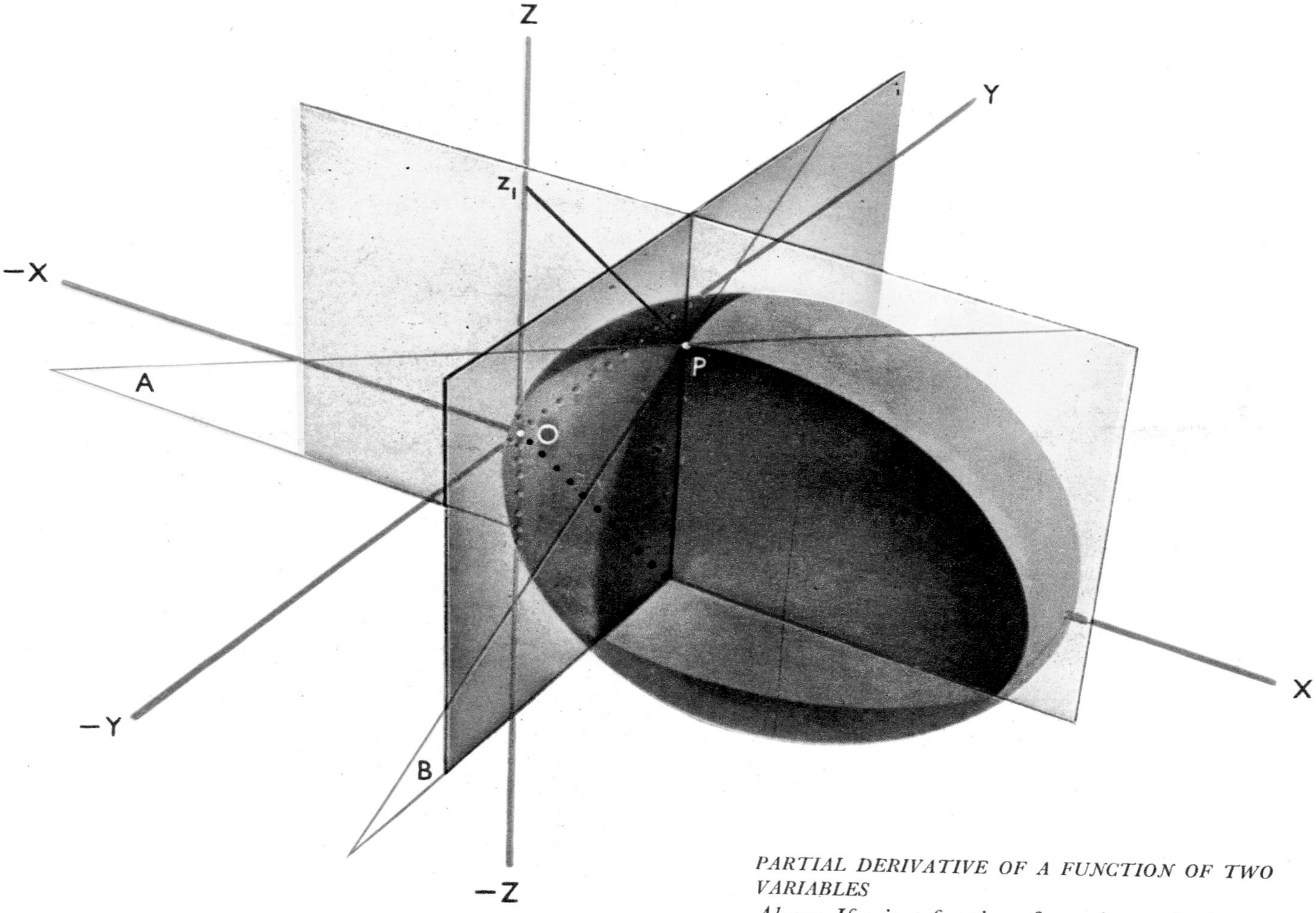

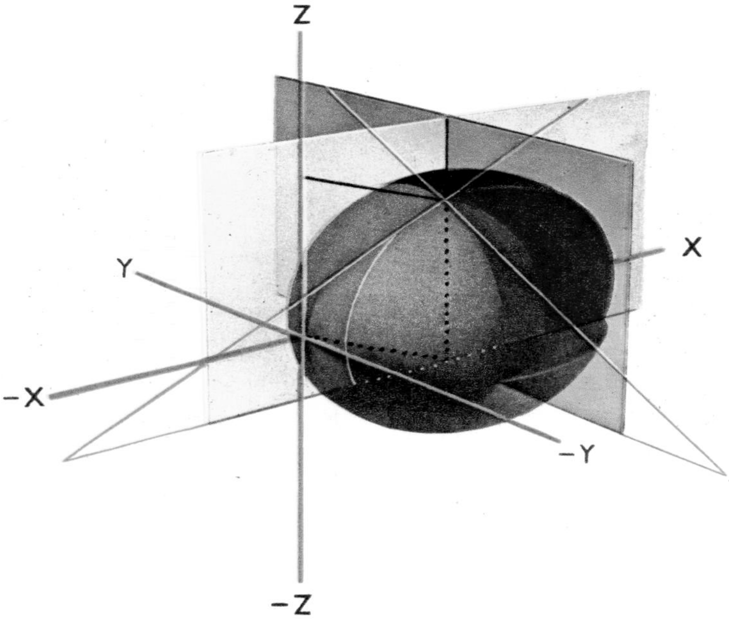

PARTIAL DERIVATIVE OF A FUNCTION OF TWO VARIABLES

Above: If z is a function of x and y, as in the equation of a surface (here an ellipsoid), we use the symbol $D_{x.y}(z)$ *or* $\frac{\partial z}{\partial x}$ *to signify the rate of change of z per unit change of x when y does not vary.*

We may visualise this as the tangent of the angle which the tangent to any point P in a plane parallel with the ZOX plane (in which y is constant, here y_1*) makes with the X-axis, i.e.* $D_{x.y}(z) = \tan A$. *Similarly we write* $D_{y.x}(z)$ *or* $\frac{\partial z}{\partial y}$ *for tan B in a plane parallel to the ZOY plane in which x (here* x_1*) is constant. Algebraically we interpret our symbols to mean* $D_{x.y}(z) \equiv D_x . z$ *when y is constant and* $D_{y.x}(z) \equiv D_y . z$ *when x is constant. If therefore* $z = 3x^3 + 2x^2y + 3y^2$, *we derive:* $D_{x.y}(z) = 9x^2 + 4xy$ *and* $D_{y.x}(z) = 2x^2 + 6y$.

The left-hand diagram shows the same figure from a different angle.

The key to this figure is:

Points	*X-Y co-ordinates*	*Lines*	*Z increment of the surface*
A	(x, y)	BE	$f(x+\triangle x, y)-f(x, y)$
B, E, F	$(x+\triangle x, y)$	BF	$f(x+\triangle x, y+\triangle y)-f(x+\triangle x, y)$
C, G, H	$(x, y+\triangle y)$	CG	$f(x, y+\triangle y)-f(x, y)$
D	$(x+\triangle x, y+\triangle y)$	CH	$f(x+\triangle x, y+\triangle y)-f(x, y+\triangle y)$

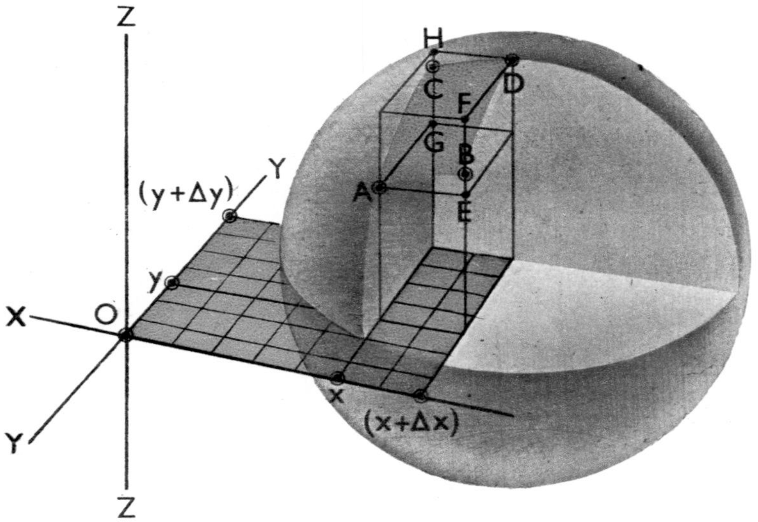

Top: The total increment of $z=f(x, y)$ is $EF=(BE+BF)=\triangle z=(CG+CH)=GH$.
In these expressions

$$\frac{BE}{\triangle x}=\frac{f(x+\triangle x, y)-f(x, y)}{\triangle x}\simeq D_{x.y}(z)$$

$$=\frac{f(x+\triangle x, y+\triangle y)-f(x, y+\triangle y)}{\triangle x}=\frac{CH}{\triangle x}$$

$$\frac{BF}{\triangle y}=\frac{f(x+\triangle x, y+\triangle y)-f(x+\triangle x, y)}{\triangle y}\simeq D_{y.x}(z)$$

$$=\frac{f(x, y+\triangle y)-f(x, y)}{\triangle y}=\frac{CG}{\triangle y}$$

$\triangle z\simeq D_{x.y}(z)\,.\,\triangle x+D_{y.x}(z)\,.\,\triangle y.$

If x and y are independent variables of t, we may write in the limit

$D_t.z=D_{x.y}(z)\,.\,D_t.x+D_{y.x}(z)\,.\,D_t.y.$

Numerical check:

Let $z=2x^2+3y^2$, so that $\triangle z\simeq 4x.\triangle x+6y.\triangle y$.
If $z=10$, $y=20$, $\triangle x=0{\cdot}1$ and $\triangle y=0{\cdot}2$, our formula then cites: $\triangle z\simeq 40(0{\cdot}1)+120(0{\cdot}2)=28$.
The exact result is: $(z+\triangle z)-z=2(10{\cdot}1)^2+3(20{\cdot}2)^2-2(10^2)-3(20^2)=28{\cdot}14$.

The importance of the result illustrated by the figure is that we may suppose z to be a function of r, itself a function of both x and y. We may then write:

$$\frac{\triangle z}{\triangle r}\simeq D_{x.y}(z)\,\frac{\triangle x}{\triangle r}+D_{y.x}(z)\,\frac{\triangle y}{\triangle r}.$$

If x and y are independent functions of the parameter r, this becomes in the limit:

$D_r.z=D_{x.y}(z)\,.\,D_r.x+D_{y.x}(z)\,.\,D_r.x.$

If r is a composite function of x and y, we may write:

$D_r.z=D_{x.y}(z)\,.\,D_{r.y}(x)+D_{y.x}(z)\,.\,D_{r.x}(y).$

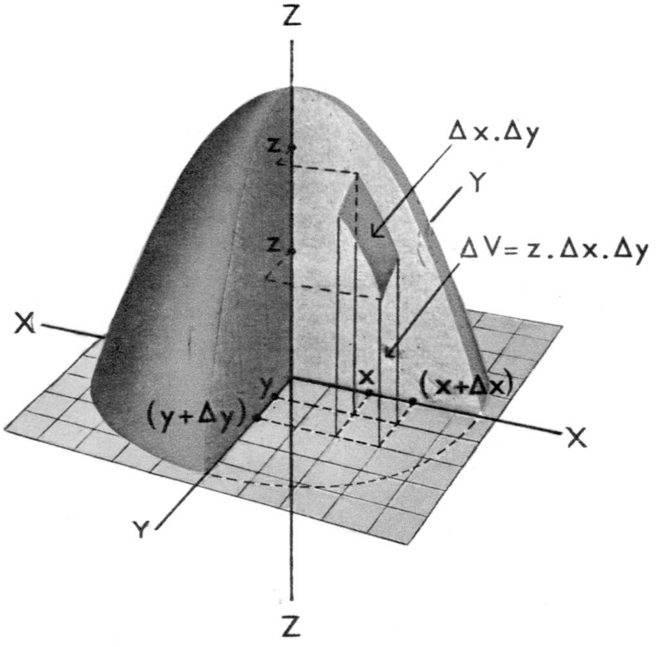

THE MULTIPLE INTEGRAL

Bottom: If $Z=f(x, y)$ is the equation of the surface of a solid figure, we may regard its volume as approximately equal to the sum of elements of height z and cross-section $\triangle x$ by $\triangle y$. In the limit, we write this for the range x from a to b and y from c to d as

$$V=\int_a^b\int_c^d z.dx.dy=\int_c^d[\int_a^b z.dx]dy.$$

Suppose for instance $z=3x^2+y$, so that $D_{x.y}{}^{-1}(z)=x^3+yx$, we proceed as follows:

$$\int_a^b\int_c^d(3x^2+y)dx.dy=\int_c^d(b^3-a^3+by-ay)dy.$$

1820 cardinal improvements in glass polishing and production of achromatic lenses led to production of microscopes, telescopes and theodolites for surveying far better than had been available heretofore. An incidental outcome was the final confirmation of the view of Aristarchus and Copernicus by Bessel's (1837–40) detection of the annual parallax of a fixed star (*61 Cygni*). In the same context, it is highly relevant to our theme that Gauss (1777–1855) directed inconclusive experiments to test whether light travels over long distances in Euclidean straight lines.

In short, many circumstances in the setting of 1800–1850 conspired to encourage a new approach to the recurrent theme of *mapping* space. One may refer to the major developments under various headings. A new tie-up between projective and co-ordinate geometry signalised the emergence of non-Euclidean geometry, of multi-dimensional geometry, of vector algebra and of topology; but this chapter can be little more than a catalogue of themes which the reader may have met in college text-books with few clues to their several inter-connexions. In one way or another, each of the foregoing signalises an invasion by algebra of the visual domain of Euclidean geometry. Accordingly, we shall here start at the stage – a little before 1850 – when grid algebra, already mentioned in Chapter 7 against the background of the Chinese counting-board as a device for

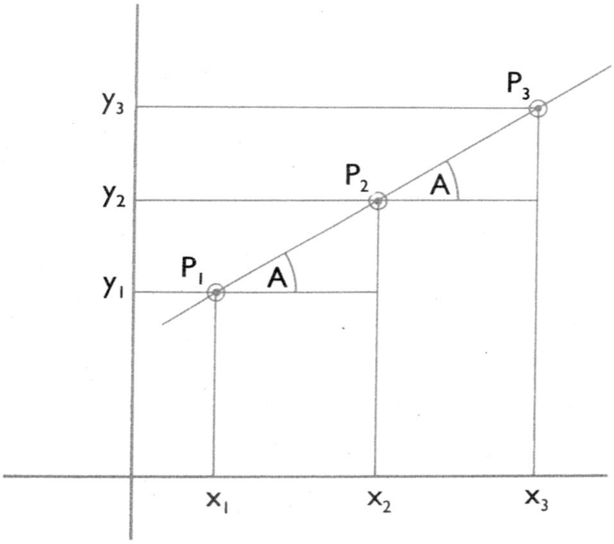

COLLINEARITY OF THREE POINTS

$$\frac{y_3-y_2}{x_3-x_2}=\tan A=\frac{y_2-y_1}{x_2-x_1}$$

$$\therefore (x_2-x_1)(y_3-y_2)=(x_3-x_2)(y_2-y_1)$$

$$\therefore x_1(y_2-y_3)-x_2(y_1-y_3)+x_3(y_1-y_2)=0$$

$$\therefore x_1\begin{vmatrix} y_2 & 1 \\ y_3 & 1 \end{vmatrix} - x_2\begin{vmatrix} y_1 & 1 \\ y_3 & 1 \end{vmatrix} + x_3\begin{vmatrix} y_1 & 1 \\ y_2 & 1 \end{vmatrix} = 0$$

$$\therefore \begin{vmatrix} x_1 & y_1 & 1 \\ x_2 & y_2 & 1 \\ x_3 & y_3 & 1 \end{vmatrix} = 0 = [x_r . y_r . 1]_3^1$$

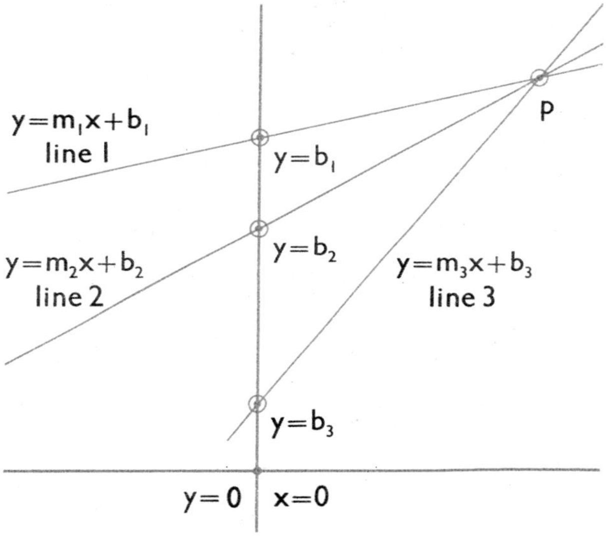

CONCURRENCE OF THREE STRAIGHT LINES IN A PLANE

If lines 1 and 2 cut at p whose co-ordinates are x_p, y_p

$$x_p=\frac{\begin{vmatrix} 1 & b_1 \\ 1 & b_2 \end{vmatrix}}{\begin{vmatrix} m_1 & 1 \\ m_2 & 1 \end{vmatrix}} \quad \text{and } y_p=\frac{\begin{vmatrix} m_1 & b_1 \\ m_2 & b_2 \end{vmatrix}}{\begin{vmatrix} m_1 & 1 \\ m_2 & 1 \end{vmatrix}}$$

If p also lies on a third line whose equation is $m_3x-y+b_3=0$: $m_3x_p-y_p+b_3=0$

$$\therefore m_3\begin{vmatrix} 1 & b_1 \\ 1 & b_2 \end{vmatrix} - \begin{vmatrix} m_1 & b_1 \\ m_2 & b_2 \end{vmatrix} + b_3\begin{vmatrix} m_1 & 1 \\ m_2 & 1 \end{vmatrix} = 0$$

$$\therefore \begin{vmatrix} m_1 & 1 & b_1 \\ m_2 & 1 & b_2 \\ m_3 & 1 & b_3 \end{vmatrix} = 0 = \begin{vmatrix} m_1 & b_1 & 1 \\ m_2 & b_2 & 1 \\ m_3 & b_3 & 1 \end{vmatrix}$$

More briefly we may write the concurrence condition:

$$[m_r . b_r . 1]=0$$

solving equations, had assumed a notational New Look in the West. This development now embraces the topics variously referred to as *determinants, matrices* and *tensors*.

In Chapter 7, we have defined the common pattern of determinants of different orders by recourse to the expansion of a determinant of order n in terms of the sum of the products of the elements of the first column (or row) and a determinant of order $(n-1)$ formed by elements of the residual rows and columns with alternating reversal of signs. We can also do so in another way, worthy of mention in passing. Just as we define a determinant of order 2 as the difference between its two 2-fold diagonal cross-products, we can define a determinant of order 3 (*etc.*) as the difference between the sum of its sets of three 3-fold (*etc.*) diagonal cross-products, if we interpret the build-up of the diagonal cross-products as a cyclic operation, *i.e.*

$$\begin{array}{ccccccccc} a_1 & \times & b_1 & \times & c_1 & \times & a_1 & \times & b_1 \\ a_2 & & b_2 & & c_2 & & a_2 & & b_2 \\ a_3 & \times & b_3 & \times & c_3 & \times & a_3 & \times & b_3 \end{array}$$

We then collect cross-products, as for the 2×2 determinant, *i.e.*

$$(a_1b_2c_3+b_1c_2a_3+c_1a_2b_3)-(a_3b_2c_1+b_3c_2a_1+c_3a_2b_1)$$
$$=a_1(b_2c_3-b_3c_2)-a_2(b_1c_3-b_3c_1)+a_3(b_1c_2-b_2c_1)$$

$$=a_1\begin{vmatrix} b_2 & c_2 \\ b_3 & c_3 \end{vmatrix} - a_2\begin{vmatrix} b_1 & c_1 \\ b_3 & c_3 \end{vmatrix} + a_3\begin{vmatrix} b_1 & c_1 \\ b_2 & c_2 \end{vmatrix} = \begin{vmatrix} a_1 & b_1 & c_1 \\ a_2 & b_2 & c_2 \\ a_3 & b_3 & c_3 \end{vmatrix}$$

The full form on the right above is unavoidable if we are using determinants as a computing device. Otherwise, we may more economically but without ambiguity express the last identity in a form such as the following:

$$a_1[b_r;\, c_r]_3^2 - a_2[b_r;\, c_r]_3^1 + a_3[b_r;\, c_r]_2^1 = [a_r;\, b_r;\, c_r]$$

We are now ready to express some important geometrical relations in a highly compact way. We have already spoken of the determinant as a grid layout with a definite numerical value; and since it has as many rows as columns, the form on the right above, *viz.* $[a_r, b_r, c_r]$ suffices to specify it without explicitly writing $r=1, 2, 3$. Henceforth we shall use the term *matrix* for a grid layout of numerical symbols without any collective numerical value defined by the cross-product rule given above. It may or may not have an equal number of rows and columns.

We may start with advantage, if we recall how much about the shape of a curve we may infer from the statement of its functional form as Ae^{-kx^2}. This has equal values for $+x$ and $-x$. It is therefore symmetrical about the Y-axis, lying wholly in the two upper quadrants. Its maximum (at $x=0$) is $y=A$. Whether or no we explicitly admit that x^{∞} and $x^{-\infty}$ mean anything definite, it is easy to see that the curve creeps in both directions nearer and nearer to the X-axis without ever quite touching it. Also, by differentiating twice and solving for $D_x^2y=0$, we find that the curve has points of inflexion at $x=\pm(2k)^{-\frac{1}{2}}$ on each side of the Y-axis, *i.e.* where it changes from being concave upwards to concave downwards or *vice versa*. However, all this is small beer, when we enlist *grid algebra* to dispense with the need to make a geometrical construction to interpret the properties of a figure.

This will be sufficiently clear if we look at three elementary problems in two dimensions:

(a) When are three points *collinear* (*i.e.* on one and the same straight line)?

(b) When are three lines *concurrent* (*i.e.* cross at the same point)?

(c) What is the *area* of a triangle in terms of the Cartesian co-ordinates of its vertices?

The rationale of the answers is in the legends of the illustrations on pp. 275 and 277, which employ an abbreviated notation, as above. Since we shall also need to refer to the co-ordinates of a point P_r as (x_r, y_r) in the plane or (x_r, y_r, z_r) in three-dimensional space, we may write $P_r=(x_r, y_r)$ or $P_r=(x_r, y_r, z_r)$ as the case may be.

Our illustrations show that:

(1) the condition that three points $P_r=(x_r, y_r)$ lie on the same straight line is (see the illustration):

$[x_r;\, y_r;\, 1]=0$

(2) in terms of the line equation $m_r x_r - y_r + b_r = 0$, the condition that *three* lines cut at one point is:

$[m_r;\ -1;\ b_r] = 0 = [m_r;\ b_r;\ 1]$

(3) the area of a triangle whose vertices are $P_r = (x_r, y_r)$ *etc.* is:

$A = \frac{1}{2}[x_r;\ y_r;\ 1]$

In the illustration, we see the derivation when all three vertices of the triangle lie in the top right-hand quadrant of the Cartesian grid. Otherwise, we can transfer the origin by a simple substitution $X = x + a$, $Y = y + b$. The reader will see that:

$$[x_r + a;\ 1]_1^2 = (x_1 + a) - (x_2 + a) = x_1 - x_2 = [x_r;\ 1]_1^2$$
$$[(x_r + a);\ (y_r + b);\ 1] = (x_r + a)\ [y_r + b; 1]$$
$$= (x_r + a)\ [y_r;\ 1]$$
$$= (x_1 + a)\ (y_2 - y_3) - (x_2 + a)\ (y_1 - y_3) + (x_3 + a)\ (y_1 - y_2)$$
$$= x_1(y_2 - y_3) - x_2(y_1 - y_3) + x_3(y_1 - y_2) = [x_r;\ y_r;\ 1]$$

The collinearity formula follows from the area equation, because collinearity of $P_1 = (x_1, y_1)$, $P_2 = (x_2, y_2)$ and $P_3 = (x_3, y_3)$ implies that the area enclosed by lines on which they lie is zero. The reader who recalls the equation of the plane $(Ax + By + Cz + D = 0)$ will find it instructive to deduce:

(a) the condition that four planes intersect at a point

$[A_r\ B_r\ C_r\ D_r] = 0$

(b) the volume of a tetrahedron in the form

$V = \frac{1}{6}[x_r;\ y_r;\ z_r;\ 1]$

The last implies the condition that four points all lie on the same plane (*i.e.* $V = 0$), *viz.*

$[x_r;\ y_r;\ z_r;\ 1] = 0$

En passant, we may recall that $y = mx + b$ for the line in the X–Y plane is only one form of the equation of the straight line in the plane. In terms of the intercepts y_0 ($= y$ when $x = 0$) and x_0 ($= x$ when $y = 0$):

$m = y_0 \div x_0$ so that $y_0 x + x_0 y - x_0 . y_0 = 0$

Whence if $A = y_0$, $B = x_0$ and $C = -(x_0 . y_0)$: $Ax + By + C = 0$. We may then write the condition of concurrence for three straight lines on the plane as:

$[A_r\ B_r\ C_r] = 0$

The foregoing uses of the determinant to describe metrical and non-metrical geometrical properties depend on its numerical value; but it – and more generally matrices which need not have the same number of rows and columns – can also disclose geometrical relations which do not depend on the numerical value we assign to it as a single entity. In this connexion the *determinant product* defined as below is of special interest. By definition we may evaluate the following determinant thus:

$$\begin{vmatrix} ae + bg & af + bh \\ ce + dg & cf + dh \end{vmatrix} = \begin{matrix} (acef + adeh + bcfg + bdgh) \\ -(acef + bceh + adfg + bdgh) \end{matrix}$$

$$= (adeh - adfg) - (bceh - bcfg) = (ad - bc)\ (eh - fg)$$

$$= \begin{vmatrix} a & b \\ c & d \end{vmatrix} \times \begin{vmatrix} e & f \\ g & h \end{vmatrix}$$

The pattern will be evident if we lay out the factors and their products thus:

$$\begin{vmatrix} a_1 & b_1 \\ a_2 & b_2 \end{vmatrix} \times \begin{vmatrix} \mathbf{A}_1 & \mathbf{B}_1 \\ \mathbf{A}_2 & \mathbf{B}_2 \end{vmatrix} = \begin{vmatrix} a_1\mathbf{A}_1 + b_1\mathbf{A}_2 & a_1\mathbf{B}_1 + b_1\mathbf{B}_2 \\ a_2\mathbf{A}_1 + b_2\mathbf{A}_2 & a_2\mathbf{B}_1 + b_2\mathbf{B}_2 \end{vmatrix}$$

AREA OF A TRIANGLE AS A DETERMINANT

Area of the triangle $P_1 . P_2 . P_3 =$

$(A + B + C + D + E + F) - (D + E + F)$

$A \quad = \frac{1}{2}(x_2 - x_1)\ (y_2 - y_1)$

$C + D + E \quad = (x_2 - x_1)\ (y_1 - y_3)$

$B + F \quad = \frac{1}{2}(x_3 - x_2)\ (y_2 - y_3)$

$D + E + F \quad = \frac{1}{2}(x_3 - x_1)\ (y_1 - y_3)$

Twice the area of the triangle is therefore

$(x_2 y_1 - x_1 y_2) - (x_3 y_1 - x_1 y_3) + (x_3 y_2 - x_2 y_3)$

$$= \begin{vmatrix} x_2 & y_2 \\ x_1 & y_1 \end{vmatrix} - \begin{vmatrix} x_3 & y_3 \\ x_1 & y_1 \end{vmatrix} + \begin{vmatrix} x_3 & y_3 \\ x_2 & y_2 \end{vmatrix}$$

$$= \begin{vmatrix} 1 & x_3 & y_3 \\ 1 & x_2 & y_2 \\ 1 & x_1 & y_1 \end{vmatrix}$$

Thus the area is $\frac{1}{2}[1 . x_r . y_r]_3^1$

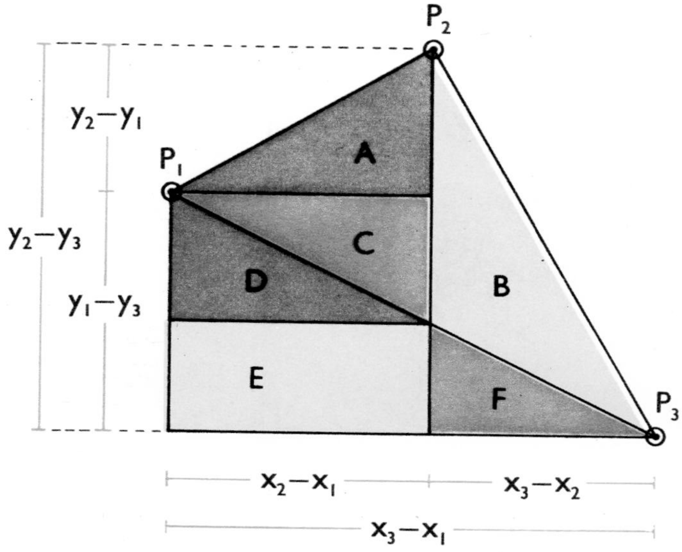

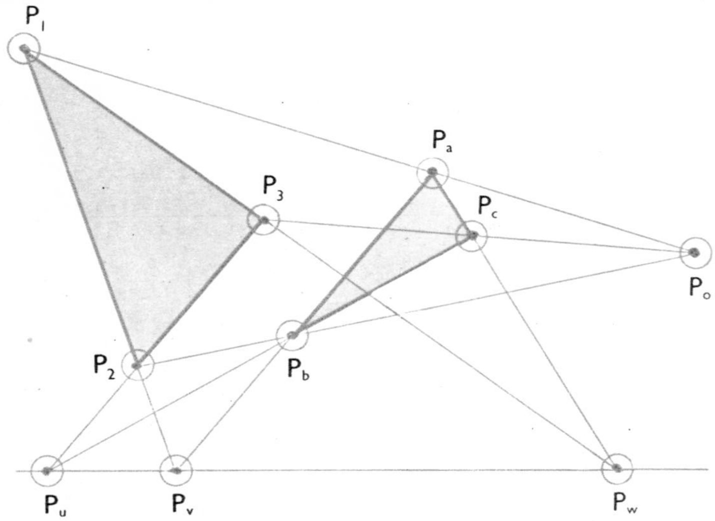

DESARGUES' THEOREM

If we place in the same plane two triangles in such a way that the three straight lines joining corresponding vertices meet at a point (here P_o), then the extensions of corresponding sides meet at three collinear points (here P_u, P_v, P_w).

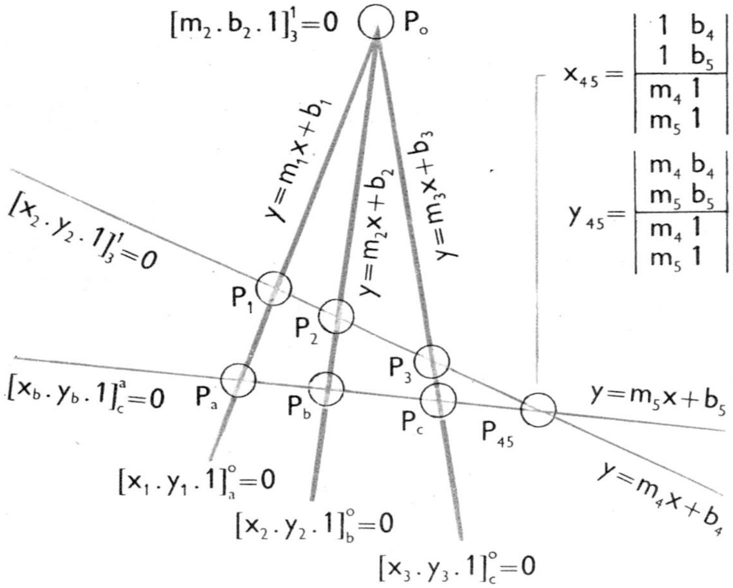

PROJECTION IN THE SAME PLANE

The third order determinants exhibited specify the condition of concurrence of the three rays at the focal point P_o and the collinearity of the points P_1, P_2, P_3, P_a, P_b, P_c, inter se *or with P_o. As for P_{45} above, the reader can define the co-ordinates of the seven other points, e.g. x_c, y_c, of P_c.*

A mnemonic for the 2×2 and 3×3 products is thus:

a_1b_1 $\mathbf{A_1A_2}$	a_1b_1 $\mathbf{B_1B_2}$
a_2b_2 $\mathbf{A_1A_2}$	a_2b_2 $\mathbf{B_1B_2}$

$a_1b_1c_1$ $\mathbf{A_1A_2A_3}$	$a_1b_1c_1$ $\mathbf{B_1B_2B_3}$	$a_1b_1c_1$ $\mathbf{C_1C_2C_3}$
$a_2b_2c_2$ $\mathbf{A_1A_2A_3}$	$a_2b_2c_2$ $\mathbf{B_1B_2B_3}$	$a_2b_2c_2$ $\mathbf{C_1C_2C_3}$
$a_3b_3c_3$ $\mathbf{A_1A_2A_3}$	$a_3b_3c_3$ $\mathbf{B_1B_2B_3}$	$a_3b_3c_3$ $\mathbf{C_1C_2C_3}$

We may write the 3×3 product in abbreviated notation as:

$$[a_r; b_r; c_r] \times [\mathbf{A}_r; \mathbf{B}_r; \mathbf{C}_r] =$$
$$[(a_r . \mathbf{A}_1 + b_r . \mathbf{A}_2 + c_r . \mathbf{A}_3); (a_r . \mathbf{B}_1 + b_r . \mathbf{B}_2 + c_r . \mathbf{B}_3);$$
$$(a_r . \mathbf{C}_1 + b_r . \mathbf{C}_2 + c_r . \mathbf{C}_3)]$$

It is here important to recall that the numerical value of a determinant does not alter if we interchange rows by columns in the same sequence, *e.g.*

$$\begin{matrix} r_1 \\ r_2 \end{matrix}\begin{vmatrix} 3 & 5 \\ 7 & 6 \end{vmatrix} = (3)(6) - (5)(7)$$
$$\quad c_1 \quad c_2$$

$$= (3)(6) - (7)(5) = \begin{matrix} c_1 \\ c_2 \end{matrix}\begin{vmatrix} 3 & 7 \\ 5 & 6 \end{vmatrix}$$
$$\quad r_1 \quad r_2$$

If our only concern were with the numerical value of the determinant, we might therefore express the product rule in an alternative way, which the reader can check. However, another use of grid algebra does not assign any numerical value to the grid as a whole, and it is then important to apply the product rule in conformity with the *preceding order* of factors *vis à vis* the prescribed row-column set-up of the product. We may thus distinguish between *determinant* multiplication which is *commutative* in the sense $\triangle_1 . \triangle_2 = \triangle_3 = \triangle_2 . \triangle_1$ and *matrix* multiplication which is not commutative, *i.e.* $\triangle_1 . \triangle_2 = \triangle_3 \neq \triangle_2 . \triangle_1$. (Translate $\neq$ as *is not equal to.*)

If the number of columns of the first factor is the same as the number of rows in the second, we may apply the product rule for determinants in the order cited to matrices which do not consist of equal numbers of rows and columns, as below:

$$\begin{vmatrix} a_1 & a_2 \\ b_1 & b_2 \\ c_1 & c_2 \\ d_1 & d_2 \end{vmatrix} \times \begin{vmatrix} \mathbf{A}_1 & \mathbf{A}_2 & \mathbf{A}_3 \\ \mathbf{B}_1 & \mathbf{B}_2 & \mathbf{B}_3 \end{vmatrix}$$

$$= \begin{vmatrix} a_1\mathbf{A}_1 + a_2\mathbf{B}_1 & a_1\mathbf{A}_2 + a_2\mathbf{B}_2 & a_1\mathbf{A}_3 + a_2\mathbf{B}_3 \\ b_1\mathbf{A}_1 + b_2\mathbf{B}_1 & b_1\mathbf{A}_2 + b_2\mathbf{B}_2 & b_1\mathbf{A}_3 + b_2\mathbf{B}_3 \\ c_1\mathbf{A}_1 + c_2\mathbf{B}_1 & c_1\mathbf{A}_2 + c_2\mathbf{B}_2 & c_1\mathbf{A}_3 + c_2\mathbf{B}_3 \\ d_1\mathbf{A}_1 + d_2\mathbf{B}_1 & d_1\mathbf{A}_2 + d_2\mathbf{B}_2 & d_1\mathbf{A}_3 + d_2\mathbf{B}_3 \end{vmatrix}$$

For memorisation, we may conveniently lay out the row-column components in tabular form as below:

$\mathbf{A}_1\mathbf{B}_1$	$\mathbf{A}_2\mathbf{B}_2$	$\mathbf{A}_3\mathbf{B}_3$
a_1a_2 b_1b_2 c_1c_2 d_1d_2	a_1a_2 b_1b_2 c_1c_2 d_1d_2	a_1a_2 b_1b_2 c_1c_2 d_1d_2

Though there is clearly little dividend from carrying out the multiplication of two numbers, *i.e.* the numerical values of the determinant factors, in such a roundabout way, the matrix

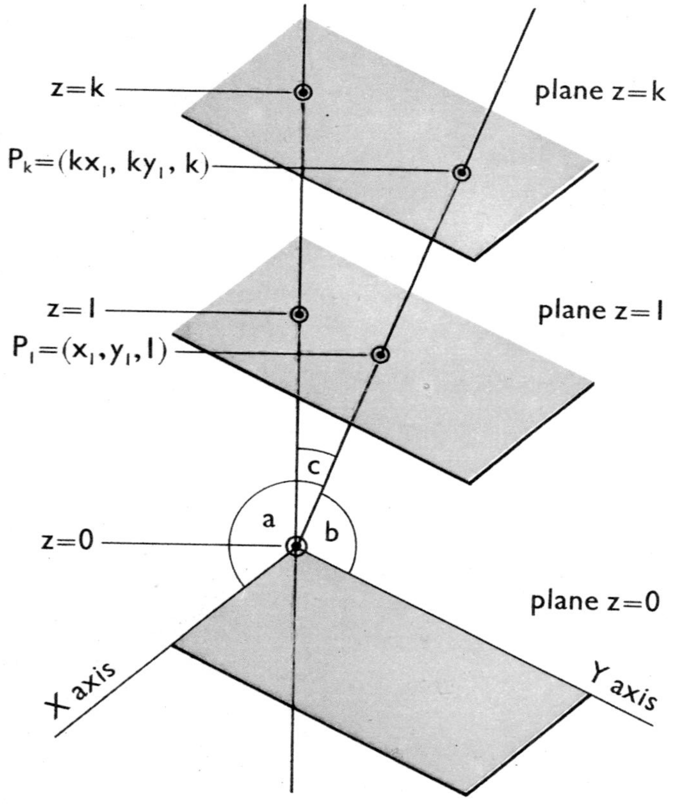

HOMOGENEOUS CARTESIAN CO-ORDINATES

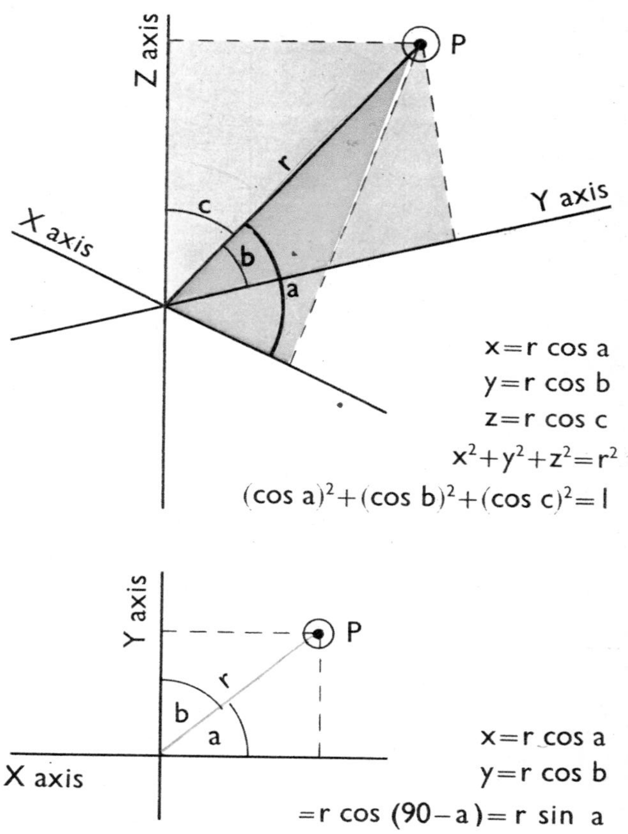

DIRECTION COSINES IN THE PLANE AND IN SOLID SPACE

pattern disclosed has a very remarkable linguistic advantage in the geometrical domain. As a simple illustration from this viewpoint, we recall that the rotation through a° of rectangular axes for one and the same figure in the plane signifies a transformation of co-ordinates:

$$x_2 = \cos a \,.\, x_1 + \sin a \,.\, y_1 \text{ and } y_2 = -\sin a \,.\, x_1 + \cos a \,.\, y_1$$

On the understanding that $\cos a = A$ and $\sin a = B$, so that $A^2 + B^2 = 1$, we may lay out the pattern of the transformation grid-wise as below:

$$\begin{aligned} x_2 &= A_1 x_1 + B_1 y_1 \\ y_2 &= -B_1 x_1 + A_1 y_1 \end{aligned} \qquad \Delta_{2.1} = \begin{vmatrix} A_1 & B_1 \\ -B_1 & A_1 \end{vmatrix} = 1$$

Similarly for a second rotation:

$$\begin{aligned} x_3 &= A_2 x_2 + B_2 y_2 \\ y_3 &= -B_2 x_2 + A_2 y_2 \end{aligned} \qquad \Delta_{3.2} = \begin{vmatrix} A_2 & B_2 \\ -B_2 & A_2 \end{vmatrix} = 1$$

By substitution of x_1 for x_2 from the above, we get:

$$\begin{aligned} x_3 &= (A_1A_2 - B_1B_2)x_1 + (A_1B_2 + B_1A_2)y_1 \\ y_3 &= (-A_1B_2 - B_1A_2)x_1 + (-B_1B_2 + A_1A_2)y_1 \end{aligned}$$

Whence we can lay out a new matrix:

$$\Delta_{3.1} = \begin{vmatrix} (A_1A_2 - B_1B_2) & (A_1B_2 + B_1A_2) \\ (-A_1B_2 - B_1A_2) & (-B_1B_2 + A_1A_2) \end{vmatrix} = 1$$

From our product formula we thus get:

$$\Delta_{3.1} = \begin{vmatrix} A_1 & B_1 \\ -B_1 & A_1 \end{vmatrix} \times \begin{vmatrix} A_2 & B_2 \\ -B_2 & A_2 \end{vmatrix} = \Delta_{2.1} \times \Delta_{3.2}$$

Here two new uses of grid notation in the service of co-ordinate geometry emerge:

(1) in rotation of rectangular axes about the origin, the numerical value of determinants which involve only the *direction cosines* ($\cos a_u = A_u$ and $\cos b_u = \sin a_u = B_u$) of the point in the plane is invariant in the sense:

$$\begin{vmatrix} A_u & B_u \\ -B_u & A_u \end{vmatrix} = 1$$

(2) without assigning any numerical value to the matrix as above, the product rule for matrix multiplication discloses the co-ordinate transformation of a rotation of axes from $P(x_1, y_1)$ to $P(x_3, y_3)$ if we know the appropriate constants for the transformations from $P(x_1, y_1)$ to $P(x_2, y_2)$ and from $P(x_2, y_2)$ to $P(x_3, y_3)$.

If we trace projective geometry to its beginnings, we learn as in Chapter 6 that the cross-ratio between the segments of a line cut by a pencil of four rays is an invariant *metrical* property of projection. Since Desargues discovered a remarkable theorem about triangles so placed that lines joining corresponding vertices are concurrent, projective geometry has been pre-eminently the study of invariant projective properties which are non-metrical. For instance, points which are collinear on a line are collinear in its projected image and straight lines which are concurrent correspond to concurrent straight lines of the projected image. Our diagrams (p. 275) show how eloquently and thriftily it is possible to express relations of collinearity and concurrence when we enlist the grid notation against the background of a co-ordinate framework.

To probe further into the tie-up between projective and co-ordinate geometry by recourse to algebraic notation, we shall now temporarily change our viewpoint. We may look at the relation between a geometrical figure and an algebraic equation naïvely in two ways. First we may ask: how does a change of our framework change the form of the equation referable to one and the same figure, as: (a) in a translation, when we shift the axes into new positions parallel with the former ones by the substitution $X = (x + a)$ and $Y = (y + b)$; (b) in a rotation of the axes about the origin by the substitution $X = Ax + By$ and $Y = -Bx + Ay$ subject to the condition $A^2 + B^2 = 1$? Alternatively, we may ask: how does a change of our framework affect the form of the figure referable to one and the same equation, as in a *scalar* change which converts an ellipse into a circle (or *vice versa*) by the substitution $X = Ax$ and $Y = By$ $(A \neq B)$? The second question suggests a new approach to projective geometry.

If we look at the tie-up between projective and co-ordinate geometry from a literally visual viewpoint, the simplest schema is to imagine the central projection as a pencil of lines radiating from the origin of an XYZ Cartesian framework and to locate the plane of the image figure as the plane parallel to the XOY plane located above it at $z=1$. We then recall that we can specify a straight line in solid space either in terms of the intersection of two planes or in terms of the parametric equations of the co-ordinates of the points which lie on it. If, as here assumed, all the rays meet at the origin, the variable parameter is the length of the radius vector (r) and the only constants involved are the so-called *direction cosines, viz.*

$$x=r\cos a \quad y=r\cos b \quad z=r\cos c$$

For any fixed value of $z=k$, we may therefore write

$$x_k=\frac{k\cos a}{\cos c};\ y_k=\frac{k\cos b}{\cos c};\ z_k=k$$

In particular, if $z=1$, as we postulate of the fundamental plane:

$$x_1=\frac{\cos a}{\cos c}=\frac{x_k}{k};\ y_1=\frac{\cos b}{\cos c}=\frac{y_k}{k};\ z_1=\frac{z_k}{k}$$

More briefly, we may write what we may appropriately call the point-equation of the ray of a pencil in central projection as $P(x_k, y_k, k)=P(kx, ky, kz)$. This signifies that any point whose co-ordinates in the framework are (x_k, y_k, k) cuts the plane of projection ($z=1$) at the points $x_1=x_k\div k$ and $y_1=y_k\div k$. Thus the specification of any point in a figure referable to the chosen three-dimensional framework suffices to specify

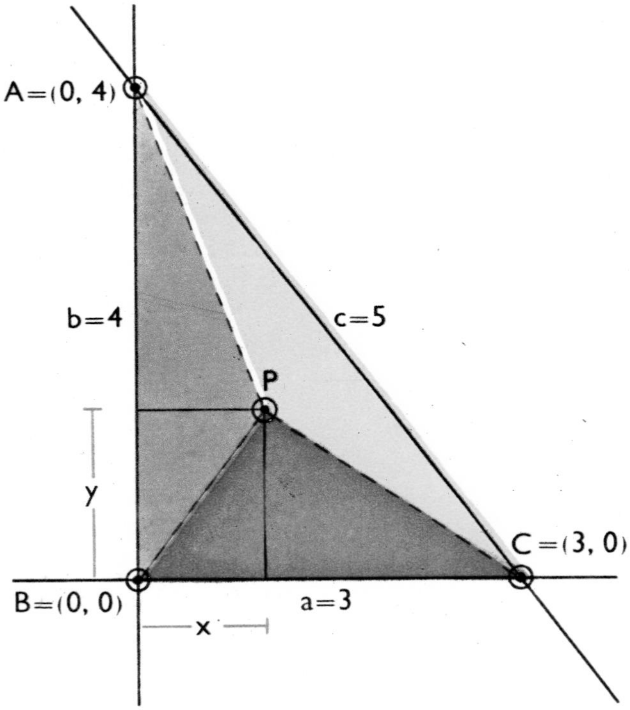

AREAL CO-ORDINATES

Here the triangle $ABC=\frac{1}{2}ab=6$

$$X_1=\frac{\triangle PAB}{\triangle ABC}=\frac{\frac{1}{2}.4x}{6}=\frac{x}{3}$$

$$X_2=\frac{\triangle PBC}{\triangle ABC}=\frac{\frac{1}{2}.3y}{6}=\frac{y}{4}$$

$\therefore x=3X_1$ *and* $y=4X_2$.

If the equation of a line in Cartesian co-ordinates is $3x+2y+12=0$

$9X_1+8X_2+12=0=9X_1+8X_2+12(X_1+X_2+X_3)$.

$\therefore 21X_1+20X_2+12X_3=0$

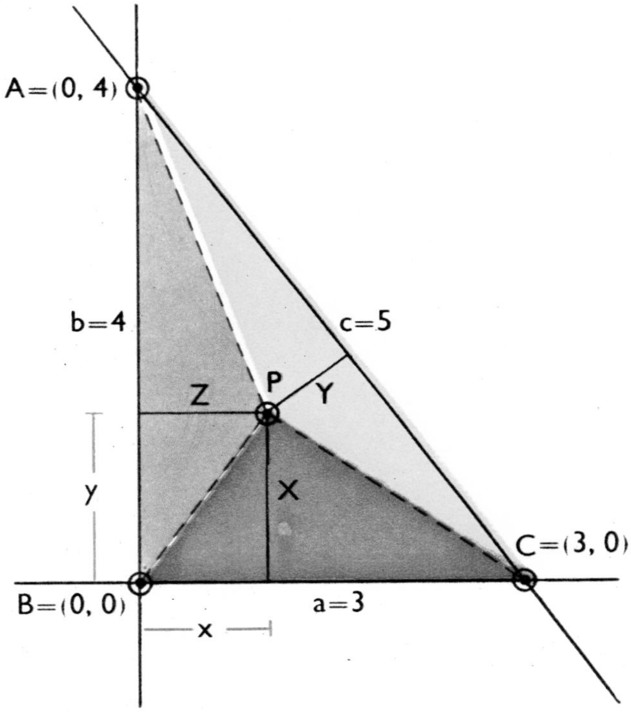

TRILINEAR CO-ORDINATES

From the figure, the triangle $ABC=\frac{1}{2}(4)(3)=6$;

$Z=x$ *and* $X=y$. *Also the triangle* $PBC=\frac{1}{2}.3X=\frac{3}{2}y$;

the triangle $PAB=\frac{1}{2}.4Z=2x$

$\frac{1}{2}.5Y=\triangle PCA=6-\triangle PBC-\triangle PAB=6-\frac{3}{2}X-2Z$.

$\therefore \frac{1}{4}X+\frac{5}{12}Y+\frac{1}{3}Z=1$.

If the Cartesian equation of a line is $3x+2y+12=0$, *the corresponding form in trilinear co-ordinates referable to the framework exhibited is*

$3Z+2X+12(\frac{1}{4}X+\frac{5}{12}Y+\frac{1}{3}X)=0$.

$\therefore 5X+5Y+4Z=0$.

also its image on the projection plane set parallel to the XOY plane at $z=1$ in the Z-axis.

In the search for an appropriate co-ordinate framework to exhibit invariant projective properties, one consideration which played a prominent role is the so-called principle of *duality*. Just as location of two points suffices to specify a straight line, it goes without saying that the intersection of two lines suffices to locate a point; and a certain family likeness (*duality* principle) occurs in many propositions about collinear points and concurrent lines in projective relationships. This suggests the need for co-ordinates which make such duality more explicit; and since the invitation emerges in the context of projection, it is natural to explore the possibility of referring points to a triangular framework of which one vertex can be the focus of projection where all the rays intersect.

We speak of one such system as *areal* co-ordinates. Of $P(x, y)$, a point inside a triangle of reference (ABC) whose area is $\triangle ABC$, we may write:

$$\triangle PAB + \triangle PBC + \triangle PCA = \triangle ABC$$

$$\frac{\triangle PAB}{\triangle ABC} + \frac{\triangle PBC}{\triangle ABC} + \frac{\triangle PCA}{\triangle ABC} = 1$$

One then defines the areal co-ordinates of P referable to the triangular framework of reference as:

$$X_1 = \frac{\triangle PAB}{\triangle ABC}; \quad X_2 = \frac{\triangle PBC}{\triangle ABC}; \quad X_3 = \frac{\triangle PCA}{\triangle ABC}$$

$$X_1 + X_2 + X_3 = 1$$

Given the Cartesian co-ordinates of the vertices $A=(x_1, y_1)$, $B=(x_2, y_2)$ and $C=(x_3, y_3)$, the determinant for the area of a triangle permits us to express the areal co-ordinates in terms of the Cartesian, thus:

$$2\triangle ABC = \begin{vmatrix} x_1 & y_1 & 1 \\ x_2 & y_2 & 1 \\ x_3 & y_3 & 1 \end{vmatrix} \qquad 2\triangle PAB = \begin{vmatrix} x & y & 1 \\ x_1 & y_1 & 1 \\ x_2 & y_2 & 1 \end{vmatrix}$$

In general, the pattern takes the form:

$$x = x_1X_1 + x_2X_2 + x_3X_3$$
$$y = y_1X_1 + y_2X_2 + y_3X_3$$

By using the third equation $X_1 + X_2 + X_3 = 1$, we can solve the above for X_1, X_2, X_3 in the form:

$$X_1 = A_1x + B_1y + C_1z, \text{ etc.}$$

Our diagram (page 281) shows the simple case of the reference triangle whose vertices are $A=(0, 4)$, $B=(0, 0)$, $C=(3, 0)$. The Cartesian equation of the line $3x+2y+12=0$ becomes $21X_1+20X_2+12X_3=0$ in which the unattached constant 12

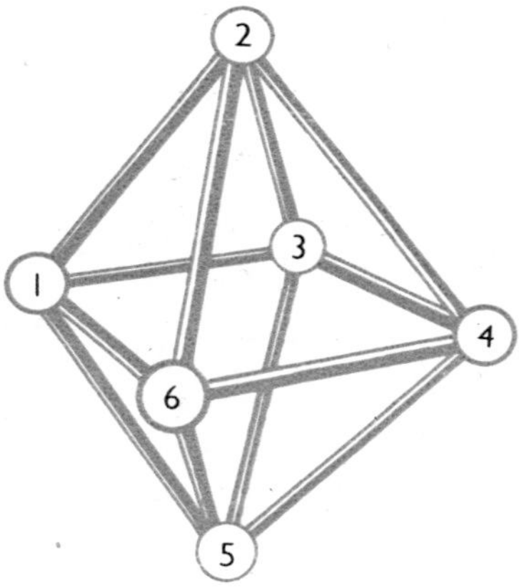

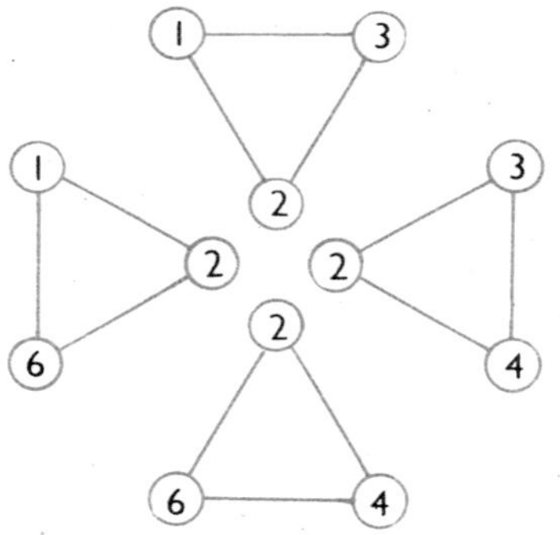

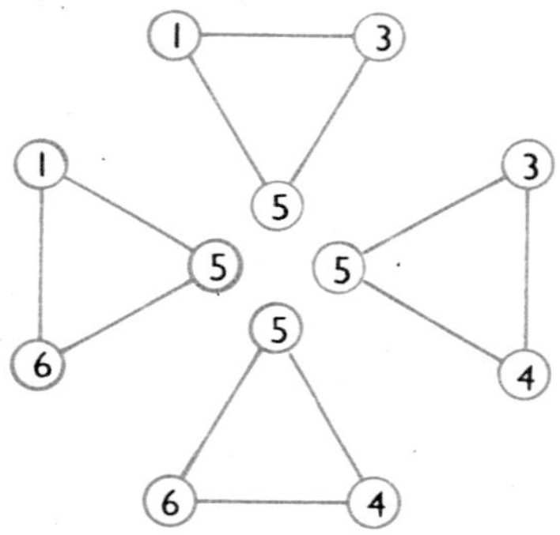

BREAKDOWN OF A POLYHEDRAL FRAMEWORK INTO RODS FOR PACKING

2 points (Dimension 0) suffice to define a line (Dimension 1). 6 points ($D=0$), 12 lines ($D=1$) and 8 faces ($D=2$) suffice to define an octahedron ($D=3$). (Faces are shown separately, above.) One approach to the geometry of four dimensions is to ask how we could unpack the figure for transport in terms of the number of (a) solid elements ($D=3$); (b) plane elements, i.e. sheets ($D=2$); (c) rods, i.e. lines ($D=1$); (c) points, i.e. joints ($D=0$).

Dimension		Points (N_0)	Lines (N_1)	Faces (N_2)	Solids (N_3)
0		$1=1_{(1)}$ A	. . .	. . .	. . .
1		$2=2_{(1)}$ A B	$1=2_{(2)}$ AB	. . .	. . .
2	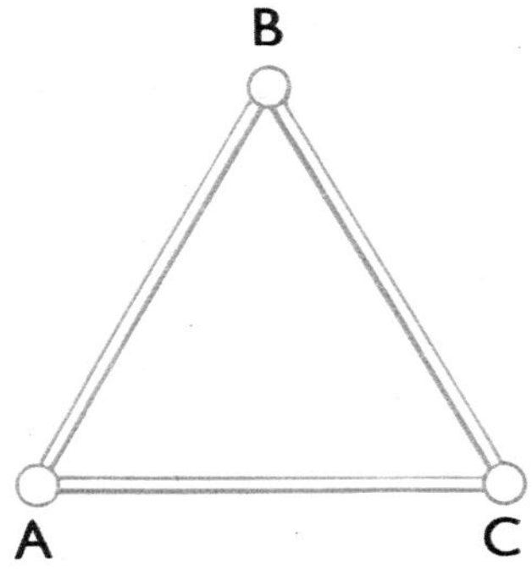	$3=3_{(1)}$ A B C	$3=3_{(2)}$ AB AC BC	$1=3_{(3)}$ ABC	. . .
3	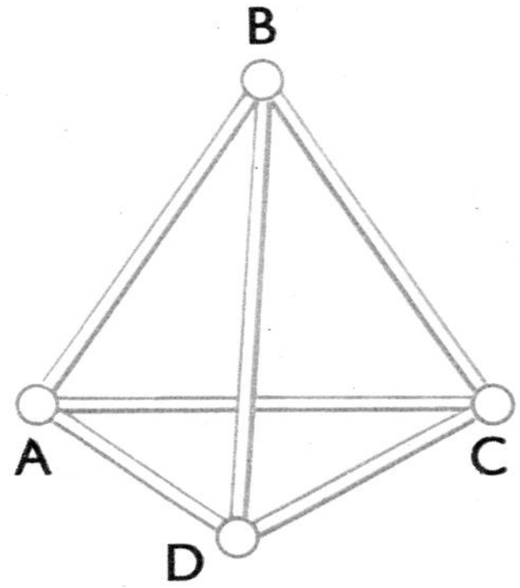	$4=4_{(1)}$ A B C D	$6=4_{(2)}$ AB AC AD BC BD CD	$4=4_{(3)}$ ABC ABD ADC BDC	$1=4_{(4)}$ ABCD
4	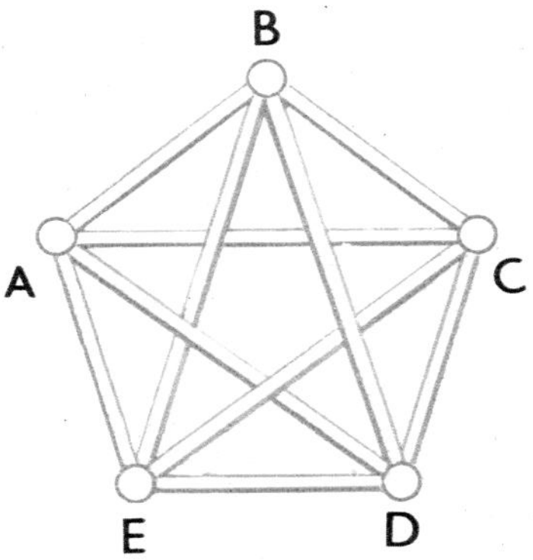	$5=5_{(1)}$ A B C D E	$10=5_{(2)}$ AB AC AD AE BC BD BE CD CE DE	$10=5_{(3)}$ ABC ABD ABE ACD ACE ADE BCD BCE BDE CDE	$5=5_{(4)}$ ABCD ABCE ABDE ACDE BCDE

THE FOUR-DIMENSIONAL TRIANGLE
The simplest problem of unravelling the four-dimensional super-solid for packing purposes arises when the two-dimensional projection is a pentagon with inscribed diagonals, which resolves into 5 tetrahedra, 10 triangles, 10 lines and 5 vertices.

disappears as such by the substitution $12 = 12(X_1 + X_2 + X_3)$ since $(X_1 + X_2 + X_3) = 1$. As an exercise the reader may choose the reference triangles whose vertices are $A\ (0, 0)$, $B\ (0, 1)$, $C\ (2, 0)$, whence $\triangle ABC = 1$, so that $X_3 = y$ and $X_2 = \frac{1}{2}x$, whence the Cartesian equation of the line $5x + 3y + 2 = 0$ becomes $2X_1 + 12X_2 + 5X_3 = 0$.

If we write the Cartesian expression in the form $5x^1 + 3y^1 + 2x^0 = 0 = 5x^1 + 3y^1 + x^0 + y^0$, we disclose an important property of areal co-ordinates. Each term in the equation of the line involves the same power of X_r; and we say that the equation is homogeneous with respect to X_1, X_2, X_3. More generally, the reader may infer the use of the term by comparing such homogeneous expressions as $x^3 + Ax^2y + Bxy^2 + Cy^3$, $x^2 + y^2$ and $x^4 + Ax^3y + Vx^2y^2$ with such non-homogeneous expressions as $x^2 + 2x + 1$, $x^3 + 3xy + y^2$ and $x^2 + y^2 + y$.

We are now ready to exhibit the principle of *duality* with reference to line and point in the plane by recourse to our criteria of concurrence, *i.e.* three lines meeting at a point, and collinearity, *i.e.* three points lying on a straight line. With this end in view, it will be more convenient to write X, Y, Z for X_1, X_2, X_3 respectively.

A rotation or translation of the Cartesian framework affects only values of the constants A, B, C, in the equation of the line $Ax + By + C = 0$, and the form of the condition of collinearity does not involve the constants. We may therefore place the reference triangle of areal co-ordinates here referred to as X, Y, Z in any convenient position with regard to the Cartesian grid. If the reference triangle is $A = (0, y_0)$, $B = (0, 0)$, $C = (x_0, 0)$, so that, as we may infer from our illustration, $x = x_0 . X$ and $y = y_0 . Y$, we may write:

$$[x_r; y_r; 1] = [x_0 . X_r; y_0 . Y_r; (X_r + Y_r + Z_r)]$$
$$= x_0 y_0 [X_r; Y_r; (X_r + Y_r + Z_r)]$$

Whence by Rule 1.b in Chapter 7:

$$[x_r; y_r; 1] = x_0 y_0 \, [X_r; Y_r; Z_r]$$
and $[x_r; y_r; 1] = 0$ if $[X_r; Y_r; Z_r] = 0$

Let us next suppose that the equations of a line referable to three points whose Cartesian co-ordinates are x_1, y_1, z_1, *etc.* and areal co-ordinates are X_1, Y_1, Z_1, *etc.* are

$$a_r x_r + b_r y_r + c_r = 0 = A_r X_r + B_r Y_r + C_r Z_r$$
$$A_r X_r + B_r Y_r + C_r (1 - X_r - Y_r) = 0$$
$$(A_r - C_r)\ X_r + (B_r - C_r)\ Y_r + C_r = 0$$

In condensed notation, the condition that there exists a point common to all three lines is:

$$[(A_r - C_r); (B_r - C_r); C_r] = 0$$

By Rule 1.b in Chapter 7:

$$[(A_r - C_r); (B_r - C_r); C_r] = [A_r; B_r; C_r]$$

Our disclosure of duality for the equations $AX_r + BY_r + CZ_r = 0$ $(r = 1, 2, 3)$ is now complete:

Condition of Concurrence:	*Condition of Collinearity:*
$[A_r\ B_r\ C_r] = 0$	$[X_r\ Y_r\ Z_r] = 0$

Areal co-ordinates of a point in the plane are not the only type of homogeneous co-ordinates we can visualise against the background of a triangular framework of reference. Another, often known as trilinear co-ordinates (X, Y, Z), is definable in terms of the perpendicular from the point $P = (x, y)$ to one of the sides (lengths a, b, c) of the triangle of reference. In the symbolism used before:

$$\triangle PBC = \frac{a}{2}X; \quad \triangle PCA = \frac{b}{2}Y; \quad \triangle PAB = \frac{c}{2}Z$$

We shall now write for brevity: $A = \dfrac{a}{2\triangle ABC}$; $B = \dfrac{b}{2\triangle ABC}$; $C = \dfrac{c}{2\triangle ABC}$, so that $AX + BY + CZ = 1$. Again, we may write in terms of arbitrary constants A_r, B_r, C_r:

$$X = A_1 x + B_1 y + C_1 \quad \text{and} \quad Y = A_2 x + B_2 y + C_2$$

Our diagram (page 281 right) shows the relation between this and the Cartesian system for a triangular framework with sides $a = 3$, $b = 4$, $c = 5$ and intercepts at $(0, 4)$, $(3, 0)$. The reader can test whether, and in what form, the principle of duality becomes explicit by recourse to this set-up.

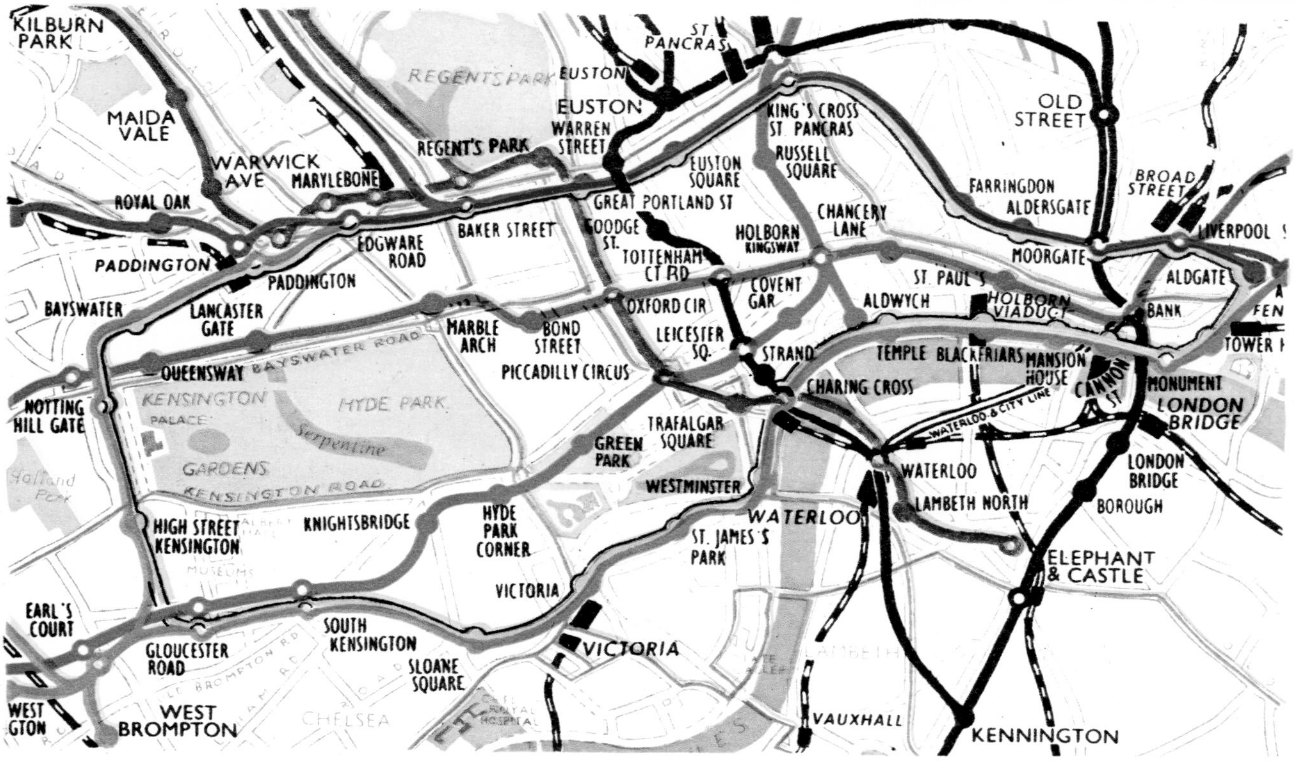

Above: Map of part of the London area in which distance remains invariant under projection and direction very nearly so.
Below: Plan of the corresponding section of the Underground system, issued by London Transport Executive for the convenience of the travelling public, in which the only property remaining invariant is the sequence of stations along each route.

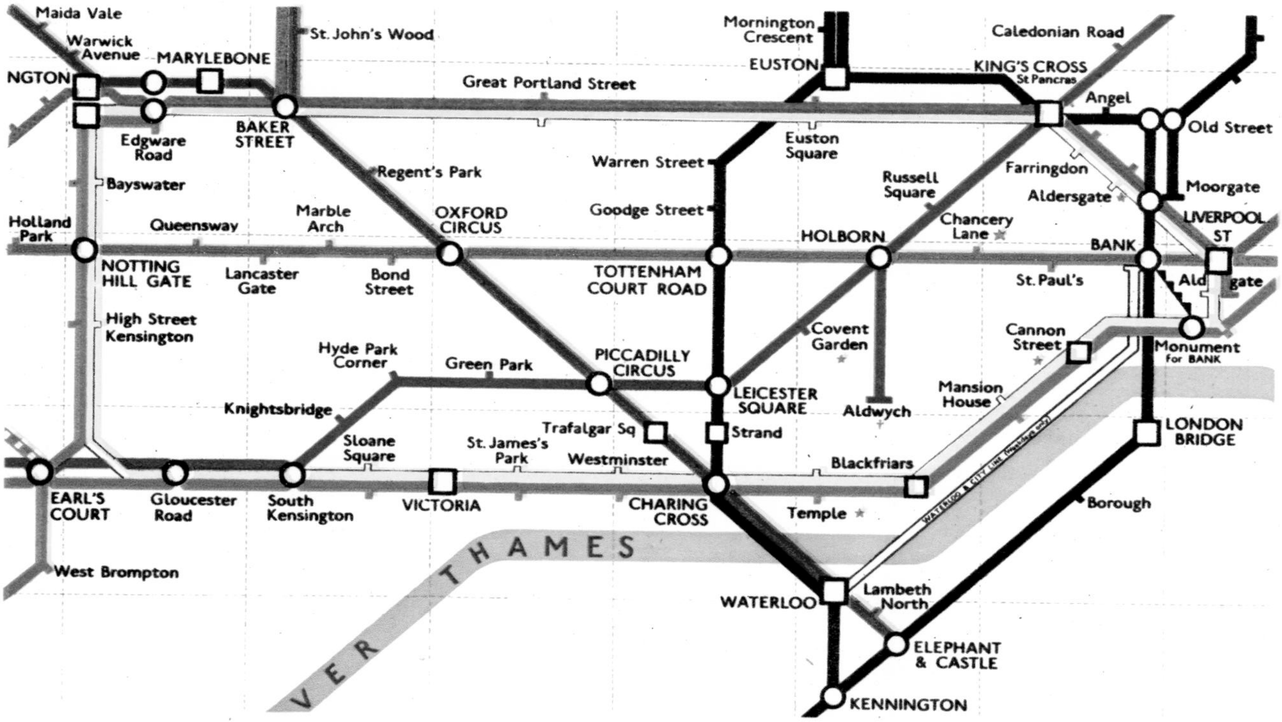

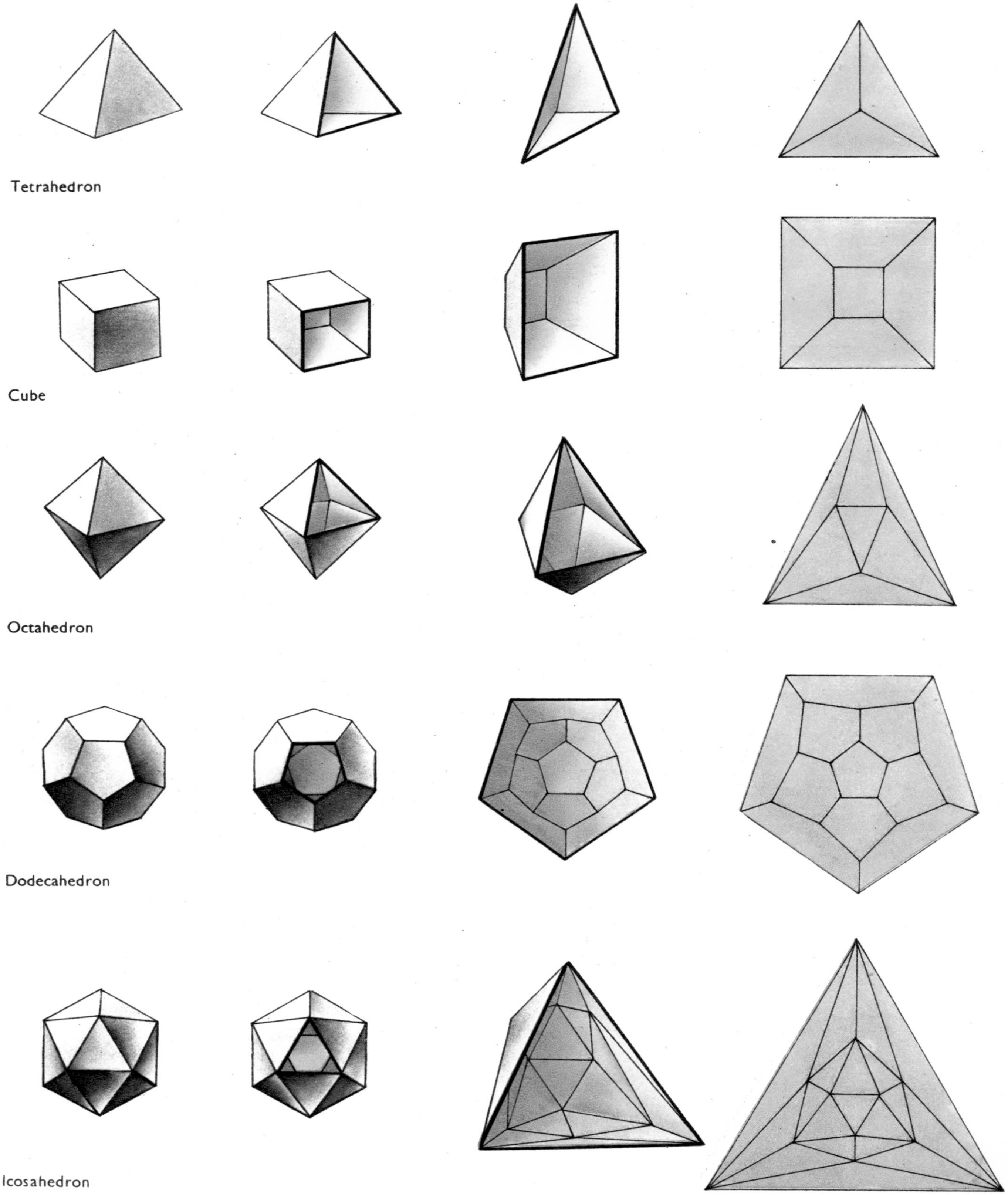

Deformation of the five regular solids as an exercise in rubber-sheet geometry. The first stage (left-hand series) is the removal of one face to permit stretching into a flat sheet of polygons one less in number than that of the original faces. The right-hand series exhibits the reduction of such sheets to single triangles by removal of the edges and vertices as explained in the text.

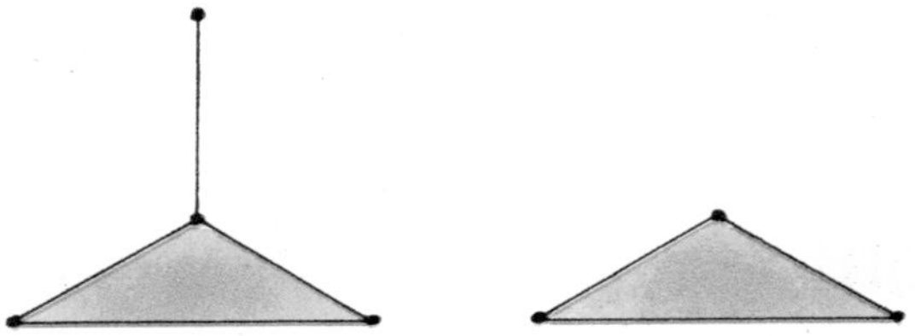

Tetrahedron

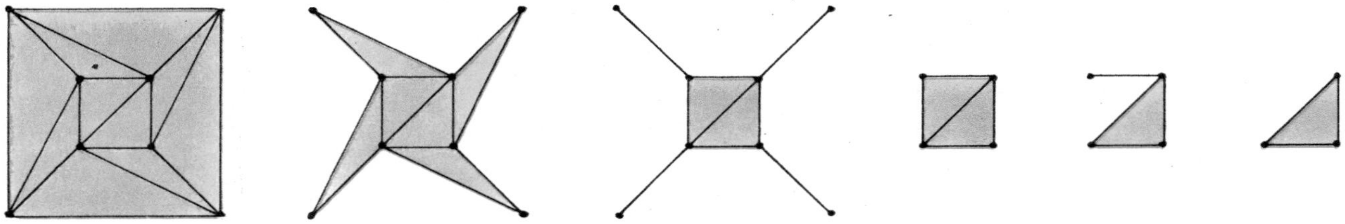

Cube

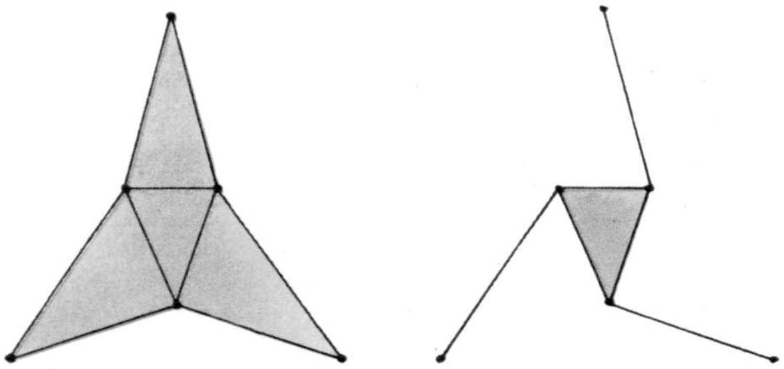

Octahedron

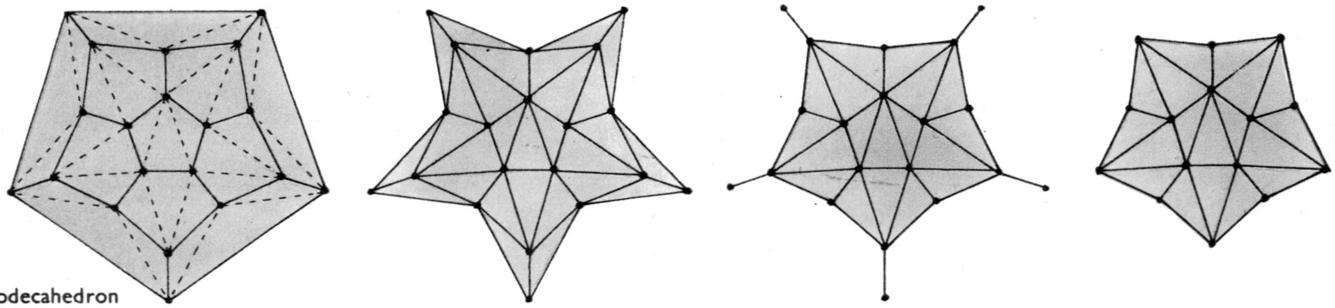

Dodecahedron

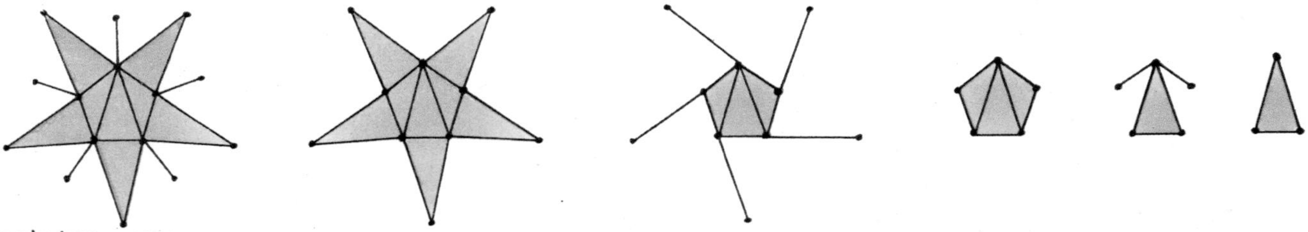

Dodecahedron, cont.

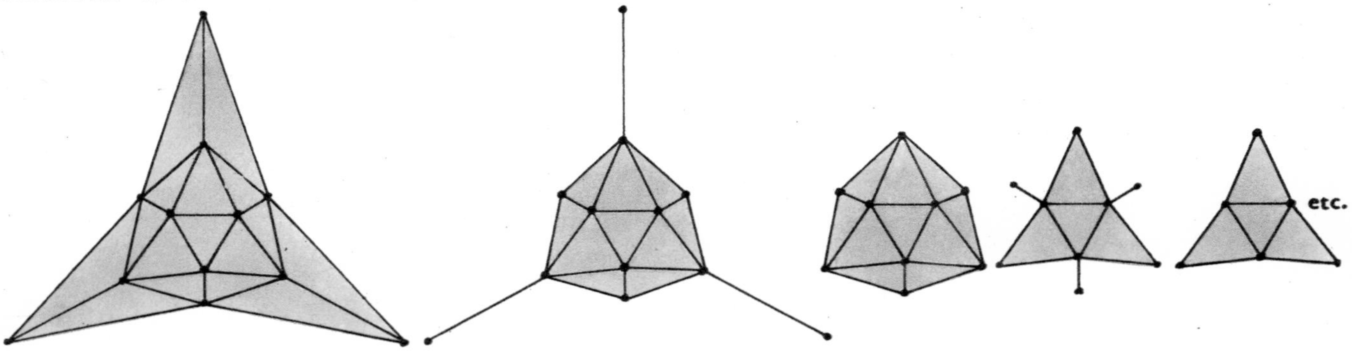

Icosahedron

About a century ago Mobius first drew attention to the existence of a surface with only one side, as when we join the ends of a strip of paper after a single twist. The illustrations show (a) what happens when we cut such a strip longitudinally into what we might expect to be separate and equal parts; (b) what happens when we do the same to a strip joined at its extremities after a double twist.

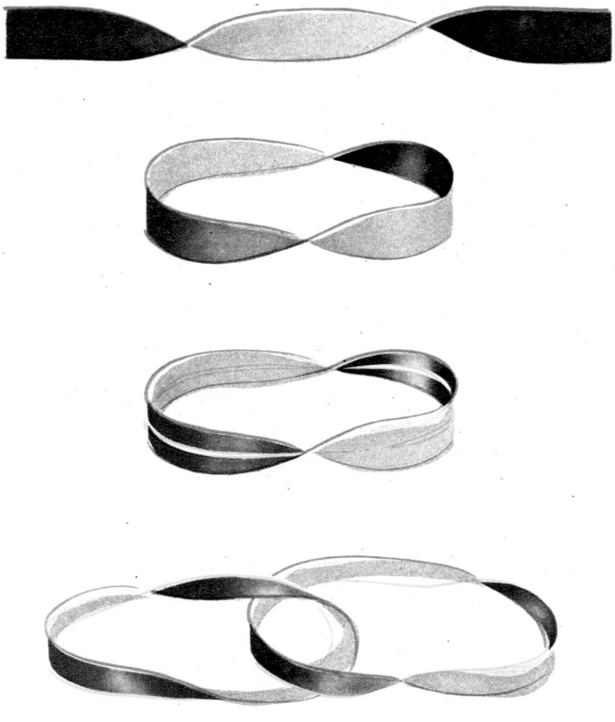

When discussing the wider implications of the duality mentioned above, it is convenient to designate such abstractions as points, straight lines, circles, triangles, cubes, spheres, *etc.*, by the term *geometrical objects*, or briefly *g*-objects. In the Cartesian system, the first of these had a special priority. One thought of extension in 0, 1, 2, 3 dimensions in terms of how many numbers suffice to specify the distance of a *point* from an arbitrarily-selected origin of reference. Actually, there is no *prima facie* reason why we should prefer to visualise a surface as a layer of minute particles (points) rather than as a closely-packed layer of fine wires (lines) pointing in the same direction; and indeed two numbers suffice to locate a point (x, y) as the intersection of two straight lines or a straight line whose intercepts are x_0, y_0 as a family of points. According to the model chosen, one is at liberty to regard either x and y or x_0 and y_0 respectively as constants or variables.

This consideration invites us to substitute for the naïve notion of absolute extension a *quasi*-space (*manifold*) which is a collection of geometrical objects. By definition, the dimensionality of a manifold depends on the *g*-objects of the collection, being the number of constants requisite to define one. Thus we speak of the plane as a two-dimensional manifold both of points and of lines. Solid space is a three-dimensional manifold of points; but it is not a three-dimensional manifold of straight lines. If we abbreviate its direction cosines as l ($=\cos a$), m ($= \cos b$), n ($=\cos c$), we may write the equation of a straight line through the origin in solid space as $x + y + z = (l + m + n)r$. In this, four fixed numbers participate. Whereas our solid space is a three-dimensional manifold with respect to points, it is therefore a four-dimensional manifold with respect to straight lines. Accordingly, duality between point and line no longer holds good; but we are free to explore the validity of the principle in relation to other pairs of *g*-objects with respect to which solid space is four-dimensional.

The concept of the manifold signalises a new freedom to enlist geometrical analogies to explore

The clover-leaf road, a major aid to the relief of traffic congestion, illustrates how topology, albeit at a very simple level, has a practical pay-off in the modern world.

a domain which is visualisable only in a very limited sense. *Inter alia*, we can extend to a hyper-space of four or more dimensions the geometrical properties of the grid notation. A few examples below exhibit explicitly what we convey metaphorically when we speak of a four-dimensional geometry, the co-ordinates of which we here write as x, y, z and t. For brevity, we shall write $x_2 - x_1 = \triangle x$, *etc.*:

(1) *Interval* (distance):

plane surface $(\triangle L)^2 = (\triangle x)^2 + (\triangle y)^2$

solid space $(\triangle L)^2 = (\triangle x)^2 + (\triangle y)^2 + (\triangle z)^2$

hyper-space

$(\triangle \mathbf{L})^2 = (\triangle \mathbf{x})^2 + (\triangle \mathbf{y})^2 + (\triangle \mathbf{z})^2 + (\triangle \mathbf{t})^2$

(2) *Orientation:*

plane $(\cos a)^2 + (\cos b)^2 = 1$

solid space $(\cos a)^2 + (\cos b)^2 + (\cos c)^2 = 1$

hyper-space

$(\mathbf{cos\ a})^2 + (\mathbf{cos\ b})^2 + (\mathbf{cos\ c})^2 + (\mathbf{cos\ d})^2 = \mathbf{1}$

(3) *Boundaries:*

line $Ax + By + C = 0$

plane surface $Ax + By + Cz + D = 0$

hyper-plane $\mathbf{Ax + By + Cz + Dt + E = 0}$

(4) *Figures:*

(a) *circle* $x^2 + y^2 - R^2 = 0$

sphere $x^2 + y^2 + z^2 - R^2 = 0$

hyper-sphere $\mathbf{x^2 + y^2 + z^2 + t^2 - R^2 = 0}$

(b) *ellipse* $\frac{x^2}{A^2} + \frac{y^2}{B^2} = 1$

ellipsoid $\frac{x^2}{A^2} + \frac{y^2}{B^2} + \frac{z^2}{C^2} = 1$

hyper-ellipsoid $\mathbf{\frac{x^2}{A^2} + \frac{y^2}{B^2} + \frac{z^2}{C^2} + \frac{t^2}{D^2} = 1}$

(5) *Areas and Volumes:*

area of triangle $\frac{1}{2!}\,[x; y; 1]$

volume of tetrahedron $\frac{1}{3!}\,[x; y; z; 1]$

hyper-volume of hyper-tetrahedron

$\mathbf{\frac{1}{4!}\,[x; y; z; t; 1]}$

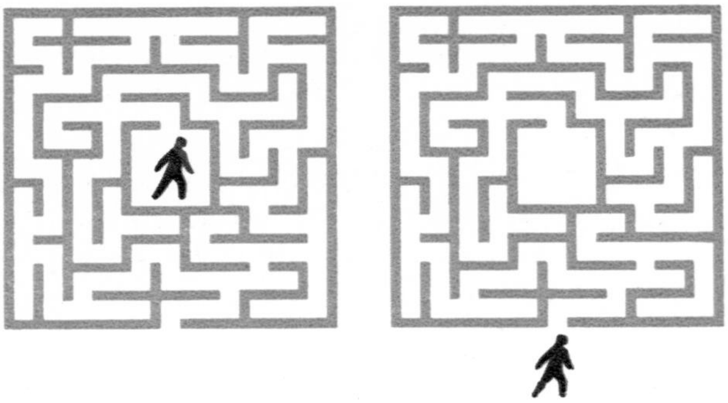

The left-hand maze above is made up of two quite distinct solid masses of wall. There is no simple escape formula. The entire wall of the right-hand maze consists of a single solid mass. One may escape from it by adhering consistently to either the wall on one's right or (consistently) to the wall on one's left.

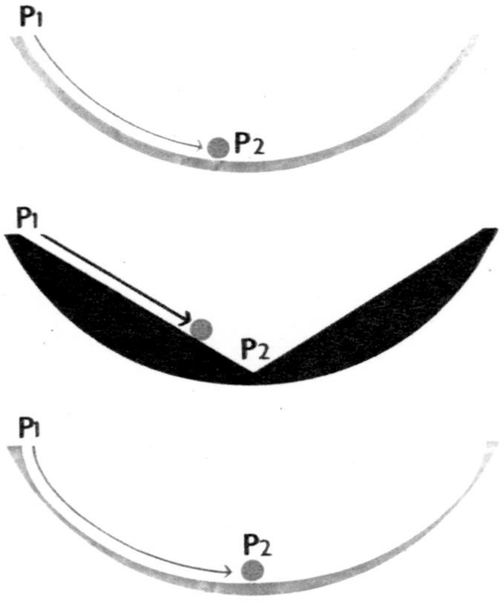

This figure shows three bowls, the initial position P_1 and the final position of a ball descending through a path with the same vertical and horizontal displacements from the initial position towards a second point (P_2) which the ball reaches in the bottom figure but fails to reach during the same time interval in the other two. The upper bowl is hemispherical and the geodesic is the arc of a great circle. The middle bowl is a cone and its geodesic is a straight line. In the plane of descent, the contour of the bottom bowl (which wins the contest) is a cycloid.

With little difficulty the reader will now think of other extensions of the algebraic pattern, *e.g.* the conditions that points are collinear, coplanar and in the same hyperplane, or that lines, planes and hyperplanes intersect. In short, every geometric relation for which we can disclose a common algebraic pattern for one, two, three dimensions suggests a corresponding theorem invoking four co-ordinates, any of which is visualisable taken not more than three at a time. For instance, we may make a solid model of a trajectory involving as one co-ordinate time (t), one co-ordinate the vertical displacement (z), the third being the horizontal displacement (either x or y).

If one's concern is with measurement (*i.e.* distances and/or angles), we can give no visual meaning to a geometry of four dimensions, Euclidean or otherwise, except in so far as the signs on the printed page have themselves a visible pattern. However, traditional geometry embraces non-metrical, though numerical, aspects of shape; and projective transformations may preserve these without retaining the metrical relations, as when we draw plausibly on paper a translucent cube with the correct number of faces (6), edges (12) and vertices (8) by making all parallel lines in the figure converge to a common focus. Now such an outline of a cube is a two-dimensional projection of a three-dimensional object in the familiar sense of the terms. To that extent, it is meaningful to think of a solid figure which is, so to say, a projection of a four-dimensional one and to project the three-dimensional projection on to the plane of the paper.

We then deem a point to be a zero-dimensional object, a line in the plane as one-dimensional, being an object specified by two points, a plane figure (two-dimensional) as an object specified by three or more lines (one-dimensional) and three or more points (zero-dimensional), a polyhedron (three-dimensional) as an object specified by four or more faces (two-dimensional), six or more lines (one-dimensional) and four or more points. In short, an object of n dimensions is specifiable by the number of objects of $(n-1)$,

($n-2$), *etc.* dimensions from which we can build it up. If we think of the edges as rods and the vertices (points) as joints of a framework comparable to steel-tube scaffolding, it is then possible to extrapolate a pattern which gives a visually meaningful projective significance to the four-dimensional tetrahedron as in the diagram on page 283. The reader may verify that the four-dimensional hyper-cube, built up of eight ordinary cubes, has 16 vertices, 32 edges and 24 faces. More generally, one may say of the family of figures whose two-dimensional representative is the square, that the number of r-dimensional objects in the build-up of the n-dimensional figure is $n_{(r)} . 2^{n-r}$.

There is none the less an important reason why we should not attempt to take this to mean that a fourth dimension is on all fours with the three dimensions of Cartesian space in a Euclidean sense. When one speaks of Euclidean geometry as the geometry of *rigid motions*, one implies that the metrical relations (angular, linear, *etc.*) of both plane and solid figures remain unchanged by translation, *i.e.* moving from one situation to another within the three-dimensional framework. As regards a plane figure, we say that we may also freely rotate it either in its own plane or in a *plane at right angles thereto*, *i.e.* we can superimpose by translation and rotation any such figure on its mirror image or *vice versa*. Thus we deem two triangles with equal corresponding sides and equal corresponding angles to be equivalent if their orientation in solid space is as mirror image to object; and this is so because we have an additional dimension in which to rotate. Now a solid figure is not equivalent to its mirror image with equivalent metrical properties. If it were so, Pasteur's interpretation of the optical properties of crystals would be meaningless; and the vast superstructure of chemical theory erected thereon would collapse. To say this implies that we have no such freedom of escape in the abstract hyper-space of four dimensions as we enjoy in the visualisable solid space of Euclid's geometry.

Non-metrical enumerative non-projective properties of figures are the theme of a branch of geometry called *topology*. From the viewpoint of mapping, this subsumes all we can usefully convey when we have abandoned any attempt to preserve metrical relations or visible shape. For instance, the familiar maps of the London Underground or of the New York Subway retain only the *ordered sequence* of other stations on one or other permissible route between any two of them. One sometimes speaks of topology as *rubber-sheet* geometry for a reason which will emerge if we examine a noteworthy non-metrical property of the simple polyhedra.

As an example of a non-metrical, but none the less numerical, property of a Euclidean figure, one thinks first and foremost of a rule discerned by Descartes and established by Euler as an isolated item of mathematical lore subsequently recognised as one of fundamental relevance to seemingly unrelated problems. The reader will readily recognise illustrations of the rule, if familiar with at least three of the regular polyhedra, *viz.* the *tetrahedron* (4 vertices, 4 faces and 6 edges), the *cube* (8 vertices, 6 faces and 12 edges) and the *octahedron* formed by fusing the bases of two pyramids (6 vertices, 8 faces and 12 edges). If we denote by V, F and E respectively the number of vertices, faces and edges, the relation $V+F-E=2$ is true of each. It happens that it is also true of any simple polyhedron, regular or otherwise.

In this context, *simple* implies that the figure is deformable into a sphere by stretching, *i.e.* it has no *hole* which makes tubular deformation possible; and the proof of the rule rests on a deformation which brings all the faces into one plane. One can imagine that the figure is hollow, that the walls are of rubber and that the edges, as also the vertices, are clearly marked on the internal surface. If one cuts along each edge of one face to make a window, what remains has still the same number of vertices and edges, but one less face, *i.e.* if $V+F-E=N$ for the complete polyhedron, $V+F-E=N-1$ for the new figure.

One now performs the following sequence of

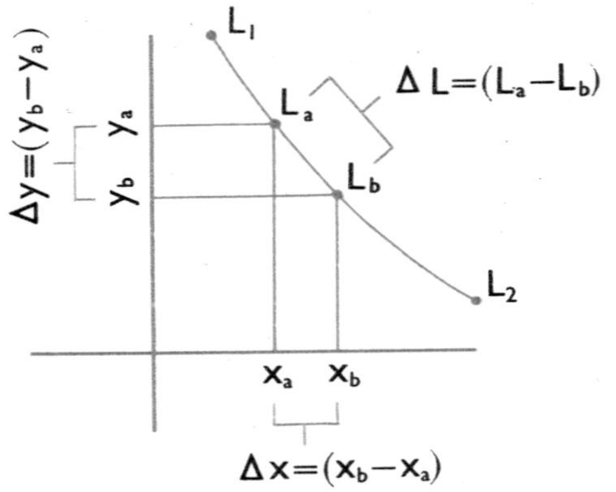

THE LINE INTEGRAL IN TWO-DIMENSIONAL RECTANGULAR CO-ORDINATES

If one approaches the meaning of the derivative of $y=f(x)$ from a geometrical viewpoint, one considers the ratio of two increments ($\triangle x$ and $\triangle y$) which are sides of a right-angled triangle whose third side is a chord (length $\triangle c$), so that $(\triangle c)^2=(\triangle x)^2+(\triangle y)^2$. As we make $\triangle x$ and $\triangle y$ smaller and smaller, the latter coincides more and more closely to a segment of the curve (length $\triangle L$) cut by the bounding ordinates $y=f(x)$ and $(y+\triangle y)=f(x+\triangle x)$, so that $\triangle L\simeq\triangle c$, whence:

$$(\triangle L)^2\simeq(\triangle x)^2+(\triangle y)^2 \text{ and } \left(\frac{\triangle L}{\triangle x}\right)^2\simeq 1+\left(\frac{\triangle y}{\triangle x}\right)^2$$

In the limit we may write:

$$D_x.L=\sqrt{[1+(D_x.y)^2]} \text{ and}$$
$$L=D_x^{-1}.\sqrt{[1+(D_x.y)^2]}+C.$$

Whence we may write for the length of a segment of the plane curve between (x_1, y_1) and (x_2, y_2):

$$L=\int_{x_1}^{x_2}\sqrt{[1+(D_x.y)^2]}.dx.$$

The corresponding expression for the line integral in polar co-ordinates is

$$L=\int_{a_1}^{a_2} r.da.$$

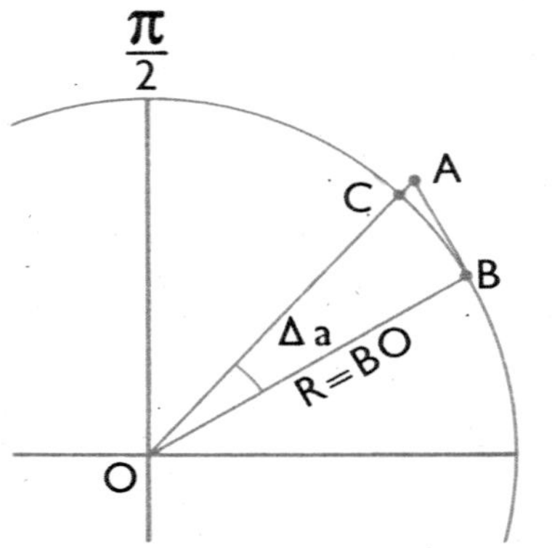

POLAR INTEGRATION OF THE CIRCLE

The length of the arc BC is $R.\triangle a$ if we measure a in radians. The area of the triangle AOB is $\frac{1}{2}AB.R=\frac{1}{2}R \tan \triangle a.R=\frac{1}{2}R^2.\tan \triangle a$. As we make $\triangle a$ smaller the area of the triangle AOB approaches more and more closely that of BOC and tan $\triangle a$ approaches $\triangle a$, so that we may write for the total area S

$$S\simeq\sum_{a=0}^{a=2\pi}\tfrac{1}{2}R^2.\triangle a.$$

In the limit:

$$S=\int_0^{2\pi}\tfrac{1}{2}R^2.da$$

Since the radius R is constant:

$$S=\tfrac{1}{2}R^2\int_0^{2\pi}da=\tfrac{1}{2}R^2(2\pi)-\tfrac{1}{2}R^2(0)$$
$$=\pi r^2.$$

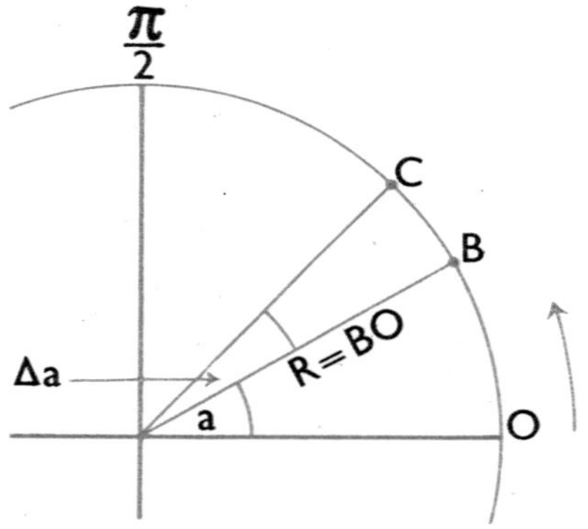

THE LINE INTEGRAL OF THE CIRCLE IN POLAR CO-ORDINATES

The arc CB is by definition $R.\triangle a$ if we measure a in radians. The length (L) of the entire boundary is

exactly $\sum_{a=0}^{a=2\pi} R\triangle a$

and this is equivalent to $\int_0^{2\pi} R.da=R\int_0^{2\pi} da=2\pi R.$

operations, which are easy to visualise as indicated on pages 286–87:

Operation 1. We stretch the edges of the window till the figure is a flat network of straight and/or curved lines, with V junctions henceforth called *vertices*, $(F-1)$ enclosures henceforth called *faces*, and E lines henceforth called *edges* linking the so-called vertices. Thus $V+F-E=N-1$ as before.

Operation 2. Unless the faces of the network are three-sided, in which event this operation is redundant, we divide each face into non-intersecting rectilinear or curvilinear triangles without changing the value of F. If the face is four-sided, we add one edge and substitute two faces for one with a net gain of zero for $V+F-E$. If the face is five-sided, we add two edges and substitute three faces for one, again with zero net gain. In general, this triangulation will leave $V+F-E$ as before, because V remains the same.

Operation 3. We remove the outer edges of the figure, thereby diminishing E and F by unity without changing the value of V at each stage. Whence again we do not change the value of $V+F-E = (N-1)$.

Operation 4. We have now a number of faceless edges, each with a single terminal point, whence the value of E is unchanged at the next step of sloughing off the loose ends. Each time we remove these, $V-E$ remains the same and $V+F-E$ remains the same, because no face change takes place.

We repeat in appropriate sequence the last two operations until we have on our hands only one residual triangle for which $V=3$, $F=1$ and $E=3$, so that $V+F-E=1$. Since the whole sequence of operations involves no change of the original equation $V+F-E=N-1$, it follows that $V+F-1=1=N-1$, whence $N=2$. From this, we can deduce what geometers of antiquity may well have suspected but, if so, were unable to prove, *viz.* that *there can be no more than five regular polyhedra*. These five include, in addition to the three mentioned, the *dodecahedron* with 12 pentagonal faces and the *icosahedron* with 20 triangular faces.

To establish the theorem, we do not need to postulate that all the edges of a regular polyhedron are of equal length. It suffices to rely on two of its properties: (a) each of its faces is a polygon with the same number (s) of sides; (b) the same number (m) of sides meet at each vertex. Hence $Fs=Vm$. Now each edge corresponds to the coalescence of two sides, whence the number of edges is half the total number of sides, *i.e.* $E=\frac{1}{2}Fs$. Since $V+F-E=2$, we derive:

$$F=\frac{4m}{2m+2s-ms};\quad V=\frac{4s}{2m+2s-ms};$$

$$E=\frac{2ms}{2m+2s-ms}$$

Now a face cannot have less than three sides, so that $s \geqslant 3$; and a vertex of a polyhedron must link at least three, so that $m \geqslant 3$. Furthermore $E>0$ must be positive. By tabulating $2m+2s-ms$ as

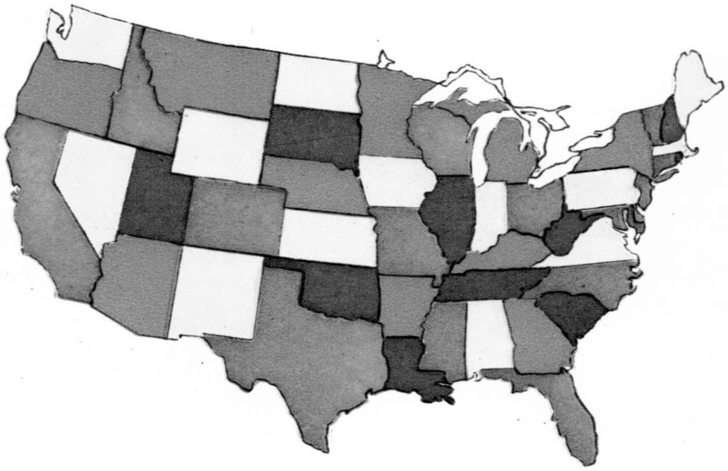

THE FOUR-COLOUR ENIGMA
Topologists suspect that four colours always suffice to distinguish any two adjacent territories on a flat or spherical map, but they have hitherto been able to prove that this must be so only when the number of territories is not greater than thirty-eight.

below, it is easy to see that this cannot be so if either s or m exceeds 5.

$s \backslash m$	3	4	5	6
3	+3	+2	+1	0
4	+2	0	−2	−4
5	+1	−2	−5	−8
6	0	−4	−8	−12

From this table we can extract all permissible values of s and m, whence all permissible sets of V, F and E, *viz.*

s	m	V	F	E	*Figure*
3	3	4	4	6	Tetrahedron
3	4	6	8	12	Octahedron
3	5	12	20	30	Icosahedron
4	3	8	6	12	Cube
5	3	20	12	30	Dodecahedron

We thus see that the five regular solids known to the Greek-speaking world are indeed the only ones which can exist.

Though Euler's Theorem is of fundamental importance in the discussion of topological themes, geometry conceived in such terms did not make much headway before a publication by Mobius (1865). He it was who first drew attention to the existence of a surface *with only one side*, as when we join the ends of a strip of paper after a single twist. Illustrations (p. 288) show what happens when we cut it longitudinally into what we might expect to be separate and equal parts, and what happens when we do the same to a strip joined at its extremities after a double twist. The properties of knots and mazes are among many other issues which involve equivalence of figures which can be transformed into one another by stretching or bending with or without cutting and restoration at the cut edge. Such problems may well have a pay-off in town-planning *vis à vis* traffic congestion. A standard topological enigma which impinges directly on our mapping theme is the famous *Four Colour Problem.*

Let us now return to geometry conceived traditionally in terms of invariant measurements as in Euclid's geometry of rigid motions, but interpreted against a background of assumptions different from those which we took for granted in our schooldays. If we seek a clue to the meaning of the term non-Euclidean, it may therefore be wise to sidestep the historical approach *via* controversies provoked by the writings of Lobachevsky (1793–1856) and Bolyai (1802–60) with reference to Euclid's parallel axiom of Chapter 4, and about what we mean by a straight line outside the restricted domain of engineering. We may then usefully start by referring to a physical problem first raised by the brothers Bernoulli, James (*alias* Jacob) and John. We imagine that two points P_1 and P_2 lie in the same vertical plane on the inner surface of a bowl. On the assumption that the vertical displacement of P_1 above P_2 and the horizontal displacement of P_2 with respect to the vertical line through P_1 have each fixed values, we can then ask: what curvature must this surface have to ensure that a ball sliding down from P_1 to P_2 will do so in least time? The answer is that the path of swiftest descent (so-called *brachistochrone*) is an arc of the same fascinating curve in which a pendulum swings with a period independent of its amplitude, *i.e.* the *cycloid.*

The solution of this problem is difficult and need not concern us except in so far as it prompts us to ask a second question which impinges on the Archimedean definition of a straight line, *viz.*, given a surface, what is the shortest path between two points on it? To formulate it in modern terms, we have to leave behind us the visual picture of a definite integral as a measure of area and tailor it to the task of measuring the length of an arc. The concept of the so-called *line integral* is simple. Its manipulation is difficult.

The figures on page 292 show the derivation of appropriate expressions for the length L of a segment of a plane curve in Cartesian and polar co-ordinates. In the same context we may exhibit for the first time the meaning of integration in polar co-ordinates, a procedure which is especially useful in connexion with the evaluation of the area of *closed* curves within whose boundary the origin of the framework lies.

$$\int_{x_1}^{x_2} \sqrt{[1+(D_{x.y})^2]} \, . \, dx = L = \int_{a_1}^{a_2} r \, . \, da$$

The formula on the right leads at once to the expression for the boundary of a circle of radius R if we set $a_2 = 2\pi$, $a_1 = 0$ and $r = R$ for all a. We leave it to the reader to obtain the same result by recourse to the Cartesian formula, by the substitution $y^2 = R^2 - x^2$.

In solid space we can express the three co-ordinates of points on a line, straight or curved, in terms of one variable parameter, which we may call p, so that

$$x = f_1(p) \quad y = f_2(p) \quad z = f_3(p)$$

We can then transform the line integral into the form:

$$L = \int_a^b [(D_p . x)^2 + (D_p . y)^2 + (D_p . z)^2]^{\frac{1}{2}} \, . \, dp$$

We have here at our disposal a tool with which we can explore the possibility of answering the question: what permissible path is the path (*geodesic*) of shortest distance between two points on a surface of definable properties? It goes without saying that different surfaces have different geodesics. All geodesics on the plane are Euclidean straight lines in the engineer's sense of the term; and all geodesics on the surface of a sphere are arcs of *great* circles. In this respect, the plane and the sphere are unique. On a conical bowl co-axial with the plumb-line, vertical geodesics will be straight lines, horizontal geodesics circular arcs with radii of different length. On our Bernoullian bowl, vertical geodesics would be cycloids, but horizontal ones will be circular arcs if the bowl is of circular cross-section.

Let us now ask what do we mean by the shortest distance between two points when imagination encompasses the outermost boundary, if any, of cosmic space? Clearly:

(a) it is meaningless to speak of a *permissible path* unless we are talking about a ray of light or other electromagnetic radiation;
(b) we cannot therefore confidently assume that a geodesic is a Euclidean straight line if it extends beyond the nearer stars whose distance our optical instruments can accurately measure by the parallax method.

Accordingly, we are free to examine the suggestion that straight lines of unrestricted length are *not* permissible paths in solid space. If not, we can conceive the geodesic between any two points P_1 and P_2 in such a space against the background of the cycloidal bowl analogy. Metaphorically speaking, we can then say that the algebraic expression for its geodesics is a measure of our space curvature. To be sure, we cannot visualise its curvature in terms of what we ordinarily mean by the curvature of a surface. What we can do is to adapt the algebraic formulae employed for the description of the latter by analogical extension comparable to the procedure implied when we referred earlier to a four-dimensional geometry.

By pursuing this line of thought, Riemann gave non-Euclidean geometry, previously developed on an arbitrary re-definition of parallelism, a new orientation destined to provide a background for the most impressive mathematical presentation of the laws of physics. Einstein's apparatus relies on far more than the differential geometry of Riemann, in particular the use of *tensors* which invoke both the notation of the matrix and the *vector* concept which we shall examine later. As every reader will know, the curvature of Einstein's universe is that of a four-dimensional space in which the interval has a more general form than as defined above for a Newtonian universe, *viz.*

$$(\Delta L)^2 = (\Delta x)^2 + (\Delta y)^2 + (\Delta z)^2 + (\Delta t)^2$$

We may write the more general form as:

$$(\Delta L)^2 = f_x(\Delta x)^2 + f_y(\Delta y)^2 + f_z(\Delta z)^2 + f_t(\Delta t)^2 + 2f_{xy} . \Delta x . \Delta y + y2f_{xz} . \Delta x . \Delta z + 2f_{xt} . \Delta x . \Delta t, \; etc.$$

The functional coefficients f_x, f_{xt} *etc.* in the above specify the curvature of the four-dimensional manifold.

Though Heaviside (1850–1925) and Gibbs tailored the notations of their predecessors to the requirements of mechanics and electromagnetism, the vector concept referred to in the last paragraph emerged in the same context as the re-examination of the credentials of the complex number by Gauss and his contemporaries in the writings of Grassman (1809–77) and Hamilton (1805–65). To at least as many freshmen as not – as to the writer of this book – vector notation is somewhat mystifying, if not repulsive, on first acquaintance. This is partly because the printed page exhibits it by recourse to typographical devices which are tiresome to transcribe, but mainly perhaps because one needs premedication in the *number field* (see next chapter) to give one the energy to grope one's way out of the semantic quagmire of what one means by number. As all readers of the foregoing chapters will sufficiently realise, the minus signs in $4-3$ and $3-4$ are not equivalent in terms of abacus subtraction; and if we wrote the second as $3+4i^2$ there would be no temptation to delude ourselves that they are. The beauty of vector algebra is that it equips us with labels to make the nature of our arithmetical postulates as explicit in three-dimensional space as does the complex number notation for the plane.

We may regard the latter as a convenient mapping device for labelling points on a plane. Hitherto we have met two ways of doing so within a rectangular framework. When we cite the co-ordinates of a point P as (x, y), we signify that its distance from the origin (O) in the XY plane is $r=\sqrt{(x^2+y^2)}$. If a is the inclination of OP to the X-axis, $x=r \cos a$, whence $a=\cos^{-1}(x \div r)$. Thus $P=(4, 3)$ signifies that $OP=5$ and that its inclination to the X-axis is $\cos^{-1}(0{\cdot}8)$. Alternatively, we may designate OP in terms of the real and imaginary axes of a *number field* as $4+3i$, meaning the same thing, *i.e.* that its *scalar* magnitude (length) is $5=\sqrt{(4^2+3^2)}$ and that its *direction* is $\cos^{-1}(4\div 5) \simeq 37°$. Either way, we conceive OP as an entity with two components, if we envisage it as the track of a point from the origin. When we do conceive it in such terms, *i.e.* as a *displacement*, we associate with it two components, a direction $(\cos^{-1}a)$ and a linear (scalar) magnitude (r) common to all points on a circle of radius r with the origin as centre. We shall henceforth speak of it as a vector. *En passant*, we may note that the alternative polar forms are more explicit, as when we write $P=(5, 37°)$ referable to the radius vector (scalar component) and the inclination to the base line or $5 \cos 37° +(5 \sin 37°)i$ in the notation of complex numbers.

The fundamental property of a vector in the plane is the rule of composition presented in the older text-books as the parallelogram (or triangle) of velocities, accelerations and forces, all of which we conceptualise in terms of displacement in unit time. If one displacement (V_1) would carry a point from the origin to $P_1 \equiv (x_1, y_1)$ and another (V_2) would carry it from the origin to $P_2 \equiv (x_2, y_2)$, the combined effect of the two displacements signifies that the length and direction of the second, which starts from P_1 if V_2 or P_2 if V_1, has the co-ordinates $x_3=x_1+x_2$ and $y_3=y_1+y_2$. In effect, this merely states that a 3-mile walk east, then 4 miles north, then 7 miles east, then 2 miles north in any order is equivalent to a walk 10 miles east and 6 miles north.

To specify displacement in three-dimensional space, we may add to our specification $z_3=z_1+z_2$, and if we write r_3 for OP_3, we derive

$$r_3=\sqrt{(x_3^2+y_3^2+z_3^2)}=\sqrt{[(x_1+x_2)^2+(y_1+y_2)^2+(z_1+z_2)^2]}$$

If a_3, b_3, c_3 are respectively the angles which OP_3 makes with the X, Y, and Z axes:

$$a_3=\cos^{-1}(x_3 \div r_3) \quad b_3=\cos^{-1}(y_3 \div r_3)$$
$$c_3=\cos^{-1}(z_3 \div r_3)$$

The additive property of the co-ordinates of a sequence of displacements enables us to carry out very simply the operation of finding the end result without recourse to a Euclidean construction in the Galileo-Newton tradition; and we may speak metaphorically of the *addition* of vectors, thereby meaning the evaluation of the resultant displacement obtained by simple numerical addition of corresponding co-ordinates. When we do so, it is convenient to substitute the signs $\oplus$ and $\ominus$ for $+$ and $-$ to avoid confusion. As is true for the simple addition of the abacus, the rule of composition is both associative and commutative, *i.e.*

$$V_1 \oplus V_2 \oplus V_4 = V_7 = V_4 \oplus V_3$$
$$V_1 \oplus V_2 = V_3 = V_2 \oplus V_1$$

Since a second displacement V_2 ($\equiv PO$) of equal scalar value in the opposite direction to V_1 ($\equiv OP$) restores the displaced point to its original position, $V_1 \oplus V_2 = 0$. In common algebra we write $x + y = 0$ if $y = -x$ on the understanding that x is not zero. Whence we label a displacement (V) in the opposite direction as $\ominus V$ in conformity with the same rule of composition, so that

$$V_3 \ominus V_1 = V_2 \text{ and } V_3 \ominus V_2 = V_1$$

The reader will doubtless recognise one result of our rule of composition, *viz.* the multiplication of a vector V whose scalar magnitude is r by a scalar (n) which may be any real number is a vector whose scalar magnitude is nr with the same inclination (a, b, c) to the axes of reference.

Because the rectangular components of a vector displacement are additive in the foregoing sense, it is essential to label them consistently in some way. The customary fixed order $P = (x, y, z)$ ensures this without making the duality of the notion of displacement so explicit as the complex number notation of the plane; but the latter (as such) is not adaptable to three-dimensional space. The beauty of vector algebra is that it does make explicit this duality (linear magnitude and direction) by conferring on each co-ordinate a scalar component (A, B, C below) and one of three

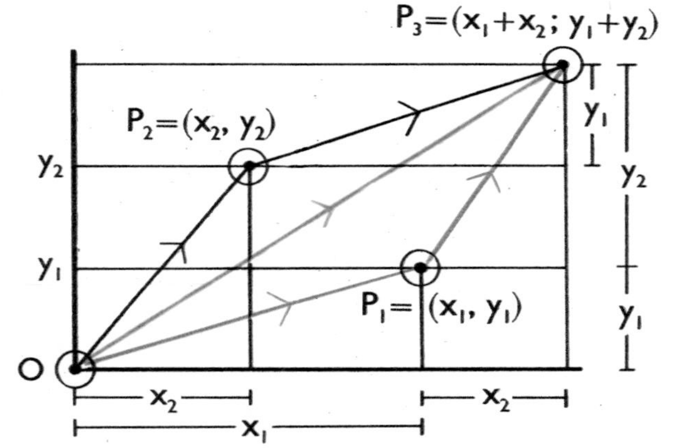

THE PARALLELOGRAM OF DISPLACEMENTS

If one displacement shifts a point from the origin (O) to $P_1 = (x_1, y_1)$ and another shifts a point from the origin to $P_2 = (x_2, y_2)$, the successive effect of both displacements in either order shifts the point to P_3 whose co-ordinates are $x_3 = (x_1 + x_2)$ and $y_3 = (y_1 + y_2)$. In vector notation one writes this:

$$V_1 = x_1 . \mathbf{i} + y_1 . \mathbf{j};\ V_2 = x_2 . \mathbf{i} + y_2 . \mathbf{j}$$
$$V_3 = V_1 \oplus V_2 = (x_1 + x_2) . \mathbf{i} + (y_1 + y_2) . \mathbf{j}$$

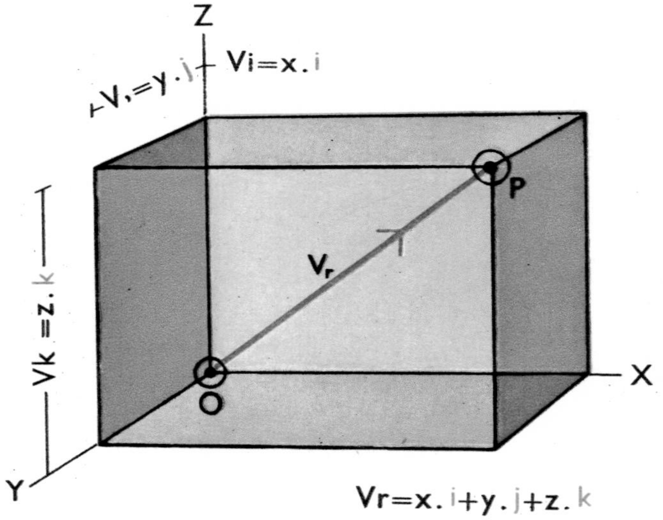

VECTOR COMPOSITION IN THREE DIMENSIONS

vectors (**i, j, k**) of unit scalar magnitude to label the axes (X, Y, Z) of reference. We may write the components in product form thus:

$$V = A.\mathbf{i} \oplus B.\mathbf{j} \oplus C.\mathbf{k}$$

Before proceeding, let us be clear that the **i** of this context is not the i we loosely refer to as $\sqrt{-1}$. If the scalar magnitude of V is N, we interpret the above to mean:

$$N = \sqrt{(A^2 + B^2 + C^2)}$$

$$a = \cos^{-1}\left(\frac{A}{N}\right) \quad b = \cos^{-1}\left(\frac{B}{N}\right) \quad c = \cos^{-1}\left(\frac{C}{N}\right)$$

Accordingly for $V_3 = V_1 \oplus V_2$, our rule of composition takes the form:

$$V_1 = A_1.\mathbf{i} \oplus B_1.\mathbf{j} \oplus C_1.\mathbf{k}$$
$$V_2 = A_2.\mathbf{i} \oplus B_2.\mathbf{j} \oplus C_2.\mathbf{k}$$
$$V_3 = (A_1 + A_2)\mathbf{i} \oplus (B_1 + B_2)\mathbf{j} \oplus (C_1 + C_2)\mathbf{k}$$

The corresponding rule for manipulating $\ominus$ needs no comment. We may now package a considerable tract of elementary mechanics, if we recall that a changing velocity is the limiting ratio of a spatial displacement V_s to time; and that a change of its direction no less than that of its scalar magnitude (speed) implies acceleration of one of its rectangular components. If v stands for velocity, we may write this in Newton-Leibniz terms retaining **i, j, k** as labels for the rate of change of one or other rectangular component of V_s, so that

$$V_s = (N_s \cos a)\mathbf{i} \oplus (N_s \cos b)\mathbf{j} \oplus (N_s \cos c)\mathbf{k}$$
$$v = D_t . V_s = D_t\ (N_s \cos a)\mathbf{i} \oplus D_t\ (N_s \cos b)\mathbf{j}$$
$$\oplus\ D_t\ (N_s \cos c)\mathbf{k}$$

Here we have exhibited the instantaneous velocity of a point in vector language. If the velocity is constant, each of the scalar components $D_t\ (N_s \cos a)$ *etc.* is a constant, and the acceleration $D_t^2 . V_s \equiv D_t . v = 0$. Otherwise we must write:

$$D_t\ (N_s \cos a) = \cos a . D_t N_s + N_s D_t \cos a$$
$$= \cos a . D_t N_s - N_s \sin a . D_t a$$

Similarly:

$$D_t\ (N_s \sin a) = \sin a . D_t N_s + N_s \cos a . D_t a$$

For the motion in the XY plane, the scalar component of k is zero and $\cos b = \sin a$. We may thus write:

$$D_t\ (N_s \cos a) = A_v \quad D_t\ (N_s \sin a) = B_v$$
$$v = A_v\mathbf{i} \oplus B_v\mathbf{j}$$

Scalar value (speed) is therefore: $s = \sqrt{(A_v^2 + B_v^2)}$; and A_v is expressible in the form $s.\cos a_v$, so that:

$$a_v = \cos^{-1}(A_v \div s) \text{ and } v = (s.\cos a_v)\mathbf{i} \oplus (s.\sin a_v)\mathbf{j}$$

If a particle rotates in a circle of radius r at fixed speed s, its displacement ($N_s = r$) from the origin is constant, whence $D_t N_s = 0$, and

$$A_v = -N_s . \sin a . D_t a \quad B_v = N_s . \cos a . D_t a$$

Since $D_t a$ is the fixed angular speed usually represented as ω and $(\sin a)^2 + (\cos a)^2 = 1$, we derive $A^2 + B^2 = N_s^2 . \omega^2 = \omega^2 . r^2$ and $s = \omega r$. This is consistent with what we might derive in one line.

Our next step introduces a real economy of effort. For any acceleration $D_t v$ we may write:

$$D_t\ (s.\cos a_v)\mathbf{i} \oplus D_t\ (s.\sin a_v)\mathbf{j}$$
$$D_t\ (s.\cos a_v) = \cos a_v . D_t s - s.\sin a_v . D_t . a_v$$
$$D_t\ (s.\sin a_v) = \sin a_v . D_t s + s.\cos a_v . D_t . a_v$$

We may now write:

$$E = D_t s \text{ and } F = s . D_t . a_v$$
$$D_t v = (E.\cos a_v - F.\sin a_v)\mathbf{i} \oplus$$
$$(E.\sin a_v + F.\cos a_v)\mathbf{j}$$

The scalar value (ft.sec^2) of this is:

$$\sqrt{[(E.\cos a_v - F.\sin a_v)^2 + (E.\sin a_v + F.\cos a_v)^2]}$$
$$= \sqrt{(E^2 + F^2)}$$

Now $(E \div \sqrt{[E^2 + F^2]})^2 + (F \div \sqrt{[E^2 + F^2]})^2 = 1$ $= (\cos q)^2 + (\sin q)^2$, so that

$$\cos q = \frac{E}{\sqrt{(E^2 + F^2)}} \qquad \sin q = \frac{F}{\sqrt{(E^2 + F^2)}}$$

$$D_t V = \sqrt{[E^2 + F^2]}\ (\cos a_v . \cos q - \sin a_v . \sin q)\mathbf{i}$$
$$\oplus \sqrt{[E^2 + F^2]}\ (\sin a_v . \cos q + \cos a_v . \sin q)\mathbf{j}$$
$$= \sqrt{[E^2 + F^2]} . \cos (a_v + q)\mathbf{i}$$
$$\oplus \sqrt{[E^2 + F^2]} . \sin (a_v + q)\mathbf{j}$$

We have now expressed acceleration in vector form entirely in terms of the vector interpretation of velocity for motion in the plane; and our notation also prescribes the appropriate form for motions in solid space. The evaluation of an acceleration (whence also of a force) thus emerges as a recipe for *motions of every sort.*

In the case of circular motion on a plane, $D_t s = 0 = E$ and $F = s\omega$, whence the scalar value of the acceleration is $\omega^2 r$. Also $\cos q = 0$ since $E = 0$ above, whence $q = \cos^{-1}(0) = 90°$. The instantaneous velocity at an inclination a_v to the x-axis is along the tangent at each point, and that of the acceleration is at an inclination of 90° to the tangent, *i.e.* to the centre. In vector notation therefore:

$$D_t v = \omega^2 r . \cos (a_v + 90°)\mathbf{i} \oplus \omega^2 r . \sin (a_v + 90°)\mathbf{j}$$

In dynamical and electromagnetic theory, we meet situations where: (a) the product of two vectors is a *scalar, e.g.* kinetic energy; (b) one field of force is at right angles to another. To tailor our algebra to these requirements, vector notation recognises products of two kinds, both of which have the associative and distributive properties of ordinary multiplication, but only one of which is commutative.

The performance of either operation recalls the arithmetic of the complex number. Just as we adopt rules $i.i = i^2 = -1$; $i^3 = -i$; $i^4 = +1$, *etc.*, in the complex domain, we adopt particular conventions for interpreting the several unit vector products as below.

Dot Product $\odot$	$\mathbf{i}^2 = \mathbf{j}^2 = \mathbf{k}^2 = 1$
(Commutative)	$\mathbf{i}.\mathbf{j} = \mathbf{j}.\mathbf{i} = \mathbf{i}.\mathbf{k} = \mathbf{k}.\mathbf{i} = \mathbf{j}.\mathbf{k}$
	$= \mathbf{k}.\mathbf{j} = 0$
Cross-Product $\otimes$	$\mathbf{i}^2 = \mathbf{j}^2 = \mathbf{k}^2 = 0$
(Non-commutative)	$\mathbf{i} \otimes \mathbf{j} = \mathbf{k} = -\mathbf{j} \otimes \mathbf{i}$
	$\mathbf{j} \otimes \mathbf{k} = \mathbf{i} = -\mathbf{k} \otimes \mathbf{j}$
	$\mathbf{k} \otimes \mathbf{i} = \mathbf{j} = -\mathbf{i} \otimes \mathbf{k}$

Whence we derive:

$$V_1 \odot V_2 = N_1 . N_2 \ (\cos a_1 . \cos a_2 + \cos b_1 . \cos b_2 + \cos c_1 . \cos c_2)$$

In the XY plane, we drop out the third component, and $\cos b_1 = \sin b_1$, *etc.*, so that

$$V_1 \odot V_2 = N_1 . N_2 \ (\cos a_1 . \cos a_2 + \sin a_1 . \sin a_2)$$
$$= N_1 . N_2 . \cos (a_1 - a_2)$$

The numerical value of the product is thus the product of its scalar factors (N_1 and N_2) with the cosine of the angle between them.

To apply common arithmetic correctly to the evaluation of the cross-product, we have to adopt some fixed convention of order to take stock of the non-commutative property of the unit vector operation. The accepted one is like this:

$$V_1 = A_1\mathbf{i} \oplus B_1\mathbf{j} \oplus C_1\mathbf{k}$$
$$V_2 = A_2\mathbf{i} \oplus B_2\mathbf{j} \oplus C_2\mathbf{k}$$
$$V_1 \otimes V_2 = A_2B_1\mathbf{j} \otimes \mathbf{i} \oplus A_2C_1\mathbf{k} \otimes \mathbf{i} \oplus B_2A_1\mathbf{i} \otimes \mathbf{j} \oplus B_2C_1\mathbf{k} \otimes \mathbf{j} \oplus C_2A_1\mathbf{i} \otimes \mathbf{k} \oplus C_2B_1\mathbf{j} \otimes \mathbf{k}$$

Whence by interpreting the meaning of $\mathbf{j} \otimes \mathbf{i}$, *etc.* as above:

$$V_1 \otimes V_2 = (B_1C_2 - B_2C_1)\mathbf{i} \oplus (C_1A_2 - C_2A_1)\mathbf{j} \oplus (A_1B_2 - A_2B_1)\mathbf{k}$$

It is most easy to memorise this if we write it in determinant form as:

$$V \otimes V = \begin{matrix} A_1 & B_1 & C_1 \\ A_2 & B_2 & C_2 \\ \mathbf{i} & \mathbf{j} & \mathbf{k} \end{matrix}$$

The pay-off from the viewpoint of the engineer depends on the fact that the cross-product of two vectors in the XY plane ($C_1 = 0 = C_2$) is the scalar product of the unit vector k being therefore in a direction at right angles to the plane, since

$$V_1 \otimes V_2 = (A_1B_2 - A_2B_1)\mathbf{k}$$

As above, we may substitute:

$$A_1 = N_1 . \cos a_1 \quad B_1 = N_1 . \sin a_1$$
$$A_2 = N_2 . \cos a_2 \quad B_2 = N_2 . \sin a_2$$

$$V_1 \otimes V_2 = N_1 . N_2 \ (\sin a_1 . \cos a_2 - \sin a_2 . \cos a_1)\mathbf{k}$$
$$= N_1 . N_2 . \sin (a_1 - a_2)\mathbf{k}$$

Chapter 14 The Great Biopsy

During the first half of the nineteenth century, mathematicians became increasingly uneasy about the foundations of a still-rising edifice. The malaise mounted in the second half, and the impulse to self-justification increasingly became an obsession in the first half of our own. At the start, the issues which provoked scrutiny were at the periphery of newly-discovered territory. Between 1750 and 1800 there had been tremendous progress in the recognition of new series. Meanwhile, there had been far too little disposition to define the circumstances in which they converge to a finite limit, and no one had successfully resolved the logical doubts expressed by the theological critics of Newton and Leibniz, more especially Berkeley, *vis à vis* their own formulation of the derivative as a limiting ratio.

A noteworthy discovery by a French mathematician put the spotlight on both issues. Fourier (*circa* 1810) discovered a remarkable series which fastidious French mathematicians of his time – including Laplace, Legendre and Lagrange – refused to recognise as such. On the other hand, physicists were not slow to recognise its usefulness as a tool. We may write Fourier's series briefly in the form:

$$y = f(x) = \sum_{n=0}^{n=\infty} B_n . \cos nx + \sum_{n=0}^{n=\infty} A_n . \sin nx$$

$$B_n = \frac{1}{\pi}\int_0^{2\pi} y . \cos nx . dx \text{ and } A_n = \frac{1}{\pi}\int_0^{2\pi} y . \sin nx . dx$$

As is true of Maclaurin's series, the derivation of the above does not enlist advanced knowledge of the Newton-Leibniz calculus; but the necessary and sufficient conditions for its convergence, if at all, raise far more sophisticated issues. These stimulated a re-examination of the notion itself by a host of Fourier's successors, in particular Cauchy (*circa* 1830), Dirichlet (*circa* 1850) and Weierstrass (*circa* 1860). However, what is most provocative about it is that it can do a job which Maclaurin's Theorem cannot accomplish. The latter can equip a function with a power series only if it is *differentiable* in the sense mentioned at the beginning of Chapter 10. Contrariwise, the sine-cosine series of Fourier can generate a series for functions which are *not* differentiable, as when the curve of the function has a cusp. This is what makes it so uniquely powerful a new tool; but the inclusion of y under the integral sign is anomalous if our only definition of the whole operation presupposes that y is a component of an anti-derivative. Seemingly, the only way out of the dilemma was to re-interpret the operation in other terms, a task successfully accomplished by Riemann.

By 1850, increasing recognition of the usefulness of complex numbers as mapping devices was another source of uneasiness which Gauss among others had sought to dispel; and controversies provoked in the birth pangs of vector algebra, which we have already noticed, heightened the tension. The creation of new geometries had weakened the pre-existing belief that the basic postulates of Euclid are in some sense laws of

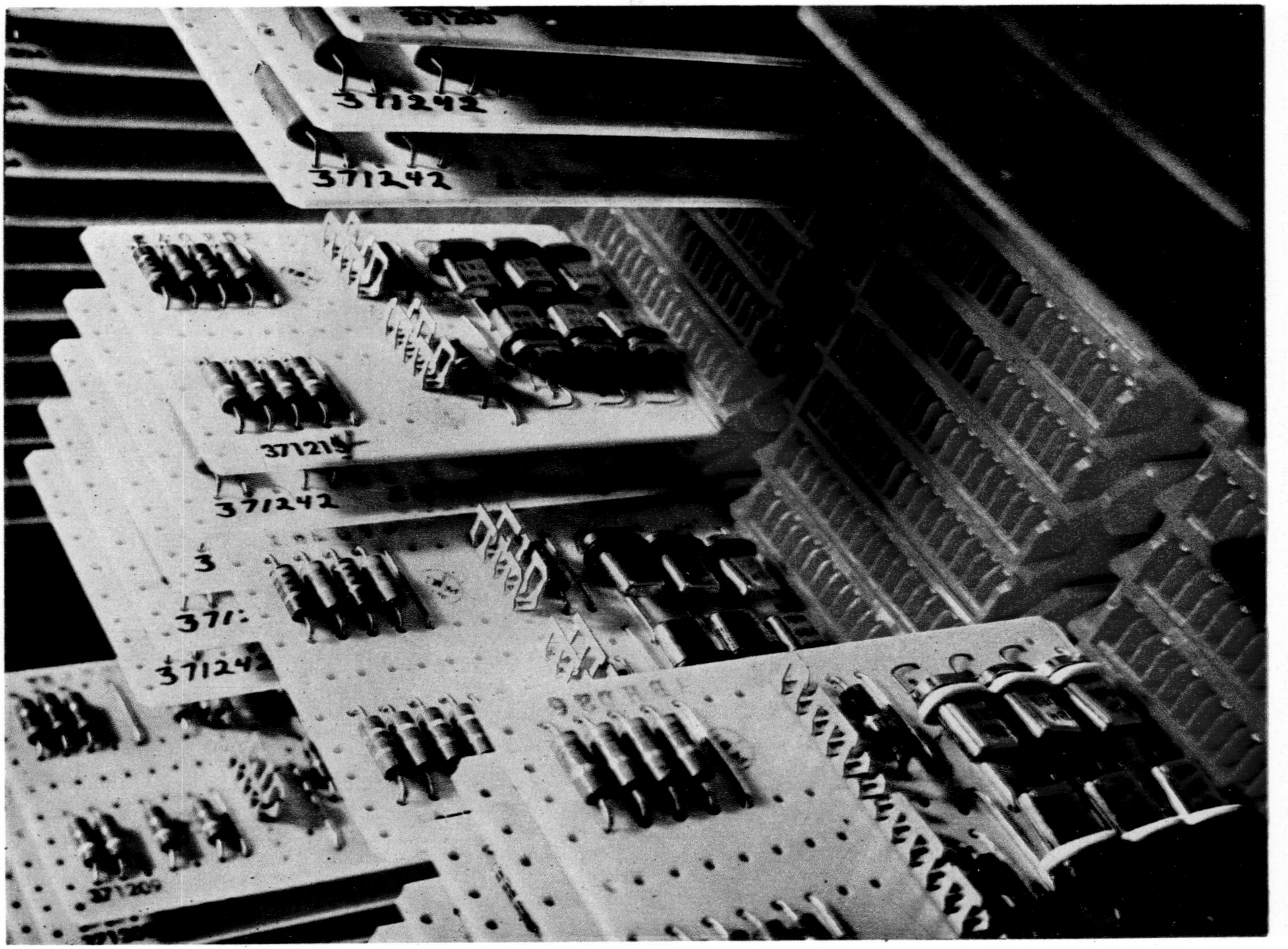

The present century has witnessed a not unnatural desire of mathematicians to validate and classify the many techniques they employ in manipulating numbers. Such attempts met with a considerable setback when Kurt Gödel demonstrated that no system of mathematical definition can contain within itself the proof of its own consistency. Today the electronic brain serves as a touchstone for the validity of mathematical reasoning. To programme it, one must issue instructions in an unequivocal idiom. Here combinations of "cards" are plugged into "gates" which form the circuits of the arithmetical, control and logical units of this transistorised IBM computer.

nature; but if they are man-made, in what sense are they true? The answer that Hilbert (1862–1943) gave is that mathematics is *the rules of a game played with meaningless marks on paper*. What we may call the ensuing Great Biopsy started with modest and diverse objectives, when few realised the full implications of the marriage of algebra and geometry. More and more, mathematicians became obsessed with the natural history of their domain, seeking, like the biologists of a previous century, some master key to classification.

Maybe, the first premonition of this is the *group* concept. It is not difficult to grasp, though its usefulness is very difficult to illustrate in a short space. Here we shall merely state what it is as a stepping-stone to other levels of classification, more especially the attempt to restate the fundamentals of computation to justify the protection of so many different entities called *numbers* by the old umbrella of abacus arithmetic. Oddly enough, the so-called *complex* numbers attracted attention long before the more overdue dilemma of how to justify the law of signs when manipulating two negative numbers.

If the intention was half jocular, Hilbert's aphorism is at least profound from one viewpoint. The first printed commercial arithmetics familiarised mathematicians with the possibility of using *marks on paper* to perform operations which their predecessors relied on the abacus to accomplish. Thenceforth there were ever newer territories of experimentation in the *paper domain*, leading to one discovery after another about manipulations of entities called *numbers* regardless of the fact that the abacus model could endorse them with no acceptable sanction. In short, the pioneers of 1550–1800 had been changing the rules of the game without either clearly conceding that they had done so or invoking any new sanctions. Precisely what form the new sanctions should take has been a battleground of controversy from 1850 onwards.

The *group* concept emerged first in connexion with the theory of equations as propounded by Galois (1846); but it has other applications in geometry, including topology. A finite group is a set of inter-related operations. What is diagnostic of their inter-relationships is fully manifest, as Cayley (1854) first pointed out, in the entries of the *group table*. An illustration based on algebraic operations expressible in functional form will make the properties of such a table sufficiently explicit. Its grid-wise construction exhibits every 2-fold repetitive permutation of the N (here $= 6$) members of the group, *i.e.* every way in which we can successively perform two operations not necessarily different. We shall define six operations on a number n as follows:

$$I(n) \equiv n \qquad C(n) \equiv 1-n^{-1}$$
$$A(n) \equiv 1-n \qquad D(n) \equiv (1-n)^{-1}$$
$$B(n) \equiv n^{-1} \qquad E(n) \equiv n(n-1)^{-1}$$

Having so defined our operations, we may define successive paired operations thus:

$$A.I(n) \equiv A(n)$$
$$A.A(n) \equiv 1-(1-n) \equiv n \equiv I(n)$$
$$A.B(n) \equiv 1-n^{-1} \equiv C(n)$$
$$A.C(n) \equiv 1-(1-n^{-1}) \equiv n^{-1} \equiv B(n)$$
$$A.D(n) \equiv 1-(1-n)^{-1} \equiv n(n-1)^{-1} \equiv E(n)$$
$$A.E(n) \equiv 1-n(n-1)^{-1} \equiv (1-n)^{-1} \equiv D(n)$$

We may write these results briefly as:

$$A.I=A;\ A.A=I;\ A.B=C;\ A.C=B;\ A.D=E;\ A.E=D$$

Similarly, we derive:

$$B.C \equiv (1-n^{-1})^{-1} \equiv n(n-1)^{-1}$$

so that $B.C \equiv E$.

Proceeding in this way, we may set out all results of two successive operations like a table of multiplication in which:

(a) each composite cell entry is equivalent to one of the six single marginal operations;
(b) every row and every column contains each of the six once only.

The reader can easily check that the table is as follows:

	I	A	B	C	D	E
I	I	A	B	C	D	E
A	A	I	C	B	E	D
B	B	D	I	E	A	C
C	C	E	A	D	I	B
D	D	B	E	I	C	A
E	E	C	D	A	B	I

That each row and each column of the foregoing table exhibits each group member implies that every operation within the group has a relation to every other in the sense that we can start with any one and accomplish the same result as with any other, if we insert an appropriate sequence of one or more operations, themselves within the group. It is essential to state that one feature of the table is *not* an essential group property. Here $P.Q=Q.P$ only if either P or $Q=I$; but one can construct groups of which this would hold good for all such pairs.

Without explicitly introducing the notion of operation or function, we may now set forth the group table properties in terms of its elements deemed to be *I, A, B, C, etc.* Where we use T, we signify any one of them.

1. In any group, each ordered couple (A, B) of its elements combines according to a rule of composition o so that $(A \, o \, B)=C$, in which C may be equal to A or to B, to both (if identical) or to neither.
2. If A and B are in the group, $C=(A \, o \, B)$ is in the group.
3. Group composition is *associative* in the sense that:
   $(A \, o \, B) \, o \, C=A \, o \, (B \, o \, C)$
4. In every group, there is a unique element I, such that
   $T \, o \, I=T=I \, o \, T$
5. On the understanding that T may be identical with T^{-1}, as is necessarily true of $I=I^{-1}$, for every T there is a T^{-1} such that
   $T.T^{-1}=I$
6. Only one class of groups (designated *Abelian* after the Scandinavian mathematician Abel) has the *commutative* property:
   $(A \, o \, B)=(B \, o \, A)$

The theory of groups had at its inception a more limited objective than a taxonomy of mathematical operations, and its reliance on a single rule of composition does not suffice to provide an abstract basis for the most elementary of all such operations, *i.e.* for numerical computation without recourse to empirical notions about number. What we may call the Programme of the *Field* has as its main objective to accomplish this aim, and by so doing to dispel anomalous features of the accepted rules of the game, some of them, *e.g.* as concerned with directed numbers, mentioned in Chapters 8 and 11. This signifies in particular the restatement of our algorithms to embrace the whole field of real and complex numbers, without arbitrarily imposing on the latter – or on manipulation with negative real numbers – concepts valid only in the proper domain of the abacus. We shall assume the intention to carry out the task as economically as possible, *i.e.* with as few axioms or postulates as need be.

Before the retreat to the Achilles-and-Tortoise tradition became a large-scale defence operation, we may trace many ingredients of what was later to be a common pattern from which the notion of *mathematical structure* takes shape; but perhaps no one foresaw more clearly the shape of things to come than did George Boole (*circa* 1850), an Englishman. Boole was undoubtedly a genius. His monograph on *Finite Differences* (the Δ operation of Chapter 9) is still a model of lucid exposition; but more important in his own view was a conviction unsuccessfully promoted by his widow with little response from his contemporaries. Boole believed that he had discovered for all time the Laws of Thought. Briefly, this signifies that one can make a logic of classes (objects) and a logic of propositions (true or false statements) on congenial marital relations with an arithmetic (base 2) which enlists only the symbols 0 and 1. This makes it possible, by reinterpreting the content of an argument in a functional notation without recourse to words listed in a dictionary: (a) to conserve the three Aristotelian canons based on a confusion of the seven different meanings of the so-called verb *be*; (b) to dispose of the inconvenience of some of the many absurdities to which syllogistic (*i.e.* lawyer-like) reasoning may lead one.

Such *symbolic* logic – more especially the part which deals with propositions – has grown immensely since Boole's time. Like himself, its practitioners stake claims which have nothing to do with mathematics as customarily conceived. In particular, it is ostensibly an apparatus for exposing verbal fallacies and inconsistencies without recourse to words. As such, we shall consider its restricted relevance to real life at a later stage. All we need here say is that a two-valued symbolic logic exploited as a tool for dissecting mathematical relations raises the question: have the rules of reasoning devised to deal with *finite* collections of objects any relevance to infinity?

Consider for instance the use G. Cantor (*circa* 1900) makes of the notion of *one-to-one correspondence*. As Galileo recognised, we can set no limit to the process of pairing the natural numbers 1, 2, 3, 4, *etc.*, with the odd ones 1, 3, 5, 7, *etc.*, with the even ones 2, 4, 6, 8, *etc.*, or with the squares 1, 4, 9, 16, *etc.* Are we to conclude that there are just as many natural and odd or even or square numbers? A cogent answer must surely take into account the fact that one-to-one correspondence in the domain of experience is a criterion of numerical identity *only in so far as the matching process terminates without a residue*. After much portentous preoccupation of Cantor's successors with so-called *transfinite* numbers, there are now indeed eminent purists who regard the late nineteenth-century excursion into the infinite as a semantic quagmire.

In the background of attempts to rehabilitate the verbal formulation of mathematical principles, a memorable bequest of Boole is the notion of the *set*. A finite set is a collection of any objects – animal, vegetable, mineral or merely marks on paper. The set (S) which is the topic of discussion is the *universal set* and two functional symbols, which invoke only the notions of *inclusion* and *identity*, subsume the relation between subsets, *i.e.* objects in the set. These are the so-called *logical sum*, which we may write $A \oplus B$ for two subsets and the *logical product* $A \odot B$ defined as follows:

(1) $A \oplus B$ signifies *all* objects which are only in

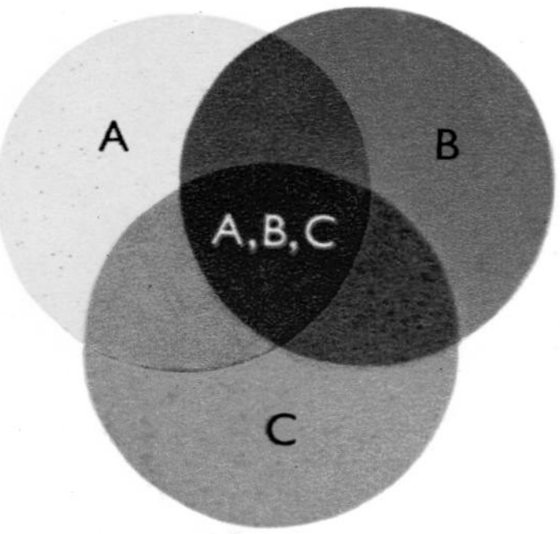

ARISTOTELIAN CLASS
A, B and C are deemed to form a single class since all share the common area of attributes A, B, C.

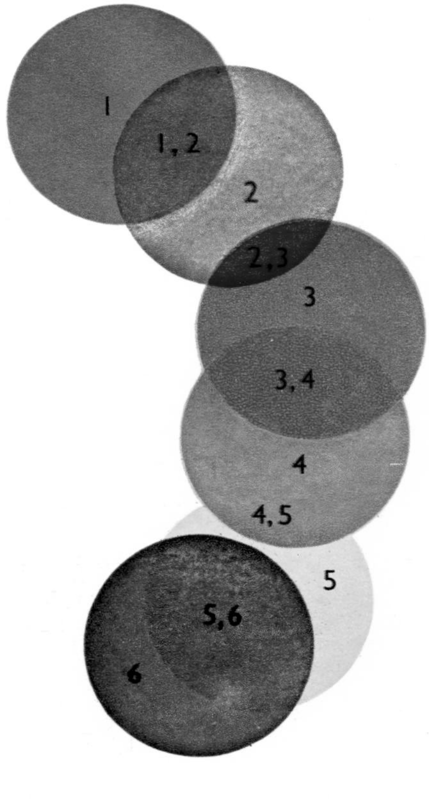

CATENARY CLASS. (*See page 311.*)
1, 2, 3, 4, 5 and 6 are deemed to form a single class since 1 shares with 2 the common area of attributes 1, 2, while 2 shares with 3 the common area of attributes 2, 3, while 3 shares with 4 the common area of attributes 3, 4, and so on.

subset A, only in subset B or also in both;

(2) $A \odot B$ signifies only objects which are both in A and in B.

For instance, S (of rivers) may be all rivers classifiable with respect to fish content, with a subset (A) of rivers which have trout in them, a subset (B) of rivers with salmon in them and a subset (C) in which there are other fish but neither salmon nor trout. Then $A \oplus B$ is the subset of rivers which have either salmon in them or trout in them or both, and $A \odot B$ is the smaller subset of rivers with both trout and salmon therein. One now adds two signs for two abstractions mentioned above, *i.e.* *inclusion* and *identity*. $A \subset B$ means: *all A are in B.* This may be true if all B are also in A, so that $A \equiv B$. Thus the combined statement $A \subset B$ and $B \subset A$ means $A \equiv B$. To complete the picture, we deem somewhat casuistically, at least on first acquaintance, that the universal set S contains a unique subset present in every other subset, A, *etc.*, of S. Here the empty subset Z is the subset which contains no rivers and is unique because all other subsets contain at least one river, whence Z is both different from all other subsets and contains nothing itself not also present in every other one. The rules of the game then take shape as the following nine tautologies of the lawyer's logic:

1. $A \subset A$ 2. $Z \subset A$ 3. $A \subset S$
4. $A \oplus A \equiv A \equiv A \odot A$ 5. $A \odot S \equiv A$ 6. $A \odot Z \equiv Z$
7. $A \oplus Z \equiv A$ 8. $A \oplus S \equiv S$
9. $A \oplus (B.C) \equiv (A \oplus B).(A \oplus C)$

Three others recall the fundamental properties common to, and distinguishing, the operations of numerical addition and multiplication, *viz.*

(*a*) *commutative* $a+b=b+a$ and $ab=ba$
(*b*) *associative* $a+(b+c)=(a+b)+c$ and $a(bc)=(ab)c$
(*c*) *distributive* $a(b+c)=ab+ac$

Thus we can derive:

10. $A \oplus B \equiv B \oplus A$ and $A \odot B \equiv B \odot A$
11. $A \oplus (B \oplus C) \equiv (A \oplus B) \oplus C$ and $A \odot (B \odot C) \equiv (A \odot B) \odot C$
12. $A \odot (B \oplus C) \equiv A \odot B \oplus A \odot C$

The formal similarity between 10–12 and (*a*)–(*c*) above is an invitation to re-define the concept of function in terms which do not explicitly invoke the number concept and to re-define the number concept in terms which do not invoke any empirical ideas about enumeration or measurement. Thence proceeds a prolific progeny of latter-day abstractions subsumed by the terms *group* (as already defined), *field*, *ring* and *lattice*. Here our sole concern will be with the *number field*. Rules 10–12 above might encourage us to hope that we can at least pack up the rules of elementary arithmetic without invoking the notion of number. If so, we have to add other ingredients to our assumptions.

What follows is not exactly like any one of several systems which express economically without appeal to experience of a matching process the rules of manipulating numbers and the relation between different sorts of *mathematical objects* severally called number. We may forgivably omit refinements on which a purist would rightly insist, if our main intention is to bring into focus how much, reluctantly, we took for granted in the algebra of the high school during the youth of any of us older than thirty. Merely for the fun of it, we may set out our autopsy in the Euclidean manner as axioms and theorems. Whether we speak of the ingredients of the concoction as *subsets* or as *elements* of the number field is a matter of taste.

We begin by defining $\odot$ and $\oplus$ without reliance on the Boolean notion of inclusion.

Axiom 1. Both $\oplus$ and $\odot$ are *commutative*, *i.e.*
$A \oplus B \equiv B \oplus A$ and $A \odot B \equiv B \odot A$

Axiom 2. Both $\oplus$ and $\odot$ are *associative*, *i.e.*
$A \oplus (B \oplus C) \equiv (A \oplus B) \oplus C$
$A \odot (B \odot C) \equiv (A \odot B) \odot C$

Axiom 3. Only $\odot$ is *distributive*, *i.e.*
$A \odot (B \oplus C) \equiv (A \odot B) \oplus (A \odot C)$

Axiom 4. If A and B are in the field, both $A \oplus B$ and $A \odot B$ are in it.

Axiom 5. $A \oplus F \equiv B \oplus F$ if, and only if, $A \equiv B$

Axiom 6. There exists in the field a unique subset Z such that for all subsets in the field $F \oplus Z \equiv F$

Axiom 7. If A and B are in the field and A is *not* Z, there is also in the field a reciprocal subset R such that
$A \odot R \equiv B$

Axiom 8. We now define a relation of order $\ogreaterthan$ such that:

(*a*) A and B are not identical if $A \ogreaterthan B$ or $B \ogreaterthan A$

(*b*) $F \ogreaterthan A$ and $A \ogreaterthan B$ implies $F \ogreaterthan B$

(*c*) $A \oplus B \ogreaterthan A \oplus C$ if $B \ogreaterthan C$.

Axiom 9. If A and $B \ogreaterthan A$ are in the field, there is also in the field a *subtrahend* subset C ($\ogreaterthan Z$) such that
$A \oplus C \equiv B$

Axiom 10. There exists in the field a unique subset $U \ogreaterthan Z$ such that for all subsets other than Z
$F \odot U \equiv F$ and $F \oplus U \ogreaterthan F$.

Before surveying what ground we have traversed so far, we may pause to cite two theorems in the Euclidean manner.

THEOREM 1. From Axioms 1, 3 and 4, we obtain:

$$\begin{aligned}(A \oplus B) \odot (C \oplus D) &\equiv [(A \oplus B) \odot C] \oplus [(A \oplus B) \odot D]\\ &\equiv [C \odot (A \oplus B)] \oplus [D \odot (A \oplus B)]\\ &\equiv (C \odot A) \oplus (C \odot B) \oplus (D \odot A) \oplus (D \odot B)\\ &\equiv (A \odot C) \oplus (A \odot D) \oplus (B \odot C) \oplus (B \odot D)\end{aligned}$$

THEOREM 2. From Axioms 3, 4 and 6, we obtain:

$$F \odot (F \oplus Z) \equiv F \odot F \equiv F \odot F \oplus Z$$
$$F \odot (F \oplus Z) \equiv (F \odot F) \oplus (F \odot Z)$$
$$F \odot Z \equiv Z$$

We shall now apply Axioms 8 and 10 to specify the ordered subsets: $P \ogreaterthan U$, $Q \ogreaterthan P \ogreaterthan U$, $R \ogreaterthan Q \ogreaterthan P \ogreaterthan U$, *etc.*, as below:

$U \oplus U \equiv P \qquad P \oplus U \equiv Q \qquad Q \oplus U \equiv R$
$R \oplus U \equiv S \qquad S \oplus U \equiv T \qquad T \oplus U \equiv V$

By applying Axioms 1, 2 and 8, we can then see that:

$P \equiv U \oplus U \qquad S \equiv P \oplus Q$ or $U \oplus R$
$Q \equiv U \oplus P \qquad T \equiv Q \oplus Q$ or $P \oplus R$ or $U \oplus S$
$R \equiv P \oplus P$ or $U \oplus Q \qquad V \equiv Q \oplus R$ or $P \oplus S$ or $U \oplus T$

We may thus build up a table such as:

$\oplus$	Z	U	P	Q	R	S	T	V
Z	Z	U	P	Q	R	S	T	V
U	U	P	Q	R	S	T	V	..
P	P	Q	R	S	T	V	..	..
Q	Q	R	S	T	V	..	..	..
R	R	S	T	V	..	..	..	..
S	S	T	V	..	..	..	..	..
T	T	V	..	..	..	..	..	..
V	V	..	..	..	..	..	..	..

So far, U, P, Q, *etc.*, are marks on paper. We are free to replace them by any other marks, *e.g.*

$Z=0 \quad U=1 \quad P=2 \quad Q=3$
$R=4 \quad S=5 \quad T=6 \quad V=7$

Our table then becomes:

+	0	1	2	3	4	5	6	7
0	0	1	2	3	4	5	6	7
1	1	2	3	4	5	6	7	..
2	2	3	4	5	6	7	..	..
3	3	4	5	6	7	..	..	..
4	4	5	6	7	..	..	..	..
5	5	6	7	..	..	..	..	..
6	6	7	..	..	..	..	..	..
7	7	..	..	..	..	..	..	..

We have now created a paper model of addition on the abacus without explicitly invoking any empirical notions about number at the outset. By applying Axiom 9, we can use the addition table to answer the question:

what is C if $A \oplus C \equiv B$?

We can thus build up a table of subtraction as below:

$\ominus$	Z	U	P	Q	R	S	T	V
Z	Z	U	P	Q	R	S	T	V
U	..	Z	U	P	Q	R	S	T
P	..	..	Z	U	P	Q	R	S
Q	..	..	..	Z	U	P	Q	R

We now apply Axiom 3 with Axiom 2 to derive a third table embodying the ordered results of the commutative operation $\odot$. It will suffice to cite the derivation of two cell entries:

$$U \odot P \equiv U \odot (U \oplus U) \equiv P, \ etc.$$
$$P \odot P \equiv P \odot (U \oplus U) \equiv (P \odot U) \oplus (P \odot U) \equiv P \oplus P \equiv R$$

Proceeding in this way, we get a table with the following entries in the first four lines, where we write $V \oplus U \equiv W$ and $W \oplus U \equiv Y$:

$\odot$	Z	U	P	Q	R	S	T	V
Z	Z	Z	Z	Z	Z	Z	Z	Z
U	Z	U	P	Q	R	S	T	V
P	Z	P	R	T	W	..	..	..
Q	Z	Q	T	Y	..	..	..	..

If we replace Z, U, P, *etc.*, as before, mark W by 8, mark Y by 9 and the mark $\odot$ by $\times$, this would look as follows:

$\times$	0	1	2	3	4	5	6	7
0	0	0	0	0	0	0	0	0
1	0	1	2	3	4	5	6	7
2	0	2	4	6	8	..	..	..
3	0	3	6	9	..	..	..	..

We have now before us the familiar table of abacus multiplication. Let us next invoke Axiom 7. This states that there is (unless $B=Z$) an answer to the question: what is D if $A \odot D \equiv B$? Two situations may arise. If A and D occur in the sequence Z, U, P, Q, *etc.*, *before* B, and D occurs as a marginal entry of our multiplication table, we have a paper model of the exact division of two integers. Otherwise, we may consider the reversal of order, *e.g.* $Q \odot D \equiv U$. By using our first three axioms and the seventh:

$$D \odot (U \oplus U \oplus U) \equiv U$$
$$(D \odot U) \oplus (D \odot U) \oplus (D \odot U) \equiv U$$
$$D \oplus D \oplus D \equiv U$$

With our alternative marking system, we may write $D=\frac{1}{3}$ and the above becomes $\frac{1}{3}+\frac{1}{3}+\frac{1}{3}=1$. We have thus provided ourselves with a paper model for the manipulation of positive rational fractions.

Let us now ask what we have *not* yet done. From the build-up of our table of addition, we know that there will be an element R such that $P \oplus R \equiv T$ if, and only if, both P and R occur earlier in the successive build-up of the U, P, Q, *etc.*, sequence. When we replace P, R and T by 2, 4 and 6, we are at liberty to write the result in arithmetical symbolism in either of three ways: (a) $6=4+2$; (b) $6-4=2$; (c) $6-2=4$. From the build-up of our subtraction table, it is clear that either of the last two implies that the mark 6 comes *after* 2 and 4 in the ordered subset of integers built up by adding units. Thus the axioms we invoke to make a self-consistent set of rules for rational positive numbers cannot give any sanction to an arithmetical identity such as $(4-6)\times(3-7)=8$. In short, we need at least one other axiom to justify the rule of signs, as in less formal language we may well have suspected when we first made its acquaintance. We can proceed in several ways. One of them is attractive because it explicitly links the sign rule to the *vector* operation $(-1)\times$ of Chapter 8. Accordingly, we add:

Axiom 11. The field contains a subset U_m which satisfies the relation:

$$U \oplus U_m \equiv Z$$

THEOREM 3. From Axioms 3 and 10, we derive $(U_m \odot U_m) \equiv U$ as follows:

$$U_m \odot (U_m \oplus U) \equiv U_m \odot Z \equiv Z \equiv U \oplus U_m$$
$$(U_m \odot U_m) \oplus U_m \equiv U \oplus U_m$$
$$\therefore U_m \odot U_m \equiv U$$

Replaced by a new mark $(U_m = -1)$, the new axiom thus gives us the two rules:

$$-1+1=0 \quad \text{and} \quad (-1)\times(-1)=1$$

THEOREM 4. For every subset $F \otimes Z$ in the field, there is a subset $F_m \equiv (F \odot U_m)$ such that $F_m \odot F_m \equiv F \odot F$. The proof is as follows:

$$(F \odot U_m) \odot (F \odot U_m) \equiv (F \odot F) \odot (U_m \odot U_m)$$
$$\equiv (F \odot F) \odot U$$
$$\therefore F_m \odot F_m \equiv F \odot F$$

The reader will easily show that:

$$Z_m \equiv Z$$

Whence by Theorem 2:

$$Z_m \odot Z_m \equiv Z \odot Z$$

If we agree to label $(P \odot U_m)$ as -2, $(Q \odot U_m)$ as -3, *etc.*, Theorem 4 generalises the rule of signs, *e.g.*

$$(-2)(-2)=4; \ (-3)(-2)=6, \ \textit{etc.}$$

Furthermore, we may write:

$$(F \odot U_m) \oplus F \equiv (F \odot U_m) \oplus (F \odot U)$$
$$\equiv F \odot (U_m \oplus U) \equiv Z$$

This is then equivalent to saying:

$$-3+3=0=-2+2, \ \textit{etc.}$$

THEOREM 5. $U \otimes Z \otimes U_m$. The proof is: Since $U \otimes Z$ by definition (Axiom 10), by Axiom 8:

$$U \oplus U_m \otimes Z \oplus U_m$$
$$Z \otimes Z \oplus U_m$$
$$Z \otimes U_m \text{ (by Axiom 6)}$$

Corollary. In conformity with the above definitions of P, Q, R:

$$P_m \otimes Q_m \otimes R_m, \ \textit{etc.}$$

By interchange of marks on paper, this signifies:

$$-1>-2>-3, \ \textit{etc.}$$

THEOREM 6. $A \oplus C_m \equiv B$ if $A \otimes B$ and $C \otimes Z$ or $C \equiv Z$

$$A \oplus C_m \oplus C \equiv B \oplus C$$
$$A \oplus Z \equiv B \oplus C$$
$$A \equiv B \oplus C$$

We may thus extend Axiom 9 in the form: if $A \otimes B$, there is in the field a subtrahend C_m such that $A \oplus C_m \equiv B$. Thus we may write:

$$Q \oplus U_m \equiv P; \ T \oplus P_m \equiv R, \ \textit{etc.}$$

By interchange of marks on paper, this is equivalent to writing:

$$3+(-1)=2; \ 6+(-2)=4, \ \textit{etc.}$$

In short, Theorem 6 extends the meaning of subtraction over the whole field of rational real numbers.

To accommodate the complex number, we shall need only one other axiom: $A \oplus (B \odot U_i) \equiv C \oplus (D \odot U_i)$ if, and only if, $A \equiv C$ and $B \equiv D$;

Axiom 12. There exists in the field a subset U_i so defined that:

$$U_i \odot U_i \equiv U_m$$

THEOREM 7. As for the derivation of Theorem 3:

$$[A \oplus (B \odot U_i)] \odot [C \oplus (D \odot U_i)]$$
$$=[(A \odot C) \oplus (B \odot D \odot U_m)] \oplus [(A \odot D \odot U_i) \oplus (B \odot C \odot U_i)]$$

By substitution of marks on paper, the reader will see that we may now write:

$$P \oplus (Q \odot U_i) = 2+3i; \ Q \oplus [U_m \odot (P \odot U_i)] = 3-2i$$
$$(2+3i)(3-2i)=12+5i$$

If we everywhere substitute $i^2=-1$, this brings complex numbers with rational components into the same framework of computation as real numbers; and if the reader now asks what we have achieved by this excursion into an abstract domain into which number as such does not intrude, the answer is that we have at least made explicit what additional postulates we need in order to re-define the fundamental operations of the abacus in a way which can accommodate without inconsistency: (a) the rules of abacus (or machine) computation; (b) the manipulation of negative real and rational numbers; (c) the arithmetic of complex numbers with rational components.

Since definition of a nonperiodic and non-terminating decimal or that of a continued fraction prescribes how to manipulate the irrational, we need not here elaborate our system further. What the reader may well ask is whether so abstract a definition of our rules of composition $\odot$ and $\oplus$ could have any meaning to us unless already familiar with the basic peculiarities of elementary addition and multiplication of numbers. In terms of sound teaching at an elementary level, this does not much matter if the outcome has forced us to face up to the failure of traditional instruction to make explicit what we are really doing. In salmon-and-trout terms, the answer we must give is that the attempt to equip the abstract notions of $\odot$ and $\oplus$ defined as axioms by 1–3 above is not interpretable in terms of the tangible notions of inclusion and identity or, seemingly, in any terms which do not rely in the last resort on our experience of an *ordered* matching process, *i.e.* of enumeration or of measurement.

It is therefore excusable to harbour a doubt about whether reasoning of the foregoing type will have any further utility when we have reformed our teaching methods to sidestep arbitrary extensions of the rules of computation. Against this, we should give weight to another consideration. Because mathematicians during the four centuries before 1850 had enlarged the concept of number without facing up to what assumptions justify prescribed manipulations in a new part of the enlarged field, the foregoing *résumé* has built upwards from what assumptions we first need to make within the framework of abacus computation. However, we can go back a step or two by contracting our number field in more than one way, as illustrated by the siphon models (page 310) which dispense with the need for Axioms 11 and 12, but impose on us the need for new ones, *e.g.* for the top model:

(a) there is a unique subset $F \circledcirc U$ such that for all $A \circledcirc Z$
$F \oplus A \equiv Z$
(b) $F \equiv U \oplus U \oplus U \oplus U$

The reader may find it entertaining to explore the formal definition of the rules of the game for this model, which is a parable of one trail pursued by mathematicians during the past century, as illustrated especially by the development of vector algebra by Heaviside (1850–1925) and Willard Gibbs (1839–1903). Instead of forcing the mathematical statement of physical laws into the framework of traditional rules, we now feel free to tailor the rules of the game to the dictates of physical reality.

Let us therefore take a fleeting glance at the ancestral pattern of symbolic logic, *i.e.* the Boolean algebra of propositions. As with the algebra of sets or classes, we recognise three rules of composition equating the outcome to unity if true and zero if false. If P and Q stand for statements:

	True	*False*
$P \oplus Q \equiv P$ or Q or both are:	1	0
$P \odot Q \equiv$ both P and Q are:	1	0
$P \rightarrow Q \equiv P$ implies Q:	1	0

Accordingly, we can lay out basic truth tables thus:

$P \oplus Q$ (columns P, rows Q)

$\oplus$	1	0
1	**1**	**1**
0	**1**	**0**

$P \odot Q$ (columns P, rows Q)

$\odot$	1	0
1	**1**	**0**
0	**0**	**0**

$P \rightarrow Q$ (columns P, rows Q)

$\rightarrow$	1	0
1	**1**	**0**
0	**0**	**1**

How we interpret these will be clear if we translate the second row of cell entries in the left-hand grid of the foregoing tables:

If P is true and Q is false, then P or Q or both are true.
If P and Q are both false, then P or Q or both are false.

One entry at the foot of the right-hand grid may puzzle the reader, being on all fours with the class axiom that the empty set Z is in all subsets. The table states that $P \rightarrow Q = 1$ when $P = 0 = Q$, *i.e.* that a false proposition implies every other proposition true or false, as when one says: *if wishes were horses, beggars could fly.*

Such is a bare outline of the fundamental conventions of Boolean logic; and the reader may well ask: (a) whether any such mechanical recipe for the reasoning process equips us with a safeguard against fallacies inherent in the customary use of words; (b) whether rules of reasoning circumscribed by a two-valued logic of classes (*A* or *not A*) and a two-valued logic of propositions (*true* or *false*) is adequate to a statement of all the laws of nature. Boole himself, and his immediate successors, subscribed with no such misgivings to what is perhaps the most vulnerable feature of traditional logic, *i.e.* the three so-called Primary Laws of Thought devised by Aristotle for the instruction of young lawyers and aspirants to

TWO UNUSUAL ARITHMETICS

The containers each siphon off when the fluid level is 4. We may set out the tables of addition *thus:*

KEY to A

Total Feed	0	1	2	3	4	5	6	7	8
Final Level	**0**	**1**	**2**	**3**	**4**	**1**	**2**	**3**	**4**

KEY to B

Total Feed	0	1	2	3	4	5	6	7	8
Final Level	**0**	**1**	**2**	**3**	**4**	**0**	**1**	**2**	**3**

A

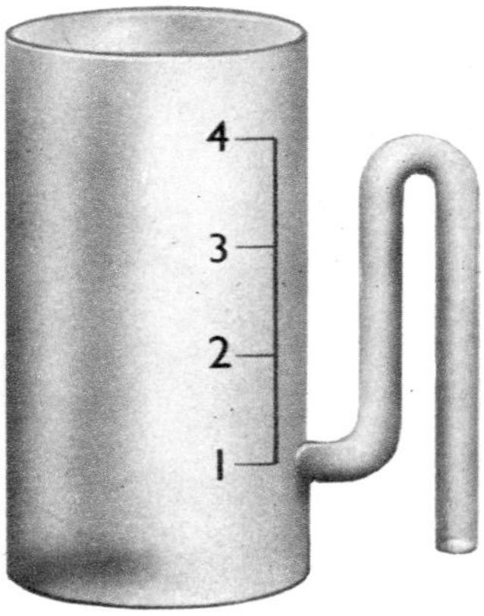

A (First Feed down, Second Feed across)

First Feed \ Second Feed	0	1	2	3	4
0	0	1	2	3	4
1	1	2	3	4	1
2	2	3	4	1	2
3	3	4	1	2	3
4	4	1	2	3	4

B (First Feed down, Second Feed across)

First Feed \ Second Feed	0	1	2	3	4
0	0	1	2	3	4
1	1	2	3	4	0
2	2	3	4	0	1
3	3	4	0	1	2
4	4	0	1	2	3

Instead of interpreting the horizontal marginal entries at the top of the foregoing tables as instructions to add a second feed, we may regard them as instructions to give 0, 1, 2, etc. successive feeds, and set out accordingly tables of multiplication *as below:*

B

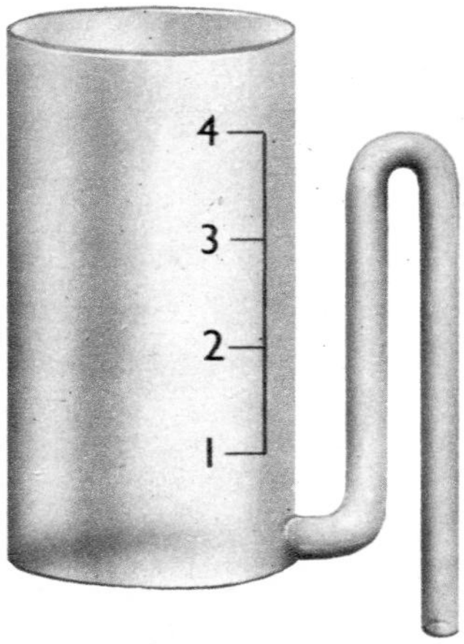

A

	0	1	2	3	4
0	0	0	0	0	0
1	0	1	2	3	4
2	0	2	4	2	4
3	0	3	2	1	4
4	0	4	4	4	4

B

	0	1	2	3	4
0	0	0	0	0	0
1	0	1	2	3	4
2	0	2	4	1	3
3	0	3	1	4	2
4	0	4	3	2	1

proficiency in political debate. For the benefit of readers unfamiliar with, or unable to remember them, they are as follows:

1. *Identity* — Whatever is, is.
2. *Contradiction* — Nothing can both be and not be.
3. *Excluded Middle* — Everything must either be or not be.

The three canons on which a two-valued logic relies are open to criticism from two different viewpoints. One is experimental in the sense that a translation of the ostensibly deep Cartesian dictum (*cogito ergo sum*) is patently silly if translated directly into Japanese and thence into Anglo-American. Actually Welsh (*inter alia*) makes a sharp distinction between the *is-be* of *identification* and the *is-be* of *predication*, as illustrated by the following sentences:

(a) John *is* our doctor=*John YW ein meddyg* (=John and our doctor are different symbols for the same object in the set).

(b) Our doctor *is* a young man=*Y MAE ein meddyg yn ddyn ifanc* (=Our doctor is one of the subset of men whose only common attribute is youth).

Among several other meanings (*e.g. live, remain*) contributed by totally different Indo-European roots to what we wrongly call the *irregular*, more properly *mixed*, verb *be*, that of identification is especially relevant to the notion of number. That of predication is closer to the notion of a mathematical function. From the viewpoint of the biologist, the first is trivial, the second of essential interest. Briefly, the justification for the last assertion is this. In a changing situation involving a limitless number of possibilities, *e.g.* the development of an embryo or the evolutionary process, it is not profitable to classify stages in any such clear-cut way as the three canons prescribe. Indeed, one often meets situations in which it is not even practicable to define a class in terms consistent with them. In its simplest form, such a definition is as follows:

> the class A is the set of all subsets whose elements share at least one of the attributes a_1, a_2, *etc.*, absent in every subset other than A of the universal set.

We leave it to our artist to dramatise (p. 304) the difference between the biological, or as the writer prefers to call it, *catenary*, and the Aristotelian class. Formally, we might define the catenary class as follows:

> class A is the set of all subsets in which every element shares with at least one other element one or other of the attributes a_1, a_2, *etc.*, absent in every subset of the universal set other than A.

Unhappily few professional mathematicians, with such conspicuous exceptions as Leibniz and Peano, have shown much interest in the pathology of verbal communication; and one may doubt whether discussion of the credentials of mathematics at a verbal level is likely to enrich mathematics unless mathematicians take a more active interest than heretofore in the semantic pitfalls of common speech. If one outcome of the Great Biopsy is to force them to do so, the sequel may be immensely beneficial to all of us, and in more than one way.

More especially during the past fifty years, three views of the credentials of mathematics have successively gained ascendancy. The first or *logicalist*, associated with the names of Peano, Whitehead and Bertrand Russell, regards mathematical reasoning as a closed system of propositions which are logically related to indisputably self-evident postulates. The *axiomatic* school led by Hilbert regards the postulates as entirely man-made and the consistency of the reasoning itself as the only criterion of validity. A third view, sometimes ineptly stigmatised as *intuitionist*, parts company with both the foregoing, but more especially with the second, in connexion with what restrictions we should impose on the admissibility of the postulates. For instance, is it permissible to erect a self-consistent system embracing the so-called transfinite numbers unless we can point to the possibility of manufacturing them in the sense that we can define the recipe for an infinite series whose sum converges to a finite limit? Some of those who take the third view also refuse to acknowledge the validity of the Euclidean proof by contradiction, *i.e.* (a) first assume a proposition is true; (b) exhibit its logical consequences; (c) exhibit an absurdity at the end of the chain of reasoning.

Inasmuch as it assumes that a proposition is either absolutely true or absolutely false, the proof by contradiction relies on Aristotle's canon of the excluded middle; and it is by no means clear that the latter is a meaningful assertion when the domain of discussion embraces *infinite* sets. In 1931 both the system of Russell's and that of Hilbert's school received a damaging setback from the development of symbolic logic, when an Austrian logician, Kurt Gödel, published a formal examination of all systems of mathematical definition. The outcome of this *tour de force* of meticulous symbolic manipulation was to show that no such system can contain within itself the proof of its own consistency. Much more recently, a new orientation towards what symbolism can usefully accomplish, including what mathematical problems are or are not soluble, has resulted from the challenge of the machine. Mathematicians, like others who converse in the vernacular, are not immune to misunderstandings inherent in the defective syntax of common speech. To programme the electronic brain, it is necessary to issue instructions in a different idiom; and the outcome will be valid only if the instructions are wholly unequivocal.

Lately, some logicians have gone as far beyond Boole as Boole beyond Aristotle. Multi-valued systems of symbolic recipes for reasoning permit one to say that a statement neither wholly true nor wholly false in the manner of lawyers may be a bit one thing and a bit the other. There are even corresponding explorations into the domain of probability, *e.g.* P is more likely to be true than Q. This line of thought would have a more promising prospect if it were easier to pin down the connexion between mathematical probability and probability as a measure of judgment.

In fairness to Boole, let us part company with the recognition that honest thinking which explores new themes is not less useful if it leads us to retrace our steps at the end of the garden path. On this understanding, the future for mathematical ingenuity is not necessarily dismal. From the global viewpoint of *Mathematics in the Making*, one may say that Man has solved the problems of power and of individual human survival so long as life is endurable. The focus of the most challenging constructive intellectual tasks of the next century may well be the unification of our means of communication. Of such, mathematics is one among several; but the accomplishment of the task depends on whether we welcome every gesture to peace on either side of the Iron Curtain.

If we persist in the predominantly paranoid temper of the present, no historian will survive to record the death of our species or the last vision of so bright a destiny. Otherwise, an Affluent Society not rooted, like that of Euclid, in the irksome toil and poverty of the many, will be well able to afford the luxury of encouraging clear thinking by those who are capable of the discomfort. Meantime, we must, as best we can, travel hopefully without the certitude of arrival. Unless the arrogance endemic among the experts confuses educational issues by new exploits of gamesmanship, one outcome of the Great Biopsy may well be a greater clarity of instruction. Another may be a new tolerance begotten of a new scepticism about what finalities are inherent in any temporarily acceptable rules of reasoning.

Quiz - Answers

Page 35 (Alexandrian addition tables)
Parts shown white on black (three patches, bottom left) complete the tables correctly.

Page 41 (Alexandrian multiplication table)
Parts shown white on black (two patches, bottom right) complete the table correctly.

Page 43 (Notable dates in world history)
(a) 1620. Voyage of the *Mayflower*.
(b) 1483. Portuguese founded Elmina (Ghana).
(c) 1912. Dr. Sun Yat Sen became first President of the Chinese Republic.
(d) 1948. State of Israel founded. Mahatma Gandhi assassinated.
(e) 1582. Pope Gregory XIII introduced his reformed (Gregorian) calendar.
(f) 1776. Declaration of American Independence.
(g) 1526. Great Mogul Empire founded.
(h) 1517. Luther protested against the sale of indulgences at Wittenberg.
(i) 1788. Sydney, Australia, founded.
(j) 1917. Russian revolution.

Page 165 (Formulae for visualisations of number series)
(a) $5n^2+4n$ (b) $5n^2$ (c) $8n^2-8n+1$ (d) $5n^2-5n+1$

Page 174 (Times around the world)
The following times are correct to within a few minutes *for the meridian of each place*, but no allowance is made for differences due to the existence of local time-zones. Singapore: 12 midnight. Hangchow: 1 a.m. Canberra: 3 a.m. Fiji: 5 a.m. New Orleans: 11 a.m. Ottawa: 12 noon. Manaos: 1 p.m. Falkland Islands: 1 p.m.

Recommended Reading

Higher Mathematics for Students of Engineering and Science. Frederick G. W. Brown. Macmillan, 1938 edn.

Mathematics: Queen and Servant of Science. Eric Temple Bell. G. Bell & Sons, London, 1952.

The Development of Mathematics. Eric Temple Bell. McGraw Hill Book Co., New York and London, 1945, 2nd edn.

What is Mathematics? Richard Courant and Herbert Robbins. O.U.P., 1941.

A Text-book on Practical Mathematics for Advanced Technical Students. Herbert L. Mann. Longmans & Co., London, 1925 edn.

Elementary Analysis. Kenneth O. May. John Wiley & Sons, New York, and Chapman & Hall, London, 1952.

Mathematics and the Imagination. Edward Kasner and James Newman. G. Bell & Sons, London, 1949.

Mathematics in Western Culture. Maurice Kline. Allen & Unwin, 1954.

Biomathematics. William M. Feldman. 3rd edn. rewritten by Cedric A. B. Smith. Griffin, 1954.

The World of Mathematics. James R. Newman (Ed.) Simon & Schuster, New York, 1956.

The Exact Sciences in Antiquity. Otto Neugebauer. Brown University Press, 1957 edn. (Copenhagen, 1951).

Science & Civilisation in China (Vol. III). Joseph Needham. C.U.P., 1959.

History of Mathematics. David Eugene Smith. Dover Publications Inc., 1958 (reprint of last-published edn. of Ginn & Co.).

Monographs on Topics of Modern Mathematics. J. W. A. Young (Ed.). Dover Publications Inc., 1951.

Elementary Mathematics from an Advanced Standpoint. Felix Klein. Dover Publications Inc., 1939.

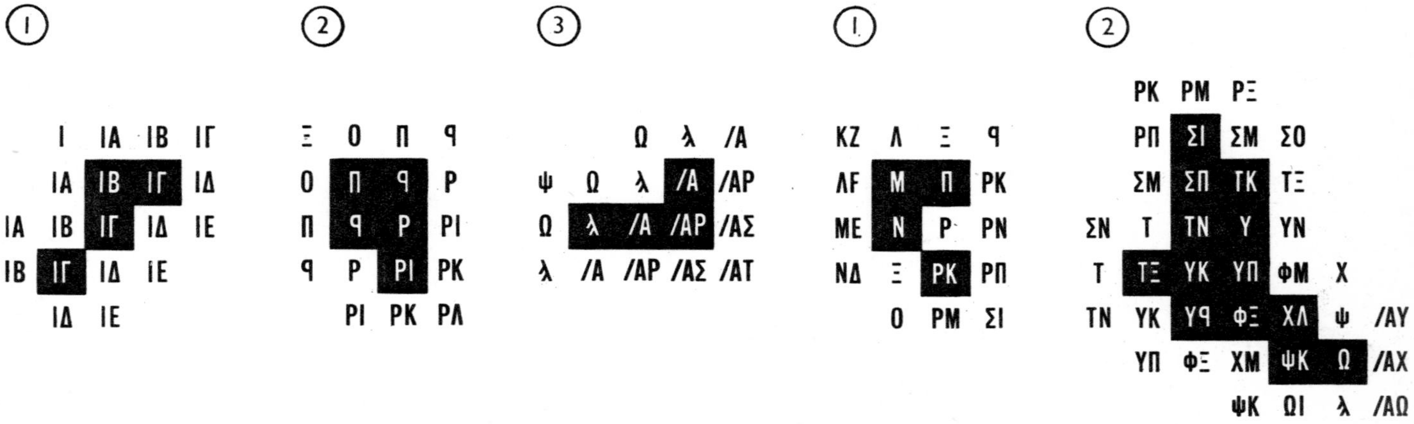

Index

Acknowledgments

Key: (B) ... bottom (M) ... middle (T) ... top
(L) ... left (R) ... right

Brit. Mus. ... courtesy of the Trustees of the British Museum.

7 (T) Wall painting from Tomb of Menna, c. 1420 B.C. Copy by Mrs. N. de Garis Davies. Brit. Mus. (B.M. – 3733)
(B) Wall painting from the Tomb of Menna. Photo: Auvergniot – Toussaint et Fautrelle. From: *Les Chefs d'oeuvre de la Peinture Egyptienne* d'André Lhote; Hachette, Paris.
11 Based on diagrams from *Algebra by Visual Aids* by G. P. Meredith (Allen & Unwin Ltd.), 1948. Visual Aid Series edited by Lancelot Hogben.
12 Mosaic from a Romano-British villa at Hemsworth, Wimborne, Dorset. Brit. Mus.
20 Diagram from *How to draw a straight line*, Kemp, 1877.
24 19-ft. quadrant from *Astronomiae Instuaratae Mechanica*, Tycho Brahe, 1598. Photo: Science Museum, London.
27 *Margaritam Philosophicam*, Gregorius Reisch, 1512. Title page to Arithmetic section. The Lewis Evans Library in the Museum of the History of Science, Oxford.
28 Stele F – Maudslay Collection. Brit. Mus.
(R) *A Guide to the Maudslay Collection of Maya Sculptures*, 1938. Brit. Mus.
29 (T) Babylonian Calendar, Brit. Mus. (B.M. – 32641)
(B) Japanese copy of a Chinese inscription. Brit. Mus.
30 (T) Egyptian Stele giving an account of expedition of Amenhotep III to N. Sudan in 1450 B.C. Brit. Mus. (B.M. – 657)
(B) Egyptian Scribe; Louvre, Paris. Photo: Mansell Collection.
31 Athenian Calendar (Cat. 12355 & 12363). Courtesy Epigraphic Museum, Athens.
32 Temple of Serpents, Mexico City. Photo: Courtesy Mexican Embassy, London.
33 Babylonian Duck Weights. Brit. Mus. (B.M. – 91438)
Alabaster Jar measure, c. 1250 B.C. Brit. Mus. (B.M. – 4659)
36 (T) Bakshali MS (Sansk. d.14, f.47r). Courtesy Bodleian Library, Oxford.
(M) From *Numbers and Numerals*, Smith and Ginsberg. Courtesy National Council of Teachers of Mathematics, Washington D.C.
(B) Folding Sundial and Compass, 1453. Brit. Mus. (B.M. – 1893, 6–16, 9)
47 Computing Machine. Photo: Courtesy International Computers & Tabulators Ltd., London.
51 Temple of Karnak. Mansell Collection.
52 (T) Minoan Script, Linear A. Photo: Ashmolean Museum, Oxford.
53 Gate of Ishtar. Reproduced with the permission of Ehemals Staatlichen Museen, Berlin.
54 (T) The Rhind Papyrus. Brit. Mus. (B.M. – 10057/8)
(B) Translation of problem 41 into hieroglyphic, from *The Rhind Mathematical Papyrus*. Courtesy Mathematical Association of America.
56 Excavations at Nippur near east corner of Ziggurat. Courtesy University Museum, Philadelphia.
57 (T) Clay Tablet Map of Nippur in possession of the Hilprecht Collection of the University of Jena. Photo: HBS – University of Jena.
(B) Clay Tablet Library Catalogue, c. 1750 B.C. (29–15–155). Courtesy University Museum, Philadelphia.
59 Calcutta Arsenal, 19th century print. Mansell Collection.
60 *Science & Civilisation in China*, Vol. III, J. Needham, C.U.P., 1959. From Shu-Hsüeh's *Shen Tao Ta Pien Li Tsung Suan Hui*, 1558.
61 Nippur Clay Tablet, c. 1800 B.C. (CBS 14233). Courtesy University Museum, Philadelphia.
62 Based on diagram from *Algebra by Visual Aids*, G. P. Meredith (Allen & Unwin, Ltd.), 1948. Visual Aid Series edited by Lancelot Hogben.
63 (T) Rhind Papyrus. Brit. Mus. (B.M. – 10057/8)
(B) Translation of problem 23 into hieroglyphic, from *The Rhind Mathematical Papyrus*. Courtesy Mathematical Association of America.
64 Babylonian Tablet. Brit. Mus. (B.M. – 15.285)
66 Pyramids at Giza. Photo: Aerofilms Ltd.
67 Chou Pei MS. Brit. Mus. (B.M. – Or. 15255.d.21)
70 (B) Step Pyramid at Sakkara. Mansell Collection.
71 Calendar from Tomb of Rameses VII. Photo: Brown University Press, U.S.A.
75 Duris Cup. Reproduced with the permission of Ehemals Staatlichen Museen, Berlin.
76-7 (B) *L'Art d'Euxdoxe* (N2325). Courtesy Dept. Egyptian Antiquities, Musée du Louvre. Photo: Séarl.
77 Cypriot Vase, c. 700 B.C. (74.51.1403). Courtesy Metropolitan Museum of Art, The Cesnola Collection; purchased by subscription 1874–6
78 Phoenician War Ship from an Assyrian relief 702 B.C. Brit. Mus. (B.M. – Stele 124772)
82 From *Corpus Vasorum*. Brit. Mus.
94 Breguet 901 on tow behind a Stamp aircraft. Photo: Dengremont, Paris.
95 Making a plane surface. Courtesy Windley Bros. Ltd., Chelmsford.
99 Fresco from Pompeii. Courtesy National Museum, Naples. Photo: Giulio Parisio, Naples.
100 Moslem MS of Euclid 1258 A.D. Brit. Mus. (Add – 23,387 f.28a)

101 *Geometry*, Hall & Stevens, 1895 edn.
102 Red-figure Kylix, Attic, c. 470 B.C. (Oldfield Bequest v517). Courtesy Ashmolean Museum, Oxford.
107 Coles Theodolite, 1586. Courtesy Museum of History of Science, Oxford. On loan from St. John's College.
109 Detail from border of the *quadratum geometricum* of Christoph Schissler 1579. Courtesy Museum of the History of Science, Oxford. On loan from Bodleian Library.
112 (T) Alexandrian coin, 188 A.D. Brit. Mus.
113 Relief from Ostia, c. 200 A.D. Mansell Collection.
115 Mosaic from Pompeii. Mansell Collection. Photo: Anderson.
116 Painted Terracotta, after 30 B.C. Brit. Mus. (B.M. – 37563)
118 *Apiaria Universae Philosophae Mathematicae*, Bettini, 1641. Brit. Mus. (B.M. – 8534.h.4, Vol. I, pt. III)
120-1 *Antifitiosi et curiosi moti spiritali di Herrone*, Battista, 1589. Brit. Mus. (B.M. – 538.g.2(1–2))
125 (T) *Cosmographia*, Ptolemy, edn. of 1482. Brit. Mus. (B.M. – IC.9303)
(B) *Cosmographia*, Ptolemy, edn. of 1470. Brit. Mus. (B.M. – Harley MS 7182)
126 Babylonian clay tablet which is possibly a primitive planisphere, c. 1200 B.C. Brit. Mus. (B.M. – K.8538)
127 *Ptolemaei Al-Magesti*, 1294 MS. Brit. Mus. (B.M. – Reg.16.A.VIII)
135 (B) Relief globe. Designed and produced by Geographical Projects Limited.
136 *Geographia*, Ptolemy, edn. of 1508, map by J. Ruysch. Brit. Mus. (B.M. – MAPS C.1.d.6)
137 *Maps*, T. W. Birch (Clarendon Press, Oxford).
138 Polar Stereographic projection, John Blagrave, 1596. Brit. Mus. (B.M. – Harley MS 5935.15)
139 Relief globe. Designed and produced by Geographical Projects Limited.
147 (T) Cave of a Thousand Buddhas. Photo: Arthaud, Paris.
(B) 13th-century Japanese painting of Hsüan Tsang (Tripitaka). Mansell Collection.
148 (T) Relief map. Designed and produced by Geographical Projects Limited.
(B) Buddhist sculptures from caves at Yun Kang, 6th century. Photo: Ecole Française d'Extrême Orient, Paris.
149 Diamond Sutra. Brit. Mus. (B.M. – Stein Collection P.2)
150 *Jāmi 'al-Tawārīkh*, by Rashīd al-Dīn, 1303. Courtesy Royal Asiatic Society, London, and Brit. Mus. (B.M. – R.A.S. loan No. 4)
151 (T) *Maqamat of Hariri*, 1337. Photo: Iraq Petroleum Co. Ltd.
(B) *Ptolemaei Al-Magesti*, 1294. Brit. Mus. (B.M. – Reg.16.A.VIII)
152 *Sūryasiddhānta*, c. 1792 (I.0.580, K.2782). Courtesy Librarian, India Office Library, Commonwealth Relations Office.
153 *Al-qānūn al-mas'ūdī*, Al-Bērūnī, 1174. Brit. Mus. (B.M. – Or.1997)
154 *Tawārīkh Guzīdah Nusrat Nāmah* (Turkish MS 1530). A history of Chingiz Khan and his descendants. Brit. Mus. (B.M. – Or.3222)
155 Mansell Collection.
156 (T) Greenwich Observatory. Photo: Science Museum.
(B) *Tabulae Ulugh Beighi*, trs. Thomas Hyde, 1665, and published at Oxford. Brit. Mus. (B.M. – 757.cc. 11(I))
157 *Science & Civilisation in China*, Vol. III, J. Needham, C.U.P., 1959. From *Suan Fa Thung Tsung*, 1593.
158 (T) *Science & Civilisation in China*, Vol. III, J. Needham, C.U.P., 1959. Left: The Lo Shu diagram. Right: The Ho Thu diagram.
159 (T) Pottery figures playing Liu-Po(?), Han dynasty (206 B.C. – 220 A.D.). Brit. Mus.
(B) Hwahaw Chinese playing cards from Shanghai, 19th century. Brit. Mus. (B.M. – Schreiber Collection No. 3, Case 54.)
160 (T) *Science & Civilisation in China*, Vol. III, J. Needham, C.U.P., 1959. From Chhêng Ta-Wei's *Suan Fa Thung Tsung*, 1593.
(B) *History of Mathematics*, D. E. Smith, 1951. From works of Seki Kōwa, c. 1661. Courtesy Dover Publications Inc., New York. ($5.00)
163 Algebra by Omar Khayyám, 6th century MS (Loth. 734–I.0.1270). Courtesy Librarian, India Office Library, Commonwealth Relations Office.
Science & Civilisation in China, Vol. III, J. Needham, C.U.P., 1959. From Chu Shih-Chieh's *Ssu Yuan Yü Chien*, 1303.
165 *Algebra by Visual Aids*, G. P. Meredith (Allen & Unwin Ltd.), 1948. Visual Aid Series edited by Lancelot Hogben.
167 *Artis Conjectandi*, Jacob Bernoulli, 1713. Brit. Mus. (B.M. – 529.f.17)
169 *Americae* 1590, engraving by T. de Bry. Brit. Mus. (B.M. – C.115.h.3(7))
170 (T) The Mosque at Cordova. Photo: Courtesy Spanish National Tourist Office.
(B) *History of Mathematics*, D. E. Smith, 1951. From a Latin MS of 1456. Courtesy Dover Publications Inc., New York. ($5.00)
171 (T) Cantino World Map, 1502. Brit. Mus. (B.M. – MAPS 7.e.8)
(B) Photo: Casa de Portugal, State Information and Tourist Office, London.
172 (B) *The Sea-Man's Kalendar*, John Tapp, 1696. Brit. Mus. (B.M. – 533.d.7)
173 *L'Atlas de Nordenskiöld*, 1889. World Map by Hakluyt in 1599 on Mercator projection. Courtesy Royal Geographical Society.
"The Draughtsman & the Lute". Woodcut by Dürer. Brit. Mus. (B.M. – 1895–1–22–731 . . . B147)
175 (L) Working reconstruction from *Horologium Oscillatorium*, 1673. (INV.1927–1981). Crown copyright, Science Museum, London.
(R) Harrison's Chronometer No. 2 from Thomas Bradley's drawing of 1840. Courtesy Royal Observatory, Greenwich.
176 (TL) *Behēde und hubsche* . . . J. Widman, 1489. Brit. Mus. (B.M. – IA.11541)
(TR) *La Disme: The Art of Tenths*. S. Stevin, 1608. (English edn. by R. Norton). Brit. Mus. (B.M. – 1393.e.15)
177 (TR) *Whetstone of witte*, Robt. Recorde, 1557. Brit. Mus. (B.M. –530.g.37)
(B) *Arithmetica Integra*, Stifel, 1544. Brit. Mus. (B.M. – 8504.cc.10)

178 Graph from *Algebra by Visual Aids* by G. P. Meredith (Allen & Unwin Ltd.), 1948. Visual Aid Series edited by Lancelot Hogben.
179 *Logarithmorum Canonis Descriptio* . . . John Napier, 1620. Brit. Mus. (B.M. – 8548.c.20)
181 *Thermometer of Lyons* (calibrated above and below zero), late 18th century (INV. 1951–581). Crown copyright, Science Museum, London.
186 *Ars Magna*, Jerome Cardan, 1545. Brit. Mus. (B.M. – 8531.g.13)
191 (T) Map of Capt. Cook's Polar Voyage, 1772–5. From a collection of maps and drawings by members of crews of *HMS Resolution* and *HMS Adventure*. Brit. Mus. (Add. MSS 15,500(2))
(B) From Capt. Cook's Log, July 1772–November 1774. Brit. Mus. (Add. MSS 27,886)
201 (L) Diagrams from *Growth and Form*, D'Arcy Thompson, C.U.P., 1942.
203 (B) *La Géometrie* (from *Discours de la Méthode* . . .), René Descartes, 1637. Brit. Mus. (B.M. – C112.c.4)
208-9 (T) *Les Grands Illustrateurs*, Georges Hirth, 1883.
208 (B) An illustration from *The Complete Gunner* (Book III of *Military & Marine Discipline* by Thos. Venn), 1672. Brit. Mus. (B.M. – 534.m.8)
213 (T) Halley's letter to Newton (Keynes MS 97(c)). Reproduced by courtesy of the Provost and Fellows of King's College, Cambridge.
(B) *Principia* trs. by Andrew Motte as *The Mathematical Principles of Natural Philosophy*, 1729. Brit. Mus. (B.M. – 233.f.22)
216 (B) "Big Dipper" at Southend Funfair. Photo: Paul Popper, London.
226 *Science & Civilisation in China*, Vol. III, J. Needham, C.U.P., 1959. From Mochinaga Toyotsugu and Ōhashī Takusei's *Kaisan-ki Kōmoku*, 1687.
228 (B) Robt. Hooke's diary, entry for 21st August 1678. From original MS. Courtesy Guildhall Library, London.
231 After an engraving from *A Treatise on the Steam Engine*, John Farey, 1827. Brit. Mus. (B.M. – 538.k.26)
234 Map reproduced from *La Figure de la terre* . . . M. Bouguer, 1749. Brit. Mus. (B.M. – TC6.a.26)
235 (T) *An Ephemeris for 9 years 1609–1617*, John Searle, 1609. Brit. Mus. (B.M. – P.P.2466.e)
(B) *Nautical Almanac & Astronomical Ephemeris for 1767* by The Commissioners of Longitude, 1766. Brit. Mus. (B.M. – P.P.2373.m)
237 (L) Illustrations from *Base du Système Metrique Decimal*, Méchain and Delambre, 1806. Brit. Mus. (B.M. – 986.e.20)
(T) As above.
(B) As above.
243 Courtesy Ford Motor Company, Dagenham, Essex.
246 (T & M) Photos taken between 1861 and 1863. Courtesy Reece Winstone Collection, Bristol.
(B) Coloured lighograph of 1831 design. Courtesy City Art Gallery, Bristol. Photo: C. & E. Photography, Bristol.
251 *le Trente-et-un* . . . coloured after an engraving by Darcis (AA_3 Darcis). Courtesy Bibliothèque Nationale Paris.
253-9 Based on diagrams from *Chance and Choice by Cardpack and Chessboard*, Vol. I, Lancelot Hogben (Max Parrish & Co. Ltd.), 1950.
262 Based on diagrams from *Algebra by Visual Aids*, G. P. Meredith (Allen & Unwin Ltd.), 1958. Visual Aid Series edited by Lancelot Hogben.
263 (B) As above.
(T) Based on diagrams from *Chance and Choice by Cardpack and Chessboard*, Vol. I, Lancelot Hogben, (Max Parrish & Co. Ltd.), 1950.
265 *Encyclopédie*, 1751, article by D'Alembert. Brit. Mus. (B.M. – 65.g.1.)
266 Posters and a Lottery Ticket reproduced from *A History of English Lotteries*, John Ashton, 1893.
267 Press Advertisement. Courtesy National Savings Committee.
268 *Observations on Reversionary Payments*, R. Price, 1771. Brit. Mus. (B.M. – 8229.bbb.36)
270 Lottery draw after a poster in Miss Bank's Collection. Brit. Mus. (B.M. – LR.301.h.8)
285 (T) Central London: detail from the Underground map. Courtesy London Transport Executive and the Controller of H.M. Stationery Office. Crown copyright reserved.
(B) Central London: detail from the diagrammatic Underground map. Courtesy London Transport Executive.
289 (T) The Harbor Freeway just south of the Four-Level Interchange, Los Angeles, California, U.S.A. Photo: courtesy California Division of Highways.
(B) The four-level Interchange, Los Angeles, California. Photo: courtesy Newsweek – Dale Healy.
301 Details of IBM Electronic Computer. Photo: courtesy IBM United Kingdom Ltd.

Artists: A. Ball, H. Bellalta, G. Brayley, K. Briggs, G. Cramp, J. Ernest, W. Greaves, N. Jones, J. Messenger, S. A. Parfitt, S. A. L. Rogers, P. Sullivan.
Photographers: K. Dustan, J. R. Freeman, G. Howson, R. Jarmain, Studio 51, D. Swann.
Miss M. Cartwright prepared the index and Miss M. Wall checked the calculations.

If the publishers have unwittingly infringed copyright in any illustration reproduced they will gladly pay an appropriate fee on being satisfied as to the owner's title.